KU-308-192

Methods in Enzymology

Volume 245
EXTRACELLULAR MATRIX COMPONENTS

METHODS IN ENZYMOLOGY

EDITORS-IN-CHIEF

John N. Abelson   Melvin I. Simon

DIVISION OF BIOLOGY
CALIFORNIA INSTITUTE OF TECHNOLOGY
PASADENA, CALIFORNIA

FOUNDING EDITORS

Sidney P. Colowick and Nathan O. Kaplan

Methods in Enzymology

Volume 245

Extracellular Matrix Components

EDITED BY

Erkki Ruoslahti

LA JOLLA CANCER RESEARCH FOUNDATION
LA JOLLA, CALIFORNIA

Eva Engvall

LA JOLLA CANCER RESEARCH FOUNDATION
LA JOLLA, CALIFORNIA

UNIVERSITY OF WOLVERHAMPTON
LIBRARY

Acc No. 869272

CLASS 572.
702
8
MET

CONTROL

DATE 25. MAY 1995   SITE RS

D22

ACADEMIC PRESS

San Diego   New York   Boston   London   Sydney   Tokyo   Toronto

This book is printed on acid-free paper. ∞

Copyright © 1994 by ACADEMIC PRESS, INC.

All Rights Reserved.
No part of this publication may be reproduced or transmitted in any form or by any means, electronic or mechanical, including photocopy, recording, or any information storage and retrieval system, without permission in writing from the publisher.

Academic Press, Inc.
A Division of Harcourt Brace & Company
525 B Street, Suite 1900, San Diego, California 92101-4495

United Kingdom Edition published by
Academic Press Limited
24-28 Oval Road, London NW1 7DX

International Standard Serial Number: 0076-6879

International Standard Book Number: 0-12-182146-3

PRINTED IN THE UNITED STATES OF AMERICA
94   95   96   97   98   99   MM   9   8   7   6   5   4   3   2   1

Table of Contents

Section I. Extracellular Matrix Proteins

Section II. Receptors

Section VI. Use of Extracellular Matrix

Contributors to Volume 245

Article numbers are in parentheses following the names of contributors.
Affiliations listed are current.

SUNEEL S. APTE (1), *Department of Cell Biology, Harvard Medical School, Boston, Massachusetts 02115*

ALEJANDRO ARUFFO (10), *Bristol-Myers Squibb Pharmaceutical Research Institute, Seattle, Washington 98121*

MINA J. BISSELL (25), *Life Sciences Division, Lawrence Berkeley Laboratory, Berkeley, California 94720*

RUEDIGER J. BLASCHKE (25), *Life Sciences Division, Lawrence Berkeley Laboratory, Berkeley, California 94720*

JON BLEVITT (26), *Telios Pharmaceuticals, Inc., San Diego, California 92121*

WAYNE A. BORDER (12), *Division of Nephrology, University of Utah Health Sciences Center, Salt Lake City, Utah 84132*

PAUL BORNSTEIN (4), *Department of Biochemistry, University of Washington, Seattle, Washington 98195*

JOHANNES BREUSS (20), *Lung Biology Center, Department of Medicine, University of California, San Francisco, San Francisco, California 94143*

IAIN D. CAMPBELL (21), *Department of Biochemistry, University of Oxford, Oxford OX1 3QU, United Kingdom*

RUTH CHIQUET-EHRISMANN (3), *Friedrich Miescher Institute, CH-4002 Basel, Switzerland*

WILLIAM S. CRAIG (26), *Telios Pharmaceuticals, Inc., San Diego, California 92121*

WILLIAM F. DEGRADO (18), *DuPont Merck Pharmaceutical Company, Experimental Station, Wilmington, Delaware 19880*

PIERRE-YVES DESPREZ (25), *Life Sciences Division, Lawrence Berkeley Laboratory, Berkeley, California 94720*

CRAIG D. DICKINSON (17), *La Jolla Cancer Research Foundation, La Jolla, California 92037*

XIAOPING DU (9), *Department of Vascular Biology, The Scripps Research Institute, La Jolla, California 92037*

JÜRGEN ENGEL (22), *Department of Biophysical Chemistry, Biozentrum, CH-4056 Basel, Switzerland*

EVA ENGVALL (5), *La Jolla Cancer Research Foundation, La Jolla, California 92037, and Department of Developmental Biology, Wenner-Gren Institute, University of Stockholm, 10691 Stockholm, Sweden*

DAVID J. ERLE (20), *Lung Biology Center, Department of Medicine, University of California, San Francisco, San Francisco, California 94143*

RANDALL J. FAULL (9), *Department of Renal Medicine, St. George Hospital, Kogarah, NSW 2217, Australia*

JOHN H. FESSLER (14), *Molecular Biology Institute and Department of Biology, University of California at Los Angeles, Los Angeles, California 90024*

LISELOTTE I. FESSLER (14), *Molecular Biology Institute and Department of Biology, University of California at Los Angeles, Los Angeles, California 90024*

NAOMI FUKAI (1), *Department of Cell Biology, Harvard Medical School, Boston, Massachusetts 02115*

WILLIAM A. GAARDE (12), *Fibrosis Research, Telios Pharmaceuticals, Inc., San Diego, California 92121*

ELIZABETH L. GEORGE (19), *Vascular Research Division, Department of Pathology, *Brigham and Women's Hospital, and Harvard Medical School, Boston, Massachusetts 02115*

FILIPPO G. GIANCOTTI (15), *Department of Pathology and Kaplan Comprehensive Cancer Center, New York University School of Medicine, New York, New York 10016*

MARK H. GINSBERG (9), *Department of Vascular Biology, The Scripps Research Institute, La Jolla, California 92037*

SCOTT GUIMOND (11), *Department of Pathology and Laboratory Medicine, University of Wisconsin, Madison, Wisconsin 53706*

JOHN R. HARPER (12), *Fibrosis Research, Telios Pharmaceuticals, Inc., San Diego, California 92121*

RONALD H. HOESS (18), *DuPont Merck Pharmaceutical Company, Experimental Station, Wilmington, Delaware 19880*

BENJAMIN G. HOFFSTROM (16), *Department of Laboratory Medicine and Pathology, University of Minnesota, Minneapolis, Minnesota 55455*

ANTHONY R. HOWLETT (25), *Life Sciences Division, Lawrence Berkeley Laboratory, Berkeley, California 94720*

RICHARD O. HYNES (19), *Center for Cancer Research and Department of Biology, Howard Hughes Medical Institute, and Massachusetts Institute of Technology, Cambridge, Massachusetts 02139*

PATRICIA J. KEELY (8), *Department of Pathology, Washington University School of Medicine, St. Louis, Missouri 63110*

DOUGLAS R. KEENE (2), *Shriners Hospital for Crippled Children, Portland, Oregon 97201*

ERKKI KOIVUNEN (17), *La Jolla Cancer Research Foundation, La Jolla, California 92037*

RANDALL H. KRAMER (7), *Departments of Stomatology and Anatomy, University of California, San Francisco, California 94143*

ALISON KRUFKA (11), *Department of Pathology and Laboratory Medicine, University of Wisconsin, Madison, Wisconsin 53706*

FABRIZIO MAINIERO (15), *Department of Pathology and Kaplan Comprehensive Cancer Center, New York University School of Medicine, New York, New York 10016*

RENÉE K. MARGOLIS (6), *Department of Pharmacology, State University of New York, Health Science Center at Brooklyn, Brooklyn, New York 11203*

RICHARD U. MARGOLIS (6), *Department of Pharmacology, New York University Medical Center, New York, New York 10016*

CURT MAZUR (26), *Telios Pharmaceuticals, Inc., San Diego, California 92121*

JOHN A. MCDONALD (24), *Department of Biochemistry and Molecular Biology, Samuel C. Johnson Medical Research Building, Mayo Clinic Scottsdale, Scottsdale, Arizona 85259*

SHAKER A. MOUSA (18), *DuPont Merck Pharmaceutical Company, Experimental Station, Wilmington, Delaware 19880*

FREDERIQUE MUSSET-BILAL (13), *Department of Molecular Biology, Princeton University, Princeton, New Jersey 08544*

ROBERT E. NELSON (14), *Department of Biology and Molecular Biology Institute, University of California at Los Angeles, Los Angeles, California 90024*

NANCY A. NOBLE (12), *Division of Nephrology, University of Utah Health Sciences Center, Salt Lake City, Utah 84132*

BJORN R. OLSEN (1), *Department of Cell Biology, Harvard Medical School, Boston, Massachusetts 02115*

BRADLEY B. OLWIN (11), *Department of Biochemistry, Purdue University, West Lafayette, Indiana 47907*

KARYN T. O'NEIL (18), *DuPont Merck Pharmaceutical Company, Experimental Station, Wilmington, Delaware 19880*

JULIAN J. O'REAR (23), *Department of Mo-*

lecular Genetics and Microbiology, Robert Wood Johnson Medical School, Piscataway, New Jersey 08854

OLE W. PETERSEN (25), Structural Cell Biology Unit, Department of Anatomy, The Panum Institute, University of Copenhagen, DK-2200 Copenhagen N, Denmark

MICHAEL D. PIERSCHBACHER (12, 26), Research and Development, Telios Pharmaceuticals, Inc., San Diego, California 92121

ROBERT PYTELA (20), Lung Biology Center, Department of Medicine, University of California, San Francisco, San Francisco, California 94143

N. RAMACHANDRAN (18), DuPont Merck Pharmaceutical Company, Experimental Station, Wilmington, Delaware 19880

ALAN C. RAPRAEGER (11), Department of Pathology and Laboratory Medicine, University of Wisconsin, Madison, Wisconsin 53706

ERKKI RUOSLAHTI (17), La Jolla Cancer Research Foundation, La Jolla, California 92037

CAROL S. RYAN (13), Department of Molecular Biology, Princeton University, Princeton, New Jersey 08544

EDWIN U. M. SAELMAN (8), Department of Pathology, Washington University School of Medicine, St. Louis, Missouri 63110

E. HELENE SAGE (4), Department of Biological Structure, University of Washington, Seattle, Washington 98195

LYNN Y. SAKAI (2), Department of Biochemistry and Molecular Biology, Oregon Health Sciences University, and Shriners Hospital for Crippled Children, Portland, Oregon 97201

RAYMOND SANDERS (15), Department of Pediatrics, New York University School of Medicine, New York, New York 10016

SAMUEL A. SANTORO (8), Department of Pathology and Internal Medicine, Washington University School of Medicine, St. Louis, Missouri 63110

SUSANNE SCHENK (3), Friedrich Miescher Institute, CH-4002 Basel, Switzerland

JEAN E. SCHWARZBAUER (13), Department of Molecular Biology, Princeton University, Princeton, New Jersey 08544

DEAN SHEPPARD (20), Lung Biology Center, Department of Medicine, University of California, San Francisco, San Francisco, California 94143

LAURA SPINARDI (15), Department of Pathology and Kaplan Comprehensive Cancer Center, New York University School of Medicine, New York, New York 10016

ROBERT C. SPIRO (12), Fibrosis Research, Telios Pharmaceuticals, Inc., San Diego, California 92121

WILLIAM D. STAATZ (8), Department of Pathology, Washington University School of Medicine, St. Louis, Missouri 63110

IVAN STAMENKOVIC (10), Department of Pathology, and Harvard Medical School, Massachusetts General Hospital, Boston, Massachusetts 02129

KIMBERLY K. STECKER (12), Fibrosis Research, Telios Pharmaceuticals, Inc., San Diego, California 92121

SHINTARO SUZUKI (20), Doheny Eye Institute, Departments of Ophthalmology and Microbiology, University of Southern California School of Medicine, Los Angeles, California 90033

RICHARD N. TAMURA (12), Desmos Incorporated, San Diego, California 92121

JAMES TOLLEY (26), Telios Pharmaceuticals, Inc., San Diego, California 92121

JUERG F. TSCHOPP (26), Telios Pharmaceuticals, Inc., San Diego, California 92121

BINGCHENG WANG (17), La Jolla Cancer Research Foundation, La Jolla, California 92037

ELIZABETH A. WAYNER (16), Department of Laboratory Medicine and Pathology, University of Minnesota, Minneapolis, Minnesota 55455

ULLA M. WEWER (5), Laboratory of Molecular Pathology, University Institute of

Pathological Anatomy, DK-2100 Copenhagen, Denmark

MICHAEL J. WILLIAMS (21), *Department of Vascular Biology/VB2, Scripps Research Institute, La Jolla, California 92047*

JUSTINA E. WU (8), *Department of Pathology, Washington University School of Medicine, St. Louis, Missouri 63110*

PETER D. YURCHENCO (23), *Department of Pathology, Robert Wood Johnson Medical School, Piscataway, New Jersey 08854*

MARY M. ZUTTER (8), *Department of Pathology, Washington University School of Medicine, St. Louis, Missouri 63110*

Preface

The extracellular matrix field has undergone remarkable expansion since the publication of Volume 82 of *Methods in Enzymology* in 1982. At that time, there was still only one laminin, one thrombospondin, and very few molecularly characterized proteoglycans. Each of the first two have become whole families, and the proteoglycans now comprise a number of families. Remarkably, there is still only one fibronectin gene. The proliferation of extracellular matrix family members reflects the progress in cloning techniques in the intervening years; PCR, in particular, has made a tremendous difference. These technical advances are covered in this volume.

The extracellular matrix receptors, which were not yet known as integrins, were just beginning to be characterized at the time Volume 82 was published; integrin research now constitutes a major portion of this book.

One of the new developments in the field concerns the many disease links that have been found for extracellular matrix proteins. These proteins as well as cell–cell adhesion proteins can clearly function as tumor suppressors, and their genes are running up as mutated tumor suppressor genes. Other mutations have clarified the molecular basis of important genetic diseases as well as helped define the function of several matrix proteins. Thus, Marfan's syndrome is caused by a mutated fibrillin, while a form of muscular dystrophy and certain skin blistering diseases have as their underlying problem mutations in newly discovered laminins. Technical advances have made it possible to disable ("knockout") genes and even introduce targeted mutations into them. The gene knockout experiments have shown that lack of fibronectin or its receptors causes an embryonal lethal phenotype but that, surprisingly, knocking out the gene for the classical tenascin has no obvious effect on the development or survival of mice. Clearly, these new techniques are extremely powerful tools in matrix research, which is why we have included a chapter on them in this volume.

A developing new focus of matrix research is matrix as a regulator of gene expression; one chapter deals with this topic. We predict that this will be one of the main topics of a future volume on extracellular matrix.

Extracellular matrix research has also found its way to biotechnology; some of the ongoing efforts to develop new drugs based on advances in matrix research are dealt with in one chapter.

We thank Drs. Abelson and Semion for their invitation to assemble this volume and the Academic Press staff for their role in making it a reality.

ERKKI RUOSLAHTI
EVA ENGVALL

METHODS IN ENZYMOLOGY

VOLUME XVII. Metabolism of Amino Acids and Amines (Parts A and B)
Edited by HERBERT TABOR AND CELIA WHITE TABOR

VOLUME XVIII. Vitamins and Coenzymes (Parts A, B, and C)
Edited by DONALD B. MCCORMICK AND LEMUEL D. WRIGHT

VOLUME XIX. Proteolytic Enzymes
Edited by GERTRUDE E. PERLMANN AND LASZLO LORAND

VOLUME XX. Nucleic Acids and Protein Synthesis (Part C)
Edited by KIVIE MOLDAVE AND LAWRENCE GROSSMAN

VOLUME XXI. Nucleic Acids (Part D)
Edited by LAWRENCE GROSSMAN AND KIVIE MOLDAVE

VOLUME XXII. Enzyme Purification and Related Techniques
Edited by WILLIAM B. JAKOBY

VOLUME XXIII. Photosynthesis (Part A)
Edited by ANTHONY SAN PIETRO

VOLUME XXIV. Photosynthesis and Nitrogen Fixation (Part B)
Edited by ANTHONY SAN PIETRO

VOLUME XXV. Enzyme Structure (Part B)
Edited by C. H. W. HIRS AND SERGE N. TIMASHEFF

VOLUME XXVI. Enzyme Structure (Part C)
Edited by C. H. W. HIRS AND SERGE N. TIMASHEFF

VOLUME XXVII. Enzyme Structure (Part D)
Edited by C. H. W. HIRS AND SERGE N. TIMASHEFF

VOLUME XXVIII. Complex Carbohydrates (Part B)
Edited by VICTOR GINSBURG

VOLUME XXIX. Nucleic Acids and Protein Synthesis (Part E)
Edited by LAWRENCE GROSSMAN AND KIVIE MOLDAVE

VOLUME XXX. Nucleic Acids and Protein Synthesis (Part F)
Edited by KIVIE MOLDAVE AND LAWRENCE GROSSMAN

VOLUME XXXI. Biomembranes (Part A)
Edited by SIDNEY FLEISCHER AND LESTER PACKER

VOLUME XXXII. Biomembranes (Part B)
Edited by SIDNEY FLEISCHER AND LESTER PACKER

VOLUME XXXIII. Cumulative Subject Index Volumes I–XXX
Edited by MARTHA G. DENNIS AND EDWARD A. DENNIS

VOLUME XXXIV. Affinity Techniques (Enzyme Purification: Part B)
Edited by WILLIAM B. JAKOBY AND MEIR WILCHEK

VOLUME XXXV. Lipids (Part B)
Edited by JOHN M. LOWENSTEIN

VOLUME 73. Immunochemical Techniques (Part B)
Edited by JOHN J. LANGONE AND HELEN VAN VUNAKIS

VOLUME 74. Immunochemical Techniques (Part C)
Edited by JOHN J. LANGONE AND HELEN VAN VUNAKIS

VOLUME 75. Cumulative Subject Index Volumes XXXI, XXXII, XXXIV–LX
Edited by EDWARD A. DENNIS AND MARTHA G. DENNIS

VOLUME 76. Hemoglobins
Edited by ERALDO ANTONINI, LUIGI ROSSI-BERNARDI, AND EMILIA CHIAN-CONE

VOLUME 77. Detoxication and Drug Metabolism
Edited by WILLIAM B. JAKOBY

VOLUME 78. Interferons (Part A)
Edited by SIDNEY PESTKA

VOLUME 79. Interferons (Part B)
Edited by SIDNEY PESTKA

VOLUME 80. Proteolytic Enzymes (Part C)
Edited by LASZLO LORAND

VOLUME 81. Biomembranes (Part H: Visual Pigments and Purple Membranes, I)
Edited by LESTER PACKER

VOLUME 82. Structural and Contractile Proteins (Part A: Extracellular Matrix)
Edited by LEON W. CUNNINGHAM AND DIXIE W. FREDERIKSEN

VOLUME 83. Complex Carbohydrates (Part D)
Edited by VICTOR GINSBURG

VOLUME 84. Immunochemical Techniques (Part D: Selected Immunoassays)
Edited by JOHN J. LANGONE AND HELEN VAN VUNAKIS

VOLUME 85. Structural and Contractile Proteins (Part B: The Contractile Apparatus and the Cytoskeleton)
Edited by DIXIE W. FREDERIKSEN AND LEON W. CUNNINGHAM

VOLUME 86. Prostaglandins and Arachidonate Metabolites
Edited by WILLIAM E. M. LANDS AND WILLIAM L. SMITH

VOLUME 87. Enzyme Kinetics and Mechanism (Part C: Intermediates, Stereochemistry, and Rate Studies)
Edited by DANIEL L. PURICH

VOLUME 88. Biomembranes (Part I: Visual Pigments and Purple Membranes, II)
Edited by LESTER PACKER

VOLUME 89. Carbohydrate Metabolism (Part D)
Edited by WILLIS A. WOOD

VOLUME 212. DNA Structures (Part B: Chemical and Electrophoretic Analysis of DNA)
Edited by DAVID M. J. LILLEY AND JAMES E. DAHLBERG

VOLUME 213. Carotenoids (Part A: Chemistry, Separation, Quantitation, and Antioxidation)
Edited by LESTER PACKER

VOLUME 214. Carotenoids (Part B: Metabolism, Genetics, and Biosynthesis)
Edited by LESTER PACKER

VOLUME 215. Platelets: Receptors, Adhesion, Secretion (Part B)
Edited by JACEK J. HAWIGER

VOLUME 216. Recombinant DNA (Part G)
Edited by RAY WU

VOLUME 217. Recombinant DNA (Part H)
Edited by RAY WU

VOLUME 218. Recombinant DNA (Part I)
Edited by RAY WU

VOLUME 219. Reconstitution of Intracellular Transport
Edited by JAMES E. ROTHMAN

VOLUME 220. Membrane Fusion Techniques (Part A)
Edited by NEJAT DÜZGÜNEŞ

VOLUME 221. Membrane Fusion Techniques (Part B)
Edited by NEJAT DÜZGÜNEŞ

VOLUME 222. Proteolytic Enzymes in Coagulation, Fibrinolysis, and Complement Activation (Part A: Mammalian Blood Coagulation Factors and Inhibitors)
Edited by LASZLO LORAND AND KENNETH G. MANN

VOLUME 223. Proteolytic Enzymes in Coagulation, Fibrinolysis, and Complement Activation (Part B: Complement Activation, Fibrinolysis, and Nonmammalian Blood Coagulation Factors)
Edited by LASZLO LORAND AND KENNETH G. MANN

VOLUME 224. Molecular Evolution: Producing the Biochemical Data
Edited by ELIZABETH ANNE ZIMMER, THOMAS J. WHITE, REBECCA L. CANN, AND ALLAN C. WILSON

VOLUME 225. Guide to Techniques in Mouse Development
Edited by PAUL M. WASSARMAN AND MELVIN L. DEPAMPHILIS

VOLUME 226. Metallobiochemistry (Part C: Spectroscopic and Physical Methods for Probing Metal Ion Environments in Metalloenzymes and Metalloproteins)
Edited by JAMES F. RIORDAN AND BERT L. VALLEE

Section I

Extracellular Matrix Proteins

[1] Nonfibrillar Collagens

By NAOMI FUKAI, SUNEEL S. APTE, and BJORN R. OLSEN

Introduction

Collagens are structural proteins that participate in the assembly of various kinds of polymers in the extracellular matrix (ECM).[1] Their molecules contain at least one triple-helical domain. Until about 15 years ago, collagens were largely known as fibril-forming or network/basement membrane-forming proteins.[2] Since then, the use of molecular biology techniques has resulted in the isolation of numerous molecules that fit the generic description of collagens, but are structurally and functionally distinct from the fibril and basement membrane-forming collagens.[3,4] Some of these molecules, such as the C1q component of complement,[5] type I and type II macrophage scavenger receptors,[6] surfactant apoproteins,[7,8] and the tail of the acetylcholinesterase receptor,[9] contain triple-helical domains, but do not participate in ECM assembly interactions and are not included in the collagen superfamily. Among the other, newer collagens (e.g., types XII and XIV collagen), non-triple-helical domains may comprise the major part of the proteins and may be of major functional significance; such "chimeric" proteins are included as collagens because they presumably interact with other ECM constituents.

Nonfibrillar collagens can be further grouped into subfamilies based on the similarities in the structure of their genes and the domain structure of their protein products. A review from our laboratory[3] described 14 collagen types encoded by at least 25 distinct genes. In the brief period since that review was published, we are now cognizant of 19 collagen

[1] E. D. Hay, ed., "Cell Biology of Extracellular Matrix," 2nd ed. Plenum, New York, 1991.
[2] K. Kühn, *in* "Immunochemistry of the Extracellular Matrix" (H. Furthmayr, ed.), Vol. 1, p. 1. CRC Press, Boca Raton, FL, 1982.
[3] O. Jacenko, B. R. Olsen, and P. A. LuValle, *Crit. Rev. Eukaryotic Gene Express.* **1,** 327 (1991).
[4] M. van der Rest and R. Garrone, *FASEB J.* **5,** 2814 (1991).
[5] K. B. M. Reid, *Biochem. J.* **179,** 367 (1979).
[6] T. Kodama, M. Freeman, L. Rhorer, J. Zabrecky, P. Matsudaira, and M. Kriger, *Nature (London)* **343,** 531 (1990).
[7] B. Benson, S. Hawgood, J. Schilling, J. Clements, D. Damm, B. Cordell, and R. T. White, *Proc. Natl. Acad. Sci. U.S.A.* **82,** 6379 (1985).
[8] A. Persson, K. Rust, D. Chang, M. Moxley, W. Longmore, and E. Crouch, *Biochemistry* **28,** 6361 (1989).
[9] T. L. Rosenberry and J. M. Richardson, *Biochemistry* **16,** 3550 (1977).

Copyright © 1994 by Academic Press, Inc.
All rights of reproduction in any form reserved.

types, and a correspondingly greater increase in the number of genes. Although data on the fibrillar collagens continue to accumulate, we have, therefore, restricted this chapter to a discussion of the newer collagen types. We have also chosen not to include new basement membrane collagen gene members [e.g., $\alpha 5$(IV) and $\alpha 6$(IV) chains] although they have been cloned,[10,11] as a discussion of basement membrane collagens and their relevance to disease represents a substantial undertaking in itself. Likewise, types VI and VII collagens have been dealt with.[3] We have, therefore, restricted the discussion to new collagens, such as the transmembrane cell surface collagen type XVII[12] and nonfibrillar gene families, the Multiplexin[13,14] and FACIT families,[15] type XIII collagen,[16] and the short-chain collagens, types X and VIII (Fig. 1). We have especially emphasized the methodological aspects of the discovery of these collagens and the analysis of their structure and function. Details of DNA clones and sequences are available from individual papers and from the GenBank/EMBL databases and chromosomal loci are listed in Table I.

A traditional feature of the nomenclature of collagenous proteins (apart from the fibrillar collagens) has been to distinguish between collagenous (COL) and noncollagenous (NC) domains, although it is more correct to refer to these as triple-helical and non-triple-helical domains. For the greater part, the numbering of these domains has been initiated at the carboxyl end of the protein. In some cases, for example, type XIII collagen, confusion has arisen from the numbering of these domains starting at the amino end. Similarly, although exons encoding fibrillar collagens were initially numbered starting at the 3' end of the genes,[17] it is now customary to number exons from the 5' end of the gene (in cases in which the complete gene structure is known).[18,19] To minimize confusion, in this

[10] S. L. Hostikka, R. L. Eddy, M. G. Myers, M. Höyhtyä, T. B. Shows, and K. Tryggvason, *Proc. Natl. Acad. Sci. U.S.A.* **87**, 1606 (1990).

[11] J. Zhou, T. Mochizuki, H. Smeets, C. Antignac, P. Laurila, A. de Paepe, K. Tryggvason, and S. T. Reeders, *Science* **261**, 1167 (1993).

[12] K. Li, K. Tamai, E. M. L. Tan, and J. Uitto, *J. Biol. Chem.* **268**, 8825 (1993).

[13] S. P. Oh, Y. Kamagata, Y. Muragaki, S. Timmons, A. Ooshima, and B. R. Olsen, *Proc. Natl. Acad. Sci. U.S.A.* **91**, 4229 (1994).

[14] M. Rehn, J. Saarela, H. Autio-Harmainen, and T. Pihlajaniemi, *Proc. Natl. Acad. Sci. U.S.A.* **91**, 4234 (1994).

[15] L. M. Shaw and B. R. Olsen, *Trends Biochem. Sci.* **16**, 191 (1990).

[16] T. Pihlajaniemi, R. Myllylä, J. Seyer, M. Kurkinen, and D. Prockop, *Proc. Natl. Acad. Sci. U.S.A.* **84**, 940 (1987).

[17] W. B. Upholt and B. R. Olsen, *in* "Cartilage: Molecular Aspects" (B. K. Hall and S. A. Newman, eds.), p. 1. CRC Press, Boca Raton, FL, 1991.

[18] E. Vuorio and B. de Crombugghe, *Annu. Rev. Biochem.* **59**, 837 (1990).

[19] L. J. Sandell and C. D. Boyd, *in* "Extracellular Matrix Genes" (L. J. Sandell and C. D. Boyd, eds.), p. 1. Academic Press, San Diego, 1990.

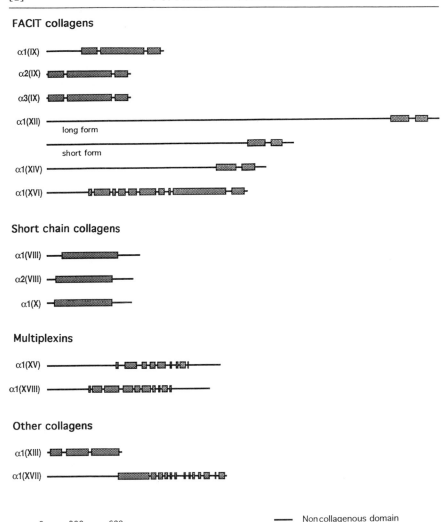

FIG. 1. Diagram showing the domain structure of the nonfibrillar collagens discussed in this chapter. The structures are based on data published for mouse[65] and human α1(IX) collagen,[68] mouse α2(IX),[66] chicken α3(IX),[62] chicken α1(XII) (long form),[50] chicken α1(XII) (short form),[83] chicken α1(XIV),[51,52] human α1(XVI),[53] rabbit[108] and mouse[126] α1(VIII), human α2(VIII),[102] chicken α1(X),[103,104] human α1(XV),[143,145] mouse α1(XVIII),[13,14] human α1(XIII),[138,140] and mouse α1(XVII) collagen.[12]

TABLE I
COLLAGEN GENES, CHAINS THEY ENCODE, AND THEIR LOCI IN HUMAN
AND MOUSE GENOMES

Gene symbol	Chain	Human locus	Mouse locus
Short-chain collagen family			
COL8A1	α1(VIII)	3q12–q13.1[149]	—
COL8A2	α2(VIII)	1p32.3–p34.4[102]	Chromosome 4[a]
COL10A1	α1(X)	6q21–q22[105]	Chromosome 10[116]
FACIT collagen family			
COL9A1	α1(IX)	6q12–q14[55,64]	Chromosome 1[55]
COL9A2	α2(IX)	1p32.3–p33[69,70]	Chromosome 4[70]
COL9A3	α3(IX)	—	—
COL12A1	α1(XII)	6q12–q14[56,b]	Chromosome 9[56]
COL14A1	α1(XIV)	—	—
COL16A1	α1(XVI)	1p34–p35[53]	—
COL19A1	α1(XIX)	6q12–q14[54]	—
Multiplexin family			
COL15A1	α1(XV)	9q21–q22[147]	—
COL18A1	α1(XVIII)	21q22.3[144]	Chromosome 10[144]
Others			
COL13A1	α1(XIII)	10q11–qter[150]	—
COL17A1	α1(XVII)	10q24.3[27]	Chromosome 19[151]

[a] Authors' unpublished data.
[b] R. W. Taylor et al., unpublished data.

chapter we use both nomenclatural conventions when referring to these proteins and their genes. We continue to refer to triple-helical and non-triple-helical domains as COL and NC domains and give the domains numbers starting at the carboxyl end, while we number exons starting at the 5′ end of the genes in cases in which the complete gene structure is known.

Type XVII Collagen: A Novel Transmembrane Cell Surface Component

Discovery

Although the existence of this protein has been known for some time, it has only recently been included as a member of the collagen family.[12] Both the manner of discovery of type XVII collagen and its structural features set it apart from other members of the collagen family.

It is the only collagen to date to have been isolated as a direct result of investigation into a human disease. In three autoimmune blistering diseases of skin, bullous pemphigoid (BP), herpes gestationis (HG), and cicatricial pemphigoid, IgG autoantibodies react with a 180-kDa protein antigen (one of two major autoantigens in these diseases) localized to the hemidesmosome.[20] Using anti-BP180 autoantibodies from BP and HG patients as molecular probes, Diaz et al.,[21] Guidice et al.,[22] and Hopkinson et al.[23] isolated overlapping human cDNA clones from keratinocyte cDNA expression libraries that encoded part of BP180. The predicted amino acid sequence of BP180 (now termed type XVII collagen[12]) was found to contain several triple-helical domains. The structure of this protein was similar to the structure predicted from a chicken cDNA clone simultaneously and independently isolated by low-stringency hybridization of a corneal cDNA library with a type XII collagen cDNA clone as probe.[24] The corneal cDNA is now recognized as representing the chicken homolog of the human BP180 or type XVII cDNA; the simultaneous discovery of type XVII collagen DNAs, using two distinct molecular genetic approaches, is a useful illustration of the manner in which various molecular methods have contributed to the collagen lexicon.

Structure

On the basis of sequencing of the $\alpha 1$(XVII) cDNAs, the complete primary structure of the human[25] and mouse[12] protein, and the greater part of the chicken[24] protein, are known. The cDNAs predict an integral membrane protein with a type II orientation (i.e., the C terminus of the protein is in the ectodomain); the triple-helical domains are in the ectodomain. The mouse[12] and human[25] $\alpha 1$(XVII) chains are 1433 and 1497 amino acid residues long, respectively, and contain identical residues at 80% of positions. The human $\alpha 1$(XVII) chain shows 64% identity with the corresponding chicken molecule.

In contrast to the fibrillar collagens, and in common with the FACIT collagens, the collagenous region of type XVII collagen consists of several interrupted triple-helical domains separated and flanked by non-triple-helical domains. The number of COL domains and consequently their numbering varies, being 16, 15, and 13 in the chicken,[24] human,[25] and mouse[12] polypeptides, respectively. The small size of these COL domains and their close spacing are reflected by the fact that the 15 COL domains

[20] P. Bernard, *Springer Semin. Immunopathol.* **13**, 401 (1992).
[21] L. A. Diaz, H. Ratrie, III, W. S. Saunders, S. Futamuta, and H. L. Squiquera, *J. Clin. Invest.* **86**, 1088 (1990).
[22] G. J. Guidice, H. L. Squiquera, P. M. Elias, and L. A. Diaz, *J. Clin. Invest.* **87**, 734 (1991).
[23] S. B. Hopkinson, K. S. Riddelle, and J. C. R. Jones, *J. Invest. Dermatol.* **99**, 264 (1992).
[24] J. K. Marchant, T. F. Linsenmayer, and M. K. Gordon, *Proc. Natl. Acad. Sci. U.S.A.* **88**, 1560 (1991).
[25] G. J. Guidice, D. J. Emery, and L. A. Diaz, *J. Invest. Dermatol.* **99**, 243 (1992).

of the human $\alpha 1$(XVII) cDNA are located within the 1007-amino acid long ectodomain. The most N-terminal triple-helical domain (COL16 in chicken, COL15 in human, and COL13 in mouse) is the largest, being 242 amino acid residues in length in all three species. Moreover, all three species contains an imperfection in the G-X-Y triplet structure at the same relative position in this large COL domain. The remaining COL domains range in size from 15 to 36 amino acid residues. Typically, the COL domains of $\alpha 1$(XVII) collagen are proline and serine rich, making it likely that even the smallest COL domains with only five G-X-Y triplets may have stable triple-helical conformations at 37°.[24]

The intervening NC domains are small, ranging from 8 to 60 amino acid residues in length. The cDNA sequences predict the presence of a transmembrane domain; the presence of possibly two transmembrane domains has been proposed for the mouse protein.[12] The mouse residues 461–508 (numbered from the amino end) are identical to the corresponding human and chicken sequences, making the transmembrane domain and the immediately adjacent intracellular domain the most highly conserved region of this molecule.[12] All four potential N-linked glycosylation sites in the human $\alpha 1$(XVII) chain are conserved in the mouse but the latter contains two additional sites (one each in the NC2 and NC4 domains) not found in the human protein.[12] It is not yet known if type XVII collagen is a homotrimer of $\alpha 1$(XVII) chains or whether there are additional, yet unknown, chains of type XVII collagen.

The mouse $\alpha 1$(XVII) polypeptide (1433 amino acid residues long) predicts a protein with a molecular mass of 144 kDa.[12] This is at variance with the estimated size of the BP180 antigen (160–180 kDa), but the difference between the observed and predicted values can, however, be explained by posttranslational modifications. There are six consensus sites for N-linked glycosylation in the mouse $\alpha 1$(XVII) polypeptide,[12] and studies by Nishizawa et al.[26] have indicated that BP180 is glycosylated; however, the extent of glycosylation is not known.

The partial structure of the $\alpha 1$(XVII) gene[12,27] is known and shows similarities to both the fibrillar and nonfibrillar collagen genes. Analysis of human genomic clones corresponding to 1.0 kilobase (kb) of cDNA sequence, encoding mostly the human COL15 domain, has shown that this region of the mRNA is encoded by 19 exons.[27] Similar to the case with fibrillar collagen genes, the exons encoding the COL domains are mostly small, and are multiples of nine nucleotides in length. The 5′ end

[26] Y. Nishizawa, J. Uematsu, and K. Owaribe, *J. Biochem.* (*Tokyo*) **113**, 493 (1993).
[27] K. Li, D. Sawamura, G. J. Guidice, L. A. Diaz, M.-G. Mattei, M.-L. Chu, and J. Uitto, *J. Biol. Chem.* **266**, 24064 (1991).

of the mouse gene has been cloned.[12] Exon 1 encodes only untranslated mRNA and exon 2 contains the translation start site. In contrast to the fibrillar collagen genes, the splice junctions in protein-coding exons split a codon.[27] Little is known of the regulatory regions of the $\alpha 1$(XVII) gene apart from the presence of a consensus TATA box 26 nucleotides upstream of the transcription start site.[12]

The identity of the BP180 antigen with $\alpha 1$(XVII) collagen has been confirmed by showing that antibodies against fusion proteins generated from cDNA clones specifically react with BP180 in keratinocyte extracts, and label hemidesmosomes on immunohistochemistry.[21] Conversely, the BP autoantibodies recognize a fusion protein encoded by the $\alpha 1$(XVII) cDNA.[21] However, direct sequence analysis of the BP180 protein is not available. Clostridial collagenase digests BP180 from epidermal extracts to yield a collagenase-resistant fragment of the expected size.[22] The predicted type II transmembrane orientation has been confirmed by immunoelectron microscopy, using extra- and intracellular domain specific antibodies.[21,23] Furthermore, tryptic digestion of keratinocytes and immunoblotting of the cell extracts with antibodies specific for the predicted ectodomain shows that this domain is removed by trypsin.[23]

Expression and Function

The $\alpha 1$(XVII) collagen chain is thus an integral membrane protein restricted in localization to the hemidesmosomes of stratified squamous epithelia, such as the cutaneous, oral, and corneal epithelium.[23] It appears to be widely conserved during evolution, based on the reactivity of anti-BP antibodies with epidermal extracts from many species.[23] The structural features of the $\alpha 1$(XVII) chain and its location in the hemidesmosome suggest that it participates in anchoring of cells to the extracellular matrix. The presence of phosphorylated residues in the intracellular domain have yet to be confirmed, but several tyrosyl residues that could be phosphorylated hold out the interesting possibility that this molecule may transduce extracellular signals to intracellular signaling pathways.[12,23] The multiple NC domains in the ectodomain may impart flexibility to the molecule (e.g., as has been shown for type IX collagen) and both the COL and NC domains may participate in interactions with other ECM components. That type XVII collagen stabilizes keratinocyte–basal lamina interactions has been indirectly demonstrated by the loss of hemidesmosomes in BP[28] and the presence of anti-BP180 autoantibodies. Direct evidence for the pathogenicity of BP180 antibodies has been published. Rabbit anti-mouse

[28] G. Shaumberg-Lever, C. E. Orfanos, and W. F. Lever, *Arch. Dermatol.* **106,** 662 (1972).

BP180 antibodies, when injected into mice, cause a subepidermal blistering disease that closely mimicks BP and HG at the clinical, histological, and immunological levels.[29] It is not known if any blistering skin diseases are due to inherited defects in this gene.

FACIT Collagens

Discovery

The FACIT (fibril-associated collagens with interrupted triple helices) group of proteins includes collagens with two or more relatively short triple-helical domains connected and flanked by non-triple-helical sequences.[15,30] The different members of the group (collagens IX, XII, XIV, XVI, and XIX) all contain a highly conserved triple-helical domain (COL1) with two cysteinyl residues at its carboxyl border. The presence of such a domain at the carboxyl end of a collagenous protein, and the unique exon structure that encodes it, is considered a necessary and sufficient structural requirement for inclusion of a gene in the FACIT group.

The "founding" member of the FACIT group, type IX collagen,[31] was first isolated as disulfide-bonded triple-helical fragments after pepsin extraction of cartilage.[32-36] It was the cloning and sequencing of cDNAs encoding two of the three distinct polypeptide subunits of type IX molecules, however, that provided the first detailed structure of the molecules.[37,38] These cDNAs, pYN1738 and pYN1731, were isolated from chicken cartilage cDNA libraries using a restriction endonuclease method to screen for cDNAs encoding collagenous proteins. As described in detail elsewhere,[39] the method is based on the fact that whenever the Gly-xxx-

[29] Z. Liu, L. A. Diaz, J. L. Troy, A. F. Taylor, D. J. Emery, J. A. Failey, and G. J. Guidice, *J. Clin. Invest.* **92,** 2480 (1992).

[30] Y. Ninomiya and B. R. Olsen, *in* "Guidebook to the Extracellular Matrix and Adhesion Proteins" (T. Kreis and R. Vale, eds.), p. 37. Oxford Univ. Press, Oxford, 1993.

[31] M. van der Rest, R. Mayne, Y. Ninomiya, N. G. Seidah, M. Chrétien, and B. R. Olsen, *J. Biol. Chem.* **260,** 220 (1985).

[32] M. Shimokomaki, V. C. Duance, and A. J. Bailey, *FEBS Lett.* **121,** 51 (1980).

[33] C. A. Reese and R. Mayne, *Biochemistry* **20,** 5443 (1981).

[34] K. von der Mark, M. van Menxel, and H. Wiedemann, *Eur. J. Biochem.* **124,** 57 (1982).

[35] S. Richard-Blum, D. J. Hartmann, D. Herbage, C. Payen-Meyran, and G. Ville, *FEBS Lett.* **146,** 343 (1982).

[36] S. Ayad, M. Z. Abedin, S. M. Grundy, and J. B. Weiss, *FEBS Lett.* **123,** 195 (1981).

[37] Y. Ninomiya and B. R. Olsen, *Proc. Natl. Acad. Sci. U.S.A.* **81,** 3014 (1984).

[38] Y. Ninomiya, M. van der Rest, R. Mayne, G. Lozano, and B. R. Olsen, *Biochemistry* **24,** 4223 (1985).

[39] G. Vasios, Y. Ninomiya, and B. R. Olsen, *in* "Structure and Function of Collagen Types" (R. Mayne and R. E. Burgeson, eds.), p. 283. Academic Press, Orlando, FL, 1987.

yyy triplets in the collagen sequence contain proline in the xxx position, the nucleotide sequence GGN-CCN-NNN contains the recognition site GGNCC for the restriction endonuclease Sau96I. Within cDNAs encoding collagenous sequences, one is therefore likely to find multiple Sau96I sites, spaced $9n$ nucleotides apart. To screen for such cDNAs, we digested DNA from randomly selected clones with Sau96I, and labeled the products with ^{32}P, using the Klenow fragment of DNA polymerase to fill in the 5' overhangs created by Sau96I. The fragments were separated by electrophoresis through 10% (w/v) polyacrylamide gels and cDNAs encoding polypeptides with triple-helical sequence domains produced a characteristic ladder of fragments with bands spaced 9 base pairs (bp) apart.[39]

The nucleotide sequence of the cDNA pYN1738 indicated that the translation product, given the designation $\alpha1$(IX) collagen, contained three triple-helical sequence domains (COL1, COL2, and COL3) separated and flanked by non-triple-helical domains (NC1, NC2, NC3, and NC4).[37] A comparison of the cDNA sequences with partial amino acid sequences of triple-helical fragments purified from pepsin extracts of chicken cartilage demonstrated that type IX collagen molecules are heterotrimers composed of $\alpha1$(IX), $\alpha2$(IX), and $\alpha3$(IX) chains.[31] Further studies showed that the NC3 domain of the $\alpha2$(IX) chain contained an attachment site for a glycosaminoglycan side chain[40] and that type IX collagen is, in fact, identical to the proteoglycan PG-Lt, previously isolated from chicken cartilage.[41–46]

Type XII collagen was first identified as a cDNA with homology to $\alpha1$(IX) by using the Sau96I screening method,[47] whereas type XIV collagen was initially identified as a short triple-helical, pepsin-resistant fragment in extracts of bovine skin.[48] Both these collagens are homotrimers of $\alpha1$(XII) and $\alpha1$(XIV) collagen chains, and their sequences are organized in homologous sequence domains. The sequence domains show similarities to domains in $\alpha1$(IX) collagen as well as domains in other extracellular matrix proteins, providing additional examples of how the principle of

[40] D. McCormick, M. van der Rest, J. Goodship, G. Lozano, Y. Ninomiya, and B. R. Olsen, *Proc. Natl. Acad. Sci. U.S.A.* **84,** 4044 (1987).
[41] A. Noro, K. Kimata, Y. Oike, T. Shinomura, N. Maeda, S. Yano, N. Takahashi, and S. Suzuki, *J. Biol. Chem.* **258,** 9323 (1983).
[42] G. Vaughan, K. H. Winterhalter, and P. Bruckner, *J. Biol. Chem.* **260,** 4758 (1985).
[43] P. Bruckner, L. Vaughan, and K. H. Winterhalter, *Proc. Natl. Acad. Sci. U.S.A.* **82,** 2608 (1985).
[44] S. Huber, M. van der Rest, P. Bruckner, E. Rodriguez, K. H. Winterhalter, and L. Vaughan, *J. Biol. Chem.* **261,** 5965 (1983).
[45] H. Konomi, J. M. Seyer, Y. Ninomiya, and B. R. Olsen, *J. Biol. Chem.* **261,** 6742 (1986).
[46] M. H. Irwin and R. Mayne, *J. Biol. Chem.* **261,** 16281 (1986).
[47] M. Gordon, D. Gerecke, and B. R. Olsen, *Proc. Natl. Acad. Sci. U.S.A.* **84,** 6040 (1987).
[48] B. Dublet and M. van der Rest, *J. Biol. Chem.* **266,** 6853 (1991).

"mix and match" of structural/functional gene cassettes has been success-fully used during the evolution of extracellular matrix components. Thus, in addition to domains that are homologous to the NC4 and COL1 domains of the cartilage form of α1(IX) collagen (see below), both α1(XII) and α1(XIV) chains contain regions of homology with von Willebrand factor A domains and fibronectin type III repeats.[49-52]

Type XVI collagen was discovered as cDNAs by cross-hybridization during screening of human fibroblast libraries.[53] The predicted polypeptide chain contains 10 triple-helical domains; the carboxyl-most of these do-mains shows homology to the type IX COL1 domain. In addition, the amino-terminal non-triple-helical domain shows limited sequence identity with the α1(IX) NC4 domain. α1(XIX) collagen, initially given the designa-tion α1(Y),[54] was identified from partial cDNA clones during screening of a human rhabdomyosarcoma cDNA library at low stringency with a chicken α1(V) collagen probe. The complete structure of α1(XIX) is not known, but partial sequences indicate that it contains regions of similarity with the NC4 and COL1 domains of α1(IX) collagen. Also, gene localization studies[54] suggest that, in humans, the α1(XIX) gene is located on chromo-some 6 relatively close to two other FACIT collagen genes, α1(IX) and α1(XII).[55,56]

Structure

The different members of the FACIT family show large variations in structure, but they are all multidomain proteins with both triple-helical and non-triple-helical regions. In type IX collagen the three triple-helical

[49] M. K. Gordon, D. R. Gerecke, B. Dublet, M. van der Rest, and B. R. Olsen, *J. Biol. Chem.* **264,** 19772 (1989).

[50] M. Yamagata, K. M. Yamada, S. S. Yamada, T. Shinomura, H. Tanaka, Y. Nishida, M. Obara, and K. Kimata, *J. Cell Biol.* **115,** 209 (1991).

[51] C. Wälchli, J. Trueb, B. Kessler, K. H. Winterhalter, and B. Trueb, *Eur. J. Biochem.* **212,** 483 (1993).

[52] D. R. Gerecke, J. W. Foley, P. Castagnola, M. Gennari, B. Dublet, R. Cancedda, T. F. Linsenmayer, M. van der Rest, B. R. Olsen, and M. K. Gordon, *J. Biol. Chem.* **268,** 12177 (1993).

[53] T.-C. Pan, R.-Z. Zhang, M.-G. Mattei, R. Timpl, and M.-L. Chu, *Proc. Natl. Acad. Sci. U.S.A.* **89,** 6565 (1992).

[54] H. Yoshioka, H. Zhang, F. Ramirez, M.-G. Mattei, M. Moradi-Ameli, M. van der Rest, and M. K. Gordon, *Genomics* **13,** 884 (1992).

[55] M. L. Warman, G. E. Tiller, P. A. Polumbo, M. F. Seldin, J. M. Rochelle, J. H. M. Knoll, S.-D. Cheng, and B. R. Olsen, *Genomics* **17,** 694 (1993).

[56] S. P. Oh, R. W. Taylor, D. R. Gerecke, J. M. Rochelle, M. F. Seldin, and B. R. Olsen, *Genomics* **14,** 225 (1992).

domains represent most of the molecular structure, whereas in types XII and XIV two small triple-helical domains constitute less than 10% of the total amino acid sequence. As mentioned above, all members contain a short triple-helical carboxyl domain (COL1) with about 100–115 amino acid residues (per chain) and with a characteristic pair of cysteinyl residues at the carboxyl border. In the middle of the domain is a conserved imperfection in the Gly-X-Y triplet structure. Although the function of this domain is presently not known, its presence in collagenous polypeptides is currently used as a sufficiently characteristic fingerprint to include them in the FACIT group. In fact, it was the presence of such a conserved COL1 domain that led to the discovery of type XIV collagen. On the basis of the similarity of the COL1 domains in types IX and XII collagen, Dublet and van der Rest[48] hypothesized that pepsin treatment of any FACIT molecule should generate a pepsin-resistant disulfide-bonded COL1 fragment, and they searched successfully for such a fragment in extracts of bovine skin. Amino acid sequence analysis of the purified material showed similarity to the sequence in the COL1 region of $\alpha 1$(XII) collagen chains, indicating the existence in skin of a type XII-like molecule. This new component was named type XIV collagen[48] and with oligonucleotide primers based on the partial amino acid sequences available, cDNAs encoding the $\alpha 1$(XIV) chain were subsequently cloned.[57]

A second conserved domain shared by the different members of the FACIT family is a non-triple-helical domain with similarity to the amino-terminal NC4 domain of $\alpha 1$(IX) collagen chains.[58] In a heterotrimeric molecule such as type IX collagen, this domain is present in only one of the three polypeptide subunits; in the homotrimeric molecules of types XII and XIV collagen all polypeptide subunits contain the domain. Consequently, the globular central region of type XII and XIV molecules appears larger in the electron microscopic images than the amino-terminal domain of type IX collagen molecules.[59–62]

With the cloning of cDNA encoding the chicken $\alpha 3$(IX) collagen

[57] M. K. Gordon, P. Castagnola, B. Dublet, T. F. Linsenmayer, M. van der Rest, and B. R. Olsen, *Eur. J. Biochem.* **201**, 333 (1991).

[58] G. Vasios, I. Nishimura, H. Konomi, M. van der Rest, Y. Ninomiya, and B. R. Olsen, *J. Biol. Chem.* **263**, 2324 (1988).

[59] L. Vaughan, M. Mendler, S. Huber, P. Bruckner, K. H. Winterhalter, M. II. Irwin, and R. Mayne, *J. Cell Biol.* **106**, 991 (1988).

[60] S. L. Watt, G. P. Lunstrum, A. M. McDonough, D. R. Keene, R. E. Burgeson, and M. P. Morris, *J. Biol. Chem.* **267**, 20093 (1992).

[61] M. Koch, C. Bernasconi, and M. Chiquet, *Eur. J. Biochem.* **207**, 847 (1992).

[62] R. G. Brewton, M. V. Ouspenskaia, M. van der Rest, and R. Mayne, *Eur. J. Biochem.* **205**, 443 (1992).

chain,[62,63] the primary structure of all three polypeptide subunits in chicken type IX collagen has now been established. In addition, cDNA and/or genomic DNA clones are available for the $\alpha1(IX)$ and $\alpha2(IX)$ chains from rat,[64] mouse,[65,66] bovine,[67] and human type IX collagen,[64,68,69] and the chromosomal locations for the corresponding genes are being determined in both the murine and human genomes.[55,64,69,70] The comparison of the $\alpha1(IX)$, $\alpha2(IX)$, and $\alpha3(IX)$ sequences shows that only the $\alpha1(IX)$ chain contributes significantly to the amino-terminal globular domain of type IX molecules in cartilage. This is because the $\alpha1(IX)$ gene is transcribed from an upstream promoter in chondrocytes.[71,72] This generates a transcript encoding a large (about 250 amino acid residues) amino-terminal NC4 domain. In tissues such as vitreous humor,[73,74] chicken primary corneal stroma,[75,76] neural retina,[77] and perinotochordal matrix,[77-79] a downstream promoter and alternative exon are used, resulting in a shorter

[63] R. Har-El, Y. D. Sharma, A. Aguilera, N. Ueyama, J.-J. Wu, D. R. Eyre, L. Juricic, S. Chandrasekaran, M. Li, H.-D. Nah, W. B. Upholt, and M. L. Tanzer, *J. Biol. Chem.* **267**, 10070 (1992).

[64] T. Kimura, M.-G. Mattei, J. W. Stevens, M. B. Goldring, Y. Ninomiya, and B. R. Olsen, *Eur. J. Biochem.* **179**, 71 (1989).

[65] I. Rokos, Y. Muragaki, M. Warman, and B. R. Olsen, *Matrix Biol.* **14**, 1 (1994).

[66] K. Elima, M. Metsaranta, J. Kallio, M. Perälä, I. Eerola, S. Garofalo, B. de Crombrugghe, and E. Vuorio, *Biochim. Biophys. Acta* **1130**, 78 (1992).

[67] Y. Ninomiya, P. Castagnola, D. Gerecke, M. K. Gordon, O. Jacenko, P. LuValle, M. McCarthy, Y. Muragaki, I. Nishimura, S. Oh, N. Rosenblum, N. Sato, S. Sugrue, R. Taylor, G. Vasios, N. Yamaguchi, and B. R. Olsen, in "Extracellular Matrix Genes" (L. J. Sandell and C. D. Boyd, eds.), p. 79. Academic Press, San Diego, 1990.

[68] Y. Muragaki, T. Kimura, Y. Ninomiya, and B. R. Olsen, *Eur. J. Biochem.* **192**, 703 (1990).

[69] M. Perälä, M. Hänninen, J. Hästbacka, K. Elima, and E. Vuorio, *FEBS Lett.* **319**, 177 (1993).

[70] M. L. Warman, M. T. McCarthy, M. Perälä, E. Vuorio, J. H. M. Knoll, C. N. McDaniels, R. Mayne, D. R. Beier, and B. R. Olsen, *Genomics* (in press).

[71] I. Nishimura, Y. Muragaki, and B. R. Olsen, *J. Biol. Chem.* **264**, 20044 (1989).

[72] Y. Muragaki, I. Nishimura, A. Henney, Y. Ninomiya, and B. R. Olsen, *Proc. Natl. Acad. Sci. U.S.A.* **87**, 2400 (1990).

[73] T. Yada, S. Suzuki, K. Kobayashi, M. Kobayashi, T. Hoshino, K. Horie, and K. Kimata, *J. Biol. Chem.* **265**, 6992 (1989).

[74] D. W. Wright and R. Mayne, *J. Ultrastruct. Mol. Struct. Res.* **100**, 224 (1990).

[75] K. K. Svoboda, I. Nishimura, S. P. Sugrue, Y. Ninomiya, and B. R. Olsen, *Proc. Natl. Acad. Sci. U.S.A.* **85**, 7496 (1988).

[76] J. M. Fitch, A. Mentzer, R. Mayne, and T. F. Linsenmayer, *Dev. Biol.* **128**, 396 (1988).

[77] R. E. Swiderski and M. Solursh, *Dev. Dyn.* **194**, 118 (1992).

[78] R. Perris, D. Krotoski, and M. Bronner-Fraser, *Development (Cambridge, UK)* **113**, 969 (1991).

[79] M. Hayashi, K. Hayashi, K. Iwata, R. L. Trelstad, T. F. Linsenmayer, and R. Mayne, *Dev. Dyn.* **194**, 169 (1992).

transcript that encodes an $\alpha 1$(IX) isoform with as short an NC4 domain as in the $\alpha 2$(IX) and $\alpha 3$(IX) chains.

As mentioned above, the NC3 domain of the $\alpha 2$(IX) chain contains an attachment site for a glycosaminoglycan side chain. The extent of glycosylation at this site is less than 100%, making type IX collagen a "part-time" proteoglycan.[80] In the vitreous of the eye the glycosylation appears more complete; in the chicken vitreous the chondroitin sulfate side chain is very large (350 kDa),[73] making type IX collagen the major proteoglycan and contributor to the gellike properties of the avian vitreous.

The interaction between type IX collagen molecules and cartilage collagen fibrils is stabilized by the formation of covalent cross-links between lysyl residues in type IX chains and lysyl residues in the amino and carboxyl telopeptides of type II collagen molecules. The identification of such cross-links has been accomplished by isolation and sequencing of cross-linked peptides.[81,82] The position of cross-link sites within the COL2 domain of type IX molecules, in relation to the relative length of the overlap and gap regions within the type II-containing cartilage fibrils, suggests that type IX and II molecules are oriented in antiparallel directions.

Although types XII and XIV collagen appear to be associated with collagen fibrils (at least at the level of resolution provided by immunogold electron microscopy) no cross-links between the two FACIT components and type I collagen molecules have been identified. Thus, rotary shadowing electron microscopy to visualize type XII/XIV molecules on fibrillar surfaces directly (as has been done for type IX/II complexes[59]) has not been accomplished, because the technique requires exposure of fibrillar preparations to denaturing solvents so that soluble fibril-associated matrix molecules can be removed.

The structure of type XII molecules has been deduced from cDNA sequencing, protein chemistry, and rotary shadowing electron microscopy of purified molecules. The molecules are homotrimers of $\alpha 1$(XII) collagen chains; two major isoforms of these chains have been described. The two isoforms result from alternative splicing of a primary transcript[83] such that two transcripts of different size are produced. Both transcripts have the same 5' untranslated sequence and encode identical signal peptides, but differ in that the short form (probably due to exon skipping) encodes a

[80] Y. Yada, M. Arai, S. Suzuki, and K. Kimata, J. Biol. Chem. **267,** 9391 (1992).
[81] M. van der Rest and R. Mayne, J. Biol. Chem. **263,** 1615 (1988).
[82] J. J. Wu, P. E. Woods, and D. R. Eyre, J. Biol. Chem. **267,** 23007 (1992).
[83] J. Trueb and B. Trueb, Biochim. Biophys. Acta **1171,** 97 (1992).

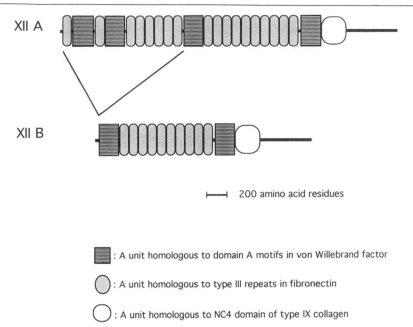

FIG. 2. Diagram of the two isoforms of α1(XII) collagen chains. The two forms have identical amino-terminal signal peptides and carboxyl regions, but the long form, XIIA, contains two von Willebrand factor A-like domains and eight fibronectin type III repeats that are not found in the short form, XIIB. The triple-helical domains at the carboxyl end of the chains are indicated by a solid line; for a detailed diagram of this region, see Fig. 1. (Modified from Trueb and Trueb.[83])

polypeptide containing 1164 fewer amino acid residues within the amino-terminal non-triple-helical domain than the long form. The long isoform, collagen XIIA, is the major form synthesized by cells in culture,[61,84,85] whereas the short form, collagen XIIB, is the major tissue form.[85-87] The two forms are identical in their carboxyl regions, but differ in the lengths of the amino-terminal non-triple-helical arms. XIIA molecules contain four von Willebrand factor A motifs and 18 fibronectin type III repeats in each of the arms, whereas XIIB molecules contain only two von Willebrand factor A-like domains and 10 fibronectin type III repeats (Fig. 2).

[84] G. P. Lunstrum, A. M. McDonough, M. P. Marinkovich, D. R. Keene, N. P. Morris, and R. E. Burgeson, *J. Biol. Chem.* **267,** 20087 (1992).

[85] S. P. Oh, C. M. Griffith, E. D. Hay, and B. R. Olsen, *Dev. Dyn.* **196,** 37 (1993).

[86] B. Dublet, S. Oh, S. P. Sugrue, M. K. Gordon, D. R. Gerecke, B. R. Olsen, and M. van der Rest, *J. Biol. Chem.* **264,** 13150 (1989).

[87] G. P. Lunstrum, N. P. Morris, A. M. McDonough, D. R. Keene, and R. E. Burgeson, *J. Cell Biol.* **113,** 963 (1991).

Collagen XIIA contains glycosaminoglycan chains and is, therefore, also a proteoglycan[60,61]; collagen XIIB does not appear to be glycosylated.[60,86,87] cDNA and/or genomic DNA clones encoding chicken,[47,49,50,83] mouse,[56,85] and human $\alpha 1$(XII)[56] collagen chains have been isolated and the chromosomal locations determined for the murine and human genes.[56]

The sequencing of cDNA clones encoding $\alpha 1$(XIV) collagen chains[51,52,57] has also led to definition of the complete primary structure of the type XII collagen homolog, collagen XIV. Comparison of the sequence with that previously published[88] for a protein called undulin[89] suggests that undulin is a proteolytic degradation product of collagen XIV.[57,90] Although the NC3 arms of collagen XIV molecules are slightly shorter than the arms of collagen XIIB, the overall domain structure of the arms is similar in the two types of molecule: von Willebrand factor A-like domains separated by several (10 in XII and 7 in XIV) fibronectin type III repeats. From cDNA cloning there is evidence that several splice variants of collagen XIV exist; variations occur both at the 5' and 3' ends.[52,90]

Expression and Function

Given the association of type IX collagen molecules with the surface of type II-containing fibrils, it is not surprising that type IX collagen expression is also seen in several noncartilaginous tissues in which type II collagen is synthesized. Type IX collagen expression has been analyzed by immunohistochemistry with both polyclonal and monoclonal antibodies.[91] Of special interest are studies of the expression of the two isoforms of $\alpha 1$(IX) mRNAs using Northern analyses, in situ hybridization, or reverse transcriptase-polymerase chain reaction (RT-PCR) with probes and primers that distinguish between the two isoforms.[75,77,92,93] These studies suggest that although the long isoform of $\alpha 1$(IX) is expressed in chondrogenic regions of developing embryos,[77,93] the short isoform has a broader distribution. For example, in the developing chick embryo limb the short form is distributed throughout the nonchondrogenic, nonmyogenic mesenchymal

[88] M. Just, H. Herbst, M. Hummel, H. Dürkop, D. Tripier, H. Stein, and D. Schuppan, J. Biol. Chem. 266, 17326 (1991).

[89] D. Schuppan, M. C. Cantaluppi, J. Becker, A. Veit, T. Bunte, D. Troyer, F. Schuppan, M. Schmid, R. Ackerman, and E. G. Hahn, J. Biol. Chem. 265, 8823 (1990).

[90] J. Trueb and B. Trueb, Eur. J. Biochem. 207, 549 (1992).

[91] M. van der Rest and R. Mayne, in "Structure and Function of Collagen Types" (R. Mayne and R. E. Burgeson, eds.), p. 195. Academic Press, Orlando, FL, 1987.

[92] T. F. Linsenmayer, E. Gibney, M. K. Gordon, J. K. Marchant, M. Hayashi, and J. M. Fitch, Invest. Ophthalmol. Visual Sci. 31, 1271 (1990).

[93] C.-Y. Liu, B. R. Olsen, and W. W.-Y. Kao, Dev. Dyn. 198, 160 (1993).

regions of the limb,[77] and it is the major form in developing chicken and mouse eyes.[75,93]

Studies of transgenic mice with mutations in type IX are providing insights into the function of this collagen. Using the technique of homologous recombination in embryonic stem cells, Fässler *et al.* have generated mice with inactivated $\alpha1$(IX) collagen alleles.[94] Surprisingly, homozygous mutant mice develop normally and appear normal at birth. However, over a period of several months after birth they develop a degenerative joint cartilage disease with histological features that resemble osteoarthritis in humans. A similar progressive age-dependent joint cartilage disease has been observed in transgenic mice expressing a dominant negative mutant $\alpha1$(IX) chain.[95] It appears, therefore, that type IX collagen is not critical for the developmental assembly of cartilage, but is important for the maintenance of cartilage (particularly articular cartilage) after birth. Because the amino-terminal NC4 domain of $\alpha1$(IX) chains is specific for the cartilage form of type IX collagen, it is possible that it plays a key role in this maintenance function, and it has been proposed that type IX molecules act as molecular bridges between cartilage collagen fibrils and other matrix components, perhaps proteoglycans.[15]

A similar bridging function has been hypothesized for the other members of the FACIT group,[15] but less direct evidence is available for such a role than in the case of type IX. As mentioned above, although type IX molecules are cross-linked to type II collagen, no cross-links that would indicate a direct association between types XII/XIV and fibrillar collagens have been identified. Attempts to demonstrate direct binding to fibrillar collagen have also been negative, but interactions have been demonstrated between type XIV collagen and decorin, heparan sulfate, and the triple-helical domain of type VI collagen.[96,97] Also, *in vitro* studies have demonstrated that the amino-terminal non-triple-helical NC3 domains of type XII and XIV collagen promote contraction of collagen gels.[98] These results, together with the immunogold localization of the two types of molecules along collagen fibrils,[99] support the bridging hypothesis, although the detailed interactions are not yet clear.

[94] R. Fässler, P. N. J. Schnegelsberg, J. Dausman, T. Shinya, Y. Muragaki, M. T. McCarthy, B. R. Olsen, and R. Jaenisch, *Proc. Natl. Acad. Sci. U.S.A.* **91,** 5070 (1994).

[95] K. Nakata, K. Ono, J.-I. Miyazaki, B. R. Olsen, Y. Muragaki, E. Adachi, K.-I. Yamamura, and T. Kimura, *Proc. Natl. Acad. Sci. U.S.A.* **90,** 2870 (1993).

[96] B. Font, E. Aubert-Foucher, D. Goldschmidt, E. Eichenberger, and M. van der Rest, *J. Biol. Chem.* **268,** 25015 (1993).

[97] J. C. Brown, K. Mann, H. Wiedemann, and R. Timpl, *J. Cell Biol.* **120,** 557 (1993).

[98] T. Nishiyama, A. M. McDonough, and R. E. Burgeson, *Mol. Biol. Cell* **3,** 74a (1992).

[99] D. R. Keene, G. P. Lunstrum, N. P. Morris, D. W. Stoddard, and R. E. Burgeson, *J. Cell Biol.* **113,** 971 (1991).

Short-Chain Collagens

Discovery

Type VIII and type X collagens are classified as *short-chain collagens* because of the relatively small size (<750 amino acid residues) of their protein products as compared with the fibrillar collagens.[100,101] These molecules consists of a single short triple-helical domain (COL1) flanked at both ends by non-triple-helical (NC1 and NC2) domains[100,102,103] (Fig. 1). Similarities between them are observed not only in their size and domain structure, but also in the genes that encode both collagens. One large 3′ exon encodes almost all of the translated region, including the entire COL1 and NC1 domains.[103,104] The COL1 domain contains eight imperfections in the Gly-X-Y repeat, with Gly-X or X-Y instead of a full triplet. The relative locations of these imperfections within the COL1 domain are remarkably similar in the different short-chain collagen chains. Moreover, the carboxyl three-quarters of the NC1 domains have a high degree of sequence similarity between the different chains; this provided the basis for molecular cloning of the human and mouse $\alpha 2$(VIII) gene[102] and the human $\alpha 1$(X) gene.[105]

Type VIII collagen was initially detected in biosynthetic studies of bovine aortic endothelial cells[106] and rabbit corneal endothelial cells,[107] and therefore given the designation EC (endothelial cell) collagen.[106] The molecular cloning of a type VIII collagen cDNA was from a library made with mRNA from rabbit corneal endothelial cells.[108] On the basis of the identity between the amino acid sequence derived from this cDNA and the amino acid sequences of cyanogen bromide peptides of type VIII collagen obtained from rabbit corneal Descemet's membrane, the cDNA was identified as encoding a type VIII collagen chain and designated $\alpha 1$(VIII).[108] However, since the sequence of the $\alpha 1$(VIII) cDNA did not predict the sequences of three additional cyanogen bromide peptides iso-

[100] N. Yamaguchi, R. Mayne, and Y. Ninomiya, *J. Biol. Chem.* **266**, 4508 (1991).

[101] M. K. Gordon and B. R. Olsen, *Curr. Opin. Cell Biol.* **2**, 833 (1990).

[102] Y. Muragaki, O. Jacenko, S. Apte, M. G. Mattei, Y. Ninomiya, and B. R. Olsen, *J. Biol. Chem.* **266**, 7721 (1991).

[103] Y. Ninomiya, M. K. Gordon, M. van der Rest, T. Schmid, T. Linsenmayer, and B. R. Olsen, *J. Biol. Chem.* **261**, 5041 (1986).

[104] P. LuValle, Y. Ninomiya, N. D. Rosenblum, and B. R. Olsen, *J. Biol. Chem.* **263**, 18378 (1988).

[105] S. Apte, M.-G. Mattei, and B. R. Olsen, *FEBS Lett.* **282**, 393 (1991).

[106] H. Sage, P. Pritzl, and P. Bornstein, *Biochemistry* **19**, 5747 (1980).

[107] P. D. Benya, *Renal Physiol.* **3**, 30 (1980).

[108] N. Yamaguchi, P. D. Benya, M. van der Rest, and Y. Ninomiya, *J. Biol. Chem.* **264**, 16022 (1989).

lated from Descemet's membrane, it was apparent that at least one additional peptide chain had to be present in type VIII collagen.[102,108] This additional chain, α2(VIII), was cloned using a PCR approach,[102] taking advantage of the high degree of sequence similarity within the NC1 region of α1(VIII) and α1(X) chains.

On the basis of the highly conserved region in the NC1 domain, 5' and 3' primers for the PCR were synthesized (Fig. 3). The sequence of the 5'-sense primer differed from the corresponding sequence in the chicken α1(X) gene by only two nucleotides and from the rabbit and human α1(VIII) genes by only one nucleotide. The 3'-antisense primer was identical to the sequence of the corresponding region in the rabbit α1(VIII) collagen gene and differed from the sequences of the human α1(VIII) and the chicken α1(X) genes in two and six nucleotide positions, respectively. Using mouse and human genomic DNAs as templates with the primers, the PCR was performed. The PCR products of expected length were cloned into λZAPII, and isolated on the basis of positive

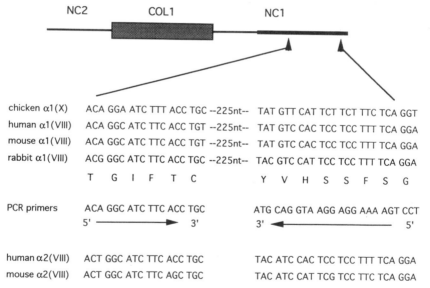

FIG. 3. Diagram showing the domain structure of short-chain collagen chains (not drawn to scale). The sequence of the carboxyl three-quarters of the non-triple-helical domain NC1 is highly conserved between the different members of the family. Below the diagram, the nucleotide sequences of two regions within the conserved portion of NC1 are shown, together with the amino acid sequences, for chicken α1(X), and human, mouse, and rabbit α1(VIII) chains. On the basis of these sequences, two primers were used for PCR with genomic DNA to isolate the human and mouse α2(VIII) genes. (Modified from Muragaki et al.[102])

hybridization with α1(VIII) and α1(X) probes. The use of a PCR product clone as probe to screen a human genomic library subsequently led to the isolation of a gene with a high degree of similarity to the α1(VIII) and α1(X) collagen genes. The amino acid sequence predicted from this gene contained the type VIII cyanogen bromide peptide sequences that were not encoded by α1(VIII). Therefore, the gene represented by the new clone was designated α2(VIII).[102]

Type X collagen, a hypertrophic cartilage-specific molecule, was first isolated as a low molecular weight collagenous protein from cultures of chicken chondrocytes.[109–111] It is a homotrimer of three α1(X) chains (each having a molecular mass of 59 kDa) that associate to form a dumbbell-shaped molecule approximately 150 nm in length.[112] Although the N-terminal protein sequence of the α1(X) chains has been obtained directly, the complete primary structure of chicken type X collagen was first determined by cDNA cloning.[103] Ninomiya et al.[103] used differential hybridization of a cDNA library from mature chondrocytes (using ^{32}P-labeled RNA from mature chondrocytes as a positive probe, and ^{32}P-labeled RNA from chick sterna as a negative probe) to obtain the first type X collagen cDNA clones. Cloning of the mammalian cDNAs encoding α1(X) collagen was slow to follow, but was eventually achieved by immunoscreening with type X antibodies (used to isolate bovine cDNAs)[113] and PCR using primers based on the compared sequences of all the members of the short-chain collagen family.[105,114–118]

Structure

Type VIII collagen has a triple-helical domain (COL1) flanked by amino (NC2) and carboxyl (NC1) non-triple-helical domains as shown in Fig. 1. The molecules are probably heterotrimers of α1(VIII) and α2(VIII)

[109] T. M. Schmid, Ph.D. Thesis, University of Illinois, Urbana (1980).
[110] G. J. Gibson and M. H. Flint, *J. Cell Biol.* **101**, 277 (1985).
[111] O. Capasso, E. Gionti, G. Pontarelli, F. S. Amesi-Impiombato, L. Nitsch, G. Tajana, and R. Cancedda, *Exp. Cell Res.* **142**, 197 (1982).
[112] T. M. Schmid and T. F. Linsenmayer, *in* "Structure and Function of Collagen Types" (R. Mayne and R. E. Burgeson, eds.), p. 223. Academic Press, Orlando, FL, 1987.
[113] J. T. Thomas, A. P. L. Kwan, M. E. Grant, and R. P. Boot-Handford, *Biochem. J.* **273**, 141 (1991).
[114] J. T. Thomas, C. J. Cresswell, B. Rash, H. Nicolai, T. Jones, E. Solomon, M. E. Grant, and R. P. Boot-Handford, *Biochem. J.* **280**, 617 (1991).
[115] S. S. Apte and B. R. Olsen, *Matrix* **13**, 165 (1993).
[116] S. S. Apte, M. F. Seldin, M. Hayashi, and B. R. Olsen, *Eur. J. Biochem.* **206**, 217 (1992).
[117] E. Reichenberger and K. von der Mark, *Dev. Biol.* **148**, 562 (1992).
[118] J. T. Thomas, C. J. Cresswell, B. Rash, H. Nicolai, T. Jones, E. Solomon, M. E. Grant, and R. P. Boot-Handford, *Biochem. J.* **280**, 617 (1991).

polypeptide chains. Analysis of bovine Descemet's membrane shows that the ratio of $\alpha1(VIII)$ to $\alpha2(VIII)$ is between 1.5 and 2, and therefore the chain composition of the molecule is likely to be $[\alpha1(VIII)]_2\alpha2(VIII)$.[119] Type X collagen molecules have simliar structural domains. A comparison of cDNA-derived sequences shows that the chicken, bovine, human, and mouse type X collagen molecules are remarkably similar.[113,115] The COL domain is 463 amino acid residues long (460 residues long in chicken), and is interrupted by eight triple-helical domains. Of these, three are of the G-X-Y-X-Y-G type, and four of the G-X-G type; each imperfection is located at the same relative location in all species studied thus far. The bovine $\alpha1(X)$ chain has a cysteine residue in the N-terminal-most of these imperfections that participates in disulfide bond formation[113]; this is not a feature of other species. The NC1 domain is 161 amino acid residues long and has 2 conserved cysteine residues that have been implicated in disulfide bond formation. It is possible that sequences in the NC1 domain may be instrumental in initiating the association and correct folding of the $\alpha1(X)$ chains.[120] The COL10A1 gene is located on chromosome 6q21[105]; the corresponding mouse gene has been mapped to chromosome 10.[116]

Expression and Function

The immunohistochemical localization of type VIII collagen has been reported using monoclonal[121–123] or polyclonal[124,125] antibodies. Different antibodies show minor differences, but immunohistochemical detection of type VIII collagen is observed in the sclera, choroid, optic nerve sheath, dura mater of the spinal cord, the perichondrium of cartilage, the meninges, periosteum, connective tissue around hair follicles, and arterioles and venules as well as Descemet's membrane. Using a fragment of genomic DNA encoding the mouse $\alpha1(VIII)$ chain as a probe, Northern blot analysis of newborn mice showed that the highest levels of mRNA were in the calvarium, eye, and skin.[126] *In situ* hybridization revealed that

[119] K. Mann, R. Jander, E. Korsching, K. Kuhn, and J. Rauterberg, *FEBS Lett.* **273**, 168 (1990).

[120] A. Brass, K. Kadler, J. T. Thomas, M. E. Grant, and R. P. Boot-Handford, *FEBS Lett.* **303**, 126 (1992).

[121] H. Sawada, H. Konomi, and K. Hirosawa, *J. Cell Biol.* **110**, 219 (1990).

[122] H. Sawada and H. Konomi, *Cell Struct. Funct.* **16**, 455 (1991).

[123] Y. Tamura, H. Konomi, H. Sawada, S. Takashima, and A. Nakajima, *Invest. Ophthalmol. Visual Sci.* **32**, 2636 (1991).

[124] J. Salonen, D. Oda, S. E. Funk, and H. Sage, *J. Periodontal Res.* **26**, 355 (1991).

[125] P. F. Davis, P. A. Ryan, R. Kittelberger, and N. S. Greenhill, *Biochem. Biophys. Res. Commun.* **171**, 260 (1990).

[126] Y. Muragaki, C. Shiota, M. Inoue, A. Ooshima, B. R. Olsen, and Y. Ninomiya, *Eur. J. Biochem.* **207**, 895 (1992).

$\alpha 1$(VIII) RNA is present in skin keratinocytes, corneal epithelial and endothelial cells, lens epithelial cells, as well as mesenchymal cells surrounding cartilage and calvarial bone, and in the meninges surrounding the brain.[126]

Type VIII collagen is considered a major component of the hexagonal lattice structure of Descemet's membrane,[121] where it may provide mechanical strength in the structure. Clinically, an altered ratio of corneal collagenous proteins in patients with pseudophakic bullous keratopathy has been reported.[127] Type VIII collagen has also been suggested to be important for cellular proliferation and angiogenesis.[128]

Type X collagen has a unique pattern of expression. As has been demonstrated in chicken,[112] mouse,[116] and human[117] cartilage, both the mRNA and type X collagen protein are synthesized by hypertrophic chondrocytes in regions of cartilage undergoing endochondral ossification. It is conspicuous by its absence in the proliferating chondrocytes of the mammalian growth plates. Thus, type X collagen is widely used as a marker of chondrocyte differentiation.

The process of chondrocyte hypertrophy occurs at the expense of the surrounding cartilage matrix, and is accompanied by the synthesis of significant amounts of type X collagen. It has been shown that, like type VIII collagen, type X collagen participates *in vitro* in the formation of supramolecular aggregates that resemble hexagonal lattices.[129] It is likely that the fine lattice structure seen in the pericellular matrix of hypertrophic chondrocytes by electron microscopy represents type X collagen aggregates.[130] Although it has been hypothesized[112] that type X collagen may be involved in matrix mineralization, or in regulating the invasion of cartilage by blood vessels, evidence suggests that the primary role of type X collagen is to provide structural support to the hypertrophic chondrocytes.

In Schmid metaphyseal chondrodysplasia, an autosomal dominantly inherited condition characterized by short stature, Warman et al.[131] have identified a 13-bp deletion in the NC1 coding domain. This frameshift mutation alters the C-terminal 60 amino acid residues and shortens the NC1 domain by 9 residues. Presumably the alterations in the NC1 domain

[127] M. C. Kenney and M. Chwa, *Cornea* **9,** 115 (1990).
[128] W. Paulus, H. Sage, K. Jellinger, and W. Roggendorf, *Acta Histochem. Suppl.* **42,** 195 (1992).
[129] A. P. L. Kwan, C. E. Cummings, J. A. Chapman, and M. E. Grant, *J. Cell Biol.* **114,** 597 (1991).
[130] C. E. Farnum and N. J. Wilsman, *J. Histochem. Cytochem.* **31,** 765 (1983).
[131] M. L. Warman, M. Abbott, S. S. Apte, T. Hefferon, I. McIntosh, D. H. Cohn, J. T. Hecht, B. R. Olsen, and C. A. Francomano, *Nat. Genet.* **5,** 79 (1993).

affect type X collagen assembly and/or function. No tissues other than the cartilage growth plates are affected in this disorder, which is characterized by the appearance of deformities and limb shortening coincidental with weight bearing. Additional mutations in the NC1 domain of type X collagen have been reported in other families with Schmid metaphyseal chondrodysplasia.[132–134]

Transgenic mice that express a chicken type X collagen transgene containing an in-frame deletion in the COL domain [and thus a truncated $\alpha1(X)$ collagen chain] also demonstrate short limbs and other skeletal deformities.[135] Presumably, the truncated chains are able to associate with normal mouse $\alpha1(X)$ collagen chains resulting in nonfunctional heterotrimers that are rapidly degraded so that there is little normal mouse type X collagen present in the growth plates. Alternatively, the truncated chicken type X collagen chains may form truncated homotrimers that disrupt the normal type X collagen aggregates in the cartilage matrix. Growth plate histology in these mutant mice demonstrates significant changes primarily in the hypertrophic zones: hypertrophic chondrocytes are decreased in number, have pyknotic nuclei, and are flattened. The number of primary metaphyseal trabeculae are also reduced in these mice. Taken together, the data from Schmid metaphyseal chondrodysplasia and the transgenic mice support a primary structural role for type X collagen in the hypertrophic zone.

Type XIII Collagen

Discovery

Type XIII collagen was initially discovered as a low molecular weight human collagen.[136] The cDNA library of a human tumor cell line, HT-1080, was screened with the mouse cDNA, pE18, encoding part of the triple-helical domain of $\alpha2(IV)$ collagen, and four overlapping positive clones that covered one-third of the corresponding mRNA were isolated. The cDNA clone did not encode any known collagen. Using the cDNA as probe, human genomic clones were isolated and the novel protein

[132] G. A. Wallis, B. Rash, W. A. Sweetman, J. T. Thomas, M. Super, G. Evans, M. E. Grant, and R. P. Boot-Handford, *Am. J. Hum. Genet.* **54,** 169 (1994).
[133] I. McIntosh, M. H. Abbott, M. L. Warman, B. R. Olsen, and C. A. Francomano, *Hum. Mol. Genet.* **3,** 303 (1994).
[134] O. Jacenko, B. R. Olsen, and M. L. Warman, *Am. J. Hum. Genet.* **54,** 163 (1994).
[135] O. Jacenko, P. A. LuValle, and B. R. Olsen, *Nature (London)* **365,** 86 (1993).
[136] T. Pihlajaniemi, R. Myllyla, J. Seyer, M. Kurkinen, and D. J. Prockop, *Proc. Natl. Acad. Sci. U.S.A.* **84,** 940 (1987).

product of the gene was designated α1(XIII).[137] The complete primary structure of the human α1(XIII) collagen chain was determined from the analysis of several overlapping cDNAs isolated from a human endothelial cell library.[138] The structure of the human α1(XIII) collagen gene has been determined from analysis of genomic clones spanning 140 kb.[139] The gene consists of 39 exons ranging in size from 24 to 133 bp. Most of these exons are multiples of 9 bp. Nine exons are utilized to generate complex, alternatively spliced transcripts.[138,139]

Structure

The type XIII collagen molecule consists of four noncollagenous domains (NC1–NC4) and three collagenous domains (COL1–COL3), as shown in Fig. 1. The polypeptide length varies between 614 and 526 amino acid residues, depending on the use of alternative splice sites.[138] Owing to alternative splicing, the lengths of the COL1 and COL3 domains vary between 57–104 and 190–235 amino acid residues, respectively, whereas the NC2 and NC4 domains vary between 12–33 and 7–18 residues, respectively. The lengths of the constant domains NC1, COL2, and NC3 are 17, 172, and 22 amino acid residues, respectively.[140] The extensive alternative splicing of the COL13A1 gene and especially the splicing in collagenous domains makes the COL13A1 gene unique among the members of the collagen gene family.[138]

At present it is not known whether this collagen contains more than one type of chain.

Expression and Function

Localization of type XIII collagen was studied mainly by Northern blot analysis and *in situ* hybridization of human fetal tissues of 15–19 weeks gestation, using a human cDNA probe.[141] Northern hybridization showed that bone, cartilage, intestine, skin, and striated muscle contain mRNAs for type XIII collagen. An intense *in situ* signal was obtained in the epidermis, hair follicles, and nail root cells of the skin.[141] It has been found that type XIII collagen mRNAs are expressed in developing muscle

[137] L. Tikka, T. Pihlajaniemi, P. Hentte, D. J. Prockop, and K. Tryggvason, *Proc. Natl. Acad. Sci. U.S.A.* **85,** 7491 (1988).

[138] T. Pihlajaniemi and M. Tamminen, *J. Biol. Chem.* **265,** 16922 (1990).

[139] L. Tikka, O. Elomaa, T. Pihlajaniemi, and K. Tryggvason, *J. Biol. Chem.* **266,** 17713 (1991).

[140] M. Juvonen, M. Sandberg, and T. Pihlajaniemi, *J. Biol. Chem.* **267,** 24700 (1992).

[141] M. Sandberg, M. Tamminen, H. Hirvonen, E. Vuorio, and T. Pihlajaniemi, *J. Cell Biol.* **109,** 1371 (1989).

and connective tissue cells in the lung and kidney[138] and fibroblastoid stromal cells of the placental villi.[142] Studies of the expression of COL13A1 are complicated by the different splice variations that have been described.[138–142] No data on the precise function of type XIII collagen or on its role in disease are available at the present time. Type XIII collagen cDNAs have also been isolated from a human endothelial cell library and human fibrosarcoma cells.[138]

Multiplexins: A Novel Family of Collagens with Multiple Triple-Helical Domains

Discovery

Members of the Multiplexin family are the most recent additions to the collagen superfamily of proteins. The first described member of the family, $\alpha1(XV)$ collagen, was discovered as a cross-hybridizing cDNA, PF19, during the screening of a human placental cDNA library with a fibrillar collagen probe.[143] Sequence analysis indicated that PF19 encoded a portion of a novel collagen chain containing multiple triple-helical domains. This was quickly followed by the discovery of another collagen chain, $\alpha1(XVIII)$, with structural homology to $\alpha1(XV)$.[13,14] Two different laboratories independently isolated and sequenced murine $\alpha1(XVIII)$ cDNAs by cross-hybridization with probes for known collagens. Oh et al.[13] used a probe encoding mouse $\alpha1(XII)$ collagen, and Rehn et al.[14] screened mouse libraries with probes for $\alpha1(XIII)$ collagen. Human $\alpha1(XVIII)$ clones have also been isolated,[144] and the complete primary structure of human $\alpha1(XV)$ chains has been established.[145,146] The Col18a1 gene has been mapped[144] close to the loci for Col6a1 and Col6a2 on mouse chromosome 10. The COL15A1 and COL18A1 genes have been mapped to the human chromosomes 9q21–q22[147] and 21q22.3,[144] respectively.

[142] M. Juvonen, T. Pihlajaniemi, and H. Autioharmainen, Lab. Invest. 69, 541 (1993).

[143] J. C. Myers, S. Kivirikko, M. K. Gordon, A. S. Dion, and T. Pihlajaniemi, Proc. Natl. Acad. Sci. U.S.A. 89, 10144 (1992).

[144] S. P. Oh, M. L. Warman, M. F. Seldin, S.-D. Cheng, J. H. M. Knoll, S. T. Timmons, and B. R. Olsen, Genomics 19, 494 (1994).

[145] Y. Muragaki, N. Abe, Y. Ninomiya, B. R. Olsen, and A. Ooshima, J. Biol. Chem. 269, 4042 (1994).

[146] S. Kivirikko, P. Heinamaki, M. Rehn, N. Honkanen, J. C. Myers, and T. Pihlajaniemi, J. Biol. Chem. 269, 4773 (1994).

[147] K. Huebner, L. A. Cannizzaro, E. W. Jabs, S. Kivirikko, H. Manzone, T. Pihlajaniemi, and J. C. Myers, Genomics 14, 220 (1992).

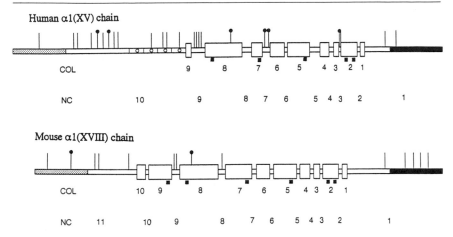

FIG. 4. Diagram comparing the domain structure of human α1(XV) and mouse α1(XVIII) collagen chains. Triple-helical (COL) regions are indicated by wide rectangular areas; non-triple-helical (NC) domains are indicated by thin rectangular areas. Homologous regions in α1(XV) and α1(XVIII) chains are indicated by dark-stippled boxes in carboxyl regions and light-stippled boxes in amino regions. Small black boxes indicate relative locations of imperfections in Gly-X-Y repeats. Vertical lines indicate potential O-linked glycosaminoglycan attachment sites and ball-and-sticks indicate N-linked glycosylation sites. Each unit of the four-amino acid sequence repeats in the NC10 domain of α1(XV) is indicated by a circle. (Modified from Muragaki et al.[145])

Structure

The cDNA sequences demonstrate that α1(XV) and α1(XVIII) chains contain multiple [9 for α1(XV) and 10 for α1(XVIII)] triple-helical (COL) domains, interrupted and flanked by non-triple-helical (NC) regions (Fig. 4). This domain structure, with a *multiple* triple-helix domain and *in*terruptions, is the basis for the proposed name, Multiplexins, for this family of collagens.[13] The way that the name was derived does not mean, however, that all collagens with multiple, interrupted triple-helical domains belong in this family. The grouping of α1(XV) and α1(XVIII) chains in a family is based on the striking similarity in the lengths of their six most carboxyl-terminal COL domains and regions of sequence homology within the amino- and carboxyl-terminal NC domains (Fig. 4).[13,14] Also, both chains contain a large number of Ser-Gly dipeptide sequences within local sequence contexts that correspond to consensus sequences for glycosaminoglycan attachment sites in proteoglycan core proteins.[13,14,144,148] It is,

[148] M. A. Bourdon, in "Extracellular Matrix Genes" (L. J. Sandell and C. D. Boyd, eds.), p. 157. Academic Press, San Diego, 1990.

therefore, possible that types XV and XVIII collagens are also, like type IX, part-time proteoglycans. Interestingly, the amino-terminal non-triple-helical domain NC10 of α1(XV) collagen contains a region (Fig. 4) with a tandem repeat structure of 45 amino acid residues repeated four times; the amino acid sequence shows a high degree of similarity to rat cartilage proteoglycan core protein.[145]

Could α1(XV) and α1(XVIII) collagen chains form heterotrimeric molecules and therefore represent two subunits of the same collagen type? This is unlikely because the amino-terminal COL domains in the two chains are different in size and number. Also, the levels of their mRNAs differ markedly in different tissues. For example, whereas human α1(XVIII) transcript levels are extremely high in liver,[144] human α1(XV) transcripts are practically undetectable in liver.[145]

Expression and Function

Northern blot analyses demonstrate that α1(XV) and α1(XVIII) collagen transcripts are expressed at high levels in internal, highly vascularized organs.[13,14,144,145] It has been suggested,[144] therefore, that they play a role in perivascular matrix assembly and/or structure. Clearly, future *in situ* hybridization and immunohistochemical studies are needed to map the distribution of these interesting new collagens in various organs and their cellular origins.[149–151]

Acknowledgments

The work from the authors' laboratory was supported by grants from the National Institutes of Health (AR36819, AR36820, HL33014, EY07334) and the Arthritis Foundation. Support was also provided by Osteoarthritis Sciences, Inc. Transgenic mice carrying a mutated type X collagen gene were generated by microinjection through the transgenic service of DNX, Inc. under a contract from the National Institutes of Health. We thank M. Jakoulov for excellent secretarial assistance.

[149] Y. Muragaki, M.-G. Mattei, N. Yamaguchi, B. R. Olsen, and Y. Ninomiya, *Eur. J. Biochem.* **197**, 615 (1991).

[150] L. Pajunen, M. Tamminen, E. Solomon, and T. Pihlajaniemi, *Cytogenet. Cell Genet.* **52**, 190 (1989).

[151] N. G. Copeland, D. J. Gilbert, K. Li, D. Sawamura, G. J. Guidice, M. L. Chu, N. A. Jenkins, and J. Uitto, *Genomics* **15**, 180 (1993).

[2] Fibrillin: Monomers and Microfibrils

By Lynn Y. Sakai and Douglas R. Keene

Introduction

Fibrillin was originally identified as the antigen recognized by monoclonal antibodies produced after immunization with a preparation of pepsin-resistant peptides extracted from human amnion.[1] Previous experience had shown that monoclonal antibody (MAb) technology could be utilized to advantage to demonstrate the molecular heterogeneity of connective tissues.[2] We have employed monoclonal antibody technology to circumvent problems related to the insoluble nature of extracellular connective tissue microfibrils: monoclonal antibodies that stained microfibrils were selected and used to isolate from the medium of cultured cells a soluble molecular component of microfibrils.

Electron microscopic immunolocalization using two different monoclonal antibodies resulted in a 54-nm periodic distribution of immunogold particles along the lengths of individual microfibrils (Fig. 1A and B). All morphologically similar microfibrils, those associated with elastin as well as bundles of microfibrils without elastin, which were described by Low in embryonic tissues[3] and in areas close to basement membranes,[4] were labeled in this periodic fashion. In addition, all tissues examined (skin, skeletal muscle, bone and cartilage, ligaments and tendons, kidney, lung, liver, blood vessels, ocular tissues, heart, and spleen) were immunolabeled. Immunoprecipitation of radiolabeled proteins secreted into the medium by cultured fibroblasts yielded a large (M_r 350,000) noncollagenous radioactive band on a gel. We named this antigen *fibrillin* because we hypothesized that it was a structural component uniting a ubiquitous class of morphologically similar microfibrils.

Fibrillin has been purified from cell culture medium in large enough quantities for visualization by rotary shadowing electron microscopic techniques (Fig. 2A) and for some biochemical characterization.[5] In addition, strings of polymeric structures that represent microfibrils have been ex-

[1] L. Y. Sakai, D. R. Keene, and E. Engvall, *J. Cell Biol.* **103**, 2499 (1986).
[2] H. Hessle, L. Y. Sakai, D. W. Hollister, R. E. Burgeson, and E. Engvall, *Differentiation* (*Berlin*) **26**, 49 (1984).
[3] F. N. Low, *Anat. Rec.* **160**, 93 (1968).
[4] F. N. Low, *Anat. Rec.* **139**, 105 (1961).
[5] L. Y. Sakai, D. R. Keene, R. W. Glanville, and H. P. Bächinger, *J. Biol. Chem.* **266**, 14763 (1991).

METHODS IN ENZYMOLOGY, VOL. 245
Copyright © 1994 by Academic Press, Inc.
All rights of reproduction in any form reserved.

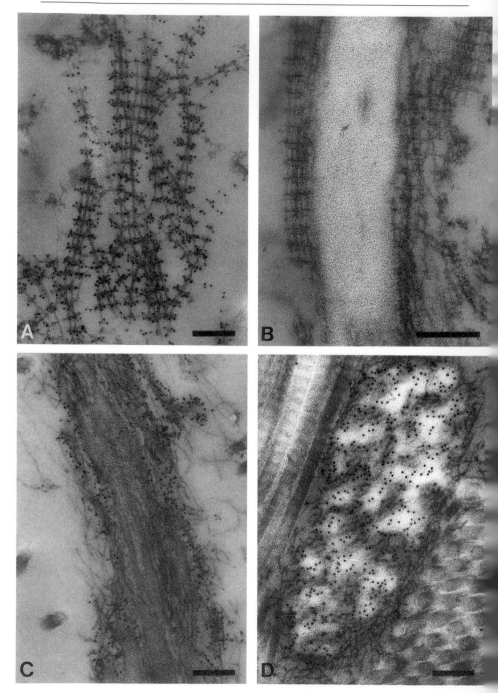

tracted from tissues by many different groups.[6-10] Rotary shadowed images of these structures revealed a "beaded string" appearance: strings of beads, variably spaced (periodicities range from 30 to 90 nm, but most are around 50 nm) and separated by a number of linear arms[6-8] (Fig. 2B). In 1991 Keene *et al.*[10] demonstrated that the structures, originally hypothesized to be microfibrils by Wright and Mayne,[6] contained fibrillin (Fig. 2C and D). Moreover, immunolocalization of a single epitope (recognized by MAb 69) of fibrillin demonstrated periodicity; labeling was just to one side of each globule ("bead").

Because monomeric fibrillin is a linear, but flexible, molecule that can extend to a length of approximately 148 nm,[5] it is thought that multiple fibrillin molecules form the individual arms and can contribute as well to the globules in the beaded string structure. Given the linear nature of the fibrillin monomer and additional immunolocalization data, it has been suggested that fibrillin molecules are arranged in a head-to-tail orientation in the microfibril.[5]

Molecular Structure

The primary structure of fibrillin has been determined by cloning and sequencing of cDNA.[11-14] Fibrillin is composed of multiple domains. The amino- and carboxyl-terminal domains are defined by unique sequences, whereas the central portion of the molecule contains multiple repeats of four types of cysteine-rich motifs (Fig. 3).

[6] D. W. Wright and R. Mayne, *J. Ultrastruct. Mol. Sruct. Res.* **100**, 224 (1988).

[7] Z. X. Ren, R. G. Brewton, and R. Mayne, *J. Struct. Biol.* **106**, 57 (1991).

[8] R. N. Wallace, B. W. Streeten, and R. B. Hanna, *Curr. Eye Res.* **10**, 99 (1991).

[9] R. Fleischmajer, J. S. Perlish, and T. Faraggiana, *J. Histochem. Cytochem.* **39**, 51 (1991).

[10] D. R. Keene, B. K. Maddox, H.-J. Kuo, L. Y. Sakai, and R. W. Glanville, *J. Histochem. Cytochem.* **39**, 441 (1991).

[11] C. L. Maslen, G. M. Corson, B. K. Maddox, R. W. Glanville, and L. Y. Sakai, *Nature (London)* **352**, 334 (1991).

[12] G. M. Corson, S. C. Chalberg, H. C. Dietz, N. L. Charbonneau, and L. Y. Sakai, *Genomics* **17**, 476 (1993).

[13] B. Lee, M. Godfrey, E. Vitale, H. Hori, M.-G. Mattei, M. Sarfarazi, P. Tsipouras, F. Ramirez, and D. W. Hollister, *Nature (London)* **352**, 330 (1991).

[14] L. Pereira, M. D'Alessio, F. Ramirez, J. R. Lynch, B. Sykes, T. Pangilinan, and J. Bonadio, *Hum. Mol. Genet.* **2**, 961 (1993).

FIG. 1. (A) and (B) *En bloc* immunolabeling of microfibrils in human skin with monoclonal antibodies specific for fibrillin. (B) Immunolabeling without secondary antibody conjugated to gold. (C and D) Immunolabeling of elastin with monoclonal antibody 10B8 using *en bloc* (C) and section surface labeling procedures (D). Bars: 200 nm.

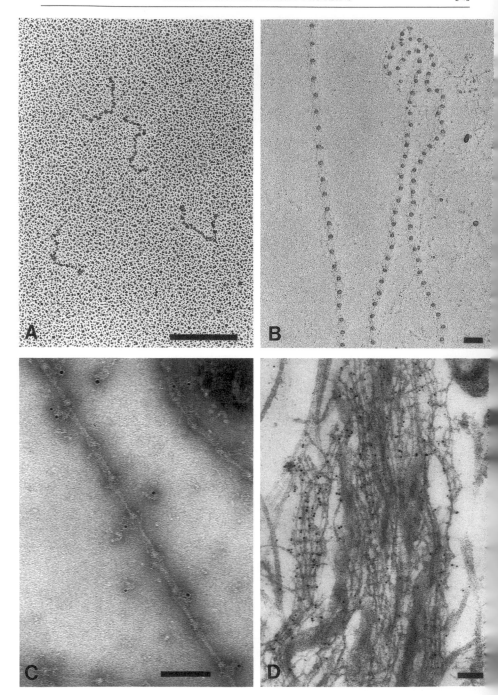

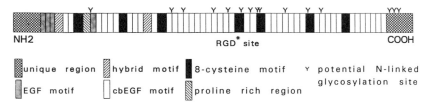

FIG. 3. Schematic representation of the molecular structure of fibrillin (FBN1).

The long central portion of fibrillin is composed of tandem arrays of two types of epidermal growth factor (EGF)-like repeats, interspersed with seven 8-cysteine repeats and two hybrid repeats. The longest tandem array of EGF-like repeats occurs in the center of the sequence and consists of 12 repeats. The cysteine-rich central portion is interrupted only once by a region composed of 47 residues, 21 of which are proline and none of which is cysteine. The amino-terminal domain contains four cysteine residues, and the carboxyl-terminal domain contains only two cysteines.

Epidermal Growth Factor-like Domains

There are two types of EGF-like repeats, which are generally distinguished by the presence of six cysteine residues with a characteristic spacing. On the basis of the structure of epidermal growth factor,[15] it is assumed that the six cysteine residues form three disulfide bonds: the first cysteine pairs with the third; the second with the fourth; and the fifth with the sixth. In fibrillin, there are 47 EGF-like repeats. Of these 47 repeats, 43 represent a special type of EGF-like motif, that is, the calcium-binding (cb) EGF-like motif. The cb EGF-like motifs in fibrillin have the following consensus sequence:

DIDEC------ C--G-C-NT-GSY-C- C--GY/F--------C-

The presence of acidic residues at the amino-terminal end of each EGF-like domain as well as the underlined portion of the sequence, which matches the consensus sequence found in proteins that undergo enzymatic

[15] C. R. Savage, J. H. Hash, and S. Cohen, *J. Biol. Chem.* **248**, 7669 (1973).

FIG. 2. (A and B) Visualization of fibrillin monomers (A) and beaded fibrils (B) after rotary shadowing and electron microscopy. (C) Beaded fibrils, labeled with monoclonal antibody specific for fibrillin, visualized after negative staining and electron microscopy. (D) Antibody-labeled beaded fibrils, visualized after conventional preparation and staining used for tissues. Bars: 100 nm.

hydroxylation of the asparagine residue,[16] have been implicated in calcium binding in other proteins that contain this type of motif.[17-19] Four of the 47 EGF-like repeats do not have these special features. Three of these occur in tandem adjacent to the amino-terminal domain. The fourth generic EGF-like domain is next to a unique region that is rich in proline residues.

It has been shown that fibrillin molecules do bind calcium, but only when disulfide bonds are intact.[12] This finding is consistent with the model of the cb EGF-like domain. The importance of calcium binding has been suggested by the identification of several different missense mutations affecting a single amino acid residue (often cysteine, but also other residues implicated in calcium binding) in a single cb EGF-like motif in different individuals with the Marfan syndrome (see Fig. 4). In addition, calcium has been shown to influence the conformation of fibrillin microfibrils[20] and to promote the deposition of fibrillin into the extracellular matrix of fibroblasts in culture.[21]

Other Cysteine-Rich Domains

A third cysteine-rich motif is repeated seven times in fibrillin. This motif contains eight cysteine residues and has the following consensus sequence in fibrillin:

D-R--- CY-------C--------- K-- CCC---G-AWG-P CE-CP---T-EF--L CP-GPGF---------

So far, this motif has been identified in fibrillin and in only one other gene, the gene for transforming growth factor (TGF)-β1-binding protein.[22] The central eight-cysteine motif contains the single RGD site that is present in the fibrillin sequence.

The fourth cysteine-rich motif occurs twice in fibrillin. It has been called the hybrid motif,[12] because its amino-terminal portion resembles the eight-cysteine motif and its carboxyl-terminal portion resembles the EGF-like motif. This motif can also be identified in TGF-β1-binding protein.

[16] J. Stenflo, A. Lundwall, and B. Dahlbäck, *Proc. Natl. Acad. Sci. U.S.A.* **84,** 368 (1987).
[17] B. Dahlbäck, B. Hildebrand, and S. Linse, *J. Biol. Chem.* **265,** 18481 (1990).
[18] P. A. Handford, M. Mayhew, M. Baron, P. R. Winship, I. D. Campbell, and G. G. Brownlee, *Nature (London)* **351,** 164 (1991).
[19] M. Selander-Sunnerhagen, M. Ullner, E. Persson, O. Teleman, J. Stenflo, and T. Drakenberg, *J. Biol. Chem.* **267,** 19642 (1992).
[20] C. M. Kielty and C. A. Shuttleworth, *FEBS Lett.* **336,** 323 (1993).
[21] T. Aoyama, K. Tynan, H. C. Dietz, U. Francke, and H. Furthmayr, *Hum. Mol. Genet.* **2,** 2135 (1993).
[22] T. Kanzaki, A. Olofsson, A. Moren, C. Wernstedt, U. Hellman, K. Miyazono, L. Claesson-Welsh, and C.-H. Heldin, *Cell (Cambridge, Mass.)* **61,** 1051 (1990).

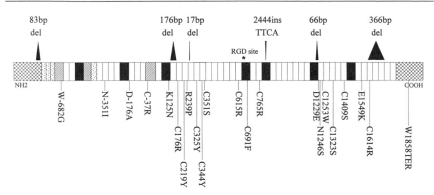

FIG. 4. Mutations identified in FBN1.

Importance of Disulfide Bonds

Ross and Bornstein[23] showed that microfibrils are structures that are stabilized by disulfide bonds. In their study, electron microscopic analysis demonstrated that morphologically normal microfibrils remain in the tissue after various methods of extraction are employed, and disappear after the tissue is extracted with disulfide bond-reducing agent. Therefore, the high concentration of cysteine residues in fibrillin (approximately 13%) is consistent with the hypothesis that fibrillin is a major component of microfibrils. Estimates of the percentage of cysteines in disulfide bonds were made by amino acid analyses before and after reduction[5] and suggested a large percentage of cysteines in intrachain disulfide bonds. Most of the available monoclonal antibodies specific for fibrillin recognize epitopes stabilized by intrachain disulfide bonds. It is likely that all 47 EGF-like domains contain cysteines in their predicted disulfide bond configurations. Whether the cysteines in the other motifs, or in the terminal domains, are paired or free for intermolecular interactions is currently unknown.

The processes whereby fibrillin becomes assembled into microfibrils and which domains may be involved in microfibril assembly are largely unknown. In cell and organ cultures, monomers of fibrillin are secreted by the cell into the extracellular space where single molecules assemble into disulfide-bonded aggregates. Pulse–chase studies[24] have suggested that the formation of intermolecular disulfide bonds occurs rapidly in tissues and precedes further insolubilization of the fibrillin monomer. In addition, single molecules of fibrillin, isolated from cell culture medium,

[23] R. Ross and P. Bornstein, *J. Cell Biol.* **40**, 366 (1969).
[24] L. Y. Sakai, *in* "Elastin: Chemical and Biological Aspects" (A. Tamburro and J. Davidson, eds.), p. 213. Congedo Editore, Galatina, Italy, 1990.

will form intermolecular disulfide bonds *in vitro,* when there are sufficient quantities of molecules in physiological solutions.[25] We are investigating which domains of fibrillin are involved in this intermolecular assembly process and whether structures that resemble microfibrils can be assembled from single molecules of fibrillin. On the basis of immunolocalization data,[5] our current working hypothesis is that the amino- and carboxyl-terminal domains participate in the formation of intermolecular disulfide bonds.

An alternative hypothesis, that amino-terminal portions of fibrillin interact with each other in the globular domains of the beaded microfibril and carboxyl-terminal portions interact with each other noncovalently, has been proposed by Kielty and Shuttleworth.[20] This hypothesis was suggested by the observation that calcium chelators (EDTA and EGTA) disturb the linear alignment of the arms between the globular domains of fibrillin microfibrils. Because most of the cb EGF-like repeats are present in the carboxyl two-thirds of the molecule, it has been proposed that calcium binding by this portion of the molecule stabilizes noncovalent interactions that would occur primarily between these portions of fibrillin molecules. This suggests that the globular domains of the beaded microfibril contain the amino termini of fibrillin molecules. Furthermore, this hypothesis implies that intermolecular disulfide bonds occur between amino-terminal portions of a limited number of fibrillin molecules and, therefore, that polymerization utilizing intermolecular disulfide bond formation is limited.

Alternative Splicing at 5' End

Three alternatively spliced exons (referred to as A, B, and C) have been identified at the 5' end of fibrillin transcripts.[12] Transcripts containing the A exon appear to be the most abundant, whereas those containing B or C are rare. The first methionine is present in the next downstream exon, exon M. Exons A and B may be in-frame noncoding regions; exon C contains stop sequences. The portion of the gene containing exons A, B, and C as well as exon M has the character of a CpG island, which may suggest the proximity of the transcription start site. If these exons are alternatively spliced first exons, they may be associated with different promoters. However, because the start site of transcription has not been identified, it is possible that an upstream initiator codon exists and that exons A and B represent coding sequences. In this case, the effect of alternative splicing would be the generation of protein products with variable amino-terminal domains.

[25] L. Y. Sakai, unpublished observations (1993).

Other Fibrillins and Fibrillin-like Molecules

A second gene that closely resembles fibrillin was identified by cDNA cloning and localized to human chromosome 5.[13] This gene, designated FBN2, is remarkably homologous to FBN1 in amino acid sequence, with the exception of one region: the proline-rich domain in FBN1 is replaced in the same position in FBN2 with a region rich in glycine. The same types of repeated motifs are arranged in FBN2 in the identical pattern of motifs found in FBN1 (see Fig. 3).[26]

Other genes have been identified that have been called "fibrillin-like."[27] These molecules are characterized by tandem arrays of cb EGF-like repeats interspersed with eight-cysteine motifs. However, unlike FBN2, these genes do not match the total size and the overall pattern of organization of the motifs found in FBN1.

Fibrillin and Marfan Syndrome

Immunolocalization of fibrillin to the ciliary zonule[1] suggested fibrillin as a candidate gene for the Marfan syndrome, a heritable disorder of connective tissue with pleiotropic manifestations.[28] One of the important features of the Marfan syndrome is the unusual weakness of the ciliary zonule, which often leads to displacement and sometimes to dislocation of the lens. The ciliary zonule is the suspensory ligament connecting the muscles at the wall of the eye to the lens, transmitting the force necessary to focus the lens. This ligament is unique in that microfibrils are the sole structural elements visible by conventional electron microscopic analysis; there are no apparent collagen fibers.

Subsequent immunofluorescence analyses of skin as well as fibroblasts from individuals with classic manifestations of the Marfan syndrome demonstrated gross visible abnormalities of fibrillin fibrils.[29] These findings further implicated fibrillin as the gene for the Marfan syndrome.

Genetic analyses confirmed the fibrillin gene (FBN1) as the Marfan gene. Identification of the Marfan gene locus on chromosome 15 was

[26] H. Zhang, S. D. Apfelroth, W. Hu, E. C. Davis, C. Sanguineti, J. Bonadio, R. P. Mecham, and F. Ramirez, J. Cell Biol. 124, 855 (1994).

[27] M. A. Gibson, E. Davis, M. Filiaggi, and R. P. Mecham, Am. J. Med. Genet. 47, 148 (abstr.) (1993).

[28] R. E. Pyeritz, in "Connective Tissue and Its Heritable Disorders" (P. Royce and B. Steinmann, eds.), p. 437. Wiley-Liss, New York, 1992.

[29] D. W. Hollister, M. Godfrey, L. Y. Sakai, and R. E. Pyeritz, N. Engl. J. Med. 323, 152 (1990).

originally accomplished with anonymous DNA markers.[30,31] Subsequently, FBN1 was localized to the same locus on chromosome 15,[13,32] and mutations in FBN1 were discovered in patients with the Marfan syndrome.[33-41] Many of these mutations are single base-pair missense mutations that affect a critical amino acid residue in one of the cb EGF-like repeats. Deletions, insertions, and frameshift mutations have also been reported. Figure 4 shows the positions of the mutations that have been identified.[33-41]

Although linkage analysis clearly identified FBN1 as the Marfan gene, only a small percentage of affected individuals has yielded defined mutations. This low success rate in screening for mutations may be due to technical problems related to the methods used for screening, the presence of null alleles, or mutations affecting noncoding regions of the gene. Elucidation of the structure of the gene, including intron/exon boundaries,[14] will facilitate further identification of mutations. In addition, characterization of multiple polymorphic markers will make molecular diagnostics possible for many individuals at risk.

Accurate diagnosis of the Marfan syndrome is still an important issue, because many individuals for whom Marfan syndrome is suspected may have no family history of the syndrome (one-fourth to one-third of individuals with the Marfan syndrome have unaffected parents[28]), not all individuals with a family history of Marfan syndrome belong to families that are

---

[30] K. Kainulainen, L. Pulkkinen, A. Savolainen, I. Kaitila, and L. Peltonen, *N. Engl. J. Med.* **323,** 935 (1990).

[31] H. C. Dietz, R. E. Pyeritz, B. D. Hall, R. G. Cadle, A. Hamosh, J. Schwartz, D. A. Meyers, and C. A. Francomano, *Genomics* **9,** 355 (1991).

[32] R. E. Magenis, C. L. Maslen, L. Smith, L. Allen, and L. Y. Sakai, *Genomics* **11,** 346 (1991).

[33] H. C. Dietz, G. R. Cutting, R. E. Pyeritz, C. L. Maslen, L. Y. Sakai, G. M. Corson, E. G. Puffenberger, A. Hamosh, E. Nanthakumar, S. Curristin, G. Stetten, D. A. Myers, and C. A. Francomano, *Nature (London)* **352,** 337 (1991).

[34] H. C. Dietz, R. E. Pyeritz, E. G. Puffenberger, R. J. Kendzior, G. M. Corson, C. L. Maslen, L. Y. Sakai, C. A. Francomano, and G. R. Cutting, *J. Clin. Invest.* **89,** 1674 (1992).

[35] H. C. Dietz, J. Saraiva, R. E. Pyeritz, G. R. Cutting, and C. A. Francomano, *Hum. Mutat.* **1,** 366 (1992).

[36] K. Kainulainen, L. Y. Sakai, A. Child, F. M. Pope, L. Puhakka, L. Ryhanen, A. Palotie, I. Kaitila, and L. Peltonen, *Proc. Natl. Acad. Sci. U.S.A.* **89,** 5917 (1992).

[37] H. C. Dietz, D. Valle, C. A. Francomano, R. J. Kendzior, R. E. Pyeritz, and G. R. Cutting, *Science* **259,** 680 (1993).

[38] D. R. Hewett, J. R. Lynch, R. Smith, and B. C. Sykes, *Hum. Mol. Genet.* **2,** 475 (1993).

[39] H. C. Dietz, I. McIntosh, L. Y. Sakai, G. M. Corson, S. C. Chalberg, R. E. Pyeritz, and C. A. Francomano, *Genomics* **17,** 468 (1993).

[40] K. Tynan, K. Comeau, M. Pearson, P. Wilgenbus, D. Levitt, C. Gasner, M. A. Berg, D. C. Miller, and U. Francke, *Hum. Mol. Genet.* **2,** 1813 (1993).

[41] K. Kainulainen, L. Karttunen, L. Puhakka, L. Y. Sakai, and L. Peltonen, *Nat. Genet.* **6,** 64 (1994).

informative for linkage analysis, and screening for mutations has had a low success rate. Therefore, efforts have been made to determine categories of protein phenotypes for individuals with the Marfan syndrome.[21,42] Whether these studies of fibrillin biosynthesis or fibrillin immunofluorescence[29] will be useful for diagnosis has not been determined, because studies have been limited so far to individuals who have defined mutations in FBN1 or otherwise fulfill the classic criteria for Marfan syndrome. Individuals for whom diagnosis is difficult may benefit from these investigations.

Other Marfan-Related Disorders

In addition to the Marfan syndrome, the FBN1 gene has been linked to ectopia lentis, a syndrome characterized by dislocated lenses and some of the skeletal, but not the cardiovascular, features of the Marfan syndrome.[43] A mutation in FBN1 in a family with ectopia lentis has been identified.[41] This mutation is a missense mutation resulting in the substitution of a glutamic acid residue with lysine at the beginning of the sixth cb EGF-like repeat from the carboxyl-terminal domain. Why this mutation, which is similar to other mutations that cause the Marfan syndrome, results in ectopia lentis is not clear.

Mutations have also been identified in a rare, severe form of Marfan syndrome, neonatal Marfan syndrome.[41] In these few cases, the mutations [K125N, DEL1EGF (176-bp deletion), and C176R] appear to cluster at the beginning of the longest string of cb EGF-like repeats in the middle of the fibrillin molecule. These data, together with the identification of the mutation in the family with ectopia lentis, suggest that different domains in fibrillin may perform different functions and that the interference of these functions by specific mutations may result in different phenotypes.

Because the Marfan syndrome is characterized by a phenotype with multiple different manifestations, there are many disorders with partial overlapping phenotypes. It may be that some of these disorders, like ectopia lentis, are also caused by abnormalities in FBN1. Alternatively, it has been suggested by the example of FBN2 that microfibril and microfibril-associated proteins may be responsible for these potentially related disorders. Although mutations have not yet been identified, FBN2 has been genetically linked to congenital contractural arachnodactyly,[13,43] a disorder characterized by joint contractures (rather than loose joints),

[42] D. McGookey Milewicz, R. E. Pyeritz, E. S. Crawford, and P. H. Byers, *J. Clin. Invest.* **89,** 79 (1992).

[43] P. Tsipouras, R. Del Mastro, M. Sarfarazi, B. Lee, E. Vitale, A. H. Child, M. Godfrey, R. B. Devereux, D. Hewett, B. Steinmann, D. Viljoen, B. C. Sykes, M. Kilpatrick, and F. Ramirez, *N. Engl. J. Med.* **326,** 905 (1992).

crumpled ears, and some of the skeletal, but not the cardiovascular or ocular, features of the Marfan syndrome.

Methods

Isolation of Fibrillin from Cell Culture Medium

Many different types of cells in culture secrete fibrillin into the medium, but some cells do not assemble fibrillin into the extracellular matrix. Fibrillin is secreted into the medium by fibroblasts, chondrocytes, endothelial cells, smooth muscle cells, ligament cells, and the established cell lines HT-1080 (human fibrosarcoma), MG-63 (male osteosarcoma), and WISH (human amnion tissue).[44] Immunofluorescence of fibrillin in the extracellular matrix has been demonstrated for most of these cells. HT-1080 and WISH are the exceptions. The following protocol is performed using medium from human skin fibroblasts or ligament cells.[5]

1. One liter/condo of serum-free medium [Dulbecco's modified Eagle's medium (DMEM)] is collected after a 24-hr incubation with cells grown in condos (Nunc, Roskilde, Denmark). Typically, each condo is seeded with $1-2 \times 10^8$ cells, and is harvested once or twice per week for 3–4 months.

2. Protease inhibitors [3 mM EDTA and 1 mM phenylmethylsulfonyl fluoride (PMSF)] are added immediately and ammonium sulfate (240 g/liter) is added gradually while stirring at 4°. Proteins are allowed to precipitate out of the medium overnight at 4°.

3. The precipitate is collected by centrifugation (15,000 g for 1 hr at 4°C), and the pellet is resuspended in 7 ml of 50 mM Tris-HCl–0.15 M NaCl–2 M urea (pH 7.4), with EDTA and PMSF included. The sample is frozen until pooled.

4. Approximately 100 ml of resuspended precipitate is thawed, treated with 10 mM diisopropyl fluorophosphate (DFP) for 3 hr on ice, and dialyzed against 50 mM Tris-HCl–2 M urea (pH 8.0) (DEAE starting buffer) overnight at 4°.

5. After clarification by centrifugation, followed by filtration through filters (0.45-μm pore size), the sample is applied to a Waters (Milford, CT) HPLC DEAE-5PW ion-exchange column (Millipore, Bedford, MA), equilibrated in 50 mM Tris-HCl–2 M urea (pH 8.0), at a flow rate of 5 ml/min. Elution is with a linear gradient of 0–0.3 M NaCl in the same buffer. Five-milliliter fractions are collected into tubes containing EDTA and

[44] L. Y. Sakai, unpublished observations (1993).

analyzed by sodium dodecyl sulfate–polyacrylamide gel electrophoresis (SDS–PAGE) using 4.5% (w/v) acrylamide gels. Fibrillin should elute beween 0.16 and 0.26 M NaCl.

6. Fractions containing fibrillin are pooled, treated again with DFP, and dialyzed against 50 mM Tris-HCl–8 M urea (pH 7.2) (Mono Q buffer), overnight at 4°.

7. The sample is applied next to an HR 10/10 Mono Q column (Pharmacia-LKB Biotechnology, Inc, Piscataway, NJ) equilibrated in 50 mM Tris-HCl–8 M urea (pH 7.2) at a flow rate of 1 ml/min. Elution is accomplished with a linear gradient of 0–0.5 M NaCl in the same buffer. One-milliliter fractions are collected into tubes containing EDTA and analyzed by SDS–PAGE. Fibrillin should elute between 0.26 and 0.30 M NaCl.

8. Fractions containing fibrillin are pooled and applied to two FPLC (fast protein liquid chromatography) HR 16/50 Superose 6 columns (Pharmacia) in tandem, equilibrated in 50 mM Tris-HCl–0.15 M NaCl–8 M urea (pH 7.2), at a flow rate of 0.25 ml/min. Fractions containing fibrillin are identified by SDS–PAGE.

9. Sieved fractions containing fibrillin are concentrated by chromatography on Mono Q.

Metabolic Labeling and Antibody Affinity Chromatography

Because fibrillin is rich in cysteine residues, [35S]cysteine is used to metabolically label fibrillin synthesized by cell or organ cultures. Cultures are fed medium that does not contain cysteine (this can be purchased or made from scratch), supplemented with [35S]cysteine (50 μCi/ml). Fibrillin can then be identified by SDS–PAGE and fluorography, or it can be isolated by antibody affinity chromatography. Antibody affinity chromatography has not been successfully utilized for large-scale purification of fibrillin. However, small amounts of radiolabeled fibrillin can be easily purified with antibody affinity chromatography.

1. Hybridomas are grown in DMEM, containing 10% (v/v) fetal calf serum (FCS) that has been depleted of IgG by repeated passage over protein G–Sepharose (Pharmacia) and then sterilized by filtration. Medium is collected by centrifugation, and the cell pellet is resuspended in fresh medium or thrown away.

2. Monoclonal antibodies are isolated from the medium by passage over protein G–Sepharose and elution with 0.1 M glycine hydrochloride, pH 2.5. Fractions are neutralized with saturated Tris, pooled, dialyzed against phosphate-buffered saline (PBS), and the protein concentration is estimated spectrophotometrically.

3. Antibodies are coupled to CNBr-activated Sepharose (Pharmacia), according to manufacturer directions.

4. Radiolabeled samples are applied to antibody–Sepharose in a small column. If medium is used, the sample is first passed over a column of gelatin–Sepharose (Pharmacia) to remove fibronectin.

5. The column is washed with PBS–0.5% (v/v) Tween 20, and fibrillin is eluted with 0.1 M glycine hydrochloide, pH 2.5. Fractions containing fibrillin are identified by scintillation counting and fluorography.

Immunofluorescence Assays

Fibrillin immunofluorescence patterns are easily recognizable and can be used to detect abnormalities of approximately 90% of individuals with the Marfan syndrome.[29] These assays were performed on samples from individuals who demonstrated clear, classic forms of the Marfan syndrome. Therefore, it is not yet known whether these assays will distinguish between individuals with only some of the phenotypic features of the Marfan syndrome. In addition, the original study yielded approximately 10% false-negatives and occasional false-positives in individuals with other types of connective tissue disorders.[29]

1. Tissues are prepared by freezing the sample in hexanes on dry ice, followed by preparation for cryomicrotomy. The sample is embedded in OCT compound (Miles, Elkhart, IN), frozen in liquid nitrogen, and cut into 8- to 10-μm sections on a cryostat. The sections are dried under a fan for 30 min.

Alternatively, cells are trypsinized, plated at very high densities (2.5 × 10^5 cells/ml) into 1-ml chamber slides, and allowed to elaborate a matrix for 2 days. On the third day, immunofluorescence is performed. Insufficient numbers of cells will yield little to no immunofluorescence.

2. Sections or cell cultures are fixed in acetone (−20°) for 10 min and then rehydrated in PBS.

3. Samples are incubated with anti-fibrillin (1 : 100 dilution of monoclonal antibodies stored at concentrations between 1.0 and 1.5 mg/ml) for 3 hr at room temperature, and then washed three times in PBS.

4. Samples are incubated with secondary antibodies conjugated with fluorescein isothiocyanate (FITC; Sigma, St. Louis, MO) (diluted 1 : 50) or with phycoerythrin (Biomeda, Foster City, CA), according to manufacturer directions, and then washed three times in PBS.

5. Nuclei in the samples can be counterstained with 0.00025% (w/v) propidium iodide.

6. A small drop of glycerol with 10% PBS is applied to the samples,

along with coverslips. If phycoerythrin is used, the samples are covered with GelMount (Biomeda, Foster City, CA).

Isolation of Fibrillin Microfibrils

Fibrillin does not normally exist as monomers in the extracellular matrix. Small quantities of monomers or disulfide bonded aggregates can be extracted from rapidly growing fetal tissues.[25,45] These molecules likely represent newly synthesized monomers or small aggregates that have not yet become fully cross-linked. However, methods for extraction of intact fibrillin from tissues usually involve treating the tissue with disulfide bond-reducing agents.[45] The disadvantage of this method is that fibrillin is denatured. Consequently, antibodies raised against reduced and denatured microfibrillar components may require treatment of tissues with reducing and denaturing reagents in order to achieve immunolocalization.[45] An alternative to extraction with reducing agents is enzymatic digestion of tissues. Large polypeptide fragments of fibrillin can be isolated from tissues that have been digested with pepsin[46] and were present in the original sample used for the production of fibrillin monoclonal antibodies.[1]

In the extracellular matrix, fibrillin molecules are components of polymeric structures that can be liberated from the matrix and visualized by electron microscopic techniques. When first observed, these structures were thought to be forms of collagen.[47,48] They were described as "beaded fibrils"[49] or "beaded filaments."[48] Wright and Mayne,[6] using rotary shadowing techniques for better visualization of the structures, first hypothesized that these beaded structures actually correspond to microfibrils. Using immunolocalization techniques, Keene et al.[10] demonstrated that beaded fibrils contain fibrillin.

Beaded fibrils may be more easily liberated from fetal tissues than from adult tissues,[48] suggesting that further cross-linking to other matrix components may occur with age in some tissues. In individuals with scleroderma, microfibrils are particularly abundant,[50] and beaded fibrils can be liberated from the tissue.[51]

Microfibrils have also been identified in the matrices of cells in culture. When cells are grown without supplementation with ascorbic acid, they

[45] M. A. Gibson, J. S. Kumaratilake, and E. G. Cleary, *J. Biol. Chem.* **264,** 4590 (1989).
[46] B. K. Maddox, L. Y. Sakai, D. R. Keene, and R. W. Glanville, *J. Biol. Chem.* **264,** 21381 (1989).
[47] A. G. Matoltsy, J. Gross, and A. Grignolo, *Proc. Soc. Exp. Biol. Med.* **76,** 857 (1951).
[48] R. L. Hayes and E. R. Allen, *J. Cell Sci.* **2,** 419 (1967).
[49] J. Gross, A. G. Matoltsy, and C. Cohen, *J. Biophys. Biochem. Cytol.* **1,** 215 (1955).
[50] R. Fleischmajer, L. Jacobs, E. Schwartz, and L. Y. Sakai, *Lab. Invest.* **64,** 791 (1991).
[51] R. L. Hayes and G. P. Rodnan, *Am. J. Pathol.* **63,** 433 (1971).

appear to be the most abundant structural element in the matrix.[52] Beaded fibrils have been extracted from cultures of dermal fibroblasts[52,53] and smooth muscle cells[53] maintained for 3 to 6 weeks at postconfluence.

Liberating Microfibrils from Tissues and Cell Culture Extracellular Matrix. Microfibrils may be liberated from tissues by a simple mechanical mechanism.

1. Excess cellular material (e.g., the epithelial layer of skin) should first be removed from the tissue with a razor blade. The tissue is then cut into pieces measuring about 1 mm on each side.

2. The tissue pieces are transferred into a small glass tube containing ice-cold 0.1 M ammonium bicarbonate, pH 7.4. With the tube immersed in ice-cold water, the tissue is sheared using a Polytron homogenizer fitted with a $\frac{1}{4}$-in. cutting blade. Homogenization is complete when tissue fragments are barely visible.

3. Discard the larger tissue fragments that settle to the bottom of the tube. The supernatant will contain some cell debris, collagen fibers, microfibrils, and elastic fibers, which are easily recognized in negative stain preparations.

Microfibrils can be more selectively isolated from tissues homogenized and treated with bacterial collagenase, using the method of Kielty *et al.*[54] These preparations contain type VI collagen microfilaments as well as fibrillin microfibrils and may also contain unknown contaminants.

1. Tissue samples (2 g wet weight) are dissected and homogenized in 50 mM Tris-HCl, pH 7.4, containing 0.4 M NaCl–0.01 M CaCl$_2$–0.01 M MgCl$_2$–2 mM PMSF–10 mM N-ethylmaleimide (NEM) (10 ml).

2. The homogenized tissue pieces are digested with bacterial collagenase (0.2 mg/ml) and DNase (0.1 mg/ml) at 4° for 18 hr. EDTA (20 mM) is added to terminate the digestion.

3. The digest is clarified by centrifugation, and the supernatant is applied to a Sepharose CL-2B gel-filtration column, equilibrated in 50 mM Tris-HCl–0.4 M NaCl (pH 7.4). Type VI collagen microfilaments and fibrillin microfibrils elute in the void volume.

A similar method, employed by Kielty and Shuttleworth,[20] can be used to extract beaded fibrils from the extracellular matrix of cell cultures.

[52] R. Fleischmajer, P. Contard, E. Schwartz, E. D. MacDonald, L. Jacobs, and L. Y. Sakai, *J. Invest. Dermatol.* **97**, 638 (1991).

[53] C. M. Kielty and C. A. Shuttleworth, *J. Cell Sci.* **106**, 167 (1993).

[54] C. M. Kielty, L. Berry, S. P. Whittaker, M. E. Grant, and C. A. Shuttleworth, *Matrix* **13**, 103 (1993).

1. Cell layers are washed in 50 mM Tris-HCl–0.4 M NaCl (pH 7.4) and then incubated for 3 hr at 20° in the same buffer with 5 mM CaCl$_2$–bacterial collagenase (0.1 mg/ml)–2 mM PMSF–5 mM NEM.

2. Extracts are clarified by centrifugation for 15 min at 7500 g at room temperature in a microfuge, and the supernatant is applied to a Sepharose CL-2B column equilibrated in 50 mM Tris-HCl–0.4 M NaCl (pH 7.4). Beaded fibrils elute in the void volume of the column.

Electron Microscopic Visualization of Fibrillin Microfibrils

Rotary Shadowing Protocol

1. Samples should be suspended in a volatile buffer such as 0.2 M ammonium bicarbonate or 0.1 M acetic acid, at a concentration of approximately 100 μg/ml.

2. Just prior to spraying, dilute the sample to 70% (v/v) glycerol.

3. From a distance of about 40 cm, spray the sample onto disks of freshly cleaved mica, cut from sheets with a paper punch and attached to glass slides with double-sided tape.

Spraying is facilitated with an airbrush (model 250; Badger, Franklin Park, IL) connected to filtered air at 35 psi. The paint delivery system of the airbrush is replaced with a threaded screw drilled through the center to accept a micropipette tip. The micropipette tip, containing 100 μl of sample, is positioned so that the end of the tip is just below the orifice from which the filtered air emerges. At this height, the flowing air will force the sample to rise through the pipette tip and be dispersed into droplets, measuring approximately 50–100 μm in diameter.

4. The sample is attached to the rotating stage of a vacuum evaporator and dried under vacuum. Freshly flamed platinum wire (2.5 cm; Fullam, Latham, NY) is coiled around the intersection of the two carbon rods (sharpened to a diameter of approximately 1 cm) of a resistance-heated carbon rod evaporator, positioned so that the angle of deposition is 6° relative to the plane of the stage and the distance between the samples and the evaporation source is 10 cm.

5. Following evacuation of the chamber into the 10^{-6}-mbar range, the carbon rods are slowly heated until the platinum wire melts, forming a molten droplet suspended from the carbon rods. Gas will emerge from the rods during heating; therefore allow the vacuum to recover into the 10^{-7}-mbar range before proceeding.

6. Rotate the sample stage at 110 rpm and slowly increase the current through the carbon rods to evaporate the metal. The evaporation should be accomplished in no less than 30 sec and at no time should the vacuum be allowed to fall into the 10^{-5}-mbar range. A combination of platinum

and carbon will be evaporated simultaneously and will color a white filter paper positioned 90° relative to the source a medium dark gray.

7. Following low-angle rotary shadowing, the sample stage is stilled and a backing film of carbon is deposited at 90° relative to the mica surface. If placed next to the samples during evaporation, a filter paper strip will be colored a very light tan after successful carbon evaporation.

8. The mica disks are removed from the evaporator and placed in a petri dish for 30 min to allow exposure to vapors arising from 1% (v/v) acetic acid, which will loosen the replicas from the mica surface. By gently inclining the mica into a small pool of distilled water, the replica will release from the mica and float onto the surface of the water. It may then be supported using a 600-mesh grid.

Negative Staining. Negative staining occurs when most of the sample itself is unstained, contrasting on a heavily stained supporting substrate. Particular detail within the sample, such as crevices, also may accumulate stain and appear dark. The sample should not absorb so much stain that it lacks contrast on a stained background. The substrate to which the sample is attached should be continuously and amorphously stained.

1. Support the sample with a continuous carbon film on an otherwise bare 600-mesh grid. Carbon films are prepared in a vacuum chamber by evaporating carbon onto freshly cleaved mica disks. A clean white filter paper positioned 90° relative to the carbon source during evaporation will be light gray once the correct thickness of carbon is obtained. The carbon film is separated from the mica surface on a pool of distilled water. The carbon film is positioned over a submersed 600-mesh grid, which is held by forceps. Excess carbon film that overhangs the grid edges is carefully fractured away.

2. To coat the hydrophobic carbon surface evenly with stain, it must be rendered hydrophilic by placing it in an ionized field, such as that afforded by either a home-made glow discharge device or one provided with vacuum evaporators. Use glow-discharged grids soon after ionization because atmospheric humidity will dissipate the charge within about 30 min.

3. Five microliters of sample is pipetted onto the grid surface, which is held in forceps. Allow the sample to settle and attach to the surface for 30 sec. Wick away excess fluid. Before the grid dries, pipette 5 μl of freshly filtered 2.0% (w/v) phosphotungstic acid (PTA, in distilled water, adjusted to pH 7.4 with NaOH) onto the surface and wait 30 sec before wicking away the PTA. If the background proves to be too densely stained, reduce the concentration of PTA. Wait until the grid is completely dry before exposing it to the vacuum of the microscope.

Electron Microscopic Immunolocalization of Elastic Fibers

Preembedding Protocol: En Bloc Technique. This technique relies on diffusion to carry antibodies to their targets. Experiments utilizing cryopreservation methods in combination with freeze-substitution demonstrate that the empty spaces between fiber bundles and cells commonly seen in conventionally fixed connective tissues are actually filled with ground substance. This material is so dense that elastic fiber microfibrils are hidden within it[55] (Fig. 5). Much of the apparently empty space seen following conventional chemical fixation probably reflects a shrinkage artifact occurring during solvent dehydration. However, there is also evidence to suggest that much ground substance is extracted from tissue even during chemical fixation.[56] Because this material is so easily extracted, it is often not necessary to digest the tissue enzymatically to facilitate diffusion of the antibody; simply soaking the tissue in PBS, pH 7.4, extracts enough ground substance to allow passage of antibody. However, in some tissues, this may not be sufficient to allow penetration of antibody. For example, in cartilage and cornea, the dense proteoglycan matrix must be digested [290 units of chondroitinase ABC (Sigma)/ml PBS overnight at 4°] before antibody can diffuse into the tissue.

1. Fresh tissue is cut into small pieces measuring no larger than 0.5 × 1.0 mm. Slice the tissue with an extremely sharp edge, such as a Schick Plus Platinum double-edge razor blade.

2. Tissue pieces are washed in PBS for 30 min (three changes of PBS). All solutions and incubations are kept at 4°. The samples are agitated (VWR, Denver, CO) during incubation in all solutions prior to fixation using an Adams Nutator, which accepts a small box with dividers capable of holding vertical a number of 10 × 75 glass tubes that contain the tissue pieces.

3. Samples are incubated in primary antibody for 14 hr. Primary antibodies, including those in ascites and serum as well as purified monoclonal antibodies (generally 1.0–1.5 mg/ml), are diluted 1:5 in PBS. A total volume of 125 μl of diluted antibody is sufficient to submerse one to five tissue pieces.

4. Samples are washed in PBS for at least 6 hr with five changes of PBS.

5. Samples are incubated for 14 hr with secondary antibody conjugated with gold.

Although the manufacturer recommends a less concentrated solution, we dilute Amersham (Arlington Heights, IL) colloidal gold products 1:3

---

[55] D. R. Keene and K. McDonald, *J. Histochem. Cytochem.* **41**, 1141 (1993).
[56] R. W. Judd and F. M. Eggert, *Histochemistry* **73**, 391 (1981).

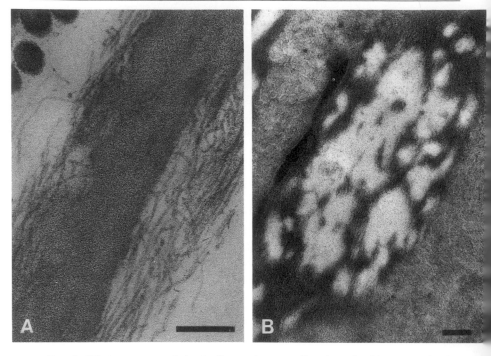

FIG. 5. (A) Appearance of elastic fiber and surrounding tissue in skin prepared using conventional fixation and staining procedures. (B) Appearance of a similar area in skin prepared using high-pressure fixation methods. Bars: 200 nm.

(v/v) in bovine serum albumin (BSA) buffer [20 mM Tris-HCl in PBS with 0.1% (w/v) bovine serum albumin and 0.05% (w/v) sodium azide, pH 8.2]. Commercial products are available in 1-, 5-, 10-, and 15-nm sizes, or larger. For penetration into dense areas, 1-nm gold may have to be used. For example, the localization of kalinin to the region between hemidesmosomes of basal keratinocytes and the subjacent lamina densa of the dermal–epidermal junction was dependent on the use of 1-nm gold conjugate.[57] Larger gold conjugates will not penetrate this region. Because 1-nm gold is difficult to recognize, it is necessary to intensify its appearance by silver precipitation methods. A 5-nm gold conjugate will penetrate approximately 100 μm into the cut edge of most connective tissue matrices and can be easily seen at a final magnification of ×60,000. Larger conjugates will penetrate less.

A secondary antibody conjugate may sometimes be omitted. If the

---

[57] P. Rouselle, G. P. Lunstrum, D. R. Keene, and R. E. Burgeson, *J. Cell Biol.* **114,** 567 (1991).

primary antibody is an immunoglobulin M (IgM), it may be possible to visualize binding due to its large size.[58] In certain circumstances (e.g., antibody binding to fibrillin microfibrils),[5] IgG may also be detected without a secondary conjugate. In longitudinal sections of fibrillin microfibrils, stained IgG, binding periodically along the lengths of individual microfibrils, forms a visible pattern (Figure 1B). However, if the antigen is positioned in a background of other stainable material, it is unlikely that it will be detected without the use of an electron-dense marker.

6. Samples are washed in PBS for at least 2 hr with five changes of PBS.

7. Samples are immersed in 0.1 M cacodylate buffer, pH 7.4, for 30 min with three changes of buffer and then fixed in 1.5% (w/v) paraformaldehyde and 1.5% (v/v) glutaraldehyde in 0.1 M cacodylate buffer, pH 7.4, for 1 hr. Samples are washed in the cacodylate buffer for 15 min with three changes of buffer, and then fixed in 1% (w/v) OsO_4 in the cacodylate buffer for 1 hr. Samples are washed in the cacodylate buffer for 15 min with three changes of buffer. All incubations are performed at 4°.

The addition of 0.1% (w/v) tannic acid to the primary aldehyde fixative will significantly increase the contrast of collagen fibrils, elastin, and microfibrils.

8. Samples are dehydrated through ethanol (30, 50, 70, 90, and 95% for 15 min each; 100% for 30 min with three changes) at room temperature. Samples are then further dehydrated in 100% (v/v) propylene oxide for 30 min, with three changes, at room temperature. The choice of solvent (acetone, ethanol with or without propylene oxide), which removes water from the tissue prior to embedding in epoxy, dramatically affects the appearance of the collagen banding pattern. If acetone is used as a dehydrant, the use of propylene oxide may be omitted because acetone and most epoxy-embedding resins are readily miscible. To dehydrate in ethanol but to avoid the use of propylene oxide, the infiltration times should be extended because ethanol and the epoxies are not so readily miscible.

9. Samples are infiltrated by Spurr's epoxy by diluting 1 : 1 (v/v) with propylene oxide for 30 min, 2 : 1 for 45 min, 3 : 1 for 60 min, and 100% Spurr's for 3 hr with three changes.

Certain embedding media, most notably Nanoplast (Biorad, Cambridge, MA), are miscible with water. Nanoplast can be infiltrated into the tissue following fixation, eliminating the exposure of the tissue to denaturing solvents. Although cell membranes are not well preserved, the appearance of the matrix resembles that following preservation by high-

[58] D. R. Keene, L. Y. Sakai, R. E. Burgeson, and H. P. Bächinger, *J. Histochem. Cytochem.* **35**, 311 (1987).

pressure freezing and freeze substitution. Therefore, omission of solvent dehydration in combination with nanoplast seems to allow retention of many matrix components normally lost without specific chemical stabilization.[59]

Surface Labeling of Antigens. For antigens that are tightly packed or surrounded by a protective coating, labeling techniques relying on *en bloc* diffusion are unlikely to succeed. For example, *en bloc* immunolabeling with anti-elastin antibodies results in labeling of only the outside surfaces of elastin cores (Fig. 1C). To label antigens in the interior of elastic fibers, the fiber interiors must be exposed on the surface of a section prior to immunolabeling.

There are many problems associated with section surface immunolabeling procedures. First, the tissue must be stabilized by chemical fixation. Even mild fixatives may affect antigens to the extent that antibodies will no longer label the tissue. In addition, solvent dehydration steps may adversely affect antigens, and polymerization of embedding media often requires heat, which may also denature the antigen. Some antigenic sites survive even the harshest treatments, whereas others seem sensitive to any treatment. Monoclonal antibody 10B8,[60] specific for elastin, recognizes an epitope that is stable after tissues are prepared for section surface labeling (Fig. 1D). Monoclonal antibodies specific for fibrillin recognize epitopes that are sensitive to surface-labeling treatments, and therefore have not been successfully used in section surface-labeling procedures. However, a polyclonal antibody specific for fibrillin[46] has been used successfully for surface labeling, indicating that there are some epitopes in fibrillin that are not affected by the preparation protocol.

The following protocol for fixation, embedding, and section surface labeling has been used successfully to localize type VII collagen in skin and cornea[61,62] and elastin in skin and aorta.

Steps 1–5 should be performed on ice.

1. Tissue pieces are fixed in 0.1% (v/v) glutaraldehyde for 30 min on ice.

2. Samples are immersed in 0.1 M cacodylate buffer, pH 7.4, for 30 min with three changes of buffer.

---

[59] D. R. Keene, G. P. Lunstrum, N. P. Morris, D. W. Stoddard, and R. E. Burgeson, *J. Cell Biol.* **113,** 971 (1991).

[60] J. M. Hurle, G. M. Corson, K. Daniels, R. S. Reiter, L. Y. Sakai, and M. Solursh, *J. Cell Sci.,* in press (1994).

[61] L. Y. Sakai, D. R. Keene, N. P. Morris, and R. E. Burgeson, *J. Cell Biol.* **103,** 1577 (1986).

[62] D. R. Keene, L. Y. Sakai, G. P. Lunstrum, N. P. Morris, and R. E. Burgeson, *J. Cell Biol.* **104,** 611 (1987).

3. Excess aldehydes are eliminated by immersing the samples in 0.15 M Tris-HCl, pH 7.4, for 4 hr with three changes of buffer.

4. Samples are dehydrated in ethanol (30, 50, 70, and 95% for 15 min each; 100% for 30 min with three changes).

5. Samples are infiltrated with Lowicryl K4M (Electron Microscopy Sciences, Fort Washington, PA), diluted 1 : 1 (v/v) with 100% ethanol for 30 min; 2 : 1 for 50 min; 3 : 1 for 1 hr; and 100% Lowicryl K4M for 3 hr with three changes of medium.

6. The embedded samples are polymerized at $-20°$ for 48 hr, using an ultraviolet (UV) light. Polymerization continues for an additional 48 hr at room temperature.

Lowicryl should be mixed in a brown bottle by bubbling nitrogen through the solution. Because oxygen will inhibit polymerization, the solution should be capped with nitrogen following exposure to atmosphere. Standard silicon flat embedding molds prevent successful polymerization. Molds manufactured from polypropylene or Teflon work well. Oxygen must be prevented from diffusing into the medium during polymerization. We assure an oxygen-poor atmosphere in the vicinity of the samples by placing them in a container that also contains dry ice, with several holes drilled in the cover (which allows a flow of CO_2 gas over the samples during polymerization). The cover is made of polypropylene, which allows the penetration of UV light. The entire container is then placed in a $-20°$ walk-in freezer, with a 15-W UV light placed 12 in. away from the embedding molds. As the dry ice sublimates, the samples are continually bathed in an atmosphere of CO_2.

7. Ultrathin sections are cut and mounted on Formvar-coated Ni, Au, or Pd grids. Avoid the use of Cu grids.

8. Grids are floated section side down, not submersed, in distilled water for 20 min at room temperature. The remaining steps proceed at room temperature.

9. Grids are floated on PBS containing 5.0% (v/v) normal goat serum and 2.0% (w/v) nonfat dry milk (Carnation) for 15 min, in order to block nonspecific binding sites, and washed by floating on PBS for 5 min.

10. Grids are floated on primary antibody, diluted 1 : 5, containing 1% (v/v) fish gelatin (Amersham) for 2 hr. Antibody incubations are done on Parafilm in a moist covered chamber.

11. Grids are washed by floating on PBS, two times over 15 min, and then immersed and washed in PBS for 1 hr. For immersion, grids are transferred to a grid box with holes drilled through the cover and back. The grid box is then suspended above a stir bar in a 600-ml beaker and washed by magnetic stirring.

12. Grids are floated on secondary antibody conjugated to gold, diluted 1 : 3 (v/v), for 80 min.

13. Grids are washed by floating on PBS for 15 min followed by immersion washing in PBS for 1 hr and distilled water for 1 hr.

14. Sections are exposed to the saturated beam of a transmission electron microscope (TEM) with the condensor lens overfocused so that the entire section is irradiated for 20 min and then stained in 2% (v/v) aqueous uranyl acetate for 15 min and Reynold's lead citrate for 3 min.

[3] Tenascins

By SUSANNE SCHENK and RUTH CHIQUET-EHRISMANN

Tenascin Family Members

At present three members of the tenascin family have been described. Erickson[1] has proposed a uniform nomenclature for these proteins, calling them tenascin-C, tenascin-R, and tenascin-X, respectively. Thus tenascin-C stands for the protein originally described as myotendinous antigen,[2] which later was renamed tenascin.[3] Because the same molecule had also been described as cytotactin,[4] it is now called tenascin-C. Tenascin-R was originally described as restrictin[5,6] or J1-160/180,[7] and tenascin-X had been discovered as human gene X.[8,9] Structural models based on the respective primary structures of the three family members are presented in Fig. 1. All of the tenascins consist at the N terminus of a central domain involved in the oligomerization, followed by differing numbers of tenascin-type epidermal growth factor (EGF)-like repeats, fibronectin type III repeats, and a C-terminal domain homologous to the globular part of β- and γ-fibrinogens. Preliminary evidence has been presented that more members of the tenascin family might exist.[10]

[1] H. P. Erickson, *Curr. Opin. Cell Biol.* **5**, 869 (1993).

[2] M. Chiquet and D. M. Fambrough, *J. Cell Biol.* **98**, 1926 (1984).

[3] R. Chiquet-Ehrismann, E. J. Mackie, C. A. Pearson, and T. Sakakura, *Cell (Cambridge, Mass.)* **47**, 131 (1986).

[4] F. S. Jones, S. Hoffman, B. A. Cunningham, and G. M. Edelman, *Proc. Natl. Acad. Sci. U.S.A.* **86**, 1905 (1989).

[5] F. G. Rathjen, J. M. Wolff, and R. Chiquet-Ehrismann, *Development (Cambridge, UK)* **113**, 151 (1991).

[6] U. Nörenberg, H. Wille, J. M. Wolff, R. Frank, and F. G. Rathjen, *Neuron* **8**, 849 (1992).

[7] B. Fuss, E. Wintergerst, and M. Schachner, *J. Cell Biol.* **120**, 1237 (1993).

[8] Y. Morel, J. Bristow, S. E. Gitelman, and W. L. Miller, *Proc. Natl. Acad. Sci. U.S.A.* **86**, 6582 (1989).

[9] J. Bristow, M. Kian Tee, S. E. Gitelman, S. H. Mellon, and W. L. Miller, *J. Cell Biol.* **122**, 265 (1993).

[10] R. Chiquet-Ehrismann, C. Hagios, and K.-I. Matsumoto, *Perspect. Dev. Neurobiol.* **2**, 3 (1994).

Copyright © 1994 by Academic Press, Inc.
All rights of reproduction in any form reserved.

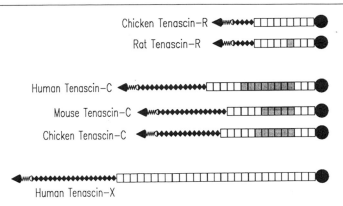

Chicken Tenascin-R
Rat Tenascin-R

Human Tenascin-C
Mouse Tenascin-C
Chicken Tenascin-C

Human Tenascin-X

FIG. 1. Structural models of the members of the tenascin family. The models are based on primary sequences for chicken tenascin-R [U. Nörenberg, H. Wille, J. M. Wolff, R. Frank, and F. G. Rathjen, *Neuron* **8**, 849 (1992)], rat tenascin-R [B. Fuss, E. Wintergerst, and M. Schachner, *J. Cell Biol.* **120**, 1237 (1993)], human tenascin-C [D. E. Nies, T. J. Hemesath, J.-H. Kim, J. R. Gulcher, and K. Stefansson, *J. Biol. Chem.* **266**, 2818 (1991); A. Siri, B. Carnemolla, M. Saginati, A. Leprini, G. Casari, F. Baralle, and L. Zardi, *Nucleic Acids Res.* **19**, 525 (1991); P. Sririmarao and M. A. Bourdon, *ibid.* **21**, 163 (1993)] mouse tenascin-C [A. Weller, S. Beck, and P. Ekblom, *J. Cell Biol.* **112**, 355 (1991)], chicken tenascin-C [F. S. Jones, S. Hoffman, B. A. Cunningham, and G. M. Edelman, *Proc. Natl. Acad. Sci. U.S.A.* **86**, 1905 (1989); J. Spring, K. Beck, and R. Chiquet-Ehrismann, *Cell (Cambridge, Mass.)* **59**, 325 (1989); R. P. Tucker J. Spring, S. Baumgartner, D. Martin, C. Hagios, P. M. Poss, and R. Chiquet-Ehrismann, *Development (Cambridge, UK)* **120**, 637 (1994)], and human tenascin-X [J. Bristow, M. Kian Tee, S. E. Gitelman, S. H. Mellon, and W. L. Miller, *J. Cell Biol.* **122**, 265 (1993)]. All tenascins consist of the same types of domains arranged in the same order, but at different numbers. At the N terminus a globular domain is located (sector), followed by heptad repeats involved in the trimerization of the molecules (wavy line), EGF-like repeats (diamonds), fibronectin type III repeats (clear rectangles, constant repeats; shaded rectangles, repeats known to be subject to alternative splicing), and a C-terminal domain homologous to the globular part of β- and γ-fibrinogens (filled circle). The exact number of fibronectin type III repeats of tenascin-X is not known yet, but at least 29 repeats have been identified in the genomic region encoding the human tenascin-X gene.

All of these tenascins are extracellular matrix proteins, showing different expression patterns in developing and adult organisms. The initial interest in tenascin-C came from its interesting tissue distribution both in normal and abnormal development. Tenascin-C is temporarily highly expressed during embryogenesis at the sites of tissue interactions during organogenesis and is reexpressed in tumors.[3] Furthermore, tenascin-C is also expressed in the nervous system.[11,12] Tenascin-R was shown to be

[11] M. Chiquet, B. Wehrle-Haller, and M. Koch, *Semin. Neurosci.* **3**, 341 (1991).
[12] M. Chiquet and B. Wehrle-Haller, *Perspect. Dev. Neurobiol.* **2**, 67 (1994).

specific for the nervous system in a more restricted fashion than tenascin-C.[5] The mRNA of tenascin-X is highly expressed in testis, gut, heart, and muscle,[9] but the tenascin-X protein has not been characterized. Our preliminary studies show that mouse tenascin-X is built up of subunits of about 500 kDa and that it is localized predominantly in heart and skeletal muscle as well as in blood vessels.[12a] Thus eventually every tissue may contain one or several types of tenascin molecules.

Tenascin Function

Tenascin-C has often been described as an antiadhesive or adhesion-modulating protein.[13–15] It might be important to counteract other adhesive proteins such as fibronectin, to enable cells to detach from the extracellular matrix and to migrate during morphogenic movements. There exists, however, no direct proof for its real *in vivo* function, because knocking out the tenascin gene in transgenic mice has not revealed any readily apparent defects.[16] In the nervous system tenascin-C and tenascin-R appear to have similar effects on neuronal cells[6] and might therefore be able to compensate for each other. Both of these tenascins also bind to the same neuronal receptors, namely F11/F3/contactin.[5,17,18] Nonneuronal cellular receptors for tenascin-C appear to be integrins[19,20] and the cell surface proteoglycan syndecan, which binds tenascin-C through its heparan sulfate side chains.[21] Tenascin-C is a heparin-binding protein[22] and also binds to other proteoglycans.[23,24] Tenascin-C and tenascin-R can also interact with each other, because tenascin-C is retained on a tenascin-R affinity column.[5] Furthermore, tenascin-C binds to fibronectin.[25] Interestingly, tenascin-C occurs

[12a] K.-I. Matsumoto, *J. Cell Biol.* **122,** 265 (1993).

[13] R. Chiquet-Ehrismann, *Curr. Opin. Cell Biol.* **3,** 800 (1991).

[14] E. H. Sage and P. Bornstein, *J. Biol. Chem.* **266,** 14831 (1991).

[15] R. Chiquet-Ehrismann, *Semin. Cancer Biol.* **4,** 301 (1993).

[16] Y. Saga, T. Yagi, Y. Ikawa, T. Sakakura, and S. Aizawa, *Genes Dev.* **6,** 1821 (1992).

[17] A. H. Zisch, L. D'Alessandri, B. Ranscht, R. Falchetto, K. H. Winterhalter, and L. Vaughan, *J. Cell Biol.* **119,** 203 (1992).

[18] P. Pesheva, G. Gennarini, C. Goridis, and M. Schachner, *Neuron* **10,** 69 (1993).

[19] M. A. Bourdon and E. Ruoslahti, *J. Cell Biol.* **108,** 1149 (1989).

[20] P. Sririmarao, M. Mendler, and M. A. Bourdon, *J. Cell Sci.* **105,** 1001 (1993).

[21] M. Salmivirta, K. Elenius, S. Vainio, U. Hofer, R. Chiquet-Ehrismann, I. Thesleff, and M. Jalkanen, *J. Biol. Chem.* **266,** 7733 (1991).

[22] M. Chiquet, N. Vruĉiniĉ Filipi, S. Schenk, K. Beck, and R. Chiquet-Ehrismann, *Eur. J. Biochem.* **199,** 379 (1991).

[23] M. Chiquet and D. M. Fambrough, *J. Cell Biol.* **98,** 1937 (1984).

[24] S. Hoffman, K. L. Crossin, and G. M. Edelman, *J. Cell Biol.* **106,** 519 (1988).

[25] R. Chiquet-Ehrismann, Y. Matsuoka, U. Hofer, J. Spring, C. Bernasconi, and M. Chiquet, *Cell Regul.* **2,** 927 (1991).

in several splicing variants, which seem to be functionally different. In Fig. 2 the three major splicing variants of chicken tenascin-C are shown together with a hypothetical chicken tenascin-C including all extra repeats presently known. The chicken tenascin-C is compared to the human counterpart, showing that the number of repeats differs slightly, both for the EGF-like repeats and the fibronectin type III repeats. The extra repeats have been proposed to be responsible for the antiadhesive effects of tenascin-C on endothelial cells.[26] In the case of fibronectin binding, the smallest tenascin variant without any extra repeats was shown to bind better to fibronectin than the larger tenascin variants.[25] The same was shown for the binding to F11, with which only the smallest tenascin-C variant was active.[27] In addition, the assembly of tenascin-C in the extracellular matrix network laid down by cells in culture differs between the tenascin-C variants. The smallest tenascin-C variant is readily incorporated into the extracellular matrix, whereas the large tenascin-C variants preferentially accumulate in the conditioned medium.[25,28] Both tenascin-C and tenascin-R can be alternatively spliced,[7] but no functional relevance has been proposed. In conclusion, the tenascins are a complex set of extracellular matrix proteins that appear to be involved in interaction with cells and in the building up of the extracellular matrix scaffold.

Isolation of Tenascin-C

Most of the tenascin research up to now has been focused on tenascin-C, because it is the longest known family member. No purification procedures based on biochemical methods have been described for the other family members. Therefore, we summarize here the isolation procedures used for tenascin-C.

Immunoaffinity Chromatography

The most commonly used, simple procedure to purify tenascin-C is by affinity chromatography to anti-tenascin-C antibody-coupled columns. Clearly this method requires the availability of sufficient anti-tenascin antibody. However, because several companies are by now selling monoclonal anti-tenascin-C antibodies and a small column can be prepared with about 1 mg of such an antibody, this method should now also be generally

[26] J. E. Murphy-Ullrich, V. A. Lightner, I. Aukhil, Y. Z. Yan, H. P. Erickson, and M. Höök, *J. Cell Biol.* **115**, 1127 (1991).
[27] A. Horwitz, K. Duggan, R. Greggs, C. Decker, and C. Buck, *J. Cell Biol.* **101**, 2134 (1985).
[28] B. Carnemolla, L. Borsi, G. Bannikov, S. Troyanovsky, and L. Zardi, *Eur. J. Biochem.* **205**, 561 (1992).

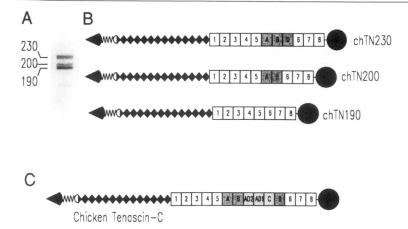

FIG. 2. Alternative splicing of tenascin-C. (A) Chicken tenascin-C isolated from fibroblast cultures contains three types of subunits of 230, 200, and 190 kDa, as seen after electrophoresis. (B) Structural models corresponding to the three major splicing variants, using the same symbols as described in Fig. 1. The structural features of the three major variants have been determined by cDNA cloning and by epitope mapping [J. Spring, K. Beck, and R. Chiquet-Ehrismann, *Cell (Cambridge, Mass.)* **59**, 325 (1989)]. (C) Three more fibronectin type III repeats of chicken tenascin have been discovered, which could lead to the production of a chicken tenascin-C molecule as diagrammed [R. P. Tucker, J. Spring, S. Baumgartner, D. Martin, C. Hagios, P. M. Poss, and R. Chiquet-Ehrismann, *Development (Cambridge, UK)* **120**, 637 (1994)]. Aligning the chicken and the human tenascin-Cs reveals that human tenascin contains one more EGF-like repeat than chicken tenascin-C, that they both contain the same number of constant fibronectin type III repeats, but that the extra repeats differ. Repeat A occurs four times in human and only once in the chicken tenascin-C. Repeat AD2 has so far been seen only in chicken tenascin-C, whereas repeat AD1 [first discovered by P. Sririmarao and M. A. Bourdon, *Nucleic Acids Res.* **21**, 163 (1993)] is present in both species.

applicable. In Fig. 3 we present the results yielded by this method. Conditioned medium from tenascin-producing cells, in this case human melanoma SK-MEL-28 cells, is collected (850 ml), concentrated by precipitation with 45% (w/v) ammonium sulfate, and the pellet dissolved in 15 ml of phosphate buffered saline (PBS) and dialyzed. This material is slowly (for about 2 hr) passed over a gelatin–Sepharose column (20-ml bed volume) that is directly connected to a Sepharose 4B column, coupled with a monoclonal antibody raised in our laboratory against human tenascin-C (bed volume, 5 ml; coupled with 2 mg of the antibody per milliliter of gel). The columns are then disconnected and the anti-tenascin-C-coupled

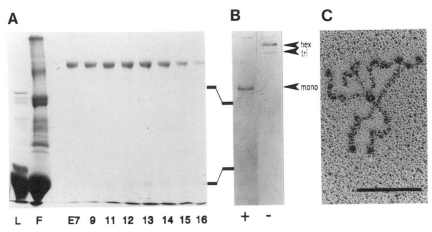

FIG. 3. Immunoaffinity isolation of tenascin-C. (A) Coomassie blue-stained sodium dodecyl sulfate (SDS)–polyacrylamide gel (6%) run under reducing conditions showing the loaded
conditioned medium (L), the flowthrough (F), and the elution fractions as indicated by their
fraction numbers (E7–16). Each fraction was 1 ml, and 20 μl was loaded per lane. Markers
of 200 and 67 kDa are indicated between (A) and (B). (B) A 3–15% gradient SDS–polyacrylamide gel run under reducing (+) or nonreducing (−) conditions of fraction 9, showing the
monomers of tenascin-C in the reduced lane and trimers and hexamers in the nonreduced lane
(marked by arrowheads labeled mono, tri, and hex, respectively). (C) Electron micrograph of
a rotary shadowed tenascin-C molecule from fraction 9 prepared by the method described in
R. Chiquet-Ehrismann, P. Kalla, C. A. Pearson, K. Beck, and M. Chiquet [*Cell* (*Cambridge,
Mass.*) **53**, 383 (1988)]. Note the central globe flanked by two T junctions, where three arms
each are assembled. Each arm has an inner, thinner part consisting of the EGF-like repeats,
followed by a thicker part consisting of the fibronectin type III repeats, and is terminated
with a globule homologous to β- and γ-fibrinogens. Bar: 100 nm.

column is extensively washed with PBS before eluting the tenascin with
50 mM diethylamine in H$_2$O (pH 11). To make sure that tenascin is not
contaminated with proteoglycans bound to it, an additional wash step
using 0.05% Triton X-100–1 M NaCl in PBS can be included before the
elution of the tenascin.[22] The tenascin isolated by this method consists of
mostly hexamers and some trimers, as can be seen in Fig. 3B, after
electrophoresis without reducing agents. After rotary shadowing this
tenascin-C preparation, hexameric tenascin molecules can be seen in electron micrographs (Fig. 3C). Each tenascin-C arm has a length of about
100 nm, and consists of an outer C-terminal fibrinogen globe, preceded
by a thicker part of the arms made up of the fibronectin type III repeats,
a thinner part of the arms consisting of the EGF-like repeats and a
T junction, where three arms are joined in a triple-coiled coil region determined by heptad repeats and flanked by interchain disulfides. In the globu-

lar region in the center of the molecule, two tenascin-C trimers are disulfide linked at the N-terminal domains to form the typical hexameric molecule. Immunoaffinity isolation is of course possible for all tenascins and has also been used successfully to isolate tenascin-R from brain extracts.[6,29] These tenascin-R preparations consisted of trimers and not of hexamers, as did the tenascin-C preparation.

Biochemical Procedures

In Fig. 4 we have summarized schematically three possible methods for purifying tenascin-C from the conditioned medium of cell cultures. These procedures are based on papers published by Oike et al.[30] (method I), Aukhil et al.[31] (method II), and Saginati et al.[32] (method III). These procedures allow the isolation of about 1–4 mg of tenascin from 1 liter of conditioned medium, with the exception of method I, for which a 20-fold lower yield was reported. (Note: There is an error in the reported yield in the article by Oike et al.,[30] which should read 500 and 450 μg of tenascin, instead of the reported 500 and 450 mg). This lower yield is most likely not due to the method applied, but to the choice of normal human fibroblasts as the tenascin-C source, because these cells seem to secrete much less tenascin-C than the cancer cell lines used in the other papers. Compared to the methods using affinity chromatography on monoclonal antibody columns these biochemical procedures have certain advantages. They allow the isolation of tenascin-C of any species and they do not require the use of harsh conditions such as high or low pH or denaturing agents as is necessary to elute tenascin-C from an antibody affinity column. However, because these biochemical procedures take about 3 days to obtain purified tenascin-C, the risk of proteolytic degradation increases.

One-Step Affinity Isolation Using Fixed Erythrocytes. This procedure has not been applied to large-scale isolation of tenascin-C. However, the analytical experiments presented in Fig. 5 demonstrate that this method could be an easy and efficient one-step procedure for the isolation of tenascin-C. Because we knew that tenascin-C hemagglutinates fixed erythrocytes,[3] we investigated whether the binding of tenascin-C to erythrocytes could be taken advantage of for its affinity isolation. In an analytical

[29] P. Pesheva, E. Spiess, and M. Schachner, J. Cell Biol. 109, 1765 (1989).
[30] Y. Oike, H. Hiraiwa, H. Kawakatsu, M. Nishikai, T. Okinaka, T. Suzuki, A. Okada, R. Yatani, and T. Sakakura, Int. J. Dev. Biol. 34, 309 (1990).
[31] I. Aukhil, C. C. Slemp, V. A. Lightner, K. Nishimura, G. Briscoe, and H. P. Erickson, Matrix 10, 98 (1990).
[32] M. Saginati, A. Siri, E. Balza, M. Ponassi, and L. Zardi, Eur. J. Biochem. 205, 545 (1992).

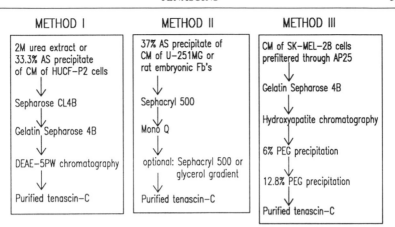

FIG. 4. Purification schemes for the isolation of tenascin-C. Method I [Y. Oike, H. Hiraiwa, H. Kawakatsu, M. Nishikai, T. Okinaka, T. Suzuki, A. Okada, R. Yatani, and T. Sakakura, *Int. J. Dev. Biol.* **34,** 309 (1990)] is for isolating human tenascin from normal umbilical vein fibroblasts; method II [I. Aukhil, C. C. Slemp, V. A. Lightner, K. Nishimura, G. Briscoe, and H. P. Erickson, *Matrix* **10,** 98 (1990)] is for isolating human and rat tenascin from the glioma cell line U-251MG or embryonic rat fibroblasts (Fbs), respectively; and method III [M. Saginati, A. Siri, E. Balza, M. Ponassi, and L. Zardi, *Eur. J. Biochem.* **205,** 545 (1992)] is for isolating human tenascin from the melanoma cell line SK-MEL-28. AS, Ammonium sulfate; CM, conditioned medium; PEG, polyethylene glycol.

batch procedure, 60 μl of packed fixed sheep erythrocytes were incubated with 50 μl of ^{35}S-labeled conditioned medium from SK-MEL-28 melanoma cells and mixed. The erythrocytes were then washed three times with 200 μl of PBS^{+} (phosphate-buffered saline including Ca^{2+}/Mg^{2+}) and then the bound tenascin-C was eluted in 75-μl fractions using either urea, NaCl, or ethylenediaminetetraacetic acid (EDTA). As can be seen in Fig. 5A, the tenascin-C bound quantitatively to the erythrocytes. The elution with urea was poor in comparison to NaCl, and in the case of the elution with salt it also seems that a small amount of fibronectin is recovered in the eluate (Fig. 5B). The best result was obtained with EDTA, which was extremely effective in eluting pure tenascin-C (Fig. 5B). It is not clear what tenascin-C is binding to on the erythrocytes. It is possible that tenascin-C binds to gangliosides[33] or sulfatides.[34] Clearly the binding is dependent on divalent cations and it is possible that the postulated Ca^{2+}

[33] U. Hofer and R. Chiquet-Ehrismann, *in* "Limb Development and Regeneration" (J. F. Fallon, ed.), p. 455. Wiley-Liss, New York, 1993.
[34] K. L. Crossin and G. M. Edelman, *J. Neurosci. Res.* **33,** 631 (1992).

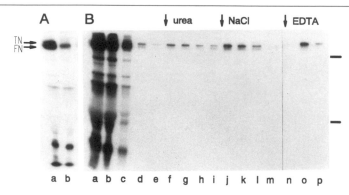

FIG. 5. Affinity isolation of tenascin-C using fixed sheep erythrocytes: autoradiographs of 10% SDS–polyacrylamide gels run under reducing conditions. Molecular mass standards of 200 and 67 kDa are marked to the right. (A) Lane a, conditioned medium of metabolically labeled SK-MEL-28 cells. The most prominent high molecular weight band, marked FN, represents fibronectin and the fainter band above it, marked TN, is tenascin-C. Lane b, supernatant after incubation of the medium with the erythrocytes. Note that the tenascin-C has been quantitatively removed. (B) Equal volumes of the fractions collected during the tenascin purification were loaded. Lane a, conditioned medium applied. Lane b, supernatant containing unbound material. Lanes c–e, wash fractions. Lanes f–i, elution fractions using consecutively increasing concentrations of urea: 0.1 M (f), 0.25 M (g), 0.5 M (h), 1 M (i). Lanes j–m, elution fractions of a parallel experiment using increasing concentrations of NaCl: 0.1 M (j), 0.25 M (k), 0.5 M (l), 1 M (m). Lanes n–p, elution fractions of a parallel experiment using EDTA: 0.1 mM (n), 5 mM (o), 10 mM (p). Note that the elution with 5 mM EDTA results in the most concentrated and purest tenascin-C fraction.

binding site in the fibrinogen globe at the end of the tenascin-C arms[35] could be involved in this binding.

Isolation of Tenascin-C from Tissues. Generally it is rather difficult to isolate tenascins from tissues, because it is not easy to solubilize tenascins in the first place. A procedure has been described to isolate tenascin-C from neutral 1 M NaCl extracts of organ cultures of embryonic cartilage.[36] Tenascin-C has been purified from embryonic or adult brain.[24,37,38] We have described the isolation of tenascin-C from chicken gizzard,[22] by extracting the tenascin, after homogenization and washing in water, using 10 mM EDTA and 20 mM 3-cyclohexylamino-1-propanesulfonic acid at

[35] K. Beck, J. Spring, R. Chiquet-Ehrismann, J. Engel, and M. Chiquet, *in* "Patterns in Protein Sequence and Structure" (W. R. Taylor, ed.), p. 229. Springer-Verlag, Berlin and New York, 1992.

[36] L. Vaughan, S. Huber, M. Chiquet, and K. H. Winterhalter, *EMBO J.* **6,** 349 (1987).

[37] M. Grumet, S. Hoffman, K. L. Crossin, and G. M. Edelman, *Proc. Natl. Acad. Sci. U.S.A.* **82,** 8075 (1985).

[38] A. Faissner and J. Kruse, *Neuron* **5,** 627 (1990).

pH 11. The dialyzed preparation is cleared from the actomyosin precipitate by centrifugation and tenascin-C is purified from the supernatant by immunoaffinity chromatography as described above. The resulting tenascin-C preparation consists of monomeric molecules, which do not appear to be degradation products, and which also include N-terminal epitopes. It is not clear whether gizzard tenascin is indeed monomeric, or whether we preferentially solubilized the monomeric molecules and were not successful in extracting oligomeric tenascin-C.

Recombinant Tenascin-C

A different approach for obtaining sufficient tenascin-C for functional studies has been the expression and isolation of recombinant tenascin-C fragments in bacteria, either as fusion proteins[39,40] or as the tenascin-C part alone.[41] Using such bacterially expressed parts of tenascin-C, many active sites have been postulated for the interaction with cells or other extracellular matrix constituents. It is too early to summarize these results, because many of the data appear to be contradictory and need to be analyzed more carefully, using alternative experimental procedures. A particularly good set of recombinant tenascin-C fragments has been produced by Aukhil et al.[41] which led them to postulate a heparin- and cell-binding site in the fibrinogen domain and a second heparin-binding site in the fourth and fifth fibronectin type III repeat. Using authentic tenascin-C and proteolytic fragments, our experiments to determine the heparin-binding site(s) localized the heparin-binding activity within the C-terminal fragment containing the fibrinogen globe exclusively,[22] thus raising the question as to whether the bacterially expressed tenascin fragments could sometimes lead to the exposure of unnatural cryptic sites. This is by now a well-known problem, and thus the expression of tenascin-C in mammalian expression systems, as has been used successfully by Aukhil et al.[41] to create artificial tenascin-C variants, may become useful in identifying active sites in an intact tenascin molecule.

Acknowledgments

We would like to thank Dr. M. Chiquet for helping to prepare some of the figures and T. Schulthess for performing the electron microscopy of the purified tenascin-C.

[39] J. Spring, K. Beck, and R. Chiquet-Ehrismann, Cell (Cambridge, Mass.) **59**, 325 (1989).
[40] A. L. Prieto, C. Andersson-Fisone, and K. L. Crossin, J. Cell Biol. **119**, 663 (1992).
[41] I. Aukhil, P. Joshi, Y. Yan, and H. P. Erickson, J. Biol. Chem. **268**, 2542 (1993).

[4] Thrombospondins

By PAUL BORNSTEIN and E. HELENE SAGE

Introduction

The characterization of platelet thrombospondin (TSP), its electron microscopic appearance, and its subcellular location were first described by Lawler and co-workers in 1978.[1] However, TSP had been identified earlier as a thrombin-sensitive protein that was released on activation of human platelets.[2,3] Thrombospondin is a disulfide-bonded trimer of three identical chains with a monomer molecular mass of about 140,000 Da. The predicted amino acid composition of human TSP1, based on analyses of DNA clones,[4,5] provides for a molecular mass of 127,500 Da. The remaining mass can be accounted for largely by the presence of about 10,000 Da per chain of carbohydrate.[6,7] TSP1 has also been reported to contain a small number of β-hydroxyasparagine residues.[8] TSP1 is an example of a modular glycoprotein, and fragments of the protein that are suitable for further biochemical and immunological analysis can be generated by limited cleavage with proteases such as chymotrypsin, thermolysin, and thrombin.[9,10] Direct visualization of amino- and carboxy-terminal globes, linked by more acicular domains in the protein, has been achieved by electron microscopic examination of rotary shadowed preparations.[11,12] The modular structure of TSP1 in part accounts for its ability

[1] J. W. Lawler, H. S. Slayter, and J. E. Coligan, *J. Biol. Chem.* **253**, 8609 (1978).

[2] N. L. Baenziger, G. N. Brodie, and P. W. Majerus, *Proc. Natl. Acad. Sci. U.S.A.* **68**, 240 (1971).

[3] N. L. Baenziger, G. N. Brodie, and P. W. Majerus, *J. Biol. Chem.* **247**, 2723 (1972).

[4] J. Lawler and R. O. Hynes, *J. Cell Biol.* **103**, 1635 (1986).

[5] S. W. Hennessy, B. A. Frazier, D. D. Kim, T. L. Deckwerth, D. M. Baumgartel, P. Rotwein, and W. A. Frazier, *J. Cell Biol.* **108**, 729 (1989).

[6] P. Vischer, H. Beeck, and B. Voss, *Eur. J. Biochem.* **153**, 435 (1985).

[7] K. Furukawa, D. D. Roberts, T. Endo, and A. Kobata, *Arch. Biochem. Biophys.* **270**, 302 (1989).

[8] C. T. Przysiecki, J. E. Staggers, H. G. Ramjit, D. G. Musson, A. M. Stern, C. D. Bennett, and P. A. Friedman, *Proc. Natl. Acad. Sci. U.S.A.* **84**, 7856 (1987).

[9] V. M. Dixit, G. A. Grant, S. A. Santoro, and W. A. Frazier, *J. Biol. Chem.* **259**, 10100 (1984).

[10] S. M. Mumby, G. J. Raugi, and P. Bornstein, *J. Cell Biol.* **98**, 646 (1984).

[11] J. Lawler, L. H. Derick, J. E. Connolly, J.-H. Chen, and F. C. Chao, *J. Biol. Chem.* **260**, 3762 (1985).

[12] N. J. Galvin, V. M. Dixit, K. M. O'Rourke, S. A. Santoro, G. A. Grant, and W. A. Frazier, *J. Cell Biol.* **101**, 1434 (1985).

Copyright © 1994 by Academic Press, Inc.
All rights of reproduction in any form reserved.

to interact with a wide variety of extracellular matrix and plasma proteins, cell surface receptors, and cations such as Ca^{2+}.

It is now recognized that TSP1 is the founding member of a small family of related proteins (Table I).[13,14] TSP1 and TSP2 are similar in overall structure but differ in their expression in the developing mouse[15] and in regulation of their expression by serum and cytokines.[16] TSP3, TSP4, and TSP5/COMP are also similar to each other in structure and differ from TSP1 and TSP2 by the absence of several of the domains present in the larger TSPs (Fig. 1). Little is known of the properties of the smaller TSPs, with the exception of TSP5/COMP, which had been identified previously as cartilage oligomeric matrix protein (COMP) in articular cartilage and in the Swarm rat chondrosarcoma.[17,18]

Our understanding of the potential for the participation of TSP in biological processes was enhanced considerably by the realization that the protein is present not only in the α granules of platelets but is also a secreted product of a wide variety of cells. These include fibroblasts, smooth muscle cells and myoblasts, epithelial cells, monocytes and macrophages, endothelial cells, osteoblasts and chondrocytes, keratinocytes and melanocytes, and many neoplastic cells (reviewed by Clezardin.)[19] It is therefore not surprising that different and sometimes conflicting functions have been attributed to TSP1 (Table II). It is possible that some of these functions are performed by some of the other TSPs whose presence was not appreciated at the time these studies were conducted and whose identity could have been confused with that of TSP1. However, in some cases divergent functions of TSP can be attributed to the multiplicity of receptors that have been described for the protein[20-22] and to the presence of different subsets of these receptors on different cells. If, as seems possible, the ligands for some of these receptors also exist on other extracellular macromolecules, the nature of the extracellular environment, together with the relative concentrations of the different TSPs that are

[13] P. Bornstein, *FASEB J.* **6**, 3290 (1992).

[14] J. C. Adams and J. Lawler, *Curr. Biol.* **3**, 188 (1993).

[15] L. Iruela-Arispe, D. J. Liska, E. H. Sage, and P. Bornstein, *Dev. Dyn.* **197**, 40 (1993).

[16] P. Bornstein, S. Devarayalu, P. Li, C. M. Disteche, and P. Framson, *Proc. Natl. Acad. Sci. U.S.A.* **88**, 8636 (1991).

[17] M. Morgelin, D. Heinegard, J. Engel, and M. Paulsson, *J. Biol. Chem.* **267**, 6137 (1992).

[18] E. Hedbom, P. Antonsson, A. Hjerpe, D. Aeschlimann, M. Paulsson, E. Rosa-Pimentel, Y. Sommarin, M. Wendel, A. Oldberg, and D. Heinegard, *J. Biol. Chem.* **267**, 6132 (1992).

[19] P. Clezardin, *in* "Thrombospondin" (J. Lahav, ed.), p. 41. CRC Press, Boca Raton, FL, 1993.

[20] A. S. Asch, J. Tepler, S. Silbiger, and R. L. Nachman, *J. Biol. Chem.* **266**, 1740 (1991).

[21] C. A. Prater, J. Plotkin, D. Jaye, and W. A. Frazier, *J. Cell Biol.* **112**, 1031 (1991).

[22] J. C. Adams and J. Lawler, *J. Cell Sci.* **104**, 1061 (1993).

TABLE I
THROMBOSPONDIN GENES AND
CHROMOSOMAL LOCATIONS

Gene	Human	Mouse
TSP1	15q15[a-c]	2F[a,d]
TSP2	6q27[e]	17A3[f]
TSP3	1q21-24[g]	3E3-F1[h]
TSP4	ND[i]	ND[i]
TSP5/COMP	ND[i]	ND[i]

[a] E. Jaffe, P. Bornstein, and C. M. Disteche, *Genomics* **7**, 123 (1990).
[b] F. W. Wolf, R. L. Eddy, T. B. Shows, and V. M. Dixit, *Genomics* **6**, 685 (1990).
[c] D. J. Good, P. J. Polverini, F. Rastinejad, M. M. Le Beau, R. S. Lemons, W. A. Frazier, and N. P. Bouck, *Proc. Natl. Acad. Sci. U.S.A.* **87**, 6624 (1990).
[d] J. Lawler, M. Duquette, P. Ferro, N. G. Copeland, D. J. Gilbert, and N. A. Jenkins, *Genomics* **11**, 587 (1991).
[e] T. L. LaBell, D. J. M. Milewicz, C. M. Disteche, and P. H. Byers, *Genomics* **12**, 421 (1992).
[f] P. Bornstein, S. Devarayalu, P. Li, C. M. Disteche, and P. Framson, *Proc. Natl. Acad. Sci. U.S.A.* **88**, 8636 (1991).
[g] H. L. Vos, S. Devarayalu, Y. de Vries, and P. Bornstein, *J. Biol. Chem.* **267**, 12192 (1992).
[h] P. Bornstein, S. Devarayalu, S. Edelhoff, and C. M. Disteche, *Genomics* **15**, 607 (1993).
[i] ND, Not determined.

available to the cell surface, could well influence the functional properties of any given member of the TSP family.

Structure of Thrombospondin 1 and Comparison with Other Thrombospondins

The structures of TSP1 and TSP2 are shown diagrammatically and are compared with the structures of TSP3, TSP4, and TSP5/COMP in Fig. 1. TSP1 and TSP2 each contain an amino-terminal domain, a region that is involved in interchain disulfide bonding, a procollagen homology, three type I (TSP) repeats, three type II (epidermal growth factor [EGF]-like) repeats, seven type III (Ca^{2+}-binding) repeats, and a carboxy-terminal

Thrombospondin 1 and 2

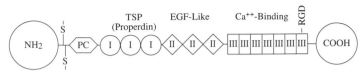

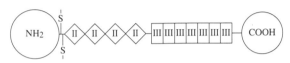

Thrombospondin 3, 4 and 5

FIG. 1. The structures of TSP1 and TSP2 in comparison with those of TSP3, TSP4, and TSP5/COMP. Models of individual chains are shown. TSP1 and TSP2 are trimeric, whereas TSP5/COMP is a pentamer. The oligomeric structures of TSP3 and TSP4 are not known, but these proteins could also exist as pentamers (see Fig. 2). TSP1 and TSP2 are composed of the following modules: (1) an NH_2-terminal globular domain, (2) an interchain disulfide knot formed by two closely spaced cysteines on each chain, (3) a region (PC) that is homologous to the α1 chain of type I procollagen, (4) three type I (TSP or properdin) repeats, (5) three type II (EGF-like) repeats, (6) seven type III (Ca^{2+}-binding) repeats, the last of which contains an RGD sequence that binds to integrins, and (7) a COOH-terminal domain. TSP3, TSP4, and TSP5/COMP lack the procollagen homology region and type I repeats and contain four type II repeats. The placement of interchain disulfide bonds following the NH_2-terminal domain in these chains is presumptive.

domain. TSP3, TSP4, and TSP5 resemble one another and differ from TSP1 and TSP2 in that they lack the procollagen domain and type I repeats and contain four, rather than three, EGF-like repeats.

TSP1 and TSP2 exist as disulfide-bonded trimers, whereas TSP5/ COMP has been shown by electron microscopy to be a pentamer.[17] The polymeric structures of TSP3 and TSP4 are not known. All five proteins contain two cysteines, separated by two or three amino acids, that could serve to form interchain disulfide bonds. In the case of TSP1, Cys-270 and Cys-274 have been implicated in such interchain bonds.[23] Mosher and co-workers have shown that the amino acid sequence surrounding these cysteines can form an α helix with a hydrophobic face and suggest that trimers could result from the formation of coiled coils.[23,24] Although the corresponding sequence in mouse, human, and chicken TSP2 differs con-

[23] J. Sottile, J. Selegue, and D. F. Mosher, *Biochemistry* **30**, 6556 (1991).
[24] D. F. Mosher, X. Sun, J. Sottile, and P. J. Hogg, *Adv. Mol. Cell Biol.* **6**, 115 (1993).

TABLE II
PROPOSED FUNCTIONS OF THROMBOSPONDIN 1

Function	Refs.[a]
Binding to activated platelets; platelet aggregation	b
	c
Neurite outgrowth; axonal regeneration	d
Interaction with components of fibrinolytic system	e
Adhesion and motility of inflammatory cells	f
Adhesive protein for normal and metastatic cells	g–i
Modulator of cell growth; sequestration and/or activation of cytokines	j
Inhibitor of angiogenesis	k–m
Chemotaxis and haptotaxis	f,n
Protease inhibitor	o,p

[a] These references are not comprehensive. In most cases, either a more recent review or reference has been cited.

[b] J. L. McGregor and H. Boukerche, in "Thrombospondin" (J. Lahav, ed.), p. 111. CRC Press, Boca Raton, FL, 1993.

[c] D. A. Walz, R. S. Zafar, and Z. Zeng, in "Thrombospondin" (J. Lahav, ed.), p. 149. CRC Press, Boca Raton, FL, 1993.

[d] K. S. O'Shea, in "Thrombospondin" (J. Lahav, ed.), p. 129. CRC Press, Boca Raton, FL, 1993.

[e] R. L. Silverstein, in "Thrombospondin" (J. Lahav, ed.), p. 165. CRC Press, Boca Raton, FL, 1993.

[f] S. J. Suchard, in "Thrombospondin" (J. Lahav, ed.), p. 177. CRC Press, Boca Raton, FL, 1993.

[g] V. Castle, J. Varani, S. Fligiel, E. V. Prochownik, and V. Dixit, *J. Clin. Invest.* **87,** 1883 (1991).

[h] D. F. Mosher, X. Sun, J. Sottile, and P. J. Hogg, *Adv. Mole. Cell Biol.* **6,** 115 (1993).

[i] G. P. Tuszynski, in "Thrombospondin" (J. Lahav, ed.), p. 209. CRC Press, Boca Raton, FL, 1993.

[j] S. Schultz-Cherry and J. E. Murphy-Ullrich, *J. Cell Biol.* **122,** 923 (1993).

[k] D. J. Good, P. J. Polverini, F. Rastinejad, M. M. Le Beau, R. S. Lemons, W. A. Frazier, and N. P. Bouck, *Proc. Natl. Acad. Sci. U.S.A.* **87,** 6624 (1990).

[l] M. L. Iruela-Arispe, P. Bornstein, and H. Sage, *Proc. Natl. Acad. Sci. U.S.A.* **88,** 5026 (1991).

[m] S. S. Tolsma, O. V. Volpert, D. J. Good, W. A. Frazier, P. J. Polverini, and N. Bouck, *J. Cell Biol.* **122,** 497 (1993).

[n] P. J. Mansfield and S. J. Suchard, *J. Immunol.* **150,** 1959 (1993).

[o] P. J. Hogg, J. Stenflo, and D. F. Mosher, *Biochemistry* **31,** 265 (1992).

[p] P. J. Hogg, D. A. Owensby, D. F. Mosher, T. M. Misenheimer, and C. N. Chesterman, *J. Biol. Chem.* **268,** 7139 (1993).

siderably, the spacing of the cysteines is preserved and the sequence also lends itself to the formation of a coiled coil. In contrast, the cysteines that could serve to form interchain disulfide bonds in TSP5/COMP (Cys-68 and Cys-71) exist in a different sequence that does not lend itself to

```
MTSP3   I R D Q V K E M S L I R N T I M E C Q V C G F H E Q R   (249-275)

RTSP5   L R H R V K E I T F L K N T V M E C D A C G M Q P A R   (51 - 77)

XTSP4   M R Q Q V K E T M F L R N T I A E C Q A C G L G P D F   (239-265)
```

FIG. 2. The amino acid sequences (single-letter code) adjacent to two cysteines (bold) that are presumed to be involved in interchain disulfide bonds in TSP3 (mouse), TSP4 (*Xenopus*), and TSP5/COMP (rat). The positions of the amino acids, with the initiating methionine at position 1, are given in parentheses. Boxed amino acids are identical or closely related.

coiled coil formation (Fig. 2). We have examined the amino acid sequences of TSP3 and TSP4 in the analogous region, immediately preceding the first type II repeat, and find that two cysteines separated by two amino acids are present in each case and that a high degree of amino acid sequence conservation exists in the 17 amino acids preceding the first cysteine (Fig.2). This conservation contrasts with little or no conservation in the remainder of the amino-terminal domain and a low degree of conservation in the first EGF repeat. Because TSP5/COMP has been shown to be a pentamer,[17] it seems likely that TSP3 and TSP4 are also pentameric.

When TSP1 and TSP2 are compared among species, it is apparent that the amino acid sequence of the carboxy-terminal domain is most highly conserved, whereas that of the amino-terminal domain is most variable.[13] Regions between the terminal domains show an intermediate degree of sequence conservation. This gradient of conservation, increasing from amino terminus to carboxy terminus, is also seen when any of the five TSPs is compared with one another within a species. In the case of the TSP3–5 subfamily, a comparison of the amino-terminal domains of these proteins with that of either TSP1 or TSP2 shows little, if any, significant conservation of sequence. A practical consequence of these observations is that antibodies directed toward the amino-terminal domains are most likely to distinguish the TSPs immunologically.

In Table III are listed sequences within each of the domains in TSP1 for which functions have been proposed, as well as proteins that show homology with one or more of the modules. Only the amino-terminal, type III repeats and carboxy-terminal domains show no homology with other proteins (with the exception of other members of the TSP family). Because the amino acid sequence of the amino-terminal domain differs so widely among members of the TSP family, it seems reasonable to suggest that the inclusion of a protein in this family should depend on the demonstration of homology with the type III repeats and carboxy-terminal domain. As can be seen in Table III, motifs representing the procollagen

TABLE III

DOMAINS AND HOMOLOGOUS MOTIFS IN THROMBOSPONDIN 1

Domain	Peptide sequence	Proposed function	Homologous proteins[a]
NH₂ terminus	BBXB[b]	Heparin and sulfatide binding[c]	None described
Procollagen	Procollagen homology[d]	Trimer assembly	Procollagen α1(I); von Willebrand factor
Type I (TSP) repeats	CSVTCG[e] WxpWSpW[j] BBXB[b]	Binding to CD36[f] Heparin and sulfatide binding Heparin and sulfatide binding	Properdin[g], f-spondin[h], complement factors C6, C7, C8 and C9[i]; antistasin (leech)[k]; unc-5 (Caenorhabditis elegans)[l]; malarial proteins[m]; several secreted proteins involved in regulation of cell growth[n,o]
Type II (EGF) repeats	EGF-like repeat	Protein–protein interactions	Many extracellular and membrane-bound proteins[p]
Type III (Ca²⁺-binding) repeats	DXDXDGXXDXXD[q] RGD	Ca²⁺-binding loop[r] Ligand for integrins[s]	None described
COOH terminus	RFYVVMWK[t]	Ligand for nonintegrin heterodimeric receptor[u]	None described

[a] Exclusive of members of the TSP gene family.
[b] B, Basic amino acid.
[c] A. D. Cardin and H. J. R. Weintraub, *Arteriosclerosis (Dallas)* **9**, 21 (1989).
[d] J. Lawler and R. O. Hynes, *J. Cell Biol.* **103**, 1635 (1986).
[e] C. A. Prater, J. Plotkin, D. Jaye, and W. A. Frazier, *J. Cell Biol.* **112**, 1031 (1991).
[f] A. S. Asch, *in* "Thrombospondin" (J. Lahav, ed.), p. 265. CRC Press, Boca Raton, FL, 1993.
[g] D. Goundis and K. B. M. Reid, *Nature (London)* **335**, 82 (1988).
[h] A. Klar, M. Baldassare, and T. M. Jessell, *Cell (Cambridge, Mass.)* **69**, 95 (1992).
[i] R. G. DiScipio and T. E. Hugli, *J. Biol. Chem.* **264**, 16197 (1989).
[j] N. Guo, H. C. Krutzsch, E. Negre, T. Vogel, D. A. Blake, and D. D. Roberts, *Proc. Natl. Acad. Sci. U.S.A.* **89**, 3040 (1992).
[k] G. D. Holt, M. K. Pangburn, and V. Ginsburg, *J. Biol. Chem.* **265**, 2852 (1989).
[l] C. Leung-Hagesteijn, A. M. Spence, B. D. Stern, Y. Zhou, M.-W. Su, E. M. Hedgecock, and J. G. Culotti, *Cell (Cambridge, Mass.)* **71**, 289 (1992).
[m] J. A. Sherwood, *in* "Thrombospondin" (J. Lahav, ed.), p. 227. CRC Press, Boca Raton, FL, 1993.
[n] P. Bork, *FEBS Lett.* **327**, 125 (1993).
[o] P. Bornstein, *FASEB J.* **6**, 3290 (1992).
[p] E. Appella, I. T. Weber, and F. Blasi, *FEBS Lett.* **231**, 1 (1988).
[q] X. Sun, K. Skorstengaard, and D. F. Mosher, *J. Cell Biol.* **118**, 693 (1992).
[r] C. V. Dang, R. F. Ebert, and W. R. Bell, *J. Biol. Chem.* **260**, 9713 (1985).
[s] J. Lawler, R. Weinstein, and R. O. Hynes, *J. Cell Biol.* **107**, 2351 (1988).
[t] M. D. Kosfeld and W. A. Frazier, *J. Biol. Chem.* **268**, 8808 (1993).
[u] R. Yabkowitz and V. M. Dixit, *Cancer Res.* **51**, 3648 (1991).

domain and type I and type II repeats are found in many other proteins. The presence of these repeats in other proteins might well indicate the existence of similar or even partially overlapping functions among these proteins. Considerations such as these are likely to assume increasing importance in the interpretation of the results of targeted disruptions of the TSP genes in the mouse. The ability of the mouse to tolerate a homozygous null mutation of one of the TSP genes could be explained not only by the presence of two closely related genes, such as TSP1 and TSP2, but also by the existence of genes whose similarity in structure and function extends to only a subset of the domains of the targeted gene.

Purification of Thrombospondin 1

The method for purification of human TSP1 from platelets generally follows procedures introduced by Lawler et al.[1] and modified by Santoro and Frazier.[25] The method used in our laboratory has been published in detail.[26] Four units of fresh human platelets are suspended in a one-fourth volume of anticoagulant citrate dextrose (ACD, 130 mM sodium citrate, 110 mM glucose) and centrifuged at 1000 g for 15 min at 20°. The platelets are suspended in 10 ml of buffer A [13 mM sodium citrate, 116 mM glucose, 20 mM Tris-HCl (pH 7.5), 145 mM NaCl, 5 mM KCl]. After a low-speed centrifugation, platelets are resuspended in 10 ml of buffer B [20 mM Tris-HCl (pH 7.5), 145 mM NaCl, 5 mM KCl, 5 mM glucose, 1 mM CaCl$_2$]. Seventy units of human thrombin (1000 units/mg; Sigma, St. Louis, MO) is added to the platelet suspension, and the suspension is stirred gently at 20° for 2 min. A 200-μl volume of 0.2 M phenylmethylsulfonyl fluoride (PMSF) in absolute ethanol is added, the platelets are stirred for 2 min, and placed on ice. After centrifugation at 27,000 g at 4° for 20 min, the supernatant, which contains the released TSP, is collected and N-ethylmaleimide (NEM) is added to a final concentration of 10 mM. This "platelet releasate" is chromatographed on a column of Sepharose CL-4B (Pharmacia, Piscataway, NJ) equilibrated with 20 mM Tris-HCl (pH 7.5), 150 mM NaCl, and 1 mM CaCl$_2$ at 4°, and the effluent is monitored by absorbance at 280 nm. Thrombospondin elutes as a broad peak within the included volume of the column. The fractions containing TSP can be ascertained by analysis of individual column fractions by sodium dodecyl sulfate-polyacrylamide gel electrophoresis (SDS–PAGE). Fractions containing TSP are then loaded onto a heparin–Sepharose column equilibrated

[25] S. A. Santoro and W. A. Frazier, this series, Vol. 144, p. 438.
[26] E. H. Sage and P. Bornstein, in "Extracellular Matrix Molecules: A Practical Approach" (M. A. Haralson and J. R. Hassell, eds.) (in press).

with 50 mM Tris-HCl (pH 7.5), 150 mM NaCl, and 1 mM CaCl$_2$. This column is eluted with the same buffer, followed by a buffer containing 0.25 M NaCl. These effluents contain little or no TSP. Thrombospondin is eluted with 30 ml of the same buffer containing 0.6 M NaCl. Fractions containing TSP are dialyzed against lower ionic strength buffers containing 1 mM CaCl$_2$ and are stored at $-20°$. Further purification can be achieved by chromatography on gelatin–Sepharose for the removal of small amounts of contaminating fibronectin, or by FPLC (fast protein liquid chromatography) with a Mono Q ion-exchange column (Pharmacia, Piscataway, NJ) or a Superose 12 column. Murphy-Ullrich et al.[27] have shown that TSP1, prepared by a combination of heparin and gel-permeation chromatography, is contaminated by a small amount of biologically active transforming growth factor β (TGF-β). TGF-β can be removed by gel-permeation chromatography in 10 mM Tris (pH 11), 0.15 M NaCl, followed by dialysis against a buffer of neutral pH. In that case, heparin–Sepharose chromatography should precede the gel-permeation chromatography step.

Thrombospondin Gene Family

A second TSP gene, encoding TSP2, was discovered in the mouse[16,28,29] and in the human[30,31] as a result of comparison of newly isolated cDNA clones with existing genomic and cDNA sequences. Similarly, a chicken TSP cDNA, characterized by Lawler et al.[32] was shown to represent chicken TSP2 on the basis of comparison with mammalian TSP1 and TSP2 sequences. The ability to identify chicken TSP as TSP2 results from the finding that in the TSP gene family, as in other gene families, the sequences of orthologous genes, that is, the same gene in two different species, bear a higher degree of nucleotide identity than the sequences of paralogous genes, that is, the sequences of two different homologous genes in a single species. More recently, two additional members of the TSP gene family were identified in the mouse (TSP3)[33,34] and in the frog and human (TSP4).[35]

[27] J. E. Murphy-Ullrich, S. Schultz-Cherry, and M. Höök, *Mol. Biol. Cell* **3**, 181 (1992).
[28] P. Bornstein, K. O'Rourke, K. Wikström, F. W. Wolf, R. Katz, P. Li, and V. M. Dixit, *J. Biol. Chem.* **266**, 12821 (1991).
[29] C. D. Laherty, K. O'Rourke, F. W. Wolf, R. Katz, M. F. Seldin, and V. M. Dixit, *J. Biol. Chem.* **267**, 3274 (1992).
[30] T. L. LaBell, D. J. M. Milewicz, C. M. Disteche, and P. H. Byers, *Genomics* **12**, 421 (1992).
[31] T. L. LaBell and P. H. Byers, *Genomics* **17**, 225 (1993).
[32] J. Lawler, P. Ferro, and M. Duquette, *Biochemistry* **31**, 1173 (1992).
[33] H. L. Vos, S. Devarayalu, Y. de Vries, and P. Bornstein, *J. Biol. Chem.* **267**, 12192 (1992).
[34] P. Bornstein, S. Devarayalu, S. Edelhoff, and C. M. Disteche, *Genomics* **15**, 607 (1993).
[35] J. Lawler, M. Duquette, C. A. Whittaker, J. C. Adams, K. McHenry, and D. W. DeSimone, *J. Cell Biol.* **120**, 1059 (1993).

A fifth member of the TSP family had previously been studied by Fife and Brandt[36,37] as a cartilage matrix protein and characterized by Morgelin *et al.*[17] and Hedbom *et al.*[18] as cartilage oligomeric matrix protein (COMP). Oldberg *et al.*[38] reported the cDNA sequence of rat COMP. This sequence reveals that COMP is a member of the TSP gene family and, more specifically, resembles TSP3 and TSP4 in its structure. All three proteins contain four rather than three EGF-like repeats, as found in TSP1 and TSP2, and lack the procollagen homology and type I repeats found in the latter two proteins (Fig. 1). However, rat COMP differs from mouse TSP3 and *Xenopus* TSP4 in containing a much smaller amino-terminal domain. In view of the obvious homology of COMP with TSP3 and TSP4, we refer to COMP as TSP5/COMP.

A sequence-based evolutionary tree of the members of the TSP family is shown in Fig. 3. This tree was constructed by multiple sequence alignment, using the algorithm described by Feng and Doolittle.[39] A similar phylogenetic tree has been created by Lawler *et al.*[40] In this dendrogram, the vertical lines that represent branch lengths are proportional to evolutionary distances. If a value of 350 million years is used for the time of divergence of amphibians, and a relatively constant rate of evolutionary divergence in the TSP gene family is assumed, the gene duplication producing TSP1 and TSP2 would have occurred about 600 million years ago. The phylogenetic tree also predicts that the gene duplication leading to the two subfamilies in the TSP gene family occurred approximately 900 million years ago.[40] Thus, a form of TSP, or a closely related protein, might exist in present-day invertebrates.

The chromosomal positions of the mouse and human TSP genes are summarized in Table I. These assignments were variously made by *in situ* hybridization, interspecific backcross mapping, and analysis of mouse–human hybrid cells. For all assigned genes, the locations in the human and mouse genomes are syntenic, that is, they confirm a region of conservation that includes other homologous genes. Although members of the two branches of the TSP gene family resemble each other in structure, we currently believe that these proteins subserve different functions. This supposition is supported by the difference in promoter structure and apparent susceptibility to induction by serum and cytokines for TSP1

[36] R. S. Fife and K. D. Brandt, *Biochim. Biophys. Acta* **802**, 506 (1984).

[37] R. S. Fife and K. D. Brandt, *J. Clin. Invest.* **84**, 1432 (1989).

[38] A. Oldberg, P. Antonsson, K. Lindblom, and D. Heinegard, *J. Biol. Chem.* **267**, 22346 (1992).

[39] D.-F. Feng and R. F. Doolittle, *J. Mol. Evol.* **35**, 351 (1987).

[40] J. Lawler, M. Duquette, L. Urry, K. McHenry, and T. F. Smith, *J. Mol. Evol.* **36**, 509 (1993).

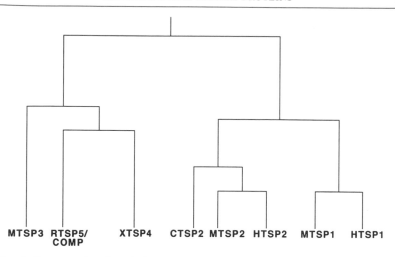

FIG. 3. Sequence-based evolutionary tree of members of the TSP gene family. Branch lengths (vertical lines) are proportional to evolutionary distances. Data sources are as follows: MTSP3 (mouse) [P. Bornstein, S. Devarayalu, S. Edelhoff, and C. M. Disteche, *Genomics* **15**, 607 (1993)]; RTSP5/COMP (rat) [A. Oldberg, P. Antonsson, K. Lindblom, and D. Heinegard, *J. Biol. Chem.* **267**, 22346 (1992)]; TSP2 (chicken) [J. Lawler, M. Duquette, and P. Ferro, *J. Biol. Chem.* **266**, 8039 (1991)]; XTSP4 (*Xenopus*) [J. Lawler, M. Duquette, C. A. Whittaker, J. C. Adams, K. McHenry, and D. W. DeSimone, *J. Cell Biol.* **120**, 1059 (1993)]; MTSP2 [C. D. Laherty, K. O'Rourke, F. W. Wolf, R. Katz, M. F. Seldin, and V. M. Dixit, *J. Biol. Chem.* **267**, 3274 (1992)]; HTSP2 (human) [T. L. LaBell, D. J. M. Milewicz, C. M. Disteche, and P. H. Byers, *Genomics* **12**, 421 (1992); T. L. LaBell and P. H. Byers, *ibid.* **17**, 225 (1993)]; MTSP1 (mouse) [C. D. Laherty, K. O'Rourke, F. W. Wolf, R. Katz, M. F. Seldin, and V. M. Dixit, *J. Biol. Chem.* **267**, 3274 (1992)]; HTSP1 (human) [J. Lawler and R. O. Hynes, *J. Cell Biol.* **103**, 1635 (1986)].

versus TSP2[13,16] and by differences in the expression of these genes in the animal. The tissue-specific expression of TSP1, TSP2, and TSP3, as judged by *in situ* hybridization in the developing mouse embryo, is distinct for each gene (see below).[15] In the postnatal rat, TSP5 expression is limited to cartilage, with a low level of expression in aorta,[38] and in the adult human TSP4 is found primarily in cardiac and skeletal muscle.[35] It is likely, however, that these two proteins are expressed more widely during embryonic development.

Analysis of Thrombospondin Expression by *in Situ* Hybridization

Thrombospondin is a secretory product of many types of cultured cells, although data pertaining to TSPs other than TSP1 are limited.[19] Localization studies *in vivo* are preferable for the identification of specific cells that express these proteins. However, there is much to be learned about the regulation, processing, chain assembly, and deposition of the TSPs; these questions are, at this time, best addressed by experiments performed *in vitro*. Quantitation of TSP1 in cell culture medium has been accomplished by radioimmunoassay (RIA),[41] and a similar technique has been used successfully to measure levels of TSP1 in human plasma and serum.[42] In some instances, however, it is important to distinguish among cells for the production of TSP, for example, in heterogeneous cultures containing more than one cell type, or in cultures in which several phenotypes are manifested (see below for a description of angiogenic cultures of endothelial cells). The technique of *in situ* hybridization *in vitro* has been developed to allow discrimination among cells with regard to the transcription of specific mRNA. This procedure has been outlined in Table IV for endothelial cells but is readily applicable to most types of adherent cells in culture.

The use of bright-field microscopy allows one to visualize the morphology and distribution of cells that hybridized with the TSP1 probe (Fig. 4). Unambiguous results can be obtained when (1) levels of expression are high, and (2) the background (due to nonspecific reactivity of the probe, nonoptimal hybridization and/or washing conditions, or poorly preserved specimens) is low. It should be stressed that this technique is not strictly quantitative; rather, it permits comparison among cells or tissues for one probe within a single experiment (i.e., identical experimental conditions that include exposure time and a probe of defined specific activity). In some instances it has been possible to perform grain counting to estimate levels of a specific mRNA,[43] and advances in the design of software for the accurate interpretation of scanning densitometry across a microscopic field are likely to resolve many of the difficulties associated with quantitation of data obtained by the *in situ* hybridization technique.

An alternative means of quantitation that can be performed concomitantly with *in situ* hybridization is the RNase protection assay. Total RNA isolated from various murine embryonic tissues by standard techniques has been hybridized with the following murine TSP riboprobes[15]: (1) TSP1, a 272-bp genomic fragment that protects 181 bp of exon 5, (2) TSP2, a 267-bp genomic fragment that protects 142 bp of exon 5, and (3) TSP3, a genomic fragment that protects 147 bp of exon 22.

Both antisense and sense-oriented (control) probes are prepared with the T3 and T7 RNA polymerases, and the levels of RNA in the assays

[41] N. R. Hunter, J. Dawes, I. R. MacGregor, and D. S. Pepper, *Thromb. Haemostasis* **52**, 288 (1984).
[42] S. D. Saglio and H. S. Slayter, *Blood* **59**, 162 (1982).
[43] R. Tucker, *Development (Cambridge, UK)* **117**, 347 (1993).

TABLE IV
PROCEDURE FOR *in Situ* HYBRIDIZATION OF THROMBOSPONDIN mRNA
IN CULTURED CELLS[a-c]

1. Cells[d] are plated at subconfluent density onto Probe-On slides (Fisher, Pittsburgh, PA) and cultivated under appropriate conditions until the desired level of confluence is attained

2. Riboprobes are transcribed with the specific RNA polymerase (Promega, Madison, WI) in the presence of $[\alpha\text{-}^{35}S]dUTP$ [specific activity, $3-4 \times 10^7$ counts per minute (cpm)/μg]

3. Slides are rinsed in serum-free culture medium; cells are fixed in 4% (w/v) paraformaldehyde (PFA) at pH 7.4 for 20 min, rinsed in phosphate-buffered saline (pH 7.4) (PBS), and treated with proteinase K (20 μg/ml) for 12 min

4. Samples are rinsed with PBS, fixed in 4% PFA for 5 min, and dipped in water containing diethyl pyrocarbonate (DEPC)

5. Samples are next treated for 10 min with 0.1 M triethanolamine (pH 8.0) containing 0.05% (v/v) acetic anhydride, rinsed in PBS for 5 min, dipped in DEPC–water, dehydrated in four solutions of ethanol (sequentially from 70 to 100%, v/v), and air dried

6. Slides are incubated overnight at 50° in prehybridization solution [PHS: 0.6 M NaCl, 0.12 M Tris-HCl at pH 8.0, 8 mM ethylenediaminetetraacetic acid (EDTA), 0.02% Ficoll 400, 0.02% polyvinylpyrrolidone, 0.1% Pentex bovine serum albumin (BSA; Sigma), salmon sperm DNA (500 μg/ml), yeast total RNA (600 μg/ml), and 50% deionized formamide]

7. Slides are subsequently incubated overnight at 50° in PHS containing 10% dextran sulfate (w/v) and 3×10^6 cpm ^{35}S-labeled riboprobe (TSP1, TSP2, or TSP3) per two slides

8. Slides are washed for 30 min at 50° in formamide buffer [FB: 50% formamide, 5% 20× SSC (1× SSC is 0.15 M NaCl, 0.015 M sodium citrate at pH 7.5), 0.01 M dithiothreitol (DTT) in DEPC–water]

9. Washes continue in (a) 0.5× SSC for 30 min at room temperature, (b) 3.5× SSC containing RNase A (40 μg/ml; Sigma) and 50 μl RNase T1 (Sigma) for 30 min at 37°, (c) 3.5× SSC for 10 min at room temperature, and (d) 0.1× SSC for 90 min at 65°

10. Samples are dehydrated in a graded series of ethanol solutions in 0.3 M sodium acetate (sequentially, from 50, 70, to 95%), and finally in 100% ethanol, and are air dried

11. Autoradiography is performed with NTB2 emulsion (Kodak, Rochester, NY) diluted to 40% with a solution of 4% Dreft detergent; slides are exposed for 4 days

12. Slides are developed in Kodak D-19 developer and fixed in GBX fixer (Kodak)

13. Samples are counterstained with toluidine blue, dehydrated in solutions of 70–100% ethanol, cleared in xylene, and mounted with Permount

---

[a] M. J. Reed, P. Puolakkainen, T. F. Lane, D. Dickerson, P. Bornstein, and E. H. Sage, *J.Histochem. Cytochem.* **41**, 1467 (1993).

[b] M. L. Iruela-Arispe, D. J. Liska, E. H. Sage, and P. Bornstein, *Dev. Dyn.* **197**, 40 (1993).

[c] P. W. H. Holland, S. J. Harper, J. H. McVey, and B. L.M.Hogan, *J. Cell Biol.* **105**, 473 (1987).

[d] Bovine aortic endothelial cells, plated at approximately 33% confluence and grown for 1–2 days. Maximal levels of TSP1 mRNA are produced by subconfluent cells.

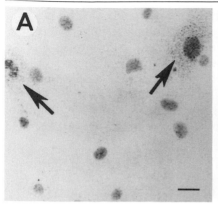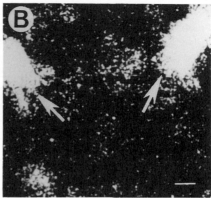

FIG. 4. Detection of TSP1 mRNA in cultured endothelial cells by *in situ* hybridization. Bovine aortic endothelial cells were grown to a subconfluent density on Probe-On slides and were prepared for *in situ* hybridization as described in Table IV. (A) Bright-field and (B) dark-field photomicrographs of fixed cultures hybridized with a human TSP1 riboprobe. Arrows indicate two cells with high levels of TSP1 mRNA. Bar: 20 μm.

are normalized to that of L32 mRNA, which encodes an RNA corresponding to a ribosomal protein. A standard curve is generated by hybridization of an antisense riboprobe to known amounts of the corresponding sense riboprobe, and quantitative analyses are performed with a phosphorimager, with normalization of signals to that of L32. It is important, both for RNase protection assays and *in situ* hybridization, to perform repeat analyses on tissues derived from several individuals and on embryos derived from a minimum of two to four different litters.

The following TSP probes have been used successfully for *in situ* hybridization studies on mouse, rat, and/or chicken tissue sections, and on bovine endothelial cell cultures[15,43,44]: (1) murine TSP1 riboprobe, 294-bp *Bgl*II/*Pst*I fragment, (2) human TSP1 riboprobe, 1.3-kb *Eco*RI/*Eco*RI fragment (61811; American Type Culture Collection, Rockville, MD), (3) murine TSP2 riboprobe, 514-bp *Pst*I/*Pst*I fragment, (4) murine TSP3 riboprobe, 1550-bp *Eco*RI/*Eco*RI fragment, (5) murine TSP3 riboprobe, 680-bp *Sal*I/*Sal*I fragment, (6) murine TSP3 riboprobe, 147-bp *Alu*I/*Alu*I fragment, and (7) chick TSP2 cDNA, 960-bp polymerase chain reaction product, corresponding to amino acids 834–1154 of the carboxy terminus of chick TSP2.

It is clearly important to minimize the frequency of cross-hybridization among the TSPs, especially because high-stringency washes of cells or

[44] M. J. Reed, P. Puolakkainen, T. F. Lane, D. Dickerson, P. Bornstein, and E. H. Sage, *J. Histochem. Cytochem.* **41,** 1467 (1993).

tissue sections are not possible without a compromise in either the morphological characteristics or the integrity of the sample. Therefore, regions of low homology have been chosen for the generation of probes that are transcript specific, and several probes for each TSP should be used to confirm the hybridization patterns. In addition, control hybridization reactions should be performed with the sense-oriented counterpart of each antisense probe. In the typical riboprobe vector (pGEM or Bluescript series; Stratagene, La Jolla, CA), sense-oriented probes can be generated by use of the opposing promoter.

In situ hybridization experiments with murine embryos have identified distinct patterns associated with the expression of TSP1, TSP2, and TSP3.[15] In general, the distribution of TSP1 mRNA from this study is concordant with that obtained by immunohistochemistry on mouse embryos with a monoclonal antibody against TSP1.[45] In contrast, it is likely that earlier studies, in which polyclonal anti-TSP antibodies were used, reported a distribution inclusive of TSP1, TSP2, and TSP3.[46,47] A procedure for the detection of TSP mRNA in tissue sections is presented in Table V.

The distribution of TSP1 and TSP2 mRNA in a 14-day mouse embryo is shown in Fig. 5. It is readily apparent that these two transcripts exhibit spatial differences in expression at this stage of embryonic development (e.g., in megakaryocytes). Some of these differences have been listed in Table VI. In summary, the expression of TSP1 mRNA was predominant over the other TSPs in mesenchyme, brain, heart, and liver from days 10–13 of embryonic development, after which constitutive expression occurred in many organs, connective tissues, and in megakaryocytes. TSP2 mRNA was prominent in the vasculature from days 10–18 and, from approximately day 14 onward, was transcribed in organized connective tissues (organ capsules, ligament, periochondrium, and periosteum). In contrast, TSP3 mRNA was observed only 2–3 days prior to birth and was confined to brain, lung, and cartilage. Although these three TSP genes are also transcribed in neonatal and adult animals, the expression of each appears to be a function of the specific cell type within a given organ or tissue. For example, TSP1, 2, and 3 mRNAs are restricted to discrete populations of neurons in the neonatal mouse brain, and the only TSP mRNA that megakaryocytes contain encodes TSP1.[15]

A further refinement of the technique of *in situ* hybridization is afforded

[45] C. L. Corless, A. Mendoza, T. Collins, and J. Lawler, *Dev. Dyn.* **193,** 346 (1992).
[46] T. N. Wight, G. J. Raugi, S. M. Mumby, and P. Bornstein, *J. Histochem. Cytochem.* **33,** 295 (1985).
[47] K. S. O'Shea and V. M. Dixit, *J. Cell Biol.* **107,** 2737 (1988).

TABLE V

PROCEDURE FOR *in Situ* HYBRIDIZATION OF THROMBOSPONDIN mRNA
IN MAMMALIAN[a,b] AND AVIAN[c] TISSUES[d]

1. Paraffin sections of tissues or whole embryos, mounted on Probe-On slides (Fisher), are hydrated in solutions of ethanol (from 100 to 70%), rinsed in PBS[e], fixed in 4% PFA for 5 min, washed in PBS, and treated for 5 min with proteinase K (20 μg/ml)[f]
2. Sections are washed again in PBS and fixed in 4% PFA for 5 min
3. Sections are incubated for 10 min in 0.1 M triethanolamine (pH 8.0) containing 0.25% acetic anhydride, washed in DEPC–water, and air dried
4. Sections are incubated in PHS for 1–4 hr at room temperature[g]
5. Slides are immediately transferred to PHS containing 10% dextran sulfate, 0.1% sodium dodecylsulfate, 10 mM DTT, and 3 × 10^6 cpm ^{35}S-labeled probe[g] per 2 slides. [The specific activity of probes is 80–180 Ci/mmol (approximately 10^9 cpm/μg). Probes larger than 200 bp are hydrolyzed to <150-bp fragments at alkaline pH[h]]
6. Hybridization is carried out at 53° for 12–16 hr in a chamber containing filter paper saturated with 4× SSC–50% formamide
7. Sections are subsequently immersed in 4× SSC–50% formamide containing 5 mM DTT for 30 min at 53°, followed by 3.5× SSC for 30 min at room temperature[i]
8. Sections are treated with RNase A (20 μg/ml) in 3.5× SSC for 30 min and are next washed in 0.1× SSC for 2 hr at 65°
9. Sections are dehydrated in solutions of ethanol containing 0.3 M ammonium acetate and are air dried
10. Slides are covered with NTB2 nuclear emulsion and are exposed for 2–4 weeks at 4°
11. After development, sections are counterstained with 1% toluidine blue, dehydrated, and mounted in Permount

---

[a] M. L. Iruela-Arispe, D. J. Liska, E. H. Sage, and P. Bornstein, *Dev. Dyn.* **197**, 40 (1993).

[b] M. J. Reed, P. Puolakkainen, T. F. Lane, D. Dickerson, P. Bornstein, and E. H. Sage, *J. Histochem. Cytochem.* **41**, 1467 (1993).

[c] R. P. Tucker, *Development (Cambridge, UK)* **117**, 347 (1993).

[d] P. W. H. Holland, S. J. Harper, J. H. McVey, and B. L. M. Hogan, *J. Cell Biol.* **105**, 473 (1987).

[e] For abbreviations, see Table IV.

[f] Alternatively, frozen sections fixed briefly in cold 4% PFA and mounted on poly-L-lysine-coated slides have been used.

[g] For *in situ* hybridization of chicken embryo tissue with a TSP2 cDNA probe labeled by random priming with [^{35}S]dCTP (Promega), the PHS was 5× SSC, 5× Denhardt's solution (1% Ficoll, 1% polyvinylpyrrolidone, and 1.1% BSA), salmon sperm DNA (100 μg/ml), 20 mM-2-mercaptoethanol (Sigma); the hybridization solution was 50% deionized formamide, 20 mM Tris, 0.1% lauryl sulfate, and 1% dextran sulfate.[c]

[h] K. H. Cox, D. V. DeLeon, L. M. Angerer, and R. C. Angerer, *Dev. Biol.* **101**, 485 (1984).

[i] Hybridization of chicken tissue with 5 × 10^6 cpm of an [^{35}S]TSP2 cDNA probe was carried out at 42°. Slides were subsequently washed in 1× SSC at room temperature followed by 42°; exposure was for 3–7 days.[c]

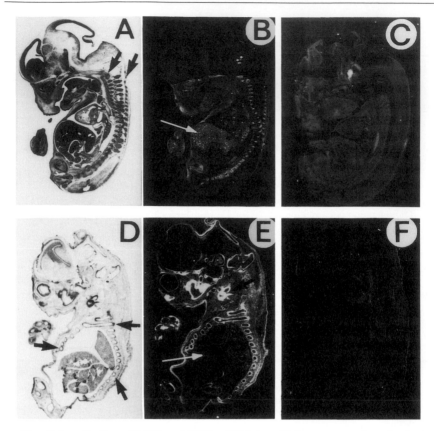

FIG. 5. Localization of TSP1 and TSP2 mRNA in a 14-day mouse embryo by *in situ* hybridization. The embryos were prepared as described in Table V, and hybridization was performed with a murine TSP1 (A,B) or TSP2 riboprobe (D,E), transcribed in an antisense orientation (A,B,D,E) or in the respective sense orientation (C,F). (A) and (D) Bright-field images of adjacent sections. and (B,C,E, and F) Dark-field images of adjacent (serial) sections of a single 14-day mouse embryo. At this stage of development, both TSP1 and TSP2 transcripts appear in the ribs (arrows, A and D); TSP1 mRNA is expressed by chondrocytes (B), whereas TSP2 mRNA is seen chiefly in the perichondrium of the ribs (E) and of the clavicle (E, dark arrow). Note the hybridization signal corresponding to TSP1 mRNA in megakaryocytes in the liver, whereas none is present for TSP2 (white arrows, B and E).

by its combination with immunohistochemistry.[44,48] This procedure permitted the simultaneous detection of TSP1 mRNA and the protein ED-1, a marker for cells belonging to the mononuclear phagocytic system, in macrophages of rat dermal wounds.[44] After hybridization of tissue sections with the TSP1 riboprobe, followed by the washing procedure and RNase

[48] S. Solberg, J. Peltonen, and J. Uitto, *Lab. Invest.* **64,** 125 (1991).

TABLE VI
EXPRESSION OF TRANSCRIPTS FOR THROMBOSPONDINS 1, 2, AND 3 IN MOUSE EMBRYOS[a,b]

Cell/tissue/organ	TSP1	TSP2	TSP3
Embryonic age: days 10 to 13 (±1)[c]			
Mesenchyme and connective tissue	+	−	−
Brain	+	+	−
Meninges	+	−	−
Heart	+	−	−
Large vessels	−	+	−
Capillaries	−	+	−
Megakaryocytes	+	−	−
Skin	+	−	−
Liver	+	−	−
Cartilage	+	−	−
Embryonic age: days 14 (±1) to 18			
Mesenchyme and connective tissue	−	+	−
Brain	+	+	+[d]
Meninges	+	+	−
Heart	+	+	−
Large vessels	−	+	−
Capillaries	+[e]	+	−
Megakaryocytes	+	−	−
Skin	−	+	−
Skeletal muscle	−	+	−
Kidney	+	+	−
Gut	+	+	−
Liver	−	+	−
Lung	+	+[e]	+[d]
Cartilage	+	+	+
Bone	+	+	−

[a] M. L. Iruela-Arispe, D. J. Liska, E. H. Sage, and P. Bornstein, *Dev. Dyn.* **197,** 40 (1993).

[b] Analyses on anterior limb, heart, liver, and brain were performed by RNase protection assays and were quantitated by phosphorimaging of autoradiograms and normalization to an internal RNA control. In all tissues, expression was noted or confirmed by *in situ* hybridization.

[c] At this stage of development, no expression was seen in gut, skeletal muscle, kidney, or lung.

[d] Embryonic ages were days 17 and 18.

[e] Days 16 and 18.

treatment as described in Table V, the samples are incubated for 90 min in 0.5× SSC (1× SSC is 0.15 M NaCl plus 0.015 M sodium citrate, pH 7.5) at 65°, rinsed twice in phosphate-buffered saline (PBS), and blocked for 1 hr at 21° in a solution of PBS containing 1% (v/v) goat serum and 0.5% (w/v) bovine serum albumin (BSA). Immunohistochemistry is performed according to protocols established for the primary antibody, and the sections are subsequently exposed to biotinylated goat antibodies against the primary immunoglobulin G (IgG) at 7.5 mg/ml (Vector, Burlingame, CA). Immune complexes are subsequently reacted with an avidin–biotin–peroxidase reagent (Vector), after which the sections are rinsed three times in tap water, dehydrated sequentially in solutions of ethanol, dried, and subjected to autoradiography as described in Table V. Controls include sections treated with preimmune antisera or IgG and with secondary antibody alone.

Thrombospondins in Angiogenesis

Possibly owing to its initial purification from platelets, numerous studies have focused on the functional properties of TSP in the context of the vascular endothelium.[49,50] In this section we discuss an activity unique to endothelial cells, that is, angiogenesis, the process by which new vessels including capillaries are formed as a result of the proliferation, migration, and biosynthetic alteration of endothelial cells that sprout from the generally quiescent monolayer that lines all vascular channels.

In vivo, TSP1 has been described in capillaries associated with healing dermal wounds.[44,51] Moreover, TSP1 mRNA was noted in capillaries of day 16–18 mouse embryos, whereas TSP2 transcripts were expressed significantly earlier in these embryos, from day 10 continuously through parturition on day 18.[15] The apparent absence of TSP2 mRNA from human umbilical vein endothelial cells *in vitro*[30] could reflect the proliferative state of these cells or the conditions under which they were initially isolated and cultured.

TSP1 is a major secretory product of most endothelial cells cultured from both the macrovasculature and the microvasculature.[19] The biosynthesis of TSP1 mRNA and/or protein has, in fact, been found to be highly dependent on the nature of the substrate and the degree of confluence and proliferation associated with endothelial cells *in vitro*.[52,53] In a model of angiogenesis *in vitro,* in which strains of cloned endothelial cells spontaneously organize into morphologically distinct cords and tubes with patent lumens, TSP1 protein was negligible in confluent monolayers and was maximal in a limited number of proliferating cells that were adjacent to

[49] W. A. Frazier, *J. Cell Biol.* **105,** 625 (1987).

[50] E. H. Sage and P. Bornstein, *J. Biol. Chem.* **266,** 14831 (1991).

[51] G. J. Raugi, J. E. Olerud, and A. M. Gown, *J. Invest. Dermatol.* **89,** 551 (1987).

[52] S. M. Mumby, D. Abbott-Brown, G. J. Raugi, and P. Bornstein, *J. Cell. Physiol.* **120,** 280 (1984).

[53] A. E. Canfield, R. P. Boot-Handford, and A. M. Schor, *Biochem. J.* **268,** 225 (1990).

the developing cords and tubes.[54] However, the overall levels of TSP1 mRNA and protein decreased in angiogenic cultures of both aortic and microvascular endothelial cells. Addition of anti-TSP1 IgG was found to enhance the production of endothelial cords by 33–50% and led to the conclusion that TSP1 acted as an inhibitor of angiogenesis *in vitro*.[54]

Several functions attributed to TSP (in most cases thought to be TSP1) are consistent with the proposed role of an angiogenesis inhibitor: (1) TSP1 destabilizes focal contacts produced by endothelial cells *in vitro*[55]; (2) TSP1 inhibits endothelial cell proliferation[56–58]; (3) TSP1 binds to several components of the extracellular matrix that modulate endothelial cell behavior[50]; (4) TSP1 modulates the motility of endothelial cells[56,58]; and (5) TSP1 mitigates the effects of known endothelial cell mitogens and motility factors.[27,56,58] Since the demonstration that TSP1 binds to active TGF-β1,[27] it has become important to reevaluate the effects of TSP1 on endothelial cell behavior under conditions in which the activity of TGF-β has been neutralized and/or TGF-β1 itself has been shown to be absent from the TSP1 preparation (e.g., TSP1 can be stripped of TGF-β1 by chromatography at alkaline pH,[27] or experiments can be performed with recombinant protein or synthetic peptides[58]). Although it is also important to demonstrate inhibition of the effect(s) mediated by TSP1 by antibodies specific for different regions of the protein, in some instances it could be argued that an antigen–antibody reaction might liberate bound TGF-β1 from TSP1; the cytokine would thus be available to the cells and, in the case of endothelium, would exert an antiproliferative effect. With regard to angiogenesis *in vitro*, however, it has been shown that TGF-β stimulated endothelial cell proliferation in cultures forming cords and tubes.[59]

A direct demonstration of the antiangiogenic property of TSP1 *in vivo* was provided by Bouck and colleagues, who isolated a fragment of TSP1 from the culture medium of hamster cells.[60] The secretion of this protein was controlled by a tumor suppressor gene.[61] Because solid tumors fail to progress without an adequate blood supply,[62] it was possible that the

[54] M. L. Iruela-Arispe, P. Bornstein, and E. H. Sage, *Proc. Natl. Acad. Sci. U.S.A.* **88,** 5026 (1991).

[55] J. E. Murphy-Ullrich and M. Höök, *J. Cell Biol.* **109,** 1309 (1989).

[56] G. Taraboletti, D. Roberts, L. A. Liotta, and R. Giavazzi, *J. Cell Biol.* **111,** 765 (1990).

[57] P. Bagavandoss and J. W. Wilks, *Biochem. Biophys. Res. Commun.* **170,** 867 (1990).

[58] T. Vogel, N. Guo, H. C. Krutzsch, D. A. Blake, J. Hartman, S. Mendelovitz, A. Panet, and D. D. Roberts, *J. Cell. Biochem.* **53,** 74 (1993).

[59] M. L. Iruela-Arispe and E. H. Sage, *J. Cell. Biochem.* **52,** 414 (1993).

[60] D. J. Good, P. J. Polverini, F. Rastinejad, M. M. Le Beau, R. S. Lemons, W. A. Frazier, and N. P. Bouck, *Proc. Natl. Acad. Sci. U.S.A.* **87,** 6624 (1990).

[61] F. Rastinejad, P. J. Polverini, and N. P. Bouck, *Cell (Cambridge, Mass.)* **56,** 345 (1989).

[62] J. Folkman, *J. Natl. Cancer Inst.* **82,** 4 (1990).

inhibition of tumor growth could be due, in part, to the inhibition of angiogenesis by TSP1. Both platelet TSP1 and the hamster protein (termed gp140) were subsequently shown to inhibit neovascularization *in vivo*, in a model utilizing a corneal implant, and the migration of endothelial cells, in a Boyden chamber assay containing the chemotactic factor basic fibroblast growth factor (bFGF).[60] Because assays of angiogenesis *in vivo* as well as *in vitro* have different limitations, their complementarity is invaluable for determination of angiogenic activity. In general, the entire process of capillary expansion is considered when a system is used *in vivo*, whereas isolated cellular activities, such as migration, protein synthesis, adhesion, and proliferation, can be quantified *in vitro*.

Given the modular structure of TSP1, it is perhaps not surprising that different regions of the protein have been shown to manifest distinct activities that contribute to the inhibition of angiogenesis. For example, Tolsma *et al.*[63] found that the central 70-kDa "stalk" region was responsible for most of the antiangiogenic activity of TSP1, because this fragment blocked DNA synthesis and migration in capillary endothelial cells in the presence of bFGF, as well as corneal neovascularization. Within this sequence, peptides from the regions of procollagen homology and from the properdin-like type 1 repeats blocked angiogenesis in the cornea and endothelial cell migration *in vitro*.[63] Consistent results were obtained by Vogel *et al.*,[58] who showed that synthetic peptides from the second type 1 repeat of TSP1 inhibited mitogenesis and migration of endothelial cells *in vitro* and augmented cell adhesion. A consensus sequence for the antiproliferative function was identified as Trp-Ser-X-Trp. Another region of TSP1, the N-terminal heparin-binding fragment, exhibited similar effects; the antiproliferative property was attributed to the binding of heparin by the peptides themselves, as well as their competitive inhibition of the binding of heparin to bFGF.[58] In contrast, trimeric recombinant TSP1 fragments containing the N-terminal heparin-binding domain displayed no activity in angiogenesis assays *in vivo* or in migration assays *in vitro* (except at high concentrations).[63] Part of the discrepancy could be due to the use of different preparations and sizes of the heparin-binding region, or to the stimulatory effect of TSP on the chemokinesis of endothelial cells.[56]

Thrombospondins in Tumor Cell Biology and Metastasis

Studies from several laboratories have demonstrated that TSP1 is involved in the adhesive and motile properties of tumor cells, as well as

---

[63] S. S. Tolsma, O. V. Volpert, D. J. Good, W. A. Frazier, P. J. Polverini, and N. Bouck, *J. Cell Biol.* **122,** 497 (1993).

in the progression of metastatic disease. By an indirect enzyme-linked immunosorbent assay (ELISA), the mean levels of TSP1 in the blood of gastrointestinal, breast, and lung cancer patients were shown, respectively, to be three-, two-, and three-fold greater than those of controls.[64] These increases, which were independent of platelet counts and were not seen in nonmalignant disease, indicate the potential utility of TSP1 as a marker for metastasis.

More direct evidence for the involvement of TSP1 in tumor growth was provided by Zajchowski et al.,[65] working with cell hybrids created by fusion between normal human mammary epithelial cells and MCF-7 human breast cancer cells. The cell hybrids exhibited a suppressed tumor cell phenotype [e.g., they failed to form tumors in nude (athymic) mice], which was correlated in part with an enhanced expression of TSP1, relative to the MCF-7 tumorigenic cells. Clearly, the known antiangiogenic properties of TSP1[60] are consistent with the lack of growth observed in tumor cells that overexpress this protein, especially because TSP1 has been shown to affect the chemokinesis and adhesion of other carcinoma and melanoma cell lines.[66,67] In contrast, Castle et al.[68] inhibited TSP1 expression by transfection of an antisense TSP1 cDNA into human squamous carcinoma cells, which normally produce high levels of this protein and are demonstrably invasive. The transfected cells exhibited reduced rates of proliferation and produced either no tumors or slowly growing tumors in nude mice. This difference could be due to the particular tumor cell studies or to the method of inhibition of TSP1 in tumor cells.

Several domains have been implicated in the interaction of TSP1 with various tumor cells. Human TSP1 contains, within the type I repeats, the related sequences CSVTCG and CSTSCG; the former interacts with the TSP receptor CD36.[69] CSVTCG and related peptides were found to support the adhesion of murine B16-F10 melanoma cells and to inhibit their metastasis in C57BL/6 mice; moreover, an anti-CSTSCG IgG specific for native TSP1 interfered with TSP1-dependent adhesion.[66] Using a metastatic line of human carcinoma cells, Yabkowitz et al.[67] found that the

[64] G. P. Tuszynski, M. Smith, V. L. Rothman, D. M. Capuzzi, R. R. Joseph, J. Katz, E. C. Besa, J. Treat, and H. I. Switalska, Thromb. Haemostasis 67, 607 (1992).

[65] D. A. Zajchowski, V. Band, D. K. Trask, D. Kling, J. L. Connolly, and R. Sager, Proc. Natl. Acad. Sci. U.S.A. 87, 2314 (1990).

[66] G. P. Tuszynski, V. L. Rothman, A. H. Deutch, B. K. Hamilton, and J. Eyal, J. Cell Biol. 116, 209 (1992).

[67] R. Yabkowitz, P. J. Mansfield, V. M. Dixit, and S. Suchard, Cancer Res. 53, 378 (1993).

[68] V. Castle, J. Varani, S. Fligiel, E. V. Prochownik, and V. Dixit, J. Clin. Invest. 87, 1883 (1991).

[69] A. S. Asch, S. Silbiger, E. Heimer, and R. L. Nachman, Biochem. Biophys. Res. Commun. 182, 1208 (1992).

thrombin-generated 140-kDa carboxy-terminal fragment of TSP1 was as effective as the intact protein for the stimulation of chemotaxis and hapto-taxis, whereas the amino-terminal heparin-binding region was inactive. Additional studies with synthetic peptides have shown that the consensus sequence WSXW, derived from the type I repeats of human TSP1, bound to heparin and diminished the binding of melanoma cells to components of the extracellular matrix.[70] This tetrapeptide was the minimal sequence that mediated both heparin and sulfatide binding, and additional basic amino-terminal amino acids flanking the maximally active octapeptide (e.g., KRFK[QDGGWSHW]SP) were found to enhance the adhesion (attachment and limited spreading) and chemotaxis of melanoma cells.[71] Because the WxpWSpW motif is conserved in TSP2 and among several other proteins associated with the regulation of cell growth and function (see Table III), the binding of cell surface heparan sulfate and/or sulfatides to TSP is likely to be fundamental to several critical aspects of cell behavior.

The presence of certain receptors for TSP on tumor cells imparts yet another tier of complexity with regard to the regulation of metastasis and tumor growth by TSP1. For example, the receptor CD36 (platelet glycoprotein IV),[69] which is expressed by both normal and neoplastic endothelial cells, has been shown to exhibit epitope heterogeneity.[72] Clearly, more studies are needed before the specificity of expression and/ or posttranslational modification as a marker for tumor progression can be fully appreciated. A second receptor for TSP, GPIIb-IIIa (the integrin $\alpha_v\beta_3$, often referred to as the vitronectin receptor), was demonstrated to mediate the adhesion of human melanoma cells to TSP1, presumably via the sequence RGD in human TSP1.[73] Because this integrin is found *in vivo* on many cells, including tumor cells,[74] and because many cells in culture appear to utilize it for attachment and/or spreading, the significance of this observation to the metastatic phenotype of melanoma cells is questionable. Despite this reservation, it should be pointed out that a major recognition sequence for many of the integrins, GRGDA or GRGDI, is conserved in all the TSP1 and TSP2 proteins for which the sequence is known.[13]

[70] N. Guo, H. C. Krutzsch, E. Negre, T. Vogel, D. A. Blake, and D. D. Roberts, *Proc. Natl. Acad. Sci. U.S.A.* **89,** 3040 (1992).

[71] N. Guo, H. C. Krutzsch, E. Nžgre, V. S. Zabrenetzky, and D. D. Roberts, *J. Biol. Chem.* **267,** 1934 (1992).

[72] E. Kudo, T. Hirose, T. Sano, and K. Hizawa, *Acta Pathol. Jpn.* **42,** 809 (1992).

[73] G. P. Tuszynski, J. Karczewski, L. Smith, A. Murphy, V. L. Rothman, and K. A. Knudsen, *Exp. Cell Res.* **182,** 473 (1989).

[74] S. Albelda, *Lab. Invest.* **68,** 14 (1992).

Acknowledgments

We thank Drs. May Reed and Luisa Iruela-Arisp for their provision of Figs. 4 and 5, and K. Doehring and B. Wood for their assistance with the manuscript. Experiments from the authors' laboratories were supported in part by National Institutes of Health Grants DE08229, HL18645, and GM40711.

[5] Laminins

By ULLA M. WEWER and EVA ENGVALL

Introduction

In writing this chapter on laminins we have considered the needs of two kinds of researchers: (1) those who do not want to buy the commercial laminins for various reasons, but want to prepare their own; and (2) those who buy laminin in small amounts and want to use it in experiments with cells. The former will need to know the sources of laminin and various purification procedures and the latter will need basic recipes for using the commercial laminin for cell culture and various biological assays.

General Structure of Laminins and Nomenclature

Laminin was first identified in 1977.[1] Mouse laminin was purified first,[2,3] followed by purification of rat laminin[4,5] and human laminin.[6-8]

[1] A. E. Chung, I. L. Freeman, and J. E. Braginski, *Biochem. Biophys. Res. Commun.* **79,** 859 (1977).
[2] A. E. Chung, R. Jaffe, I. L. Freeman, J.-P. Vergnes, J. E. Braginski, and B. Carlin, *Cell (Cambridge, Mass.)* **16,** 277 (1979).
[3] R. Timpl, H. Rohde, P. Gehron Robey, S. I. Rennard, J.-M. Foidart, and G. R. Martin, *J. Biol. Chem.* **254,** 9933 (1979).
[4] U. Wewer, R. Albrechtsen, and E. Ruoslahti, *Cancer Res.* **41,** 1518 (1981).
[5] E. Engvall, T. Krusius, U. Wewer, and E. Ruoslahti, *Arch. Biochem. Biophys.* **222,** 649 (1983).
[6] L. Risteli and R. Timpl, *Biochem. J.* **193,** 749 (1981).
[7] M. Ohno, A. Martinez-Hernandez, N. Ohno, and N. A. Kefalides, *Biochem. Biophys. Res. Commun.* **112,** 1091 (1983).
[8] U. Wewer, R. Albrechtsen, M. Manthorpe, S. Varon, E. Engvall, and E. Ruoslahti, *J. Biol. Chem.* **258,** 12654 (1983).

Copyright © 1994 by Academic Press, Inc.
All rights of reproduction in any form reserved.

The number of publications about laminin has gradually increased every year and in 1992 was approximately 470 (source, Silverplatter-Medline).

Initially thought to be a single molecule, laminin has been redefined as a multigene family of related proteins. Thus, eight different subunits (A, B1, B2, S, M, K, B2t, and B1k) have been identified so far (see Table I) but more are likely to be added in the future. The most recent reviews on laminins include those by Yurchenco and Schittny,[9] Engel,[10] Yamada and Kleinman,[11] Engvall,[12] Kleinman *et al.*,[13] and Tryggvason.[14]

Laminins are large (10^6 Da), heterotrimeric glycoproteins arranged in a cruciform structure (Fig. 1). The molecule is generally composed of three related but different chains: one heavy chain (200–400 kDa) and two distinct lighter chains (150–200 kDa) held together by coiled-coil interactions and disulfide bonds (for a review on structure, see Beck *et al.*[15] and Engel[10]). We know at present three homologous heavy chains (A,[16–19] M,[20,21] and K[22,23]) and five homologous lighter chains (B1[24–26] and

---

[9] P. D. Yurchenco and J. C. Schittny, *FASEB J.* **4**, 1577 (1990).

[10] J. Engel, *Biochemistry* **31**, 10643 (1992).

[11] Y. Yamada and H. K. Kleinman, *Curr. Opin. Cell Biol.* **4**, 819 (1992).

[12] E. Engvall, *Kidney Int.* **43**, 2 (1993).

[13] H. K. Kleinman, B. S. Weeks, H. W. Schnaper, M. C. Kibbey, K. Yamamura, and D. S. Grant, *Vitam. Horm. (N.Y.)* **47**, 161 (1993).

[14] K. Tryggvason, *Curr. Opin. Cell Biol.* **5**, 877 (1993).

[15] K. Beck, I. Hunter, and J. Engel, *FASEB J.* **4**, 148 (1990).

[16] M. Sasaki, H. K. Kleinman, H. Huber, R. Deutzmann, and Y. Yamada, *J. Biol. Chem.* **263**, 16536 (1988).

[17] M. Nissinen, R. Vuolteenaho, R. Boot-Handford, P. Kallunki, and K. Tryggvason, *Biochem. J.* **276**, 369 (1991).

[18] T. Haaparanta, J. Uitto, E. Ruoslahti, and E. Engvall, *Matrix* **11**, 151 (1991).

[19] M. Kuche-Gullberg, K. Garrison, A. J. MacKrell, L. I. Fessler, and J. H. Fessler, *EMBO J.* **11**, 4519 (1992).

[20] K. Ehrig, I. Leivo, W. S. Argraves, E. Ruoslahti, and E. Engvall, *Proc. Natl. Acad. Sci. U.S.A.* **87**, 3264 (1990).

[21] R. Vuolteenaho, M. Nissinen, K. Sainio, M. Byers, R. Eddy, H. Hirvonen, T. B. Shows, H. Sariola, E. Engvall, and K. Tryggvason, *J. Cell Biol.* **124**, 381 (1994).

[22] M. P. Marinkovich, G. P. Lunstrum, and R. E. Burgeson, *J. Biol. Chem.* **267**, 17900 (1992).

[23] M. P. Marinkovich, G. P. Lunstrum, D. R. Keene, and R. E. Burgeson, *J. Cell Biol.* **119**, 695 (1992).

[24] M. Sasaki, S. Kato, K. Kohno, G. R. Martin, and Y. Yamada, *Proc. Natl. Acad. Sci. U.S.A.* **84**, 935 (1987).

[25] T. Pikkarainen, R. Eddy, Y. Fukushima, M. Byers, T. Shows, T. Pihlajaniemi, M. Saraste, and K. Tryggvason, *J. Biol Chem.* **262**, 10454 (1987).

[26] D. J. Montell and C. S. Goodman, *Cell (Cambridge, Mass.)* **53**, 463 (1988).

S^{27-29}, $B2^{30-33}$, $B2t$,[34] and B1k).[35] B1k and B2t are shorter than the other B chains and lack large portions of the short arm structure.[34] Another short A-like chain has been described.[36,37] Laminin-related proteins include kalinin,[22,38] BM600/nicein,[39,40] and epiligrin.[41]

Cells assemble these laminin chains into different laminin variants/isoforms. In this chapter we use the word *chain* when talking about the individual laminin subunits and *laminin variants* when referring to the heterotrimeric molecule. Our nomenclature of the laminin variants, schematically shown in Fig. 2 is as follows.

Laminin variant		Composition
Classic laminin	is composed of	A-B1-B2 chains
S-Laminin	is composed of	A-S-B2 chains
M-Laminin/merosin	is composed of	M-B1-B2 chains
S-Merosin	is composed of	M-S-B2 chains
K-Laminin	is composed of	K-B1-B2 chains
Kalinin/nicein/epiligrin	is composed of	K-B1k-B2t chains

Sources for Laminin Purification

The primary source of laminins is any basement membrane-rich material. Basement membranes may be recognized morphologically as the

[27] D. D. Hunter, V. Shah, J. P. Merlie, and J. R. Sanes, *Nature (London)* **338**, 229 (1989).

[28] D. D. Hunter, B. E. Porter, J. W. Bulock, S. P. Adams, J. P. Merlie, and J. R. Sanes, *Cell (Cambridge, Mass.)* **59**, 905 (1989).

[29] U. M. Wewer, E. A. Wayner, B. G. Hoffström, B. Meyer-Nielsen, I. Damjanov, E. Engvall, and R. Albrechtsen, *Lab. Invest.* (in press).

[30] M. Sasaki and Y. Yamada, *J. Biol. Chem.* **262**, 17111 (1987).

[31] M. E. Durkin, B. B. Bartos, S.-H. Liu, S. L. Phillips, and A. E. Chung, *Biochemistry* **27**, 5198 (1988).

[32] T. Pikkarainen, T. Kallunki, and K. Tryggvason, *J. Biol. Chem.* **263**, 67751 (1988).

[33] H.-C. Chi and C.-F. Hui, *J. Biol. Chem.* **264**, 1543 (1989).

[34] P. Kallunki, K. Sainio, R. Eddy, M. Byers, T. Kallunki, H. Sariola, K. Beck, H. Hirvonen, T. B. Shows, and K. Tryggvason, *J. Cell Biol.* **119**, 679 (1992).

[35] D. R. Gerecke, D. W. Wagman, M.-F. Champliaud, and R. E. Burgeson, *J. Biol. Chem.* **269**, 11073 (1994).

[36] Y. Aratani and Y. Kitagawa, *J. Biol. Chem.* **263**, 16163 (1988).

[37] Y. Tokida, Y. Aratani, A. Morita, and Y. Kitagawa, *J. Biol. Chem.* **265**, 18123 (1990).

[38] P. Rouselle, G. P. Lunstrum, D. R. Keene, and R. E. Burgeson, *J. Cell Biol.* **114**, 567 (1991).

[39] P. Verrando, A. Pisani, and J.-P. Ortonne, *Biochim. Biophys. Acta* **942**, 45 (1988).

[40] P. Verrando, C. Blanchet-Bardon, A. Pisani, L. Thomas, F. Cambazard, R. A. J. Eady, O. Schofield, and J.-P. Ortonne, *Lab. Invest.* **64**, 85 (1991).

[41] W. G. Carter, M. C. Ryan, and P. J. Gahr, *Cell (Cambridge, Mass.)* **65**, 599 (1991).

TABLE I
CHROMOSOMAL ASSIGNMENT OF LAMININ CHAINS[a]

Laminin chain	Mouse	Human
α1 (previous A)	17[b]	18p11.3[c]
α2 (previous M)	10[d]	6q22→q23[21]
α3 (previous K)	18[e]	18q11[f,g]
α4	Not known	6q21[h]
β1 (previous B1)	12[i,j]	7q31[k]; 7q22[25]
β2 (previous S)	9[l,m]	3p21[n]
β3 (previous B1k)	1[e]	1q32[o]
γ1 (previous B2)	1[p]	1q25→q31[q]; 1q31[r]
γ2 (previous B2t)	1[e]	1q25→q31[34]

[a] R. E. Burgeson, M. Chiquel, R. Deutzmann, P. Ekblom, J. Engel, H. Kleinman, G. R. Martin, G. Meneguzzi, M. Paulsson, J. Sanes, R. Timpl, K. Tryggvason, Y. Yamada, P. Yurchonco, *Matrix Biology* **14,** 209 (1994).

[b] N. W. Kaye, A. E. Chung, P. A. Lalley, M. E. Durkin, S. L. Phillips, and R. L. Church, *Somatic Cell Mol. Genet.* **16,** 599 (1990).

[c] T. Nagayoshi, M. G. Mattei, E. Passage, R. Knowlton, M. L. Chu, and J. Uitto, *Genomics* **5,** 932 (1989).

[d] Y. Sunada, S. M. Barzler, C. A. Kozak, Y. Yamada, and K. P. Campbell, *J. Biol. Chem.* **269,** 13729 (1994).

[e] D. Aberdam, M. F. Galliano, M. G. Mattei, A. Pisani-Spadafora, J. P. Ortonne, G. Meneguzzi, *Mamm. Genome* **5,** 229 (1994).

[f] C. Baudoin, M. F. Galliano, P. Verrando, J. Vailly, J. P. Ortonne, G. Meneguzzi, *J. Invest Dermatol.* **102,** 549a (1994).

[g] C. Baudoin, D. Aberdam, C. Miquel, M. G. Mattei, J. P. Ortonne, G. Meneguzzi, submitted 1994.

[h] A. J. Richards, L. A. Imara, N. P. Carter, J. C. Lloyd, M. A. Leversha, and F. M. Pope, *Genomics* **22,** 237 (1994).

[i] R. W. Elliott, D. Barlow, and B. L. Hogan, *In Vitro Cell. Dev. Biol.* **21,** 477 (1985).

[j] R. W. Elliott, *Mouse News Lett.* **78,** 74 (1987).

[k] M. Jaye, W. S. Modi, G. A. Ricca, R. Mudd, I. M. Chiu, S. J. O'Brien, and W. N. Drohan, *Am. J. Hum. Genet.* **41,** 605 (1987).

[l] B. E. Porter, M. J. Justice, N. G. Copeland, N. A. Jenkins, D. D. Hunter, J. P. Merlie, and J. R. Sanes, *Genomics* **16,** 278 (1993).

[m] D. Aberdam, M. G. Mattei, J. P. Ortonne, G. Meneguzzi, *Mamm. Genome* **6,** 393 (1994).

[n] U. M. Wewer, D. R. Gerecke, M. E. Durkin, K. S. Kurtz, M. G. Mattei, M.-F. Champliaud, R. E. Burgeson, and R. Albrechtsen, *Genomics,* submitted.

[o] J. Vailly, P. Szepetowski, F. Pedeutour, M. G. Mattei, F. Depeuteur, R. E. Burgeson, J. P. Ortonne, and G. Mcneguzzi, *Genomics* **21,** 286 (1994).

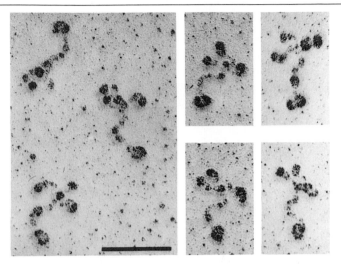

Fig. 1. Rotary shadowing electron microscopy of M-laminin isolated from human term placenta. (Reproduced from Ehrig et al.[20]) Bar: 100 mm.

structure separating the basal part of epithelial cells from the surrounding interstitial connective tissue. In the mesenchymal tissue the individual skeletal, cardiac, or smooth muscle, fat, nerve, and decidual cells are completely surrounded by a basement membrane. Specialized basement membrane structures are present, for example, in the glomerulus and at the neuromuscular junction.

Most basement membranes are approximately 100 nm thick and therefore cannot be resolved at the level of light microscopy in sections prepared with routine hematoxylin–eosin staining. At the ultrastructural level, with classic electron microscopy, three parts can be distinguished: the *lamina lucida* (closest to the cell membrane), *lamina densa* (located beneath), and the *pars fibroreticularis,* which forms an incomplete layer merging into the connective tissue and consists of the anchoring fibrils. This nomenclature is recommended by the International Anatomical Nomenclature

[p] M. F. Seldin, H. C. Morse, R. C. LeBouef, and A. D. Steinberg, *Genomics* **2,** 48 (1988).

[q] Y. Fukushima, T. Pikkarainen, T. Kallunki, R. L. Eddy, M. G. Byers, L. L. Haley, W. M. Henry, K. Tryggvason, and T. B. Shows, *Cytogenet. Cell. Genet.* **48,** 137 (1988).

[r] M. G. Mattei, D. Weil, D. Pribula-Conway, M. P. Bernard, E. Passage, N. Van-Cong, R. Timpl, and M. L. Chu, *Hum. Genet.* **79,** 235 (1988).

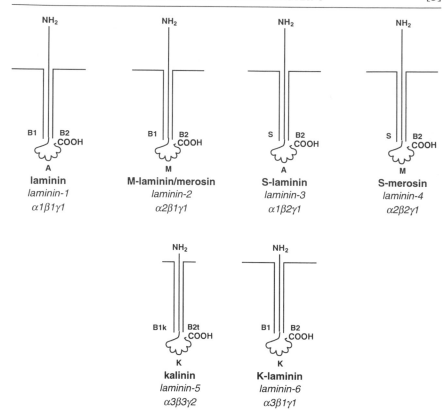

FIG. 2. Schematic representation of members of the laminin family and their characteristic tissue distribution. The new nomenclature by Burgeson *et al.* (*Matrix Biology* **14**, 209 (1994)) is shown in italics.

Committee[42] and more recently has been discussed and modified in a review on the ultrastructure of basement membranes.[43]

The small amounts of basement membrane material present in normal tissue together with the relative insolubility of this structure hampered the initial isolation and characterization of its individual constituents. Major progress was made possible after it was discovered that mouse tumors known as parietal yolk sac carcinomas produce large amounts of basement membrane material.[44] Such tumors imitate the biosynthetic pattern of normal parietal yolk sac endodermal cells of rodents. The parietal yolk sac tumors have been and are still instrumental in the progress

[42] International Anatomical Nomenclature Committee, *in* "Nomina Anatomica." Williams & Wilkins, Baltimore and London, 1983.
[43] S. Inoue, *Int. Rev. Cytol.* **117**, 57 (1989).
[44] G. B. Pierce, T. F. Beals, J. Sri Ram, and A. R. Midgley, *Am. J. Pathol.* **45**, 929 (1964).

of basement membrane research, with examples discussed below. The major variant of laminin produced by parietal yolk sac carcinomas (mouse and rat) is what we now call classic laminin, composed of A–B1–B2 chains.

Mouse parietal yolk sac carcinoma cell lines were originally derived from retransplantable testicular teratocarcinomas identified in strain 129 mice by L. C. Stevens.[45] These spontaneous and subsequently experimentally produced murine teratocarcinomas have been extensively reviewed.[45,46] Embryonal carcinoma cells, the stem cells of these tumors, have the capacity to differentiate into various tissues including cells equivalent to the visceral and parietal yolk sac endoderm of mouse choriovitelline placenta. A number of cell lines have been established including parietal yolk sac endoderm cell lines PYS 1 and 2,[47] M1536-B3, and PFHR-9.[48] Laminin was first identified in the material secreted by these parietal yolk sac cell lines.[1] These and other cell lines of similar phenotype are widely used for studying the classic form of laminin.[49,50]

The Engelbreth-Holm–Swarm (EHS) tumor is a spontaneous tumor of an ST/Eh mouse strain that arose in the mouse colony at the University Institute of Pathological Anatomy (Copenhagen, Denmark). It was initially transplanted by J. Engelbreth-Holm and later made available to the National Cancer Institute (NIH, Bethesda, MD). Originally designated by Swarm in 1963 as a chondrosarcoma,[51] it was properly identified in 1977 when Orkin et al. characterized it as a basement membrane-producing tumor.[52] On the basis of morphology and biochemistry the EHS tumor represents a spontaneous mouse parietal yolk sac tumor. For biosynthesis studies short-term cultures from the EHS tumor can be used[53] and continuous cell lines derived from the EHS tumor have been established.[54] The EHS tumor, grown subcutaneously, is one of the most widely used sources for laminin isolation and purification studies. Procedures are described by Yurchenco and O'Rear.[55]

[45] L. C. Stevens, Adv. Morphog. **6,** 1 (1967).
[46] L. C. Stevens, in "Mechanisms of Sex Differentiation in Animals and Man" (C. R. Austin and R. G. Edwards, eds.), p. 301. Academic Press, New York, 1981.
[47] J. M. Lehman, W. C. Speers, D. E. Swartzendruber, and G. B. Pierce, J. Cell. Physiol. **84,** 13 (1974).
[48] A. E. Chung, L. E. Estes, H. Shinozuka, J. Braginski, C. Lorz, and C. A. Chung, Cancer Res. **37,** 2072 (1977).
[49] M. Kurkinen, D. P. Barlow, J. R. Jenkins, and B. L. M. Hogan, J. Biol Chem. **258,** 6543 (1983).
[50] A. Damjanov, U. M. Wewer, B. Tuma, and I. Damjanov, Differentiation (Cambridge, UK) **45,** 84 (1990).
[51] R. L. Swarm, J. Natl. Cancer Inst. (U.S.) **31,** 953 (1963).
[52] R. W. Orkin, P. Gehron, E. B. McGoodwin, G. R. Martin, T. Valentine, and R. Swarm, J. Exp. Med. **145,** 204 (1977).
[53] S. R. Ledbetter, B. Tyree, J. R. Hassell, and E. A. Horigan, J. Biol. Chem. **260,** 8106 (1985).
[54] K. G. Danielson, A. Martinez-Hernandez, J. R. Hassell, and R. V. Iozzo, Matrix **11,** 22 (1992).
[55] P. D. Yurchenco and J. J. O'Rear, this volume [23].

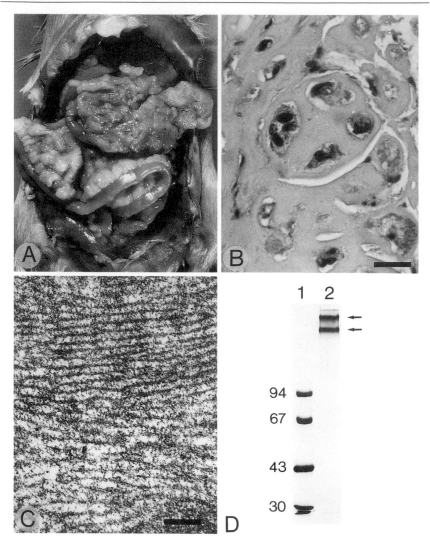

FIG. 3. The L2 rat yolk sac carcinoma. (A) Gross appearance of L2 tumor masses from a Lewis rat. (B) The tumor masses are composed of tumor cell islands surrounded by large amounts of extracellular matrix. Hematoxylin–eosin stain. Bar: 25 μm. (C) At the ultrastructural level the extracellular matrix is composed of repeating layers reminiscent of lamina lucida and lamina densa of the basement membrane. Bar: 0.27 μm. (D) Purified intact classic laminin A–B1–B2 on SDS–PAGE under reducing conditions showing the characteristic 400- and 200-kDa bands (arrows). Markers to the left represent molecular weights ($\times 10^{-3}$).

Parietal yolk sac carcinomas can be experimentally induced in the rat[56] (reviewed by Damjanov[57]). The L2 rat yolk sac carcinoma was induced by puncturing the uterine wall of a pregnant Lewis rat during midgestation.[4,58] It was established as a transplantable tumor line in Lewis rats and also as the continuous L2 cell line in culture.[58] We have used the L2 tumor and cell line extensively as a laminin source (see Fig. 3).

Parietal yolk sac tumors may not necessarily represent normal basement membranes, and the need to isolate human laminin has stimulated the search for a readily available source of human laminin. Human placenta has been useful as a source because large amounts of normal tissue can be readily obtained. Thus, placental tissue has been used for the isolation of intact and truncated forms of human laminin.[6–8,20,59,60] Using the protocol described in this chapter it was shown that purified intact placental laminin consists primarily of M-laminin (M–B1–B2) with small amounts of M–S–B2 laminin also present.[61] Classic laminin (A–B1–B2) is released after treatment with proteases. Other normal tissues that have been used for laminin purification are mouse and bovine heart.[62] These preparations also contain predominantly M-laminin.[62,63]

Serum and other body fluids may contain laminin in small amounts. In serum, the concentration of laminin is too low[64] to be adequate for biochemical purification. The amniotic fluid, another body fluid, is relatively rich in classic laminin (A–B1–B2) and is almost devoid of M chain-containing laminins (M. P. Marinkovich and R. E. Burgeson, personal communication, 1994). Detailed analyses of laminins in other body fluids have not yet been performed.

The heterogeneity of laminin variants has been explored immunohistochemically using monoclonal antibodies against extracts of basement membrane-containing material. This approach resulted in the identification of new laminin chains S, M, and K. For example, Sanes' group isolated a series of monoclonal antibodies that reacted with the basement membrane of the neuromuscular junction.[65] These immunological reagents were subsequently used for screening a cDNA library and led to the identification of the S chain of laminin.[27] Monoclonal antibodies raised against partially purified placental laminin[8] led to the identification of the M chain of laminin.[20,60] Using the monoclonal antibody strategy it was possible

[56] I. Damjanov, N. Skreb, and S. Sell, *Int. J. Cancer* **19,** 526 (1977).

[57] I. Damjanov, *Int. J. Dev. Biol.* **37,** 39 (1993).

[58] U. Wewer, *Dev. Biol.* **93,** 416 (1982).

[59] S. N. Dixit, *Connect. Tissue Res.* **14,** 31 (1985).

[60] I. Leivo and E. Engvall, *Proc. Natl. Acad. Sci. U.S.A.* **85,** 1544 (1988).

[61] E. Engvall, D. Earwicker, T. Haaparanta, E. Ruoslahti, and J. R. Sanes, *Cell Regul.* **1,** 731 (1990).

[62] M. Paulsson and K. Saladin, *J. Biol. Chem.* **264,** 18726 (1989).

[63] M. Paulsson, K. Saladin, and E. Engvall, *J. Biol. Chem.* **266,** 17545 (1991).

[64] J. Risteli, H. Rohde, and R. Timpl, *Anal. Biochem.* **113,** 372 (1981).

[65] J. R. Sanes and A. Y. Chiu, *Cold Spring Harbor Symp. Quant. Biol.* **48,** 667 (1983).

to identify the newly described laminin-related proteins such as kalinin,[22,38] BM600/nicein,[39,40] epiligrin,[41] which may be all related,[66] and K-laminin.[22,23] In addition, laminin-like domains are found in many proteins such as perlecan,[67] agrin,[68,69] and neurexins.[70]

Techniques for Laminin Purification

Protocols for the isolation and purification of laminin from the parietal yolk sac tumors have been described.[3,5] Mouse and rat laminin can be obtained from a variety of commercial companies that may use similar protocols. Here we describe (1) a protocol (adapted from Engvall et al.[5]) for the isolation and purification of biologically active classic rat laminin (A–B1–B2) in milligram amounts, using the L2 parietal yolk sac carcinomas as a source, (2) a protocol (adapted from Wewer et al.,[8] Ehrig et al.,[20] and Paulsson and Saladin[62]) for purification of intact M-laminin (M–B1–B2) from human term placenta, and (3) the sodium dodecyl sulfate–polyacrylamide gel electrophoresis (SDS–PAGE) pattern of the various laminin variants.

Purification of Classic Laminin (A–B1–B2) from Rat L2 Tumors

The L2 rat yolk sac carcinoma cell line is available from the author (U.W.) and will soon be available from the American Type Culture Collection (ATCC, Rockville, MD). These cells grow in most culture media, for example, Dulbecco's modified Eagle's medium (DMEM) with high glucose supplemented with 10% (v/v) fetal bovine serum gives consistently good results. The amount of laminin synthesized in the cell culture medium is 10–50 $\mu g/ml$. It is convenient to start a "transplantable tumor line" by injecting 5×10^7 cultured cells in 1–2 ml of phosphate-buffered saline (PBS) (the number is not critical) intraperitoneally in 8-week-old Lewis rats. After approximately 6 weeks the rats form ascites (25–30 ml/rat) and the multiple tumor masses (5–10 g/rat) are located most often in the omental tissue at the lower curvature in proximity to the gastric sac. The amount of tumor masses varies from 5–10 g/rat. In some rats the L2 cells

[66] M. P. Marinkovich, P. Verrando, D. R. Keene, G. Meneguzzi, G. P. Lunstrum, J. P. Ortonne, and R. E. Burgeson, Lab. Invest. 69, 295 (1993).

[67] D. M. Noonan, E. A. Horrigan, S. R. Ledbetter, G. Vogeli, M. Sasaki, Y. Yamada, and J. R. Hassell, J. Biol. Chem. 263, 8106 (1988).

[68] F. Rupp, D. G. Payan, C. Magill-Solc, D. M. Cowan, and R. H. Scheller, Neuron 6, 811 (1991).

[69] K. W. K. Tsim, M. A. Ruegg, G. Escher, S. Kröger, and U. J. McMahan, Neuron 8, 677 (1992).

[70] Y. A. Ushkaryov, A. G. Petrenko, M. Geppert, and T. C. Südhof, Science 257, 50 (1992).

seem to grow mostly in ascites form with less solid tumor tissue. The tendency to form ascites rather than solid tumors increases following consecutive serial transplantations. The ascites fluid is rich in clumps of tumor cells and can be used for tumor propagation using 1–2 ml of ascites fluid per rat. Once the rats have developed ascites to a certain stage (approximately 30 ml), they quickly die. Therefore, it is important to check the development of ascites and euthanize the rats in due time. The ascites fluid is also rich in soluble laminin (usually in excess of 0.5 mg/ml), which can also be used for purification. The tumor and ascites materials are best stored at −80° until further use.

The purification procedure should be performed at 4° and in the presence of protease inhibitors. Once started, the purification should proceed as quickly as possible through all the steps to avoid unnecessary proteolytic degradation. A typical purification procedure requires approximately 10 g of tumor tissue and the expected yield of purified classic laminin A–B1–B2 is in the range of 3–5 mg.

Procedure

1. Thaw solid L2 tumors rapidly and mince into smaller pieces with razor blades.

2. Homogenize tumor tissue in 50 ml of 0.05 M Tris-HCl, pH 7.4 (five times the tumor volume is adequate).

3. Spin the homogenate in a tabletop centrifuge at 3000 rpm for 20 min and discard the supernatant (use two 50-ml tubes).

4. Extract the pellet with 50 ml of 0.01 M ethylenediaminetetraacetic acid (EDTA) in 0.05 M Tris-HCl, pH 7.4, overnight. Use either an end-over-end or a magnetic stirrer for this step.

5. Spin at 10,000 rpm for 20 min and save the supernatant for chromatography. The pellet can be reextracted for residual laminin.

6. Gel filtration is used to eliminate low-molecular-mass-contaminating proteins. Use a long wide column (i.e., 50–60 cm long and 5–6 cm wide; approximate bed volume, 500 ml), pack with 6B-CL beads (Pharmacia, Piscataway, NJ), equilibrate with 0.05 M Tris-HCl, pH 7.4, and 0.025 M NaCl. Load the column with 50 ml of extract and collect 6- to 7-ml samples. Read the A_{280} of the fractions and examine by SDS–PAGE. Laminin should be in the void volume, the first peak coming off the column.

7. Ion-exchange chromatography can be used as a step for further laminin purification. The laminin-containing sample is loaded at low ionic strength and eluted at high ionic strength. We use Whatman (Clifton, NJ) DE-52 cellulose, approximately 1 ml of cellulose per calculated 10–12 mg of protein. Do not overload or else too much laminin will remain in the

nonbound fraction. Equilibrate with 0.05 M Tris-HCl, pH 7.4, and 0.025 M NaCl. If the NaCl concentration is too high, too much laminin will remain in the nonbound fraction. Flush the column with at least four column volumes of buffer and then elute with four column volumes 0.05 M Tris-HCl, pH 7.4, and 0.35 M NaCl. Collect fractions throughout, read the A_{280}, and examine by SDS–PAGE. Laminin should be eluted as a peak.

8. Heparin–Sepharose can be used as the final purification and concentration step. The sample is loaded at low ionic strength and eluted at high ionic strength. If no ion-exchange chromatography has been performed one can load the sample from the Sepharose 6B-CL column directly but if ion exchange has been performed it is necessary to dilute the sample with 0.05 M Tris-HCl, pH 7.4, until the NaCl concentration is approximately 0.05 M. Load the heparin–Sepharose column with the sample, wash with 4 vol of 0.05 M Tris-HCl, pH 7.4, and 0.05 M NaCl and elute with 0.05 M Tris-HCl and 0.05 M NaCl. Collect fractions throughout, and read the A_{280}. An example of purified classic laminin (A–B1–B2) is shown in Fig. 3D.

Purification of Intact Laminin (M-Laminin) from Human Term Placenta

Procedure

1. Frozen or fresh placental tissue is cut into pieces and rinsed thoroughly with cold H_2O before blending in 0.05 M Tris-HCl, pH 7.4 (five times tissue volume).

2. Spin the sample in a tabletop centrifuge at 3000 rpm for 20 min at 4° and discard the supernatant. If the supernatant is a very dark red, consider washing the pellet once with Tris buffer.

3. Extract the pellet with 0.01 M EDTA in 0.05 M Tris-HCl, pH 7.4, 0.05 M NaCl overnight at 4°.

4. Spin the sample at 10,000 rpm and save the supernatant.

5. Add NaCl to give a final concentration of 4 M NaCl and let it sit overnight at 4° to form a precipitate.

6. Spin the sample at 10,000 rpm, dissolve the pellet in 50 mM Tris-HCl, pH 7.4, 10 mM EDTA and respin before chromatography.

7. Eliminate contaminating proteins by gel filtration (see step 6 of the preceding section).

8. Purify further by ion exchange (see step 7 of the preceding section).

9. The sample may be applied to a heparin–Sepharose column (see step 8 of the preceding section).

Steps 7–9 generally follow the same principles as described above for rat laminin. Monoclonal antibody chromatography can also be used as a step in the purification of laminins, but because these antibodies may not always be available we omit the detailed protocol here.

SDS–PAGE Migration Pattern of Various Laminin Variants

Metabolic labeling followed by single or double immunoprecipitations can be a useful technique by which to elucidate which laminin variants a given cell type may synthesize. Another interesting question is which chains of laminin may preferentially assemble and if or how different cells may regulate this process. Consequently, we thought it would be useful to illustrate schematically the expected migration pattern on SDS–PAGE of subunits of biosynthesized laminin variants from cell culture as well as of biochemically purified laminins (Fig. 4).

Laminin Fragments

Fragments of laminin have been used to investigate the functions of various domains of the molecule.[15] Fragments containing the globular domains at the ends of the short arms have been shown to participate in

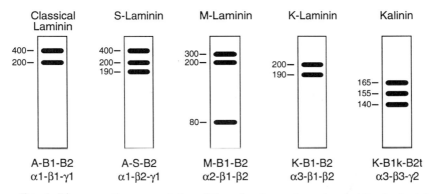

FIG. 4. Diagrammatic representation of the migration pattern observed for the laminin variants when electrophoresed under reducing conditions in SDS–PAGE. Markers to the left represent molecular weight ($\times 10^{-3}$). Classic laminin has a 400-kDa A chain band and a broader, 200-kDa band consisting of B1 and B2 chains. The S chain of laminin is a 190-kDa band. The heavy chain of M-laminins appears as two bands, one 300 kDa and the other 80 kDa. K-laminin consists of the B1 and B2-containing 200-kDa band and the 190-kDa band of the K chain.[22,23] The SDS–PAGE migration pattern of kalinin is complicated because the chains are differentially processed in cell culture. Thus, on the basis of Rouselle et al.[38] and Marinkovich et al.,[22] kalinin exists in three "forms," namely, the initial transcript [the cellular KC form (200; 155; and 140-kDa bands)] and two secreted forms [the KM1 (165; 155; and 140-kDa bands) shown here, and the KM2 (165; 140; and 105-kDa bands)]. The 190-kDa chain of K laminin may or may not be identical to the 200-kDa chain of kalinin (KC).

the polymerization of laminin.[55] Fragments containing either the center of the cross or the large globular domain at the C terminus of the heavy chain have similarly been shown to contain the major cell-binding activities of laminin. Most of these studies have been done with fragments of the mouse EHS laminin, because it has been the only laminin available in quantities large enough for such purposes. It is possible that laminin A–B1–B2 from other species could be used to prepare similar fragments corresponding to the same domains. Laminin M–B1–B2 or "heart laminin" was digested and compared with laminin A–B1–B2.[61] The M chain is cleaved in different positions than the A chain and fragments corresponding to the E8 fragment of A laminin are not obtained. A fragment with a long arm but with no short arms is the main product obtained by purifying laminins from placenta after pepsin digestion.[8] However, at least some fragments of M-laminin and A-laminin are similar, for example, the fragments consisting of repeats 4 and 5 in the G domains.[20]

Activities of Laminins

We describe here *in vitro* assays that can be used to analyze the interaction of cells with laminin. The assumption is that these *in vitro* interactions closely mimic *in vivo* interactions and thus have biological significance.

Laminin is secreted from cells in a soluble form. To become incorporated into the distinct structures that are typical of basement membranes, laminin must self-assemble and connect with other basement membrane components such as type IV collagen, entactin, and heparan sulfate proteoglycan.[55,71] *In vivo* cells then interact with the assembled basement membrane including laminin, probably via several mechanisms. It is not necessarily so simple or even possible to mimic *in vitro* the conformation and activities of laminin in basement membranes *in vivo*. Therefore, certain approximations must be made. Either one can study the interaction of cells with purified laminin on artificial laminin-containing matrices and perhaps miss an activity, or one can use preparations of basement membrane material, for example, the so-called Matrigel (see below), to try to mimic the situation *in vivo*. However, in the latter case, the interactions of cells with laminin may be obscured by the interaction of the cells with other basement membrane proteins and by the presence of growth factors in basement membranes.

Preparation of Substrates with Purified Laminin: General Considerations

Glass or polystyrene can be coated with laminin, and the polystyrene may be untreated or treated for tissue culture. However, if isolated laminin

[71] R. Timpl, *Eur. J. Biochem.* **180,** 487 (1989).

alone is used to coat the substrate, then untreated polystyrene is the choice, because most of what is known about protein binding is from the use of untreated plastic. Various proteins coat the plastic in similar amounts, mostly depending on the size of the protein.[72] Laminin is a large protein and binds well to the hydrophobic plastic. Such substrates are most suitable for cell attachment assays. The binding of proteins to plastic causes some unfolding/denaturation of the protein, but it is not known if the denaturation is random or affects a specific part of the protein. It has been consistently observed that even slightly denatured laminin may be inactive in cell attachment.

Attachment of Nonneuronal Cells to Laminin

Wells in 96-well microtiter plates [enzyme-linked immunosorbent assay (ELISA) or "assay" plates] are incubated with laminin in PBS or other physiological buffer for 2 hr at 37° or for 4–5 hr at room temperature. If the laminin solution is titrated, the optimal concentration will be about 3 μg/ml and little cell attachment will be seen below 0.3 μg/ml. The laminin-coated wells are washed with PBS and then incubated with 3% (w/v) bovine serum albumin (BSA) in PBS for 30–60 min at 37°. This "blocking step" is important, because cells will otherwise bind nonspecifically to remaining hydrophobic sites on the plastic and a high background will be obtained.

Cells requiring solid substrate for their growth are most often examined for attachment to extracellular matrix. Leukocytes and platelets also interact with laminin and can be used in cell–substrate attachment assays.[73,74] To prepare a cell suspension from monolayers of adherent cells, the cells may be dissociated with EDTA or with trypsin with or without EDTA. If trypsin is used, care should be taken to inactivate the trypsin before adding the cells to the laminin-coated wells. The safest method is to use crystalline trypsin for dissociation, and inactivate it with soybean trypsin inhibitor immediately after the cells have detached. The tissue culture grade of trypsin often contains additional proteolytic enzymes that may not be inactivated by any one trypsin inhibitor or even by serum. By adding serum one not only inactivates trypsin but also adds cell attachment proteins such as fibronectin and vitronectin, and these may possibly influence the assay and confuse the results.

After the cell suspension has been prepared, it is added immediately to the laminin-coated wells and the cells are allowed to attach for 30–90 min. To record the degree of cell attachment and/or the morphology of

[72] E. Engvall, this series, Vol. 70A, p. 419.
[73] M. E. Hemler, *Annu. Rev. Immunol.* **8**, 365 (1990).
[74] M. H. Ginsburg, J. C. Loftus, and E. F. Plow, *Thromb. Haemostasis* **59**, 1 (1988).

Laminin   Fibronectin

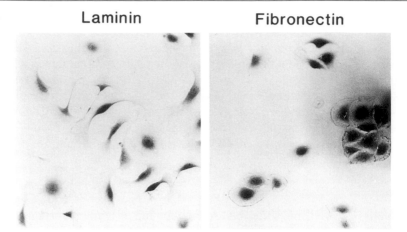

FIG. 5. Attachment of Rugli (rat glioblastoma) cells to microtiter wells coated with purified laminin or fibronectin. Note the different morphologies that the cells attain on these two proteins.

the cells, the unattached cells are washed away, and the attached cells are fixed with glutaraldehyde or other fixative, and stained (Fig. 5). Amido black, Coomassie blue, and crystal violet are all suitable stains. For routine quantitative cell attachment assays, the microtiter plates with stained cells may be read in an ELISA reader. The highest and most reproducible readings are obtained if the stain is first solubilized, for example, by adding 1% (w/v) SDS to the wells.

Neurite Outgrowth/Neuronal Culture on Laminin Substrates

Neuronal cells do not attach well to the laminin-coated polystyrene described above, although they do extend neurites on it, suggesting that the growth cone but not the cell body contains the active receptors. The poor attachment of the cells means the cell cultures are difficult to handle at the end of an assay. The cells detach from the substrate easily when fixed. Therefore, for assays using neuronal cells, the laminin is coated on tissue culture plastic that has been precoated with polylysine or polyornithine.[75,76] The polyamine binds to the negatively charged tissue culture plastic and laminin binds to the polyamine and perhaps also to any remaining negative charges. The neurons presumably use the polyamine for attachment and laminin for neurite outgrowth.

[75] M. Manthorpe, E. Engvall, E. Ruoslahti, F. M. Longo, G. E. Davis, and S. Varon, *J. Cell Biol.* **97,** 1882 (1983).

[76] E. Engvall, G. E. Davis, K. Dickerson, E. Ruoslahti, S. Varon, and M. Manthorpe, *J. Cell Biol.* **103,** 2457 (1986).

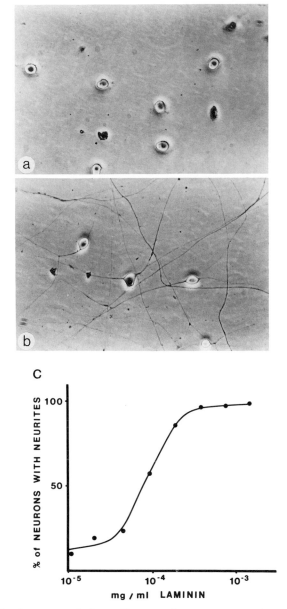

FIG. 6. Neurite outgrowth from ciliary ganglion neurons on substrates without (a) or with (b) laminin, and a laminin titration curve (c).

Wells in 96-well culture plates are incubated overnight at room temperature with 50 μl of polyornithine (0.1 mg/ml). After three washes with distilled water or PBS, the precoated wells are incubated with 50 μl of laminin in a physiological buffer. If laminin is serially diluted, the optimal concentration is usually 0.1 to 1 μg/ml (Fig. 6). Activity of laminin may still be seen in the range of 0.03 μg/ml. Thus, one needs less laminin for this type of assay than for the cell attachment assay described above. This may be partly because laminin may be presented more optimally in this configuration, perhaps owing to laminin retaining a more native conformation. Another reason is that neurite outgrowth assay is exquisitely sensitive and the response may be seen at a low concentration of laminin. After washing with PBS containing 1% (w/v) BSA, cells may be seeded in the wells. Neurite outgrowth on laminin can be detected under the microscope after 1–2 hr, but extensive neurite outgrowth facilitates evaluation of the assay and may require longer incubations, such as overnight. Controls for this type of assay should include the omission of the laminin-coating step, because some types of neuronal cells extend neurites on polyamines only.

Culture of Cells on Matrigel

Matrigel is a soluble preparation of basement membrane components and is easily prepared from EHS tumors.[77] Briefly, the tumor tissue is extracted with 2 M urea in 0.05 M Tris-HCl, pH 7.4, and the extract is dialyzed to remove the urea and kept frozen or at 4°. On warming to 37°, the extract forms a gel of reconstituted basement membrane. Matrigel can also be purchased ready to use from GIBCO-BRL (Gaithersburg, MD). Matrigel contains a number of growth factors in addition to laminin, type IV collagen, and entactin.[78] Cells can be placed on top of the gel or incorporated into the gel for growth in three dimensions. Matrigel promotes differentiation of cells *in vitro* as seen from several cell culture systems (see, e.g., Streuli *et al.*[79]). *In vivo,* Matrigel promotes the growth in nude mice of transformed but normally nontumorigenic cells and may even transform normal cells into tumorigenic cells.[80] On the other hand, Matrigel may promote the differentiation of tumor cells rather than make

[77] H. K. Kleinman, M. L. McGarvey, J. R. Hassell, V. L. Star, F. B. Cannon, G. W. Laurie, and G. R. Martin, *Biochemistry* **25,** 312 (1986).
[78] S. Vukicevic, H. K. Kleinman, F. P. Luyten, A. B. Roberts, N. S. Roche, and A. H. Reddi, *Exp. Cell Res.* **202,** 1 (1992).
[79] C. H. Steuli, N. Bailey, and M. J. Bissell, *J. Cell Biol.* **115,** 1383 (1991).
[80] R. Fridman, T. M. Sweeney, M. Zain, G. R. Martin, and H. K. Kleinman, *Int. J. Cancer* **51,** 740 (1992).

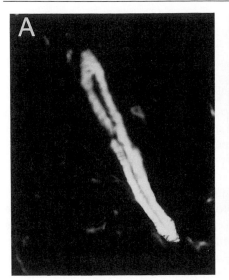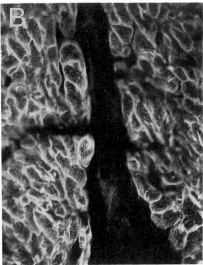

FIG. 7. Immunohistochemistry: staining of adjacent sections of rabbit heart tissue with monoclonal antibody to the laminin A chain (A) or to the M chain (B) in indirect immunofluorescence. The A chain is found predominantly in the blood vessel wall, whereas the M chain is localized around the individual striated muscle cells.

them invasive and metastasizing.[81] Matrigel has been used to grow tumors from primary patient tumors including difficult-to-grow tumor types such as acute lymphoblastic leukemia cells.[82] The content of growth factors in Matrigel or the combination of extracellular matrix and growth factors may be responsible for its growth-promoting effects.

Molecular Mechanisms of Cell Attachment and Neurite Outgrowth on Laminin

Isolated laminin binds to integrin-type cell surface receptors and cell surface proteoglycans and perhaps also to other cell surface components. Of these, the binding of laminin to a large number of different integrins has received the most attention.[83,84] However, despite a large body of work in this area, relatively little is known about the nature of the interactions and exactly where the binding sites on the laminin molecule are located. One reason for this is that the complicated structure of laminins

[81] A. C. Noël, A. Calle, H. P. Emonard, B. V. Husgens, L. Simar, J. Foidart, C. M. Lapiere, and J.-M. Foidart, *Cancer Res.* **51,** 405 (1991).

[82] K. Sterling-Levis, L. White, A. E. Trickett, C. Gramacho, S. M. Pittman, and V. Tobias, *Cancer Res.* **53,** 1222 (1993).

[83] R. O. Hynes, *Cell (Cambridge, Mass.)* **69,** 11 (1992).

[84] E. Ruoslahti, *J. Clin. Invest.* **87,** 1 (1991).

has prevented the use of straightforward molecular biology techniques that have been used so successfully to delineate active domains in other proteins such as fibronectin.

Tissue Distribution Pattern and Changes Associated with Diseases

Most of the earlier immunohistochemical work was performed using polyclonal antibodies raised against classic laminin (A–B1–B2) and therefore could not take into account the possible heterogeneity in the structural composition of the laminins. However, chain-specific monoclonal antibodies and corresponding cDNA are now available for dissecting the heterogeneity of laminin chain expression in normal tissue and during disease. Such an approach has already led to new information about the relative expression of the individual chains. Thus, in most normal adult tissue laminin chains B1 and S seem to be mutually exclusive, as do laminin chains A and M (Fig. 7).[60,85] It is also apparent that during differentiation the expression changes, and transient or permanent changes in the composition of laminin may contribute to important developmental events.[60,86,87]

Disruption of the normal composition and/or assembly of laminin molecules is likely to be associated with disease. There are already several studies supporting this. In junctional epidermolysis bullosa (Herlitz syndrome), in which blisters arise within the basement membrane area, the laminin-related BM600/nicein molecule is absent.[40] Another striking example is the deficiency of the laminin M chain in Fukuyama congenital muscular dystrophy.[88] Consequently, animal models with diseases attributable to basement membrane defects have become interesting to investigate for any changes that may occur in the laminin expression pattern. We have found that mice with congenital muscular dystrophy due to the mutation *dy* do not produce a detectable M chain polypeptide as observed by immunohistochemistry and immunoblotting.[89]

Acknowledgments

The original work was supported by the Danish Medical Research Council and the Danish Cancer Society (U.M.W.) and by the National Cancer Institute, NIH (E.E.).

[85] J. R. Sanes, E. Engvall, R. Butkowski, and D. D. Hunter, *J. Cell Biol.* **111,** 1685 (1990).

[86] M. Ekblom, G. Klein, G. Mugrauer, L. Fecker, R. Deutzmann, R. Timpl, and P. Ekblom, *Cell (Cambridge, Mass.)* **60,** 337 (1990).

[87] U. M. Wewer, E. Engvall, M. Paulsson, Y. Yamada, and R. Albrechtsen, *Lab. Invest.* **66,** 378 (1992).

[88] Y. K. Hayashi, E. Engvall, E. Arikawa-Hirasawa, K. Goto, R. Koga, I. Nonaka, H. Sugita, and K. Arahata, *J. Neurol. Sci.* **119,** 53 (1993).

[89] X. Hong, H. Xu, P. Christmas, X.-R. Wu, K. Arahata, U. Wewer, and E. Engvall, *Proc. Natl. Acad. Sci. USA* **91,** 5572 (1994).

[6] Aggrecan–Versican–Neurocan Family of Proteoglycans

By RICHARD U. MARGOLIS and RENÉE K. MARGOLIS

Introduction

Cartilage aggrecan was one of the first proteoglycans to be cloned,[1] and it was soon found to be homologous to a human fibroblast proteoglycan that was subsequently named versican.[2,3] It was not until several years later, with the cloning of neurocan from brain,[4] that this proteoglycan family began to expand beyond its original two members. However, nervous tissue has proved to be a surprisingly diverse source of proteoglycans capable of forming macromolecular aggregates with hyaluronic acid (some, and perhaps all, of which are link protein stabilized). Besides neurocan, which is specific to nervous tissue, these include aggrecan[5,6] and versican themselves, brevican[6a], and two chondroitin sulfate proteoglycans (whose primary structures have not yet been reported) recognized by the monoclonal antibodies Cat-301[7] and T1.[8] These latter proteoglycans are biochemically and/or immunochemically related to aggrecan and are also capable of interacting with hyaluronic acid.

The isolation and properties of aggrecan, the prototypical cartilage proteoglycan whose biochemistry, cell biology, and pathology have been intensively studied for over 25 years, were described in previous volumes of this series[9–12] and are therefore not covered here. The first portion of this chapter summarizes the biochemical properties and structural relationships of versican and neurocan, with particular emphasis on methodological considerations insofar as they may affect the interpretation of some-

---

[1] W. B. Upholt, L. Chandrasekaran, and M. L. Tanzer, *Experientia* **49**, 384 (1993).
[2] T. Krusius, K. R. Gehlsen, and E. Ruoslahti, *J. Biol. Chem.* **262**, 13120 (1987).
[3] D. R. Zimmermann and E. Ruoslahti, *EMBO J.* **8**, 2975 (1989).
[4] U. Rauch, L. Karthikeyan, P. Maurel, R. U. Margolis, and R. K. Margolis, *J. Biol. Chem.* **267**, 19536 (1992).
[5] L. Hao and N. B. Schwartz, *FASEB J.* **7**, Abstr. 486 (1993).
[6] P. Milev, M. Flad, R. K. Margolis, and R. U. Margolis, unpublished results (1993).
[6a] H. Yamada, K. Watanabe, M. Shimonaka, and Y. Yamaguchi, *J. Biol. Chem.* **269**, 10119 (1994).
[7] H. J. L. Fryer, G. M. Kelly, L. Molinaro, and S. Hockfield, *J. Biol. Chem.* **267**, 9874 (1992).
[8] M. Iwata, T. N. Wight, and S. S. Carlson, *J. Biol. Chem.* **268**, 15061 (1993).
[9] D. Heinegård and Y. Sommarin, this series, Vol. 144, p. 305.
[10] D. Heinegård and Y. Sommarin, this series, Vol. 144, p. 319.
[11] J. H. Kimura, T. Shinomura, and E. J.-M. A. Thonar, this series, Vol. 144, p. 372.
[12] D. Heinegård and M. Paulson, this series, Vol. 145, p. 336.

Copyright © 1994 by Academic Press, Inc.
All rights of reproduction in any form reserved.

times conflicting results, whereas the following sections provide procedures for the isolation and biochemical analysis of neurocan and related proteoglycans, their immunocytochemical localization, demonstration of their cellular sites of synthesis by *in situ* hybridization histochemistry, and a brief discussion of methods for studying their interactions with other proteins and with cells.

Versican

The primary structure of the C-terminal portion of versican was obtained by screening a human fibroblast cDNA library with an antiserum made against a proteoglycan fraction from human fetal membranes, and antibodies to synthetic peptides based on the deduced amino acid sequence indicated that these clones encoded the 400-kDa core (glyco)protein of a chondroitin sulfate proteoglycan.[2] The complete sequence revealed that this 2389-amino acid core protein contained an N-terminal domain homologous to the hyaluronic acid-binding domains of link protein and cartilage aggrecan, a C-terminal domain containing two epidermal growth factor (EGF)-like repeats, a lectin-like sequence, and a complement regulatory protein-like domain, and that these were separated by a region containing up to 15 potential attachment sites for chondroitin sulfate chains.[3] The versican gene has been mapped to the long arm of human chromosome 5 (5q12–5q14).[13] The evolutionary relationship of the hyaluronic acid-binding domain of link protein to homologous sequences in aggrecan, versican, and neurocan has been described,[14] and the primary structures and homology domains of aggrecan, versican, and neurocan are summarized in Fig. 1.

Little has been published concerning the biochemical properties of versican purified from tissues. However, recombinant versican and a 78-kDa truncated form of versican containing its N terminus bind to hyaluronic acid (but not to chondroitin sulfate or heparin) with a dissociation constant (4 nM) in the same range as that reported for aggrecan, suggesting that versican could form a bridge between pericellular hyaluronic acid and cell surfaces.[15] The estimated molecular mass of the recombinant versican core protein (270–290 kDa) was somewhat less than that of versican produced by MG-63 cells (~400 kDa), and the intact recombinant proteoglycan migrated as a broad band immediately above the position of the core protein, whereas intact versican immunoprecipitated from MG-63 cells did not enter a 4–12% gradient gel.[15]

[13] R. V. Iozzo, M. F. Naso, L. A. Cannizzaro, J. J. Wasmuth, and J. D. McPherson, *Genomics* **14,** 845 (1992).
[14] E. Barta, F. Deák, and I. Kiss, *Biochem. J.* **292,** 947 (1993).
[15] R. G. LeBaron, D. R. Zimmermann, and E. Ruoslahti, *J. Biol. Chem.* **267,** 10003 (1992).

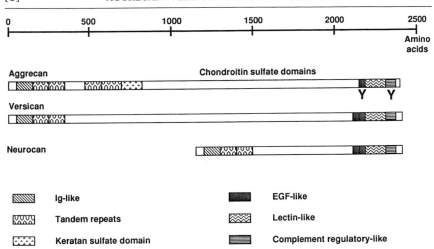

FIG. 1. Summary of the primary structures and homology domains of aggrecan, versican, and neurocan. **Y**, Aggrecan domains that may be deleted by alternative splicing.

In aggrecan mRNA, the exons encoding the EGF-like and complement regulatory protein (CRP)-like sequences may be deleted by alternative splicing.[16,17] In aggrecan of human articular cartilage and cultured chondrocytes, the CRP-like domain is usually present whereas most of the transcripts lack the EGF-like domain.[18] Versican mRNA is also present in human articular cartilage, although at a much lower level than that of aggrecan, and its two EGF-like domains and one CRP-like domain are predominantly present in all transcripts.[18] Neurocan resembles versican in containing two EGF-like domains and, as in the case of versican mRNA from cartilage, only a single neurocan transcript was detected in early postnatal or adult brain.[4]

A chondroitin sulfate proteoglycan (PG-M) with a 388-kDa core protein that is expressed during chondrogenesis in chick limb buds has been cloned and appears to be the chick homolog of versican.[19] This proteoglycan, which occurs in two alternatively spliced forms, the larger of which is ~100 kDa greater in size than human versican, has 68–82% amino acid sequence identity with human versican in its three N-terminal domains, 68–84% identity in the two EGF-like domains, and 93–96% identity in the lectin-like and complement regulatory protein-like sequences, but only

[16] C. T. Baldwin, A. M. Reginato, and D. J. Prockop, *J. Biol. Chem.* **264**, 15747 (1989).
[17] K. J. Doege, M. Sasaki, T. Kimura, and Y. Yamada, *J. Biol. Chem.* **266**, 894 (1991).
[18] J. Grover and P. J. Roughley, *Biochem. J.* **291**, 361 (1993).
[19] T. Shinomura, Y. Nishida, K. Ito, and K. Kimata, *J. Biol. Chem.* **268**, 14461 (1993).

38–44% identity in the central chondroitin sulfate attachment region (where the alternative splicing occurs). The N- and C-terminal homology with human versican and the lack of similarity of the central portion (which accounts for ~80% of PG-M) with other reported proteins are similar to the case for neurocan and may reflect a general pattern in this proteoglycan family.

Glial Hyaluronic Acid-Binding Protein and Hyaluronectin

Many reports have appeared describing the properties and localization of various low molecular size (usually 60- to 70-kDa) hyaluronic acid-binding proteins detected in brain and other tissues.[20] Two of these hyaluronic acid-binding proteins have also been used for the histochemical detection of hyaluronic acid,[21,22] in a manner similar to that earlier employed for the light and electron microscopic localization of hyaluronic acid in developing brain using the biotinylated hyaluronic acid-binding domain isolated after clostripain or trypsin treatment of rat chondrosarcoma proteoglycan.[23,24] Because hyaluronectin, the first of these proteins to be described, and the closely related or identical "glial hyaluronic acid-binding protein" (GHAP) are both isolated from brain homogenates in the absence of protease inhibitors and under acidic conditions that would favor lysosomal enzyme activity, it was suggested that they may be proteolytic degradation products of larger proteoglycan core proteins.[25] The GHAP isolation procedure[26] was based on and closely follows the original method described for the preparation of hyaluronectin.[27] It includes initial homogenization of tissue in 10 mM HCl,[26] and in the case of hyaluronectin a brain supernatant fraction is extensively dialyzed at pH 5.[27]

It was suggested by Zimmermann and Ruoslahti[3] that GHAP is derived from versican because of the finding that peptide sequences in the 60-kDa glial hyaluronic acid-binding protein of brain[26] are virtually identical to sequences within the amino-terminal hyaluronic acid-binding domain of versican. This origin of GHAP/hyaluronectin has been supported by the

[20] R. K. Margolis and R. U. Margolis, Experientia 49, 429 (1993).
[21] A. Bignami and R. Asher, Int. J. Dev. Neurosci. 10, 45 (1992).
[22] N. Girard, M. N. Courel, A. Delpech, G. Bruckner, and B. Delpech, Histochem. J. 24, 21 (1992).
[23] J. A. Ripellino, M. M. Klinger, R. U. Margolis, and R. K. Margolis, J. Histochem. Cytochem. 33, 1060 (1985).
[24] J. A. Ripellino, M. Bailo, R. U. Margolis, and R. K. Margolis, J. Cell Biol. 106, 845 (1988).
[25] R. K. Margolis, Ciba Found. Symp. 143, 221 (1989).
[26] G. Perides, W. S. Lane, D. Andrews, D. Dahl, and A. Bignami, J. Biol. Chem. 264, 5981 (1989).
[27] B. Delpech and C. Halavent, J. Neurochem. 36, 855 (1981).

isolation from human brain, in a yield of ~0.2 mg/100 g fresh weight, of a 365-kDa chondroitin sulfate proteoglycan with a 345-kDa core protein that binds to hyaluronic acid and contains N-terminal amino acid sequences previously reported for versican and GHAP.[28] The isolation procedure involves homogenization of brain in 3 vol of 10 mM Tris–acetate buffer, pH 7.4, containing protease inhibitors [5 mM ethylenediaminetetraacetic acid (EDTA), 1 mM phenylmethylsulfonyl fluoride (PMSF), 0.1 M aminohexanoic acid, and 5 mM benzamidine]. The homogenate is centrifuged for 10 min at 15,000 g, and the supernatant is recentrifuged under the same conditions and made to 7 M urea and 0.2% (v/v) Triton X-100 at pH 7.4. The extracted proteins are adsorbed on DEAE-Sepharose (1 ml/g brain), which is then washed with 0.22 M NaCl, and the proteoglycan-containing fraction is eluted with 0.5 M NaCl. The pooled fractions are concentrated by ultrafiltration and the proteoglycan is further purified by gel filtration on Sephacryl S-500. Fractions are monitored by sodium dodecyl sulfate–polyacrylamide gel electrophoresis (SDS–PAGE). Although only a single component, with a molecular size of ~365 kDa, is recognized by the 5D5 monoclonal antibody to versican, it is not clear whether other proteoglycans of larger molecular size that do not enter the 5–15% gradient gel may also be present in the purified proteoglycan fraction, and account for the significant amount of silver-staining material seen at the top of the gel.

Certain inconsistencies in the immunocytochemical localization of hyaluronectin, GHAP, and the 365-kDa proteoglycan have led to some confusion in this area and were originally used to support the contention that GHAP and hyaluronectin were not derived from a larger proteoglycan core protein. Although GHAP/hyaluronectin may also be present in brain together with versican, either as a result of *in vivo* proteolytic processing or conceivably as a small alternatively spliced product of the versican gene, the most convincing evidence for the presence of the smaller N-terminal fragment in tissues is that GHAP but not versican immunoreactivity is removed from tissue sections by hyaluronidase treatment.[28] However, because none of the monoclonal antibodies raised against and reactive with human GHAP recognized human versican on immunoblots,[28,29] the exact specificity of these "anti-GHAP" monoclonal antibodies remains unclear. To help resolve these questions, it would be important to demonstrate that GHAP can also be isolated under conditions designed to preclude proteolysis, and that purified versican added to brain homogenates

[28] G. Perides, F. Rahemtulla, W. S. Lane, R. A. Asher, and A. Bignami, *J. Biol. Chem.* **267,** 23883 (1992).
[29] A. Bignami, G. Perides, and F. Rahemtulla, *J. Neurosci.* **34,** 97 (1993).

can be recovered intact under such conditions. It should also be noted that the presence of N-terminal *in vivo* proteolytic processing fragments of versican would be consistent with the demonstration of similar hyaluronic acid-binding fragments derived from neurocan (see N-Terminal Fragments of Neurocan, below).

Neurocan

Neurocan is an ~300-kDa chondroitin sulfate proteoglycan that was originally isolated from rat brain using the 1D1 monoclonal antibody (MAb), and is developmentally regulated with respect to its molecular size, concentration, carbohydrate composition, sulfation, and immunocytochemical localization.[30] The proteoglycan isolated from 7-day postnatal brain contains an average of three 22-kDa chondroitin sulfate chains, and after chondroitinase digestion yields a major core glycoprotein with an apparent molecular size of 245 kDa on 5% SDS–PAGE. A smaller proteoglycan species with a 150-kDa core glycoprotein is also present at 7 days, and by 2–3 weeks postnatal this becomes the major or exclusive component, containing a single 32-kDa chondroitin 4-sulfate chain. The concentration of proteoglycan isolated with the 1D1 monoclonal antibody decreases during development, from 20% of the total chondroitin sulfate proteoglycan protein (0.1 mg/g brain) at 7 days postnatal to 6% in adult brain. During development, chondroitin 6-sulfate, which accounts for 20% of the chondroitin sulfate at 1 week postnatal, is replaced almost exclusively by chondroitin 4-sulfate, comprising >97% of the total glycosaminoglycan in the "adult form" of the 1D1 proteoglycan, which represents the C-terminal half of neurocan (see below). This developmental change in chondroitin sulfate 4- vs 6-sulfation is seen to an even greater extent in the case of the 3F8 proteoglycan, whereas the structurally related 3H1 chondroitin/keratan sulfate proteoglycan contains exclusively chondroitin 4-sulfate in both early postnatal and adult brain.[30] 3-Sulfated HNK-1 carbohydrate epitopes, which occur in a number of neural cell adhesion molecules, are present at all developmental stages, but the novel O-glycosidic mannose-linked oligosaccharides that have previously been characterized in the chondroitin sulfate proteoglycans of brain[31,32] are essentially undetectable in neurocan from early postnatal brain but increase significantly during development. The biochemical properties of neurocan are summa-

[30] U. Rauch, P. Gao, A. Janetzko, A. Flaccus, L. Hilgenberg, H. Tekotte, R. K. Margolis, and R. U. Margolis, *J. Biol. Chem.* **266,** 14785 (1991).

[31] T. Krusius, J. Finne, R. K. Margolis, and R. U. Margolis, *J. Biol. Chem.* **261,** 8237 (1986).

[32] T. Krusius, V. N. Reinhold, R. K. Margolis, and R. U. Margolis, *Biochem. J.* **245,** 229 (1987).

rized in Table I. Immunocytochemical studies demonstrated that like the total brain population of soluble chondroitin sulfate proteoglycans, the localization of which was previously examined using polyclonal antibodies,[33,34] staining with the 1D1 monoclonal antibody is generally most intense in the prospective white matter and absent from the external granule cell layer of early postnatal cerebellum, whereas in adult brain staining is strongest in the molecular layer and also present in the granule cell layer.

Primary Structure of Neurocan

A 5.2-kb composite sequence of overlapping cDNA clones contained an open reading frame of 1257 amino acids that encodes a 136-kDa protein containing 10 peptide sequences present in the adult and/or early postnatal forms of the proteoglycan.[4] The deduced amino acid sequence revealed a 22-amino acid signal peptide followed by an immunoglobulin domain, tandem repeats characteristic of the hyaluronic acid-binding region of aggregating proteoglycans, and an RGDS sequence. However, the RGDS sequence may not be functionally active insofar as a GRGDSP peptide (at 100 μg/ml) does not inhibit the binding of chick neurons to neurocan in a centrifugation assay.[35]

The C-terminal portion (amino acids 951–1215) has ~60% identity to regions in the C termini of versican and aggrecan, including two epidermal growth factor-like domains, a lectin-like domain, and a complement regulatory protein-like sequence (Fig. 1). The central 595-amino acid portion of neurocan has no homology with other reported protein sequences. The proteoglycan that can be isolated from adult brain with the 1D1 monoclonal antibody, and represents the C-terminal half of neurocan, contains a single 32-kDa chondroitin 4-sulfate chain linked at Ser-944, whereas three additional potential chondroitin sulfate attachment sites (only two of which are utilized) are present in the N-terminal portion of the larger proteoglycan species. A probe corresponding to a region of neurocan having no homology with versican or aggrecan hybridized with a single band at 7.5 kb on Northern blots of mRNA from both 4-day and adult rat brain and from PC-12 pheochromocytoma cells, but not with muscle, kidney, liver, or lung mRNA. These findings indicate that the 1D1 MAb-reactive proteoglycan of adult brain, containing a 68-kDa core protein, probably originates by a developmentally regulated *in vivo* proteolytic processing of the 136-kDa species, which is predominant in early postnatal brain, because the results

---

[33] D. A. Aquino, R. U. Margolis, and R. K. Margolis, *J. Cell Biol.* **99**, 1117 (1984).
[34] D. A. Aquino, R. U. Margolis, and R. K. Margolis, *J. Cell Biol.* **99**, 1130 (1984).
[35] M. Flad and R. K. Margolis, unpublished results (1993).

TABLE I
BIOCHEMICAL PROPERTIES OF NEUROCAN ISOLATED USING 1D1
MONOCLONAL ANTIBODY

Property	7-Day brain	Adult brain
Average molecular size (gel filtration)	300 (180)[a] kDa	180 kDa
Core glycoprotein size (SDS–PAGE)	245 (150) kDa	150 kDa
Chondroitin sulfate chain size	22,000	32,000
Percentage of total CSPG[b] protein	20	6
Carbohydrate (% by weight)		
Chondroitin sulfate	22%	22%
4-Sulfate	80[c]	97
6-Sulfate	20	<3
Glycoprotein oligosaccharides	20%	17%
HNK-1 epitopes	+	+
Yield of mannitol after NaOH/NaBH$_4$[d]	<1	15

[a] Values in parentheses are for the less abundant component.
[b] CSPG, Chondroitin sulfate proteoglycan.
[c] As a percentage of total chondroitin sulfate.
[d] As a percentage of total mannose (reflecting proportion of mannosyl-O-serine/threonine-linked oligosaccharides).

of earlier studies had precluded the possibility that it was generated by proteolysis during the isolation of neurocan from adult brain.[30]

Interactions of Neurocan

Neurocan copurifies with a 45-kDa link protein and forms aggregates with hyaluronic acid similar to those known to occur with aggrecan and versican (Fig. 2). It also binds with high affinity ($K_d \sim 0.6$ nM) to the neural cell adhesion molecules Ng-CAM and N-CAM, inhibits their homophilic interactions, and inhibits neuronal adhesion and neurite outgrowth on both Ng-CAM and anti-Ng-CAM IgG substrates.[36,37] These findings indicate that interactions between neurocan and neural cell adhesion molecules may play important roles in developmental processes. *In situ* hybridization histochemistry demonstrated that the proteoglycan is synthesized by neurons, and that neurocan mRNA is widely distributed in pre- and postnatal rat nervous tissue but not in other organs.[38]

[36] M. Grumet, A. Flaccus, and R. U. Margolis, *J. Cell Biol.* **120**, 815 (1993).
[37] D. R. Friedlander, P. Milev, L. Karthikeyan, R. K. Margolis, R. U. Margolis, and M. Grumet, *J. Cell Biol.* **125**, 669 (1994).
[38] M. Engel, R. U. Margolis, and R. K. Margolis, unpublished results (1993).

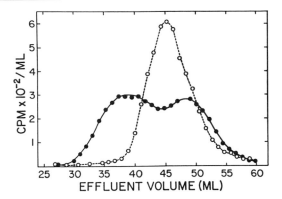

FIG. 2. [^{35}S]Sulfate-labeled neurocan isolated from 7-day brain and eluted from Sepharose CL-2B (0.9 × 90 cm column) with 0.5 M sodium acetate (pH 7.0) in the presence (●) and absence (○) of hyaluronic acid. To test aggregation with hyaluronic acid, 50 μg of proteoglycan is mixed for 18 hr at 4° with 3 mg of hyaluronic acid at a concentration of 2 mg/ml of 0.5 M sodium acetate, pH 7.0. The V_0 (elution of blue dextran) is at 21.5 ml and the V_i (glucose) elution is at 60 ml. No difference in the extent of aggregation is seen when the ratio of hyaluronic acid to proteoglycan is increased or decreased to twice or one-half of that used above. The proteoglycan that does not aggregate represents the C-terminal half of neurocan, which lacks a hyaluronic acid-binding region.

Proteolytic Fragments of Neurocan

Neurocan fragments that are resolved by the usual SDS–PAGE procedures can be produced by CNBr treatment or the use of specific proteases.[4,30] Some of these were originally used for peptide mapping and for obtaining internal amino acid sequences. When prepared from ^{125}I-labeled neurocan and employed in a radioligand-binding assay, a mixture of such fragments has also proved useful for localizing regions involved in the binding of neurocan to other proteins such as neural cell adhesion molecules, after elution of the labeled glycopeptide(s) followed by SDS–PAGE.[39]

CNBr treatment of chondroitinase-digested neurocan yields three major glycopeptides with apparent molecular sizes of 70, 24, and 13 kDa on SDS–PAGE.[4] The 70-kDa glycopeptide is derived from the N-terminal portion of the core protein (N terminus at Leu-39; numbering according to the complete sequence including the signal peptide,[4] whereas the N termini of the 24- and 13-kDa glycopeptides are Tyr-971 and Val-1129, respectively. A 65-kDa endoprotease Lys-C glycopeptide begins at almost the same position as the 70-kDa CNBr glycopeptide (after Lys-40), and

[39] P. Milev, M. Grumet, and R. U. Margolis, unpublished results (1993).

two other endo Lys-C glycopeptides (35 and 47 kDa) both begin after Lys-927. The smaller of these is almost identical to a 35-kDa endoprotease Asp-N glycopeptide (N terminus at Asp-925).

Identification of N-Terminal Fragments of Neurocan Using Monoclonal Antibody 1F6

Initial studies of neurocan involved the full-length core glycoprotein with an apparent molecular size on SDS–PAGE of 245 kDa, and the C-terminal half with an apparent molecular size of 150 kDa, both of which were isolated by immunoaffinity chromatography with the 1D1 monoclonal antibody.[4,30] Another monoclonal antibody (1F6) raised to the chondroitin sulfate proteoglycans of brain recognizes an N-terminal fragment of neurocan that appears to be derived by *in vivo* proteolytic processing, as described above.

On immunoblots of the total soluble chondroitin sulfate proteoglycans of brain after chondroitinase treatment, the 1F6 monoclonal antibody recognizes bands at 130 kDa and (in proteoglycans from early postnatal brain) 245 kDa. When the 1F6 monoclonal antibody is used for the immunoaffinity chromatographic isolation of proteoglycans from such a population of proteoglycan species, additional proteins with apparent molecular sizes of 45 and 90 kDa are seen on Coomassie blue-stained gels.

N-Terminal amino acid sequences of the 90- and 130-kDa bands correspond to the N-terminal sequence of neurocan. The proteoglycan fraction isolated by immunoaffinity chromatography using the 1F6 monoclonal antibody also contained significant amounts of hyaluronic acid (identified by the presence of glucosamine-containing unsulfated disaccharides after chondroitinase treatment), as well as a 45-kDa band with the N-terminal amino acid sequence of link protein, and which is reactive with the 8A4/5, but not the 1F6, monoclonal antibodies.[40] The rat brain link protein corresponds to the smaller, predominant form of rat chondrosarcoma link protein,[41] rather than to the alternatively spliced transcript encoding an additional 53 amino acids that was later identified in rat chondrosarcoma cDNA clones.[42] These results demonstrate that link protein does not contain 1F6 epitopes, but was present in the mixture of total chondroitin sulfate proteoglycans purified by ion-exchange chromatography only because it was bound to hyaluronic acid.

---

[40] R. K. Margolis and R. U. Margolis, unpublished results (1993).

[41] P. J. Neame, J. E. Christner, and J. R. Baker, *J. Biol. Chem.* **261**, 3519 (1986).

[42] C. Rhodes, K. Doege, M. Sasaki, and Y. Yamada, *J. Biol. Chem.* **263**, 6063 (1988).

TABLE II
REACTIVITY OF MONOCLONAL ANTIBODIES WITH NEUROCAN, NEUROCAN-DERIVED
PEPTIDES, AND RAT BRAIN LINK PROTEIN[a]

MAb	Amino acids	Immunoreactive bands (kDa)				
		45 (link)	**90**	**130**	**150**	**245**
5C4	213–216	+	+	−	−	+
3B1	310–313	+	+	+	−	+
8A4/5	218–224	+	+	+	+	+
	316–321					
	725–729					
1F6		−	−	+	−	+
1D1		−	−	−	+	+

[a] The total soluble chrondroitin sulfate proteoglycans of rat brain, or proteins isolated using MAb 1F6, were electrophoresed and transferred to nitrocellulose. Apparent molecular sizes of bands reactive with the monoclonal antibodies are listed. The amino acids column indicates the location (where known) of monoclonal antibody epitopes/mimotopes in the neurocan sequence (see text). The 90-, 130-, 150-, and 245-kDa bands (indicated in bold) are derived from portions of the neurocan core protein that contain chondroitin sulfate attachment sites and appear only after chrondroitinase treatment. The 45-kDa band corresponds to link protein.

1F6 is therefore an anti-neurocan monoclonal antibody that recognizes the 245-kDa core protein as well as a 130 kDa N-terminal fragment, but not the C-terminal 150-kDa neurocan core glycoprotein. Conversely, whereas the 1D1 monoclonal antibody recognizes an epitope in the C-terminal half of neurocan, and which is present in the 150-kDa core glycoprotein, it does not recognize N-terminal proteolytic processing fragments of neurocan. When used in conjunction with 1D1, the 1F6 monoclonal antibody should therefore be useful for the immunocytochemical or immunochemical identification of N-terminal fragments of neurocan that are generated by developmentally regulated proteolytic processing,[30] but are not recognized by the 1D1 monoclonal antibody.

Other monoclonal antibodies to cartilage proteoglycans produced and characterized by Caterson and co-workers[43,44] are also useful in identifying neurocan peptides and associated proteins with which they share epitopes

[43] B. Caterson, in "Biology of Proteoglycans" (T. N. Wight and R. P. Mecham, eds.), p. 1. Academic Press, Orlando, FL, 1987.
[44] B. Caterson, personal communication (1993).

or mimotopes. These reactivities are summarized in Table II. 3B1 recognizes a DGSV sequence present in link protein and neurocan. We have found that in neurocan the 8A4 monoclonal antibody (and the 8A5 MAb, which has a similar specificity[44]) recognizes a PISGP mimotope[4] of the PITKP and PISRP epitopes in link protein. The 5C4 monoclonal antibody recognizes an RTVR mimotope in the 90-kDa band of neurocan. It also recognizes a (G)SVQ mimotope in link protein and weakly stains link protein in chondroitin sulfate proteoglycans of brain.

Isolation and Biochemical Characterization of Neurocan

Monoclonal antibodies are purified from ascites fluid by hydroxyapatite chromatography on HA-Ultrogel (IBF Biotechnics, Inc., Savage, MD). The ascites fluid is diluted with 9 vol of 10 mM phosphate buffer (pH 6.8) containing 0.02% (w/v) sodium azide, centrifuged, and applied at a rate of <1 column volume/hr to a column of HA-Ultrogel in a ratio of 1 ml of diluted ascites per milliliter of bed volume. The purified antibody is then eluted by a step gradient of 50 and 200 mM phosphate buffer (pH 6.8) containing 0.02% sodium azide, and elution is monitored by absorbance at 280 nm. The fractions eluted with 50 mM buffer are discarded, and the 200 mM eluate is used for preparation of immunoaffinity matrices of the 1D1 and 1F6 antibodies. After elution, the column is washed with 500 mM buffer before reuse for purification of the same antibody. Antibody eluates are reduced in volume (to a protein concentration of ~10 mg/ml) by pressure ultrafiltration on a Diaflo PM30 membrane (Amicon, Danvers, MA), exchanged into 50 mM phosphate-buffered saline (PBS, pH 7.2) containing 0.02% sodium azide, and stored at 4°.

CNBr-activated Sepharose 4B (Pharmacia-LKB, Piscataway, NJ) is allowed to swell in 1 mM HCl and washed successively with 1 mM HCl and with coupling buffer (0.1 M NaHCO$_3$, pH 8.3, containing 0.5 M NaCl) on a sintered glass funnel according to manufacturer instructions. After dialysis into coupling buffer, the antibody is gently mixed for 2 hr at room temperature with the activated Sepharose at a ratio of 5 mg of antibody protein per milliliter of swollen gel. After removal of uncoupled antibody by filtration, the gel is washed with 1 M ethanolamine, pH 8.5, and gently mixed for 18 hr at 4°. The gel is then washed with four cycles of 0.1 M sodium acetate buffer, pH 4, followed by 0.1 M Tris-HCl buffer, pH 8, both containing 0.5 M NaCl, and stored at 4° in PBS containing 0.02% sodium azide.

For immunoaffinity isolation of neurocan, chondroitin sulfate proteoglycans isolated from a PBS extract of brain[45] are dissolved in 50 mM PBS (pH 7.2) containing 0.1% (v/v) 3-[(cholamidopropyl)-dimethyl-ammonio]-1-propanesulfonate (CHAPS) and mixed gently overnight at 4° with 0.5 to 1 vol of Sepharose beads, using a ratio of 1 mg of proteoglycan protein per milliliter of beads. After washing the beads several times by centrifugation to remove unbound proteoglycan, they are poured into a column and washed further with 50 mM PBS, followed successively by phosphate buffer containing 0.5 M NaCl, 50 mM PBS, 10 mM phosphate buffer (pH 8), and 50 mM diethylamine, pH 11.5. All solutions contain 0.1% CHAPS. The progress of the elution is followed by measurement of [^{35}S]sulfate radioactivity or absorbance at 280 nm, and the pH 11.5 eluates are immediately neutralized by addition of 1 M monobasic sodium phosphate. Proteoglycan-containing fractions are pooled, concentrated by pressure ultrafiltration on a PM30 membrane, and dialyzed thoroughly against water before lyophilization. Unbound proteoglycans can be concentrated and subsequently used for the isolation of other species by means of different antibody–Sepharose beads.

The core proteins are identified by 5% (w/v) SDS–PAGE after chondroitinase treatment of the proteoglycans. Digestion is performed for 45–60 min at 37° with protease-free chondroitinase ABC (Seikagaku America Inc., Rockville, MD) in 100 mM Tris-HCl buffer (pH 8.0 at 37°C) containing 30 mM sodium acetate, using a ratio of 0.5 mU chondroitinase/μg proteoglycan protein. For quantitation of chondroitin sulfate, proteoglycans are digested with chondroitinase ABC (chondroitin ABC lyase, EC 4.2.2.4) and filtered on a Centricon-30 membrane (Amicon) in 0.1 M NaCl. An aliquot of the released disaccharides is desalted by gel filtration on Sephadex G-15 and hydrolyzed for galactosamine analysis as described below. Aliquots of these samples can also be used for determination of chondroitin sulfate disaccharide isomers by high-performance liquid chromatography on an Ultrasil amino column (Beckman Instruments, San Ramon, CA) according to the procedure of Yoshida et al.[46] The elution positions of the unsulfated, and mono-, di-, and trisulfated chondroitin sulfate disaccharides are identified by use of standards obtained from Seikagaku America, Inc. (Rockville, MD). Chondroitin sulfate chain sizes can be determined after alkaline-borohydride treatment of proteoglycans followed by gel filtration on Sepharose CL-6B eluted with 0.1 M NaCl, with sizes calculated from the K_{av} values published by Wasteson.[47]

[45] W.-L. Kiang, R. U. Margolis, and R. K. Margolis, J. Biol. Chem. **256**, 10529 (1981).

[46] K. Yoshida, S. Miyauchi, H. Kikuchi, A. Tawada, and K. Tokuyasu, Anal. Biochem. **177**, 327 (1989).

[47] Å. Wasteson, J. Chromatogr. **59**, 87 (1971).

The monosaccharide composition of neurocan can be analyzed by high-performance anion-exchange chromatography with pulsed amperometric detection according to the general procedure of Hardy et al.[48] Sugars are separated on a column (4.6 × 250 mm) of Carbopac PAI pellicular anion-exchange resin equipped with a PA guard column (Dionex, Sunnyvale, CA), using a flow rate of 1 ml/min at ambient temperature. The analysis of monosaccharides is carried out at an isocratic NaOH concentration of 17 mM for 23 min. The column is regenerated for 5 min with 300 mM NaOH and equilibrated for 10 min with the starting eluant. The separated monosaccharides are detected by pulsed amperometric detection, with addition of 300 mM NaOH to the postcolumn effluent via a mixing tee (at a flow rate of 1 ml/min) to minimize baseline distortion.

Sialic acid is determined after hydrolysis in 0.1 N HCl for 1 hr at 80°, neutral sugars after hydrolysis in 2 N trifluoroacetic acid for 5 hr at 100°, and hexosamines after hydrolysis at 100° for 3 hr in 6 N HCl or 8 hr in 4 N HCl. After cooling, the acid is removed by rotary evaporation at 45° followed by several rinses with water, and the samples are dissolved in water for analysis. Because of the multiple hydrolyses required for quantitation of sialic acid, neutral sugars, and hexosamines, and for replicate chromatographic analysis of each hydrolysate, it is advantageous to begin with samples containing 5–10 nmol of each sugar, if possible, so that they can be redissolved in a convenient volume for column injection. However, a single determination requires <100 pmol.

To determine the yield of mannitol from O-glycosidic mannose-linked oligosaccharides,[31,32] proteoglycan samples are treated with 1 M NaBH$_4$ in 0.05 N NaOH for 16 hr at 45°. Excess NaBH$_4$ is destroyed with acetic acid, the sample is passed through a column of Dowex 50-X4 (H$^+$) (Bio Rad) to remove Na$^+$ ions, and boric acid is removed by repeated evaporations with HCl–methanol (1 : 1000, v/v). Samples are then hydrolyzed and analyzed for mannitol (which elutes at the beginning of the chromatogram just before fucose) as described above.

Immunocytochemical Localization of Proteoglycans

Neurocan can be specifically localized in tissue sections using the 1D1 or 1F6 monoclonal antibodies, whereas polyclonal antisera will cross-react with versican, brevican, and aggrecan. We describe here a general procedure for immunostaining Vibratome sections of postnatal brain using peroxidase-conjugated second antibodies in conjunction with diaminoben-

---

[48] M. R. Hardy, R. R. Townsend, and Y. C. Lee, *Anal. Biochem.* **170**, 54 (1988).

zidine. An alternate procedure for embryonic tissue is immunofluorescence staining of cryostat sections prepared as described in the later description of *in situ* hybridization histochemistry.

Sprague-Dawley rats are anesthetized with sodium pentobarbital (6.5 mg/100 g body weight) and perfused through the left ventricle, first briefly with 0.1 M phosphate buffer (pH 7.4), followed by the picric acid–paraformaldehyde–glutaraldehyde fixative described by Somogyi and Takagi.[49] The tissue is removed and allowed to stand in fixative for 2 hr at room temperature before being placed at 4° overnight. After washing several times with PBS, the tissue is sectioned with a Vibratome to a thickness of 15–30 μm for light or electron microscopy. In certain cases, sections for electron microscopy are frozen in liquid nitrogen, and thawed slowly (2 hr at −20° followed by 2 hr at 4°) before being brought to room temperature. All sections are washed with 50 mM Tris-buffered saline (TBS, pH 7.6) before antibody staining.

For light microscopy, sections are washed for 30 min in TBS containing 0.5% (v/v) H_2O_2, three times (15 min each) in TBS, followed by 1 hr in TBS containing 5% (w/v) BSA, 1% (v/v) normal rabbit serum, and 0.2% (v/v) Triton X-100. They are then incubated overnight at 4° in TBS containing monoclonal antibody, 0.5% BSA, 0.1% normal rabbit serum, and 0.2% Triton X-100. After washing for 1 hr at room temperature in several changes of TBS without Triton X-100, the sections are incubated for 1.5 hr at room temperature in TBS containing 0.5% BSA, 0.1% normal rabbit serum, 0.2% Triton X-100, and peroxidase-conjugated rabbit anti-mouse immunoglobulins (1 : 200, Jackson Immunoresearch, West Grove, PA). The sections are then washed for 1 hr in several changes of TBS without Triton X-100, followed by a 12-min incubation with 0.05% (v/v) 3,3′-diaminobenzidine–0.01% (v/v) H_2O_2 and three 10-min washes in TBS.

The staining procedure for electron microscopy is similar to that described above, except that the incubation with 5% BSA–1% normal rabbit serum is performed overnight at 4°, the incubation with monoclonal antibody is for 48 hr at twice the concentration used for light microscopy, followed by 24 hr with second antibody and washing overnight with several changes of TBS. Triton X-100 is also omitted from all solutions. The peroxidase reaction product may be intensified by treatment of sections for 15 min with silver methenamine.[50]

Sections for light microscopy are mounted on gelatin-coated slides, dehydrated through a series of graded ethanols, cleared with xylene, and

[49] P. Somogyi and H. Takagi, *Neuroscience* **7,** 1779 (1982).
[50] E. M. Rodriquez, R. Yulis, B. Peruzzo, G. Alvial, and R. Andrade, *Histochemistry* **81,** 253 (1984).

sealed in Permount (Fisher Scientific Co., Pittsburgh, PA). For electron microscopy, sections are fixed for 2 hr at room temperature in 0.1 M phosphate buffer (pH 7.4) containing 2% (v/v) glutaraldehyde, washed for 1–2 hr or overnight in buffer, then treated for 2 hr at 4° with 1% (v/v) osmium tetroxide, washed again for 1 hr with several changes of buffer, dehydrated through a series of graded ethanols, and embedded in Epon for thin-sectioning.

Localization of mRNA by *in Situ* Hybridization Histochemistry

The cellular sites of synthesis of proteoglycans such as neurocan, which are secreted into the extracellular space, cannot be reliably inferred from their immunocytochemical localization in tissue sections, or from their association with particular cell types in mixed primary cultures. Association with cultured cells may occur by specific or nonspecific binding of secreted proteoglycan, and anatomical gradients of proteoglycan expression during development could similarly result in extensive redistribution of proteoglycan to areas of tissue sections where the mRNA is not present. These questions can therefore be answered only by examining the biosynthesis of proteoglycan by homogeneous and well-characterized primary cultures of individual cell types (e.g., neurons, astrocytes, and oligodendroglia) or, in the case of cloned proteoglycans, by *in situ* hybridization histochemistry. We have chosen this latter approach both because of the greater amount of information that can be obtained, and in view of uncertainty as to how well isolated and often immature cultured cells reflect the biosynthetic capabilities of neurons and glia *in vivo*.

We have used a digoxigenin-labeled riboprobe because it allows a more precise cellular localization of mRNA and yields results more rapidly (a total of 3–4 days) as compared to radioactive probes. Single-stranded probes of this type also offer other advantages over double-stranded cDNA probes, including their higher hybridization efficiency and the possibility of using sense RNA for control of specificity. Because digoxigenin is produced only by digitalis and is not present in animal tissues, it does not lead to nonspecific background labeling of the type seen with other non-radioactive probes such as those employing biotin. Our *in situ* hybridization studies have demonstrated that neurocan is synthesized predominantly or exclusively by neurons, such as cerebellar granule cells.[38]

Rat embryos are fixed for 4–6 hr at room temperature in 4% (v/v) paraformaldehyde–0.1 M PBS, washed twice for 5 min in PBS, and cryoprotected by gentle shaking overnight at 4° in 15% (w/v) sucrose–0.1 M PBS before freezing on powdered dry ice, whereas postnatal brain is frozen directly after dissection. Cryostat sections (15–20 μm) are thaw-mounted on poly(L-lysine)-subbed slides and quickly dried under a stream of cool air. The tissue is fixed for 5 min in 3% paraformaldehyde in 0.1 M PBS (pH 7.4), and rinsed twice for 1 min in PBS and twice for 1 min in 2× SSC (1× SSC is 0.15 M NaCl plus 0.015 M sodium citrate). All solutions are made with diethyl pyrocarbonate-treated water. Sections are then treated for 3 min at 37° with proteinase K (1 μg/ml in 0.1 M Tris-HCl, pH 8.0, containing 50 mM EDTA), rinsed twice with water, and acetylated for 10 min with 0.25% (v/v) acetic anhydride in 0.1 M triethanolamine, pH 8.0. After rinsing for 1 min each in 2× SSC and 0.1 M PBS, sections are incubated for 30 min in glycine–PBS (2 mg/ml, pH 7.0), rinsed for 1 min in 2× SSC, dehydrated through graded ethanols, and air dried.

We have used a riboprobe corresponding to a 307-bp PstI/BamHI restriction fragment of neurocan (nucleotides 2624–2930, just preceding the EGF-like domains homologous with versican, brevican, and aggrecan) subcloned into pGEM3Zf. The plasmid is linearized with HindIII and transcribed into digoxigenin-labeled antisense RNA with T7 RNA polymerase, using the Genius 4 RNA-labeling kit (Boehringer Mannheim, Indianapolis, IN). Control sections are treated with either a labeled sense probe transcribed from the same plasmid, or with labeled antisense probe diluted with a 10-fold excess of unlabeled probe.

Sections are covered with digoxigenin-labeled riboprobe, which is diluted to a concentration of 0.2–1 ng/ml in 30 μl of hybridization solution [40% (v/v) formamide, 10% (w/v) dextran sulfate, 1× Denhardt's solution, 4× SSC, 10 mM dithiothreitol, yeast RNA (1 mg/ml), and sheared salmon sperm DNA (1 mg/ml)] and denatured for 10 min at 65°. Hybridization is carried out overnight at 52° in an atmosphere of 40% formamide, after which sections are washed for 5 min at 52° with 50% formamide–2× SSC followed by a further 20-min wash using fresh solution. Slides are rinsed twice for 1 min at room temperature with 2× SSC, treated for 30 min at 37° with RNase A (20 μg/ml 2× SSC), and rinsed twice for 1 min in 2× SSC. After incubation for 5 min at 52° in 50% formamide–2× SSC, sections are treated for 2 hr at room temperature with 2% blocking reagent (Cat. No. 1096176; Boehringer Mannheim) in 0.1 M sodium maleate buffer (pH 7.5) containing 0.15 M NaCl.

The blocking solution is removed and sections are incubated for 1 hr at room temperature with alkaline phosphatase-conjugated sheep anti-digoxigenin Fab fragments (Boehringer Mannhein) diluted 1 : 500 in 2% (v/v) blocking solution containing 0.3% (v/v) Triton X-100. Slides are washed for 10 min in 0.1 M TBS, pH 7.5, followed by 30 min in 0.1 M

Tris-HCl (pH 9.5) containing 0.1 M NaCl and 50 mM MgCl$_2$. Sections are covered with chromogen solution [nitro blue tetrazolium (338 μg/ml), 5-bromo-4-chloro-3-indolylphosphate (175 μg/ml), and levamisole (240 μg/ ml) in 0.1 M Tris-HCl, pH 9.5, containing 0.1 M NaCl and 50 mM MgCl$_2$] and incubated in the dark at room temperature in a humidified chamber until color development is complete, after which slides are washed twice for 5 min in 10 mM Tris-HCl (pH 8.1) containing 1 mM EDTA. The sections are then dehydrated through a series of graded ethanols, cleared in xylene, coverslipped, and sealed in Permount (Fisher Scientific Co.).

Interactions of Proteoglycans with Other Proteins and Cells

Chondroitin sulfate proteoglycans are known to have effects on cell adhesion, cell migration, and neurite growth.[20,51,52] Several types of interactions have been demonstrated between chondroitin sulfate proteoglycans of brain and neurons, or neural cell adhesion and extracellular matrix molecules.[36,37,53] These studies suggest that such processes may provide a general mechanism for modulating homophilic and heterophilic binding between neurons and glia and to extracellular matrix molecules during nervous tissue development. This section briefly summarizes some of the methods that can be employed for investigating the interactions of proteoglycans with other proteins and with cells.

Covasphere Aggregation Assays

The Covasphere aggregation assay is relatively simple and sensitive, employs protein-coated fluorescent microbeads (Covaspheres; Duke Scientific Corp., Palo Alto, CA) that are available in different colors, and is useful for demonstrating homophilic or heterophilic interactions between proteins, and for investigating potential effects of other molecules that may perturb such interactions.

Proteins (50 μg) are covalently coupled to 200 μl of 0.5-μm Covaspheres, washed twice in PBS containing BSA (1 mg/ml)–10 mM NaN$_3$, and resuspended in the original volume of buffer (200 μl) as described[54]; Covaspheres as supplied by the manufacturer are at a concentration of 850 cm^2 of surface area per milliliter. Quantitative measurements indicate that under these conditions ~20% of proteins such as the neuron–glia

[51] E. Ruoslahti, *J. Biol. Chem.* **264,** 13369 (1989).
[52] T. N. Wight, M. G. Kinsella, and E. E. Qwarnstrom, *Curr. Opin. Cell Biol.* **4,** 793 (1992).
[53] M. Grumet, P. Milev, T. Sakurai, L. Karthikeyan, M. Bourdon, R. K. Margolis, and R. U. Margolis, *J. Biol. Chem.* **269,** 12142 (1994).

cell adhesion molecule (Ng-CAM) is bound to the Covaspheres. For Covasphere aggregation experiments, prior aggregates in the bead preparations are first dissociated by sonication for 10–20 sec, and 6-μl aliquots (containing ~0.3 μg of protein) are mixed with varying amounts of soluble protein in 54 μl of PBS, such as to give a final protein concentration in the range of 0–30 μg/ml. After a 30-min incubation on ice, the samples are resonicated and aggregation is monitored at 25°. The appearance of superthreshold aggregates of Covaspheres can be measured using a Coulter counter (Coulter, Hialeah, FL) fitted with a 100-μm aperture set at an amplification of 0.17, aperture current of 0.33, threshold 10–100; these settings allow detection of particles >~4 μm.[54,55] Superthreshold aggregates are measured in samples of 20 μl that are diluted into 20 ml of PBS. Microscopic observations of Covasphere preparations that had been measured in the Coulter counter indicate that populations containing aggregates of 5–10 Covaspheres are subthreshold, inasmuch as they gave measurements that are not significantly greater than the background levels for buffer alone (range, 300–1000). It is difficult to resolve the numbers of individual Covaspheres present in larger aggregates.

Radioligand-Binding Assay

The binding of proteoglycans to other proteins can be quantitated using a radioligand-binding assay.[37] Proteoglycans are labeled with ^{125}I by the lactoperoxidase–glucose oxidase method, using Enzymobeads (Bio-Rad, Richmond, CA), typically using 50 μg of protein and 1 mCi of ^{125}I per reaction, and free iodine is removed by gel filtration on a PD-10 column (Pharmacia, Piscataway, NJ). One to 30 μg of soluble protein in 16 mM Tris (pH 7.2)–2 mM CaCl$_2$–2 mM MgCl$_2$–0.02% NaN$_3$ (binding buffer, which also contains 50–150 mM NaCl) is adsorbed to removable Immulon-2 wells (Dynatech, Chantilly, VA) by overnight incubation at room temperature. Unbound proteins are removed by three washes in binding buffer–0.02% (v/v) Tween 20, and the wells are blocked by incubation with heat-treated bovine serum albumin (BSA; 1 mg/ml) in binding buffer. Wells are then emptied and 50 μl/well of labeled proteins or mixtures of labeled and unlabeled proteins (in binding buffer containing 1 mg of BSA/ml) are incubated for 2 hr at room temperature. Unbound proteoglycan is removed by four washes with TBS [50 mM Tris (pH 7.2)–150 mM NaCl–0.02% (v/v) Tween 20], and radioactivity bound to wells is

---

[54] M. Grumet and G. M. Edelman, *J. Cell Biol.* **106**, 487 (1988).
[55] S. Hoffman and G. M. Edelman, *Proc. Natl. Acad. Sci. U.S.A.* **84**, 2533 (1987).

determined with a γ counter. Scatchard plots can be generated and the K_d determined using the Ligand program.[57]

Gravity Cell-Binding Assay

The gravity cell-binding assay is useful for measuring the effects of proteoglycans on cell adhesion and can be modified in many ways for specific purposes.[37,58] In a typical application, neurons are prepared from 9-day chicken embryo brains[54] and cell binding to protein-coated dishes is measured. Two-microliter aliquots of protein in PBS are placed in a circular array near the center of a polystyrene petri dish. After 30 min, the dishes are washed three times with PBS containing BSA (10 mg/ml). After a 1-hr incubation, the BSA blocking solution is removed and replaced with 5×10^5 cells in 0.25 ml of Eagle's minimal essential medium with spinner salts (GIBCO Laboratories, Grand Island, NY) containing deoxyribonuclease I (50 μg/ml), and incubated for 1 hr at 37°. The unbound cells are removed by four gentle washes in PBS and a fifth wash in medium. The cells are fixed with 3.7% formalin for later examination. Dishes are observed by phase-contrast microscopy, and cells are counted in four fields (2.2 mm^2) within the protein-coated area.

The effects of proteoglycans on neurite growth can also be determined using an extension of this assay. Brain cells (10^5) are incubated for 2 days under the same conditions used for the cell adhesion assay, and are fixed with formalin. Neurite length is defined as the distance between the furthest removed neurite tip and the cell body. Quantitation is performed under phase-contrast microscopy with the help of a measuring eyepiece.

Centrifugation Cell-Binding Assay

To determine whether cells can bind directly to proteoglycans one can use a centrifugation assay that has previously been employed to demonstrate binding of cells to tenascin/cytotactin, an extracellular molecule that can inhibit cell adhesion and migration.[58,59] Because of its short duration, the centrifugation assay reflects more closely the strength of the initial molecular interactions than does the gravity assay, which is highly

[56] Deleted in proof.

[57] P. J. Munson and D. Rodbard, *Anal. Biochem.* **107,** 220 (1980).

[58] D. R. Friedlander, S. Hoffman, and G. M. Edelman, *J. Cell Biol.* **107,** 2329 (1988).

[59] K. L. Crossin, A. L. Prieto, S. Hoffman, F. S. Jones, and D. R. Friedlander, *Exp. Neurol.* **109,** 6 (1990).

sensitive to the ability of the immobilized molecule to allow cell spreading. Thus, proteins that bind to cells but cause a "repulsive" reaction (as is the case with neurocan and other brain proteoglycans such as phosphacan[53,59a]) may demonstrate cell binding in the centrifugation assay even though no binding can be observed in the gravity assay.[36]

For the centrifugation assay, proteins are adsorbed for 1 hr to U-shaped wells of 96-well polyvinyl chloride microtiter plates, which are then washed and blocked with BSA.[58] Two hundred microliters of a cell suspension containing 5×10^4 cells is placed in each well and the plate is centrifuged at ~250 g for 2 min at room temperature. The pattern of cells in each well is observed under dark-field illumination, and reflects a balance between the centrifugal force and the adhesivity of the substrate. On nonadhesive substrates, the cells are centrifuged into a pellet at the bottom of the well and, as the adhesivity of the substrate increases, more cells bind to the substrate along the wall of the well, which can be detected in a circular area with a measurable diameter. On strongly adhesive substrata, cells are distributed more or less uniformly on the well.

Conclusions

Studies of nervous tissue, including the recent characterization of brevican[6a], suggest that the aggrecan–versican–neurocan family of proteoglycans may still be growing, as will become clearer after the primary structures of the Cat-301 and T1 proteoglycans mentioned earlier are known. It would also be desirable to have well-established methods for the isolation of versican and brevican, which will be necessary in order to obtain a more complete picture of their biochemical properties and how they may differ between tissues, as well as for comparison with aggrecan and neurocan in cell biological studies. Because chick brain aggrecan appears to differ in its size, glycosylation, and other properties from chick cartilage aggrecan,[60] it will probably also be necessary to utilize the tissue-specific forms of these proteoglycans in such studies in order to draw meaningful conclusions concerning their probable biological functions. In view of their multidomain primary structures and developmental regulation, it can be expected that there will be increasing interest in exploring the functional roles of native aggrecan, versican, neurocan, brevican, and related proteoglycans, of their core glycoproteins obtained by chondroitinase treatment,

[59a] P. Maurel, U. Rauch, M. Flad, R. K. Margolis, and R. U. Margolis, *Proc. Natl. Acad. Sci. U.S.A.*, **91**, 2512 (1994).

[60] R. C. Krueger, A. K. Hennig, and N. B. Schwartz, *J. Biol. Chem.* **267**, 12149 (1992).

and of recombinant protein domains in such processes as cell interactions and neural histogenesis.

Acknowledgments

We thank Dr. Bruce Caterson for unpublished information concerning epitopes recognized by his monoclonal antibodies to aggrecan. Research cited in this chapter that was performed in the authors' laboratories was supported by grants from the National Institutes of Health (NS-09348, NS-13876, and MH-00129).

Section II

Receptors

[7] Characterization of Laminin-Binding Integrins

By RANDALL H. KRAMER

This chapter addresses the approaches used to study the interaction of integrin adhesion receptors with laminin and its isoforms. Sample protocols are given for studies of integrin binding to laminin or its fragments that should be readily applicable to other laminin variants. Methods are outlined for adhesion assays in which the interaction of surface receptors can, with immobilized laminin or laminin fragments, be probed with available anti-integrin blocking antibodies or peptides. Protocols for the identification and isolation of integrin receptors using ligand-affinity chromatography are described. Although much of the work to date has concerned studies with the Engelbreth-Holm–Swarm (EHS)-derived murine laminin containing the A–B1–B2 chains, future work will undoubtedly examine the interaction of integrins with the newly discovered laminin isoforms.

Overview of Laminin Receptors

Laminin is a major adhesive glycoprotein found in all basement membranes.[1,2] A number of integrin receptors have been shown to mediate adhesion to laminin. In addition to integrins, numerous nonintegrin receptors have also been identified. This diverse group of macromolecules includes the 67-kDa family of elastin-binding receptors, mammalian lectins, galactosyltransferase, and several apparently related 110- to 120-kDa proteins (cranin, LBP-120, and dystroglycan). Several reviews deal with these groups of nonintegrin receptors.[3–5]

The integrins are an important group of transmembrane heterodimeric adhesion receptors that mediate attachment of cells to extracellular matrices, including basement membranes, and form linkages between the cytoskeleton and the extracellular environment.[6,7] Integrins are formed by the noncovalent association of α and β subunits. This pairing of α and β partners forms the ligand-binding site at the amino-terminal globular re-

[1] R. Timpl, *Eur. J. Biochem.* **180,** 487 (1989).
[2] K. Beck, I. Hunter, and J. Engel, *FASEB J.* **4,** 148 (1990).
[3] R. P. Mecham, *Annu. Rev. Cell Biol.* **7,** 71 (1991).
[4] R. P. Mecham, *FASEB J.* **5,** 2538 (1991).
[5] M. Mercurio and L. M. Shaw, *BioEssays* **13,** 469 (1991).
[6] R. O. Hynes, *Cell (Cambridge, Mass.)* **48,** 549 (1987).
[7] E. Ruoslahti and M. D. Pierschbacher, *Science* **238,** 491 (1987).

METHODS IN ENZYMOLOGY, VOL. 245
Copyright © 1994 by Academic Press, Inc.
All rights of reproduction in any form reserved.

gions of the integrin subunits.[8,9] Several of the β1 series of integrins are important receptors for laminin. Figure 1 summarizes the current information on the integrin-binding sites for EHS laminin.

The first laminin to be extensively characterized was isolated from rodent tumor cells and was found to be composed of A, B1, and B2 chains and is found in various tissues such as skin and blood vessels.[10,11] This prototypic laminin is a high molecular weight trimer with multiple domains that are involved in the interaction with other basement membrane components such as entactin/nidogen, type IV collagen, and heparan sulfate proteoglycan (reviewed by Beck *et al.*[2] and Engel[12]). Several other laminin isoforms have been identified and characterized to varying degrees and include merosin, S-laminin, S-merosin, kallinin/epiligrin, and K-laminin (see [5] in this volume). These isoforms of laminins usually are composed of a heavy chain (A or M) and two homologous light chains (B1, B2, and S). Studies have confirmed that several forms of laminin have truncated A-like chains. Most likely there are other variants yet to be identified.

Laminin-rich basement membranes are sheets of extracellular matrix that provide anchorage for various cell types, underlie epithelia and endothelium, and surround fat and muscle cells.[13] In addition to contributing to the structure, stability, and physical characteristics of basement membranes, laminin also has important biological properties. Most laminin isoforms have been shown to be adhesive and can induce a multitude of cellular responses that can include migration, cell polarization, and neurite outgrowth, and influence proliferation and differentiation.[1-3,14] Furthermore, the different laminin isoforms are tissue specific and are expressed in a developmentally regulated pattern. The biological response to laminin appears to be cell type specific and this appears to be due in part to the specific receptors that are expressed by individual cells.

On the basis of work from the well-characterized classic laminin isolated from the murine EHS tumor, it is established that laminins have multiple functional domains. An important approach to identify the multiple sites with adhesive or inductive activity has been to use specific frag-

[8] E. Ruoslahti, *J. Clin. Invest.* **87**, 1 (1991).

[9] R. O. Hynes, *Cell (Cambridge, Mass.)* **69**, 11 (1992).

[10] A. E. Chung, R. Jaffe, I. L. Freeman, J. P. Vergues, J. E. Graginski, and B. Cartin, *Cell (Cambridge, Mass.)* **16**, 277 (1979).

[11] R. Timpl, H. Rohde, P. Gehron Robey, S. I. Rennard, J. M. Foidart, and G. R. Martin, *J. Biol. Chem.* **254**, 9933 (1979).

[12] J. Engel, *in* "Molecular and Cellular Aspects of Basement Membranes" (D. H. Rohrbach and R. Timpl, eds.), p. 239. Academic Press, San Diego, 1993.

[13] R. Vracko, *in* "New Trends in Basement Membrane Research" (K. Kuehn, ed.), p. 1. Raven Press, New York, 1982.

[14] G. R. Martin and R. Timpl, *Annu. Rev. Cell Biol.* **3**, 57 (1987).

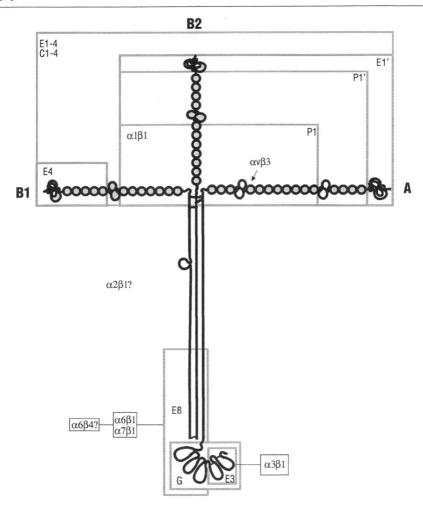

FIG. 1. Location of integrin-binding sites in the mouse A-chain laminin. Current proposed binding sites on laminin for the major laminin-binding integrins. Schematic of laminin and its proteolytically derived fragments is modified from Beck *et al.*[2] and Engel.[12] $\alpha 1\beta 1$ has been shown to bind to the inner-cross region of laminin.[22,24,26] The site for binding of $\alpha 2\beta 1$ on laminin has not yet been determined. For $\alpha 3\beta 1$, the binding site is located in the globular end of the long arm of laminin[19] and one report has identified an active peptide from the G5 domain found in the E3 fragment.[20] The $\alpha 6\beta 1$ and $\alpha 7\beta 1$ integrins have been shown to bind to the E8 fragment,[21–25] the $\alpha v\beta 3$ integrin has been shown to bind to a cryptic site exposed in the P1 fragment,[23,27,28] and possibly other non-RGD-sensitive sites.[36,55,56] Although inactive in many cell types, $\alpha 6\beta 4$ appears to bind to laminin and indirect evidence suggests it binds to the E8 fragment.[63]

ments generated by limited proteolytic digestion (e.g., with elastase or pepsin). These isolated fragments have then been used to study their adhesive and other biologically inductive properties.[15–18]

Using this approach, a dominant adhesion- and migration-promoting region in the A–B1–B2 forms of laminin has been located in the long arm near the globular end. In EHS laminin this activity is found in the elastase-derived E8 fragment.[15–17] For both human and murine A-chain laminin, several integrin receptors have been shown to bind to this region of laminin, including $\alpha 3\beta 1$,[19,20] $\alpha 6\beta 1$,[21–23] and $\alpha 7\beta 1$.[24,25]

Another major cell attachment site is present in the short arms of laminin that comprise a cross structure and it appears to be the major binding site for the $\alpha 1\beta 1$ integrin.[22,24,26] In the same inner-cross region, there is a cryptic RGD sequence contained in the A chain that becomes exposed after pepsin digestion and promotes attachment via the $\alpha v\beta 3$ receptor.[23,27,28]

Sources of Laminin and Laminin Isoforms

Laminin isolated from the mouse EHS tumor matrix has been extensively used for studies of receptor binding and for the preparation of laminin fragments.[11,29] This has been a major source for laminin because

[15] R. Timpl, S. Johansson, V. van Delden, I. Oberbaumer, and M. Hook, *J. Biol. Chem.* **258**, 8922 (1983).

[16] S. L. Goodman, R. Deutzmann, and K. von der Mark, *J. Cell Biol.* **105**, 589 (1987).

[17] D. Edgar, R. Timpl, and H. Thoenen, *EMBO J.* **2**, 1463 (1987).

[18] V. Nurcombe, M. Aumailley, R. Timpl, and D. Edgar, *Eur. J. Biochem.* **180**, 9 (1989).

[19] K. R. Gehlsen, L. Dillner, E. Engvall, and E. Ruoslahti, *Science* **241**, 1228 (1988).

[20] K. R. Gehlsen, P. Sriramarao, L. T. Furcht, and A. P. N. Skubitz, *J. Cell Biol.* **117**, 449 (1992).

[21] M. Aumailley, R. Timpl, and A. Sonnenberg, *Exp. Cell Res.* **188**, 55 (1990).

[22] D. E. Hall, L. F. Reichardt, E. Crowley, B. Holley, H. Moezzi, A. Sonnenberg, and C. H. Damsky, *J. Cell Biol.* **110**, 2175 (1990).

[23] A. Sonnenberg, C. J. Linders, P. W. Modderman, C. H. Damsky, M. Aumailley, and R. Timpl, *J. Cell Biol.* **110**, 2145 (1990).

[24] R. H. Kramer, M. P. Vu, Y. P. Cheng, D. M. Ramos, R. Timpl, and N. Waleh, *Cell Regul.* **2**, 805 (1991).

[25] H. von der Mark, J. Durr, A. Sonnenberg, K. von der Mark, R. Deutzmann, and S. J. Goodman, *J. Biol. Chem.* **266**, 23593 (1991).

[26] S. L. Goodman, M. Aumailley, and H. von der Mark, *J. Cell Biol.* **113**, 931 (1991).

[27] M. Aumailley, M. Gerl, A. Sonnenberg, R. Deutzmann, and R. Timpl, *FEBS Lett.* **262**, 82 (1990).

[28] A. Sonnenberg, K. R. Gehlsen, M. Aumailley, and R. Timpl, *Exp. Cell Res.* **197**, 234 (1991).

[29] R. Timpl, H. Rohde, L. Ristelli, U. Ott, P. Gehron Robey, and G. R. Martin, this series, Vol. 82, p. 831.

it is possible to isolate gram quantities of pure laminin per kilogram of EHS tumor. Several different protocols have been used for the purification of the EHS laminin (see [5] in this volume). A convenient method for preparation of biologically active laminin is the chelation protocol of Paulsson *et al.*,[30] which allows a mild extraction and purification of the laminin–entactin (nidogen) complex. Studies have indicated that this preparation of laminin yields an active ligand for receptor analysis. The more rigorous isolation methods originally described by Timpl *et al.*,[11,29] in which much of the contaminating entactin/nidogen is removed, also yields highly active laminin. Both laminin and the laminin–nidogen complex have given similar results in cell adhesion assays and in ligand-affinity chromatography assays.[31] However, the report that the $\alpha 3 \beta 1$ integrin can bind recombinant nidogen through an RGD-sensitive mechanism[32] should be considered when using laminin–nidogen complex for cell adhesion studies.

A major complication in the study of cell interaction with laminin is that there are multiple forms of this ubiquitous adhesive glycoprotein. It is likely that certain cell types may interact differently with each laminin isoform and there are indications that there may be species-specific differences in the structural and biological properties of each laminin type. An example of the different behavior observed for laminin isoforms is seen with human skin keratinocytes that adhere poorly to murine EHS laminin yet attach with high efficiency to human A-chain laminin and to human merosin (Fig. 2), or to human epiligrin/kalinin.[33,34] Analysis of adhesion to human A-chain laminin with specific anti-integrin blocking antibodies indicates that both $\alpha 2 \beta 1$ and $\alpha 3 \beta 1$ receptors contribute to keratinocyte adhesion (Fig. 2B). Apparently, these receptors are unable to bind to EHS laminin efficiently. In the case of $\alpha 3 \beta 1$, these results are similar to that previously reported for laminin affinity columns or in adhesion assays where this receptor binds well to human A-chain laminin or to epiligrin/kallinin but weakly or not at all to EHS laminin.[19,28,35-37] Similarly, $\alpha 6 \beta 1$

[30] M. Paulsson, M. Aumailley, R. Deutzman, R. Timpl, K. Beck, and J. Engel, *Eur. J. Biochem.* **123**, 63 (1987).

[31] M. Aumailley, V. Nurcombe, D. Edgar, M. Paulsson, and R. Timpl, *J. Biol. Chem.* **262**, 11532 (1987).

[32] S. Dehar, K. Jewell, M. Rojiani, and V. Gray, *J. Biol. Chem.* **267**, 18908 (1992).

[33] W. G. Carter, M. C. Ryan, and P. J. Gahr, *Cell (Cambridge, Mass.)* **65**, 599 (1991).

[34] P. Rousselle, G. P. Lunstrum, D. R. Keene, and R. E. Burgeson, *J. Cell Biol.* **114**, 567 (1992).

[35] M. J. Elices, L. A. Urry, and M. E. Hemler, *J. Cell Biol.* **112**, 169 (1991).

[36] R. H. Kramer, Y. F. Cheng, and R. Clyman, *J. Cell Biol.* **111**, 1233 (1990).

[37] J. B. Weitzman, R. Pasqualini, Y. Takada, and M. E. Hemler, *J. Biol. Chem.* **268**, 8651 (1993).

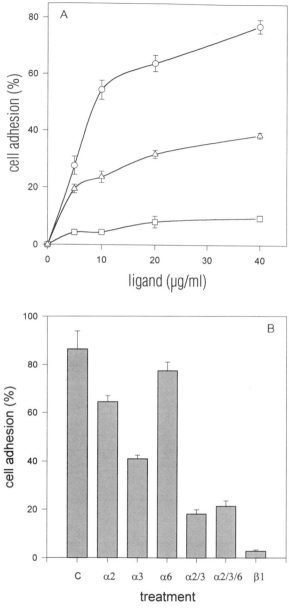

Fig. 2. Adhesion of human skin keratinocytes to laminin isoforms and inhibition by anti-integrin antibody. (A) Keratinocytes were plated on polystyrene microtiter wells coated with (□) mouse EHS laminin,[30] (○) human A-chain laminin,[40] and (△) human merosin[41] at the indicated coating concentrations. For attachment, cells were incubated for 30 min;

binds with relatively high affinity to ligand columns of human A-chain laminin but weakly to EHS laminin.[28,36,38] Nevertheless, $\alpha6\beta1$ promotes efficient adhesion of many cell types to immobilized EHS laminin. A study by Brown and Goodman[39] reported that the adhesion of different cell lines to human placental laminin purified by the chelator method is mediated by a set of receptors distinct from $\alpha6\beta1$. Another difference between the EHS laminin and human A-chain laminin is that the RGD sequence exposed following pepsin digestion of murine laminin is not present in the A chain of human laminin but is replaced by RAD.[12] Taken together these results indicate that more work is needed to identify potential differences between the activities of different laminin isoforms.

Although the EHS laminin is readily available, other isoforms of laminin are more difficult to isolate in pure form. Procedures are now available for the purification of human placental A-chain laminin (solubilized by mild pepsin digestion)[40] or human merosin [solubilized by ethylenediaminetetraacetic acid (EDTA) extraction][41]; both are commercially available. The epithelium-associated laminin (variously called epiligrin, kalinin, GB3 antigen, and nicein) and another variant, K-laminin, are not as yet completely characterized at the molecular level.[33,42]

Modulation of Integrin Activity

Studies have shown that frequently many integrin receptors require activation before they can bind ligand (reviewed by Hynes[9] and Ginsberg et al.[43,44]). The mechanisms that modulate receptor activity and specificity

[38] D. M. Ramos, E. D. Berston, and R. H. Kramer, *Cancer Res.* **50,** 728 (1990).

[39] J. C. Brown and S. L. Goodman, *FEBS Lett.* **282,** 5 (1991).

[40] U. Wewer, R. Albrechtsen, M. Manthorpe, S. Varon, E. Engvall, and E. Rouslahti, *J. Biol. Chem.* **258,** 12654 (1983).

[41] K. Ehrig, I. Leivo, W. S. Argraves, E. Ruoslahti, and E. Engvall, *Proc. Natl. Acad. Sci. U.S.A.* **87,** 3264 (1990).

[42] R. E. Burgeson, *in* "Molecular and Cellular Aspects of Basement Membranes" (D. H. Rohrbach and T. Timpl, eds.), p. 239. Academic Press, San Diego, 1993.

[43] M. H. Ginsberg, T. E. O'Toole, J. C. Loftus, and E. F. Plow, *Cold Spring Harbor Symp. Quant. Biol.* **57,** 221 (1992).

[44] M. H. Ginsberg, X. Du, and E. F. Plow, *Curr. Opin. Cell Biol.* **4,** 766 (1992).

background adhesion to control BSA substrates was less than 10% of mean values and has been subtracted from all datapoints. (B) Adhesion to A-chain human laminin. Anti-integrin antibodies against the following subunits were used at optimal inhibitory concentration, either alone or in combination: anti-$\alpha2$ (VM-1), anti-$\alpha3$ (P1B5), anti-$\alpha6$ (GoH3), and anti-$\beta1$ (AIIB2). Note that inhibition was maximal when antibodies to both $\alpha2$ and $\alpha3$ were present. For (A) and (B), values are the mean and SD of triplicate wells.

are not completely understood but certain evidence indicates that receptor function is controlled posttranslationally at different levels. In particular, integrin activity and ligand specificity appear to be under the control of the cellular microenvironment. In addition, postligand-binding events are regulated by the α and β cytoplasmic domains.[45] Numerous factors can alter receptor function including divalent cations, lipids, status of the cytoplasmic domain, and the binding of certain anti-β1 antibodies.[44] For example, the cell type-specific activity of $\alpha2\beta1$, which is (1) inactive or (2) binds collagen or (3) binds collagen and laminin, can be modified by phorbol esters and by activating antibodies to the β1 subunit.[46] Finally, work by Hogervorst et al.[47] indicates that the phosphorylation of the α6A cytoplasmic tail can downregulate $\alpha6A\beta1$ adhesion to laminin.

The type of divalent cation available to the integrin complex has a significant impact on the affinity and specificity of receptor for its ligand. There are numerous examples of modulation of integrin receptor activity by the specific divalent metal salts present in assays of cell adhesion and migration,[48] ligand-affinity chromatography,[36,48,49] or in solubilized receptor–ligand-binding assays.[50,51]

For ligand-affinity chromatography of laminin-binding integrins, optimal conditions usually include Mn^{2+} as the divalent cation. For many integrins the presence of Mn^{2+} will produce higher yields of receptor by increasing the affinity of the integrin for the ligand. However, as in the case of the $\alpha2\beta1$ receptor binding to laminin, Mg^{2+} and not Mn^{2+} yields the optimal assay conditions.[36] On the other hand, Ca^{2+} by itself is usually inhibitory for laminin-binding integrins, apparently by lowering the receptor–ligand affinity.[36] In fact, this inhibitory effect of Ca^{2+} can be used as a mild method to elute specific integrin selectively from a complex mixtures of integrins. In the case of the K1735 melanoma, $\alpha1\beta1$, $\alpha6\beta1$, and $\alpha7\beta1$ all bind to laminin; elution with low salt (0.2 M NaCl) and Ca^{2+} (10 mM) can be used to elute the different integrins selectively (Fig. 3). The $\alpha7\beta1$ receptor, a high-affinity integrin, is partially resistant to Ca^{2+} elution.

[45] M. E. Hemler, P. D. Kassner, and M. M. C. Chan, Cold Spring Harbor Symp. Quant. Biol. **57,** 213 (1992).
[46] M. M. C. Chan and M. E. Hemler, J. Cell Biol. **120,** 537 (1992).
[47] F. Hogervorst, I. Kuikman, E. Noteboom, and A. Sonnenberg, J. Biol. Chem. **268,** 18427 (1993).
[48] J. J. Grzesiak, G. E. Davis, D. Kirchhofer, and M. D. Pierschbacher, J. Cell Biol. **117,** 1109 (1992).
[49] R. Pytela, M. D. Pierschbacher, S. Argraves, S. Suzuki, and E. Ruoslahti, this series, Vol. 144, p. 475.
[50] J. Gailit and E. Ruoslahti, J. Biol. Chem. **263,** 12927 (1988).
[51] I. F. Charo, L. Nannizzi, J. W. Smith, and D. A. Cheresh, J. Cell Biol. **111,** 2795 (1990).

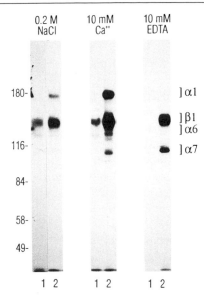

FIG. 3. Differential elution of laminin-binding integrins from laminin–Sepharose by Ca^{2+}. K1735 melanoma cells were surface labeled with ^{125}I, and the cell lysate was applied to a laminin–Sepharose column. After washing with running buffer, the column was sequentially eluted with 0.2 M NaCl, 10 mM $CaCl_2$, and finally 10 mM EDTA. The fractions were then subjected to immunoprecipitation with antibody to the α6 subunit, GoH3 (lanes 1), or anti-β1 antibody (lanes 2) and the recovered immunoprecipitates were separated by SDS–PAGE under reducing conditions. The locations of individual integrin chains and the apparent molecular masses (in kilodaltons) are indicated. Note that α6 is mostly eluted with salt whereas α1 is mostly resistant to the salt but is completely recovered with Ca^{2+}. α7 is completely resistant to salt, mostly resistant to Ca^{2+}, but is eluted with EDTA.

Studies with Laminin Fragments

Laminin fragments have been a powerful tool for identifying specific domains that contain receptor-binding sites. These fragments are gener-ated by selective proteolytic digestion followed by purification and are then used in cell adhesion assays and other studies. However, for some of these fragments the yield can be low. A number of useful fragments can be produced from EHS laminin by limited elastase or pepsin digestion. Commonly, the E1-4, E3, E4, E8, and P1 fragments (or similar fragments) have been isolated and characterized for adhesion-promoting activity (re-viewed by Engel[12]). The fragments can also be coupled to Sepharose and then used in ligand-affinity chromatography procedures to identify domain-specific receptors. For example, using ligand-affinity chromatography, α7β1 and α1β1 integrins selectively bind to immobilized E8 and P1 fragments, respectively (Fig. 4). Additionally, antibodies can be generated to specific laminin fragments and then tested for their adhesion-neutralizing activity on intact laminin. Results from such different approaches can provide strong evidence for identifying the binding site of the receptors on laminin.

One difficulty that can limit the success of using laminin fragments is that receptor-binding sites may be destroyed or altered following proteolytic digestion of laminin. For example, the binding site for the $\alpha6\beta1$ and $\alpha7\beta1$ integrins has been shown to bind to the E8 fragment of the long arm of laminin.[24,27] A slightly smaller subfragment derived from the E8 fragment by trypsin digestion also retains activity.[25] Attempts to further resolve the receptor-binding site in the E8 fragment have not been successful. Apparently, the E8-binding site for this receptor requires an intact quaternary structure composed of the A, B1, and B2 chains.[52,53] The E8 domain contains a major part of the globular (G) domain of the COOH-terminal part of the laminin A chain.[2] There is no efficient proteolytic digestion protocol for the preparation of the G domain. However, it has been possible to generate this region by recombinant approaches and then to use this domain in studies of adhesion.[54]

On the other hand, as mentioned above, proteolytic digestion of laminin yields fragments that may contain cryptic sites not normally exposed in the intact molecule. This is the case for the RGD site in the P1 fragment isolated from the EHS laminin following digestion of laminin with pepsin. Both $\alpha v\beta3$ and $\alpha IIb\beta3$ have been shown to bind this cryptic RGD sequence.[23,27,28] Additionally, $\alpha v\beta3$ appears to interact with laminin apparently through a non-RGD-sensitive mechanism.[36,55,56]

$\alpha3\beta1$ has been reported to recognize multiple ligands such as fibronectin, collagen, laminin (human and murine A chain), and epiligrin/kallinin.[19,20,33–35,57] Binding of $\alpha3\beta1$ to laminin occurs at the globular end of the long arm, where it effectively competes with $\alpha6\beta1$ for available sites.[28] The ability of the $\alpha3\beta1$ receptor to bind to different ligands by RGD-dependent and -independent means points to the possibility that multiple sites on the receptor complex interact with different ligand recognition

[52] R. Deutzmann, M. Aumailley, H. Wiedemann, W. Pysny, R. Timpl, and D. Edgar, *Eur. J. Biochem.* **191,** 513 (1990).

[53] U. Sung, J. J. O'Rear, and P. D. Yurchenco, *J. Cell Biol.* **123,** 1255–1268 (1993).

[54] P. D. Yurchenco, U. Sung, M. D. Ward, Y. Yamada, and J. J. O'Rear, *J. Biol. Chem.* **268,** 8356 (1993).

[55] R. I. Clyman, F. Mauray, and R. H. Kramer, *Exp. Cell Res.* **200,** 272 (1992).

[56] D. G. Stupack, C. Shen, and J. A. Wilkins, *Exp. Cell Res.* **203,** 443 (1992).

[57] M. E. Hemler, M. J. Elices, B. M. Chan, B. Zetter, N. Matsuura, and Y. Takada, *Cell Differ. Dev.* **32,** 229 (1990).

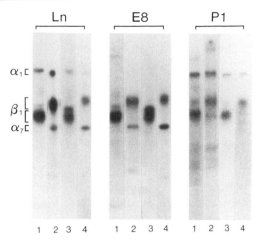

FIG. 4. $\alpha7\beta1$ binds to the E8 subfragment of laminin. MeWo melanoma cells were surface radioiodinated, solubilized, and the cell extract applied to either a laminin (Ln)–Sepharose (*left*), E8–Sepharose (*center*), or P1–Sepharose (*right*) column. Specifically bound material that was eluted with EDTA (lanes 1 and 2) was subjected to immunoprecipitation with antibody (AIIB2) to the $\beta1$ subunit (lanes 3 and 4) and separated by SDS–PAGE under nonreducing (lanes 1 and 3) and reducing (lanes 2 and 4) conditions. Note that the $\alpha7$ subunit is retained and eluted from both the intact laminin and the E8–Sepharose columns but not from the P1 column. Low levels of $\alpha1\beta1$ were recovered from the P1–Sepharose column. (Reproduced from Kramer *et al.*,[24] with permission of the publisher.)

signals.[35,57] However, it was reported that $\alpha3\beta1$ transfected into K562 cells fails to bind fibronectin, collagen, or EHS laminin but will bind an epithelium-derived matrix (kallinin/epiligrin?).[37]

These results with $\alpha3\beta1$ are difficult to reconcile. It is possible that either the activation state of the receptor can confer different ligand specifities or, alternatively, that different forms of the receptor exist. In fact, $\alpha3$, $\alpha6$, and $\alpha7$ represent a subset of integrin that may be alternatively spliced in the cytoplasmic domain; each α chain can exist as the A or B form.[58–62] More recently, $\alpha6$ and $\alpha7$ and probably $\alpha3$ have been shown to be alternatively spliced in the extracellular domain near the ligand-binding

[58] H. M. Cooper, R. N. Tamura, and V. Quaranta, *J. Cell Biol.* **115**, 843 (1991).
[59] R. N. Tamura, H. M. Cooper, G. Collo, and V. Quaranta, *Proc. Natl. Acad. Sci. U.S.A.* **88**, 10183 (1991).
[60] F. Hogervorst, I. Kuikman, A. G. Van Kessel, and A. Sonnenberg, *Eur. J. Biochem.* **199**, 425 (1991).
[61] G. Collo, L. Starr, and V. Quaranta, *J. Biol. Chem.* **268**, 19019 (1993).
[62] B. L. Ziober, M. P. Vu, N. Waleh, J. Crawford, C. S. Lin, and R. H. Kramer, *J. Biol. Chem.* **268**, 26773 (1993).

site.[62] Similarly, $\alpha6\beta4$ was unable to bind laminin under conditions conducive to $\alpha6\beta1$ binding.[23,28] However Lee *et al.*[63] reported that $\alpha6\beta4$ can bind to EHS laminin. The apparently divergent behavior of both $\alpha3$ and $\alpha6$ toward different laminins could be due to alternative splicing of the α subunits in the extracellular domain.

Cell Attachment Assays: General Considerations

Adhesion assays provide a simple approach for the identification of individual receptors that bind laminin and other ECM ligands. When such assays are used in conjunction with a panel of blocking monoclonal antibodies, now available against most major human integrins, it is possible to identify which integrin or subgroup of integrins are important in mediating adhesion. Frequently, however, multiple integrin receptors contribute to the adhesion process, which can complicate the analysis. A supplemental approach employs isolated fragments of laminin with restricted receptor specificity (Fig. 1) as the target ligand in adhesion assays.

Adhesion assays can be conveniently performed in uncharged polystyrene 96-well microtiter plates available from several manufacturers. The 96-well format for adhesion assays has many advantages that have already been described.[64,65] Detection of adherent cells can be measured by several different methods using a plate reader. However, because of the small diameter of the wells, vigorous washing procedures during the assay can result in the detachment or lysis of adherent and even well-spread cells by the shear force generated at the cell–air meniscus. Consequently a method should be used that produces a mild and reproducible shear force for effectively removing weakly attached cells while leaving strongly attached cells in place.

In any adhesion assay, it is important to determine the possible contribution of adhesion proteins that are synthesized and deposited during the assay. Protein synthesis can usually be inhibited by preincubating cell cultures with cycloheximide (2.5 μg/ml) for 2 hr before cell harvest; cycloheximide should be present during the assay as well; the level of protein synthesis should be monitored to confirm that there is substantial inhibition by the inhibitor. In some cells, even cycloheximide may not sufficiently block the synthesis and deposition of certain matrix proteins during even short-term adhesion assays.

[63] E. C. Lee, M. M. Lotz, G. D. Steele, and A. M. Mercurio, *J. Cell Biol.* **117,** 671 (1992).
[64] F. Grinnell, D. G. Hays, and D. Minter, *Exp. Cell Res.* **110,** 175 (1977).
[65] E. Ruoslahti, E. D. Hayman, M. D. Pierschbacher, and E. Engvall, this series, Vol. 82 [46].

Adhesion Assay: Detailed Procedure[24,66]

Ligand Coating. Laminin, laminin fragments, or other extracellular matrix macromolecules are immobilized in wells of virgin polystyrene 96-well plates (not treated for cell culture) at predetermined concentrations (usually between 1 and 50 μg/ml) by coating for 1 hr or overnight at 4°. Generally, highly active fragments of laminin, such as E8, are used at lower coating concentrations. The efficiency of cell adhesion to matrix proteins will vary considerably depending on the specific cell type. It is best initially to determine the final density of absorbed protein by measuring the fraction of bound substrate with ligand that is trace labeled with ^{125}I. After coating with ligand, the wells are blocked with 1% (w/v) bovine serum albumin (BSA) in phosphate-buffered saline (PBS) for 1–2 hr; sometimes BSA that has been heat denatured at 80° for 10 min will increase the effectiveness of the blocking procedure. Extended coating times with BSA should be avoided because it progressively removes the bound ligand. Control wells coated with BSA alone should always give negligible levels of cell attachment; this level of background attachment is subtracted from the raw adhesion data.

Harvesting of Culture Cells. For consistency it is best to use cultured cells in log-phase growth because it has been shown that integrin adhesion profiles can vary depending on the time at confluency.[67] For harvesting cultured cells, use a mild treatment such as EDTA to detach adherent cells for use in adhesion assays. However, for certain cells (particularly epithelial cell lines), EDTA may not be effective in detaching cells and protease treatment may be required. Exposure to protease can, of course, alter receptor levels or their activity. An alternative method is to harvest cells by brief trypsin treatment, then replate the cells 12–24 hr prior to the adhesion assay in order for the cells to recover. This usually facilitates subsequent detachment of the cells with EDTA. Note that integrins are somewhat resistant to trypsin digestion in the presence of stabilizing divalent cations; trypsin–EDTA is not recommended for the detaching of cells.

The cultured cells are removed from the culture plates by incubation for 10–20 min at 37° with 2 mM EDTA–0.05% (w/v) BSA in PBS, followed by washing twice with PBS. The BSA is added to minimize cell damage during harvesting. The cell pellet is resuspended in cold serum-free medium such as Dulbecco's minimum essential medium (DMEM) supplemented with 20 mM N-2-hydroxyethylpiperazine-N'-2-ethanesulfonic acid (HEPES) and 0.1% (w/v) BSA, and adjusted to a final concentration of

[66] R. H. Kramer, K. A. McDonald, E. Crowley, D. M. Ramos, and C. H. Damsky, *Cancer Res.* **49**, 393 (1989).

$2–4 \times 10^5$ cells/ml and kept in an ice bath until the time of assay (to minimize cell clumping). The type and concentration of divalent cation present in the attachment medium will influence adhesion. Ca^{2+} and Mg^{2+} at 1 mM each are commonly used.

Adhesion Assay. For the assay, 100 μl of the cell suspension is added to each well of a 96-well plate that is precooled in an ice bath. At least three 100-μl aliquots of the original cell suspension are saved and collected by centrifugation for measurement of total input of cells. If synthetic peptides or blocking antibodies (to integrin or laminin) are to be tested, then an appropriate concentration of the reagent is added to wells in a volume of 50 μl, followed by 50 μl of a 2× cell suspension, and mixed by gently pipetting. If reagents or cells are precious, a 50-μl total volume can be used instead of 100 μl. After a 30-min incubation at 0–4°, the assay is initiated by floating the plate on a 37° water bath in a humidified CO_2 chamber for 15–30 min or longer. Care is taken not to jar the plate because this will unevenly disperse the cells. It is usually best to use short adhesion periods (15–30 min) in order to study the early phase of cell attachment and to minimize deposition of cell-derived matrix or degradation/redistribution of adsorbed ligand. If the kinetics of adherence are to be analyzed, then at various time points the incubation is interrupted and the number of attached cells is determined; for this, multiwell strips are available.

The assay is terminated by shaking the plates on an orbital shaker (such as Lab Line model 3520, Melrose Park, IL); plates are subjected to a series of six rotational pulses each at 350 rpm over a 60-sec period; then 200 μl of DMEM containing 0.1% BSA is added to each well with a multichannel pipettor and the wells are aspirated, using an eight-well manifold (such as the Accutran; Schleicher & Schuell, Keene, NH), down to a volume of about 50 μl. The plates are subjected to two more rotational pulses for 20 sec each at 350 rpm, then are washed and completely aspirated as described above. The shear forces generated by this procedure are sufficient to remove cells from control BSA-coated wells but will not detach strongly adhering cells from ECM-coated wells. By varying the speed and duration of the rotational pulses, it is possible to select for more or less strongly attached cells.

Measurement of Attached Cells. At the end of the adhesion assay, adherent cells can be measured using one of several methods that detect radioactivity (^{125}I, ^{51}Cr, or [^3H]thymidine), enzymatic activities (hexosaminidase or phosphatase),[68,69] or bound dye (crystal violet, trypan blue,

[67] E. Fingerman and M. Hemler, *Exp. Cell Res.* **177,** 132 (1988).
[68] U. Landegren, *J. Immunol. Methods* **67,** 379 (1984).
[69] C. A. Prater, J. Plokin, D. Jaye, and W. A. Frazier, *J. Cell Biol.* **112,** 1031 (1991).

or toluidine blue).[55] For the enzymatic assays, such as that for hexosamini-dase, it should be noted that the level of activity of this enzyme can vary substantially between different cell lines. Colorimetric assays that measure enzymatic activity or dye binding have the advantage that the signal can be conveniently read on an enzyme-linked immunosorbent assay (ELISA)-type plate reader and the results can be directly downloaded into computer-based programs for data analysis.

If the [^{125}I]iodo-2′-deoxyuridine ([^{125}I]IUdR) labeling method is used, cells are labeled for 16–20 hr in 1 μCi of [^{125}I]IUdR per milliliter and are washed two times in culture medium to remove excess radiolabel. At the end of the assay, any unincorproated radiolabel is extracted from the attached cells by exchanging the medium for 70% ethanol at 4° for 2 hr or overnight. The residue is then solubilized with a solution of 1 N NaOH and 0.1% (w/v) sodium dodecyl sulfate (SDS) at 37° for 1 hr and radioactiv-ity is then measured in a γ counter. Alternatively, the radioactivity in each well is measured directly with a Molecular Dynamics Phosphorimager (Sunnyvale, CA) or similar device.

Isolation of Integrin-Binding Laminins: General Considerations and Precautions

In many respects, the protocols and precautions previously described for other integrin-affinity chromatographic assays[49] hold for the study of laminin-binding receptors. When using laminin–Sepharose columns it is important to realize the potential for ligands present in the detergent cell extract eventually contaminating the affinity matrix. Once adsorbed to the column, they are difficult to remove. In crucial experiments, it is probably a good practice not to reuse ligand–Sepharose columns. Cell-associated fibronectin can adsorb onto laminin–Sepharose columns, re-sulting in the binding and recovery of the fibronectin receptor $\alpha5\beta1$. In this case it is probably a good idea to pass the cell lysate over an anti-fibronectin antibody column to remove any fibronectin present and thus minimize contamination of the resin. If laminin–Sepharose chromatogra-phy is to be used for large-scale isolation of receptors, then the presence of albumin in the sample should be avoided because it tends to bind nonspecifically to the affinity matrix. Care should be taken to wash cul-tured cells or tissue samples thoroughly with buffer prior to detergent extractions and solubilization. However, in the case of analytical runs, in which limited numbers of cells are used, the inclusion of small amounts of BSA (10 μg/ml) can help produce cleaner elution profiles.

As for other ligand-affinity chromatographic procedures, a variety of conditions during the chromatographic run will necessarily influence receptor binding and subsequent detection. The type of detergent used, the flow rate, the presence and type of divalent cation, the salt concentration, and the method of elution can all significantly affect the characteristics of chromatographic assay. In particular, retention and detection of low-affinity integrins on immobilized ligand columns are problematic; low-affinity receptor can easily be lost in the wash fractions. In general, integrin–ligand binding typically does approach the high-affinity interactions observed in receptor–peptide hormone binding. Integrin–ligand binding is estimated to be in the micromolar range whereas hormone–receptor interactions are in the nanomolar range.

Laminin-binding integrins can be readily purified from cultured cells or tissues. The availability of tissues as a source for the receptor of interest is a definite advantage when large amounts of the receptor are required for biochemical characterization. However, in some cases, higher levels of integrins may be expressed in cultured cells. For example, in certain tumors, integrins may be present at low or nondetectable levels *in vivo,* but high levels of the receptor may be expressed in the corresponding cultured tumor cells.[70] Also, it should be noted that tissues usually are a composite of different cell types; this may increase the diversity of the receptors present.

For most cells or tissues, laminin-binding integrins are usually present as a complex mixture of different receptors and this may result in a competition for available binding sites on the ligand–resin columns during purification. This has been demonstrated especially by the higher affinity receptors and it may be necessary to deplete one integrin in order to observe binding of a second. For example, Elices *et al.*[35] found that to observe binding of the lower affinity $\alpha3\beta1$ to fibronectin, the higher affinity $\alpha5\beta1$ had to be depleted. Similarly, Sonnenberg *et al.*[28] found that to detect $\alpha6\beta1$ binding to human laminin columns, it was necessary to remove $\alpha3\beta1$. Depletion of a competing integrin can be achieved either by passing the cell extract over several ligand columns or by extracting the cell lysate with an anti-integrin antibody column.

For analytical studies of laminin receptors, large amounts of starting material are normally not required. However, integrin receptors are usually expressed in trace amounts and certain precautions are needed to reduce nonspecific binding of irrelevant cellular proteins. For small-scale studies, a minimum of several million cells is needed and larger numbers $(10^7–10^8)$ usually yield cleaner results. Iodination by the glucose oxidase-coupled lactoperoxidase method is a technique commonly used to label surface receptors. Alternatively, the NHS(N-hydroxysuccinimide)-biotin

---

[70] S. M. Albelda and C. A. Buck, *FASEB J.* **4,** 2868 (1990).

labeling system has also been used. When specific antibodies are available it is feasible to perform Western blot analysis to detect integrin subunits in the eluted fractions, particularly if column fractions are concentrated prior to sodium dodecyl sulfate-polyacrylamide gel electrophoresis (SDS–PAGE).

A typical elution pattern of fractions from an EHS laminin–Sepharose column is shown in Fig. 5. Assignment of integrin subunit positions in the SDS gel is based on analysis of fractions by immunoprecipitation with specific antibodies. In this cell line (human MeWo melanoma), only trace amounts of the $\alpha6\beta1$ receptor are present and eluted in the 0.2 M NaCl-eluted fractions. The EDTA initially eluted small amounts of $\alpha1\beta1$ followed by large amounts of the high-affinity $\alpha7\beta1$ complex. The $\alpha7$ and $\beta1$ subunits comigrate as a single broad band under nonreducing conditions, but following reduction they separate into the $\alpha7$ and $\beta1$ bands at 140 and 90 kDa, respectively.[71]

Laminin–Sepharose Affinity Chromatography: Detailed Procedures[24,36,72]

Coupling of Laminin and Laminin Fragments. EHS laminin is isolated by the method of Timpl *et al.*[29,72] or Paulsson *et al.*[30] and coupled to CNBr-activated Sepharose to yield 1–2 mg/ml of packed gel. For analytical studies, small columns are used [0.8 cm diameter× 3 cm (Bio-Rad, Richmond, CA); bed volume, ~1.5–2 ml] and equilibrated with buffer A [50 mM octyl-β-D-glycopyranoside–50 mM Tris-HCl (pH 7.4), 1 mM MnSO$_4$, 0.02% NaN$_3$, 1 mM phenylmethylsulfonyl fluoride (PMSF)]. Laminin-derived proteolytic fragments such as P1 and E8 are prepared[27,70] and are coupled at 0.5–1.0 mg/ml of packed activated Sepharose gel.

Cell Surface Labeling and Solubilization of Receptors. Cultured cells are harvested from culture dishes at preconfluency with EDTA, thoroughly washed with PBS, and then iodinated with ^{125}I by the method of Hubbard and Cohn.[73] Alternatively, cells are labeled with NHS-biotin.[74] Labeled cells are extracted with buffer A but containing 100 mM octyl-β-D-glycopyranoside, for 30–60 min at 4°. The cell lysate is centrifuged first at 2000 g (10 min, 4°) to remove nuclei, then at 20,000 g (30 min, 4°). If tissues are used, the integrins present may be strongly bound to endogenous laminin-containing matrices. More complete solubilization can be achieved by first extracting the tissue in the presence of EDTA, which will help dissociate receptor–ligand complexes. The tissue is first

---

[71] R. H. Kramer, K. A. McDonald, and M. P. Vu, *J. Biol. Chem.* **264**, 15642 (1989).
[72] R. Timpl, M. Paulsson, M. Dziadek, and S. Fujiwara, this series, Vol. 145, p. 363 (1987).
[73] A. L. Hubbard and Z. A. Cohn, *J. Cell Biol.* **64**, 438 (1975).
[74] R. R. Isberg and J. M. Leong, *Cell* **60**, 861 (1990).

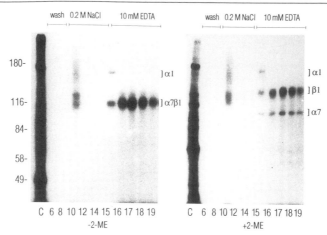

FIG. 5. Identification of laminin receptors, using ligand-affinity chromatography. ^{125}I-Surface-labeled human MeWo melanoma cells were extracted with octyl-β-pyranoside detergent in 1 mM Mn^{2+} (lane C) and applied to an EHS laminin–Sepharose column as described.[72] The column was washed with five column volumes of running buffer [50 mM octyl-β-D-glycopyranoside–50 mM Tris-HCl (pH 7.4), 1 mM MnSO$_4$, 0.02% NaN$_3$, 1 mM PMSF] (lanes 6–8) and then washed with 200 mM NaCl in running buffer (lanes 10–14), followed by elution with 10 mM EDTA in running buffer lacking divalent cation (lanes 15–19). Samples from the indicated fractions were analyzed by SDS–PAGE under nonreducing (−2-ME) and reducing conditions (+2-ME). The positions of the integrin α and β1 chains are indicated.

sectioned into 1- to 2-mm slices with a scalpel, quickly rinsed with divalent cation-free PBS, and treated with 1 mM EDTA, 150 mM NaCl, 50 mM Tris-HCl (pH 7.4), 1 mM PMSF for 1 hr at 4°. The tissue sections are recovered by low-speed centrifugation (50 g at 4° for 5 min) and are then extracted with detergent as described above for cultured cells.

Ligand-Affinity Chromatography. The detergent lysate is loaded onto a laminin–Sepharose column, and incubated for at least 30–60 min (2–4 hr or longer is required for maximum recovery). The extract can also be effectively applied by recirculating through the column. Alternatively, the laminin–Sepharose can be incubated with the extract in a batchwise manner to increase the efficiency of receptor binding. The chromatography is conveniently performed at 20° although at 4° it may be possible to obtain more stable integrin–ligand binding. When nonspecific adsorption of cellular proteins on the column is a problem, it may be advantageous to first pass the cell lysate through a short column of plain Sepharose or BSA–Sepharose to remove the contaminating material.

After application of the cell extract the column is carefully washed with three to five column volumes of the low-salt running buffer. Receptors that bind laminin with low affinity can be lost during this stage, therefore it is best to perform the initial column wash as rapidly as possible. Next, a salt wash of the column is performed with the application of 0.2 M NaCl in the running buffer; up to five column volumes is applied. This step will usually elute weakly bound integrins (e.g., $\alpha 6\beta 1$). If high-affinity receptors are present, then the column can be washed next with running buffer containing 0.5 or even 1 M NaCl. Medium- to high-affinity receptors can be eluted in running buffer without divalent cation containing either 10 mM CaCl$_2$ or 10 mM EDTA (for more effective elution). If EDTA is used then the eluted integrin should be stabilized by immediately adjusting with excess Ca^{2+} or Mg^{2+}.

Analysis of Elution Profile. Fractions containing eluted proteins are analyzed by SDS–PAGE with either silver or Coomassie blue staining for large-scale runs or by other detection methods for analytical experiments. For the latter, either surface ^{125}I-labeling by lactoperoxidase followed by autoradiography or NHS-biotin labeling followed by detection with blotting using avidin-coupled reporter systems is used. Alternatively, if suitable antibodies are available (e.g., polyclonal anti-cytoplasmic chain antibodies), then column fractions containing eluted integrins can be readily detected by immunoblotting. For further characterization, fractions containing the eluted integrin are subjected to immunoprecipitation with specific anti-integrin antibodies, either individually or sequentially, followed by SDS–PAGE and autoradiography.

Acknowledgments

This work was supported by NIH Grants CA33834, CA51884, DE10564, and DE10306, and by the TRDRP of the University of California.

[8] Analysis of Collagen Receptors

By Samuel A. Santoro, Mary M. Zutter, Justina E. Wu, William D. Staatz, Edwin U. M. Saelman, and Patricia J. Keely

The collagens represent major constituents of both the interstitial extracellular matrix and basement membranes. Although in earlier years the more purely structural roles of collagenous molecules were emphasized, we have now come to appreciate that many cells interact directly with collagen through cell surface receptors in a dynamic manner. Cellular

Copyright © 1994 by Academic Press, Inc.
All rights of reproduction in any form reserved.

interactions with collagens play major roles in processes such as cell migrations during development, organogenesis, wound healing, tumor cell invasion and metastasis, hemostasis, and may influence the expression of genes characteristic of differentiated cells. The role of the $\alpha_1\beta_1$ and $\alpha_2\beta_1$ integrins as collagen receptors has now been well established as a result of studies by many investigators. Although other classes of molecules may also contribute to cell–collagen interactions and the methods described below may be adapted to probe the function of these other receptors, the techniques as described below were all initially applied to the study of collagen receptors of the integrin family.

Purification of Collagen Receptors

The most efficient and effective schemes for the purification of collagen receptors, especially of the integrin class, exploit the divalent cation-dependent binding of the receptors to collagen. Detergent extracts of cells are subjected to affinity chromatography on collagen–Sepharose in the presence of the divalent cations Mg^{2+} and Mn^{2+}. Elution of bound receptor is accomplished by chelation of divalent cations with ethylenediaminetetraacetic acid (EDTA). A detailed procedure for isolating the $\alpha_2\beta_1$ integrin from platelets is given below.[1] Slight variations of the procedure have been employed to isolate the $\alpha_1\beta_1$ and/or $\alpha_2\beta_1$ integrins from endothelial cells,[2] HT-1080 fibrosarcoma cells,[3] MG-63 osteosarcoma cells,[3] hepatocytes,[4] fibroblasts,[4] melanoma cells,[5] PC-12 neural cells,[6] and smooth muscle cells.[7]

The preparation of type I collagen–Sepharose has been previously described in detail in this series.[8] It is, therefore, not reiterated here. Other collagenous molecules may be coupled via the same procedure.

Preparations of platelet-derived receptor may be carried out with platelets derived from 3 to 25 units of platelet concentrates. Platelets are washed by centrifugation as described below for use in adhesion assays or as previously described in this series for the isolation of RGD-dependent

[1] W. D. Staatz, S. M. Rajpara, E. A. Wayner, W. G. Carter, and S. A. Santoro, *J. Cell Biol.* **108**, 1917 (1989).

[2] D. Kirchhofer, L. R. Languino, E. Ruoslahti, and M. D. Pierschbacher, *J. Biol. Chem.* **265**, 615 (1990).

[3] P. M. Cardarelli, S. Yamagata, I. Taguchi, F. Gorcsan, S. L. Chiang, and T. Lobl, *J. Biol. Chem,*. **267**, 23159 (1992).

[4] D. Gullberg, L. Terracio, T. K. Borg, and K. Rubin, *J. Biol. Chem.* **264**, 12686 (1989).

[5] R. H. Kramer and N. Marks, *J. Biol. Chem.* **264**, 4684 (1989).

[6] K. J. Tomaselli, C. H. Damsky, and L. F. Reichardt, *J. Cell Biol.* **107**, 1241 (1988).

[7] V. M. Belkin, A. M. Belkin, and V. E. Koteliansky, *J. Cell Biol.* **111**, 2159 (1990).

[8] R. Timpl, this series, Vol. 82, p. 472.

receptors.[9] The platelet pellet is lysed by the addition of an equal volume of cold isotonic buffer [0.05 M phosphate, N-2-hydroxyethylpiperazine-N'-2-ethanesulfonic acid (HEPES), or Tris buffers adjusted to pH 7.4 and containing 0.14 M NaCl all are effective in our hands] containing 100 mM octylglucoside or 50 mM octylthioglucoside. Although octylglucoside and octylthioglucoside have become standard for these types of receptor solubilizations, we have successfully employed Triton X-100, Nonidet P-40 (NP-40), and Lubrol. The lysis buffer should also include a cocktail of protease inhibitors such as 2 mM phenylmethylsulfonyl fluoride (PMSF), 2 mM Trasylol, and 100 μM leupeptin. After resuspension in lysis buffer, the receptors are extracted at 4° for 20 min. The lysate is subjected to centrifugation at 30,000 g at 4° for 30 min and the supernatant collected for use in subsequent steps. Although in our initial description of the purification of the $\alpha_2\beta_1$ integrin we next found it useful to purify the receptor partially by chromatography on concanavalin A–Sepharose,[1] the step has subsequently been omitted by many investigators.

The lysate should be made 2 mM in MgCl$_2$ and 1 mM in MnCl$_2$ before chromatography on collagen–Sepharose. Buffers should contain no added Ca^{2+} because the binding of $\alpha_2\beta_1$ to collagen is inhibited by Ca^{2+}. The presence of Mn^{2+} is now recommended to enhance binding of integrin collagen receptors to ligand and therefore receptor yield. The extract is applied to a 2- to 5-ml column of collagen–Sepharose equilibrated with buffer containing 200 μM octylglucoside, 2 mM MgCl$_2$, and 1 mM MnCl$_2$ at 4°. The extract is allowed to flow slowly through the column and the column is washed extensively with buffer. Receptor is eluted by the substitution of 2–5 mM EDTA for the divalent cations in the column buffer. Column fractions are monitored for purity by sodium dodecyl sulfate–polyacrylamide gel electrophoresis (SDS–PAGE). A second round of collagen affinity chromatography may be required to remove trace contaminants. Receptor purified in this manner is suitable for biochemical and functional analysis by, for example, reconstitution into phosphatidylcholine liposomes.[1,9]

Analysis of Collagen Receptors in Tissues

Immunohistochemistry

Recognition of the widespread distribution and the potential importance of the integrins in cell differentiation and development has come in

[9] R. Pytela, M. D. Pierschbacher, S. Argraves, S. Suzuki, and E. Ruoslahti, this series, Vol. 144, p. 475.

part from immunohistochemical studies.[10] Many papers characterizing the cell type and differentiation-dependent expression of integrin receptor subunits have broadened our understanding of the complex roles the integrin receptors play.

Several techniques have been used to localize the integrin subunits in either cell lines or in tissues. The methods include immunofluorescence and a variety of immunohistochemical techniques, using different developing reagents such as alkaline phosphatase or horseradish peroxidase. The advantages and disadvantages of the different techniques are discussed below.

For optimal identification of the integrins in tissues, the specimens are best obtained immediately after surgical resection or animal sacrifice, cut into blocks of approximately 1.0 × 1.0 × 0.5 cm, embedded in optimal cryopreserving tissue compound (OCT compound; Miles Laboratories, Elkhart, IN), snap-frozen in liquid nitrogen-cooled isopentane, and stored at −70°. Tissue blocks can be stored for prolonged periods of time if thawing does not occur. Frozen sections are then cut at 4–6 μm using a cryostat microtome, placed on slides coated with either poly(L-lysine) or on Superfrost Plus slides (Fisher Scientific, Pittsburgh, PA), and fixed with acetone at −20° for 5 min. Frozen sections that have been briefly fixed as outlined above can be held at −70° for several months before staining.

Prior to staining, frozen sections are allowed to defrost quickly and then are washed in phosphate-buffered saline (PBS), pH 7.4–7.5, for a total of 15 min with gentle rocking. The pH of the buffers used is critical. The tissues are fragile at this time, and caution must be used to prevent loss of frozen tissue from the slides. The slides are carefully dried with a tissue to remove most of the PBS from the slide and from the tissue, again taking care not to disturb the tissue. Sections are then incubated with nonimmune horse or goat serum (1.5%, v/v) in PBS for 30 min to minimize nonspecific binding of the primary and secondary reagents. Tissues with high-level endogenous peroxidase activity, such as bone marrow or gastrointestinal tract, should be treated for 30 min with 0.3% (v/v) H_2O_2 in either absolute ethanol or methanol to quench the background peroxidase activity. The optimal dilution of primary antibody diluted in the blocking serum [PBS plus 1.5% (v/v) horse serum] is added to the slide and the primary antibody incubated for approximately 1–2 hr. Longer periods of time may increase nonspecific binding, whereas shortened incubation times will decrease sensitivity. The optimal dilution of primary antiserum must be empirically established in pilot experiments. We find that 5–10 μg of monoclonal anti-integrin antibodies per milliliter typically

---

[10] M. M. Zutter and S. A. Santoro, *Am. J. Pathol.* **137**, 113 (1990).

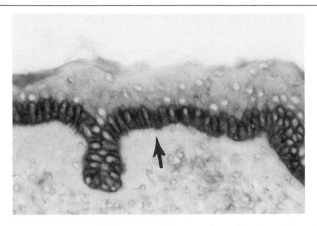

FIG. 1. Immunohistochemical detection of the $\alpha_2\beta_1$ integrin. The α_2 integrin subunit is expressed at high levels along the basal cell layer (arrow), and at lower levels in the more superficial layers of the keratinizing stratified squamous epithelium of the skin. Sections were stained immunohistochemically with monoclonal antibody P1E6 (GIBCO) directed against the α_2 integrin subunit, using the ABC technique, and counterstained with methyl green–Alcian blue. Magnification: ×400.

gives satisfactory results. Sections are washed with PBS three times for 5 min to remove the unbound primary antibody. We prefer to detect the primary antibody by the avidin–biotin complex method.[11] Detection is achieved with two additional steps, which include a biotinylated secondary antibody of appropriate species for the primary antibody [i.e., rabbit anti-mouse immunoglobulin G (IgG)] and a tertiary avidin–biotin–peroxidase complex (Vector Laboratories, Burlingame, CA). Incubations with the secondary antibody and the avidin–biotin complex (ABC) proceed for 1–2 hr each and are followed by three washes of 5 min each in PBS at room temperature. Each reagent is reconstituted and used according to the specific instructions of the supplier.

Detection of the antibody reactivity is accomplished by the addition of 0.06% (w/v) 3,3'-diaminobenzidine containing 0.03% (v/v) H_2O_2. During the development process, which can vary between 1 and 10 min, the sections should be carefully monitored and the reaction stopped when a positive control yields an intense brown signal. The sections may then be either counterstained with methyl green–Alcian blue or hematoxylin or left unstained, then dehydrated through graded ethanol, cleared in xylene, and mounted. Figure 1 shows the intense staining of the $\alpha_2\beta_1$ integrin expressed by the basal cell layer of the stratified squamous epithelium of

[11] S. M. Hsu, L. Raine, and H. Fanger, *J. Histochem. Cytochem.* **29,** 577 (1981).

skin. A control in which the primary antiboby is omitted should always be carried out.

Several modifications of this basic protocol have been employed for detection of collagen receptors and other integrin receptor subunits in tissue. A two-step technique using peroxidase-conjugated secondary antibody decreases the time required for the entire procedure and is adequate for detection of most integrin subunits that are present at a high level. However, use of the avidin–biotin–peroxidase complex does increase the sensitivity of detection. Although the majority of immunohistochemistry reports have utilized 3,3'-diaminobenzidine, which forms a dark-brown precipitate, some authors prefer development with 0.05% (w/v) 3-amino-9-ethylcarbazole and 0.01% (v/v) H_2O_2 in 0.05 M acetate buffer, pH 4.9, which results in a bright red stain at the site of immunoreactivity.[12,13]

The technique of immunofluorescence microscopy is similar to that described above. Specimen preparation, sectioning, fixation, and the steps through incubation with primary antibody and subsequent washes are identical. For localization by immunofluorescence, the secondary antibody is conjugated to either rhodamine isothiocyanate, fluorescein isothiocyanate, or other fluorophores. The reagents are widely available from any of several vendors. Specimens are incubated for 30–60 min, rinsed in PBS, and mounted in aqueous embedding compound with p-phenylenediamine to inhibit quenching of the fluorescence signal. Slides may be analyzed immediately or may be stored at $-20°$ for several days in the dark.

The major advantages of the immunohistochemical over the immunofluorescence techniques is the permanence of the slides and the ability to identify specific anatomical structures by light microscopy. The slides can be evaluated repeatedly, and a permanent photographic record is not required. Immunofluorescence yields greater sensitivity and better cellular resolution for identification of single cells or cell structures. The disadvantages are the rapid quenching of the fluorescent signal, and the need to take photographs at the time of initial analysis.

In Situ Hybridization

Although the expression patterns of the integrin collagen receptors can be elucidated by immunocytochemistry as described above, the technique has some limitations. Antibodies that recognize integrins for study in animal models are scarce. Furthermore, epitopes on integrins are often sensitive to fixation with standard fixatives such as paraformaldehyde or

[12] K. Koretz, P. Schlag, L. Boumsell, and P. Moller, *Am. J. Pathol.* **138**, 741 (1991).
[13] R. Volpes, J. J. van den Oord, and V. J. Desmet, *Am. J. Pathol.* **142**, 1483 (1993).

formalin. Unfortunately, many specimens of human clinical material are fixed in this manner and embedded in paraffin. Occasionally, it may be difficult to distinguish which cells are expressing the integrin when a mixture of different cell types are in close apposition, such as in the sympathetic or sensory ganglia. At the resolution of the light microscope, cell surface immunopositivity on neuronal cell bodies may resemble cell surface immunopositivity of satellite cells that surround and interdigitate between the neurons or vice versa. An alternative technique to examining integrin expression in tissues is *in situ* hybridization. *In situ* hybridization can be done not only on postfixed fresh frozen tissue sections but also on aldehyde-fixed tissue embedded in paraffin.[14,15] The technique can also clarify which cell types actually express the integrin in tissues such as the sympathetic and sensory ganglia (Fig. 2). In addition, the technique is a useful complement to immunocytochemistry. However, the technique of *in situ* hybridization also has its disadvantages. The technique can only indirectly reveal the production of integrin subunits by revealing the production of their mRNAs. In addition, the technique involves cumbersome RNase-free techniques, extensive preparation, and long turnaround times if radioactive probes are used. Nevertheless, *in situ* hybridization is a powerful technique for studying the expression of integrin collagen receptors in developmental, tissue repair, and disease processes as exemplified by studies of the $\alpha_2\beta_1$ integrin.[15,16]

The *in situ* hybridization technique employs antisense and sense riboprobes transcribed from plasmids using T3, T7, or SP6 promoters 5' and 3' to a cDNA fragment encoding the α_1, α_2, or β_1 integrin subunits. Sense riboprobes, which should not be able to hybridize specifically to mRNAs, serve as controls for nonspecific background hybridization (Fig. 2C) while the hybridization of antisense riboprobes generates the signal (Fig. 2B). To avoid runaround transcripts that encode the entire plasmid and cause high backgrounds, the plasmids need to be linearized with restriction enzymes that cut at either the 5' or the 3' ends of the cDNA insert prior to *in vitro* transcription of the sense or antisense probes. Restriction enzymes that leave 3' overhangs should be avoided to encourge the T3, T7, or SP6 promoters to initiate transcription specifically at the appropriate promoter sites rather than at the cut ends. Usually riboprobes are generated so that they range from 400 to 1500 bases (b) in size. The riboprobes can be labeled with either ^{35}S or digoxigenin-conjugated ribonucleotides.

[14] I. W. Prosser, K. R. Stenmark, M. Suther, E. C. Crouch, R. P. Mecham, and W. C. Parks, *Am. J. Pathol.* **135,** 1073 (1989).

[15] M. M. Zutter, H. R. Krigman, and S. A. Santoro, *Am. J. Pathol.* **142,** 1439 (1993).

[16] J. E. Wu and S. A. Santoro, *Developmental Dynamics* **199,** 292 (1994).

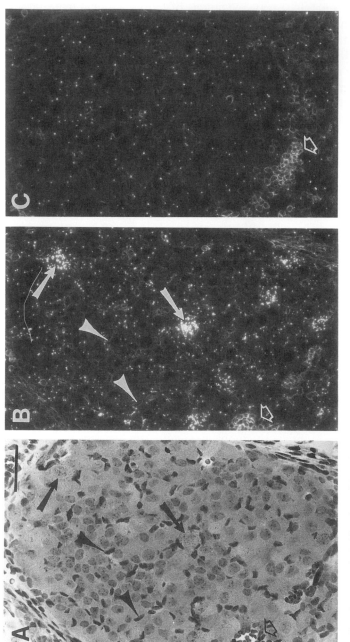

FIG. 2. *In situ* hybridization with α_2 integrin subunit antisense (A and B) and sense (C) riboprobes of a dorsal root ganglion (DRG) taken from a mouse embryo after 16.5 days of gestation. The bright-field image (A) shows a hematoxylin- and eosin-stained DRG containing large neuronal cell bodies (arrows) with euchromatic nuclei and small satellite cells (arrowheads) with dark nuclei interdigitating between the neurons. The outline arrow shows a blood vessel filled with erythrocytes that are refractile under dark-field illumination (B and C). (B) In contrast to the satellite cells (arrowheads), a subset of neurons (arrows) shows strong signal for α_2 mRNA expression. (C) *In situ* hybridization with control sense riboprobes of an adjacent section shows the background level of nonspecific hybridization. Bar: 50 μm.

After *in situ* hybridization of riboprobes to tissue or cells, detection of probes can be accomplished either autoradiographically or immunocytochemically with enzyme-conjugated antibodies directed against digoxigenin. Kits available for transcribing and detecting digoxigenin-labeled riboprobes are commercially available (Boehringer Mannheim, Indianapolis, IN) and are not described here.

RNase-free techniques (e.g., using gloves) are needed throughout the procedure and RNase-free reagents are required. Generally, solutions should be treated with 0.1% (v/v) diethyl pyrocarbonate (DEPC) and autoclaved whenever possible or prepared with DEPC-treated and autoclaved doubly distilled H_2O and sterile filtered. All glassware and metal tools should be baked for at least 8 hr at $\geq 160°$. The procedure for *in situ* hybridization can be dissected into four parts: generation of riboprobes, tissue preparation, hybridization of riboprobes and stringency washes, and detection of hybridized riboprobes by autoradiography.

Reagents and Materials

T3, T7, or SP6 RNA polymerases

Nuclease-free (phenol–chloroform and chloroform extracted), linearized plasmid containing integrin subunit cDNA with T3, T7, or SP6 promoter sites upstream and downstream of the cDNA

Transcription buffer (5×): 200 mM Tris (pH 8.0), 40 mM MgCl$_2$, 10 mM spermidine, and 250 mM NaCl

Dithiothreitol (DTT)

RNasin

rATP, rCTP, rGTP

[^{35}S]rUTP

RNase-free DNase I

Phenol–chloroform solution: 50% (v/v) phenol equilibrated to 0.1 M Tris (pH 8.0) and 50% (v/v) chloroform

Chloroform

Yeast tRNA

Sodium acetate (3 M), pH 5.2 (pH adjusted with glacial acetic acid)

Formamide

Dextran sulfate

Denhardt's solution

2-Mercaptoethanol (2-ME)

Triethanolamine (0.1 M), pH 8.0 (pH adjusted with 12 N HCl)

Acetic anhydride

Coverslips siliconized by dipping into Sigmacote (Sigma, St. Louis, MO) and autoclaved

Humidifying chamber

Generation of Riboprobes. Generally, radioactively labeled full-length riboprobes can be generated by using cold rUTP as well as [³⁵S]rUTP in the transcription reaction. However, if the mRNA studied is made in low abundance, high specific activity riboprobes would be more appropriate. To generate these riboprobes, cold rUTP is omitted from the transcription reaction. In such reactions, the concentration of radioactive nucleotide limits the reaction, and the generation of full-length and large quantities of riboprobes becomes more difficult.

For high specific activity riboprobes, 15 μl or 150 μCi of [³⁵S]rUTP (>1000 Ci/mmol) dried in a desiccator is used for each transcription reaction. For low specific activity riboprobes, only 5 μl of [³⁵S]rUTP is used for each reaction with 12.5 μM cold rUTP. The transcription reaction should also contain 1× transcription buffer, 10 mM DTT, RNasin (1.6 U/μl), 0.5 mM rATP, rCTP, and rGTP, 0.5 μg of linearized plasmid DNA template, and 5 units of RNA polymerase in a total volume of 10 μl for generation of high specific activity probes and a total volume of 20 μl for low specific activity probes. The reaction is allowed to proceed for 2 hr at 37° with an addition of 5 additional units of RNA polymerase after 1 hr of incubation. The transcription reaction is terminated by degrading the template plasmid by the addition of 1 unit of RNase-free DNase I. The riboprobes are purified by a phenol–chloroform and chloroform extraction. To separate the riboprobes from the free [³⁵S]rUTP, the riboprobes are precipitated by adding 20 μg of tRNA as carrier, 0.1 vol of 3 M sodium acetate solution, and 2.5 vol of 100% ethanol followed by incubation of the riboprobes at −70° for 30 min. The riboprobes are pelletted by centrifugation for 10 min at 10,000 g at 4°. The pellet is washed with ice-cold 70% ethanol and redissolved in 100 μl of 20 mM DTT. The riboprobes can be aliquoted to avoid repeated freeze-thawing and stored at −20° for up to 2 months.

Tissue Preparation. Conventionally, tissue is fixed in 10% (w/v) formalin in PBS, pH 7.0, or 4% (w/v) paraformaldehyde in PBS, pH 7.2. Tissue should be well fixed for 4 to 24 hr at room temperature, depending on the size of the specimen. Both underfixation and overfixation can decrease signal. The fixed tissue can be processed for either paraffin sections or cryosections. For paraffin sections, the tissue is slowly dehydrated through a graded series of alcohols, cleared in xylenes, and embedded in paraffin preferably under vacuum for complete tissue penetration. For cryosectioning, the tissue is incubated at 4° in 30% (w/v) sucrose in PBS until the tissue sinks and snap-frozen in OCT or M-1 (Shandon Lipshaw, Pittsburgh, PA) embedding medium. Alternatively, fresh-frozen tissue sections can be postfixed with 4% paraformaldehyde or 10% formalin in PBS for 20 min at room temperature prior to *in situ* hybridization. Both paraffin and cryosections should be cut at a thickness of 5 μm and collected onto Superfrost Plus-treated slides (Fisher). The use of conventional gelatin-, albumin-, or polylysine-coated slides is not recommended to avoid tissue loss during the *in situ* hybridization procedure. Collection of paraffin sections must be done with DEPC-treated and autoclaved doubly distilled H_2O. The sections should be stored with desiccant at 4° for the paraffin sections and at $-20°$ or colder for cryosections.

Before hybridization of sections to riboprobes, paraffin sections need to be deparaffinized with several washes in xylenes, rehydrated through a graded series of alcohols, and washed in PBS. Cryosections need to be treated with 95% ethanol for 5 min and washed with PBS. The sections should be pretreated with proteinase K to make mRNAs more accessible to hybridization and blocked to avoid high backgrounds. The optimal duration for proteinase K digestion can vary and should be derived empirically for each tissue. Generally 1 μg of proteinase K per milliliter in PBS at 37° for 10 to 30 min is used. After proteinase K treatment, the sections are equilibrated with 0.1 M triethanolamine, pH 8.0, and then blocked for 10 min with 0.25% (v/v) acetic anhydride (freshly made) in 0.1 M triethanolamine, pH 8.0. The sections are washed twice with 2× SSC (1× SSC is 0.15 M NaCl plus 0.015 M sodium citrate) and dehydrated up to 70% ethanol and air dried prior to hybridization.

Hybridization and Stringency Washes. The hybridization buffer consists of 2× SSC, 50% (v/v) formamide, 20 mM Tris (pH 8.0), 1× Denhardt's solution, 1 mM EDTA (pH 8.0), 10% (w/v) dextran sulfate, 100 mM DTT, yeast tRNA (500 μg/ml) and riboprobe (12,000 cpm/ml), and should be heat denatured for 10 min at 75° and then chilled on ice prior to hybridization. In addition, RNasin should be added to the hybridization buffer at 0.4 U/μl after chilling on ice. The hybridization buffer is placed on the tissue sections and gently covered with siliconized and autoclaved coverslips. Generally, 25 μl of hybridization buffer is used for every 400-mm^2 area of coverslip. The sections are then incubated for 4 to 24 hr at 15–20° below the melting temperature of the riboprobe, usually 55 to 60°, in a moist chamber containing 2× SSC, 50% formamide, 20 mM Tris (pH 8.0), and 1 mM EDTA (pH 8.0).

The initial washes consist of 15-min washes in 2× SSC then 0.5× SSC supplemented with 25 mM 2-ME at 55–60°. The first stringency wash is done with 0.1× SSC at 55–60°. Sections are then treated with boiled RNase A (20 μg/ml) in RNase A buffer (0.5 M NaCl, 1 mM EDTA, 10 mM Tris base, pH 8.0) for 30 min at 37°. This step is absolutely essential for decreasing background hybridization, for it will degrade single-stranded portions of riboprobes that have partially and nonspecifically hybridized to mRNAs within the tissue. The sections are then washed for 15 min with RNase A buffer alone at 37°, 15 min with 2× SSC at 55°, and 5 min with 0.5× SSC at 55°. The last stringency wash consists of a 15-min wash in 0.1× SSC at 55°. The slides are then again dehydrated through a graded series of alcohols beginning with 50% ethanol and ending at 95% ethanol and air dried before autoradiography.

Autoradiography. If desired, the slides may be autoradiographed against autoradiographic film for 1 to 4 days, depending on the intensity of the signal, to assess the success of the *in situ* hybridization. Generally, the incubation time for autoradiography with emulsion is approximately 5 to 10 times longer than that needed to see a clear signal on autoradiographic film. To obtain resolution at the cellular level, NTB emulsion (Kodak, Rochester, NY) is used. NTB emulsion is melted at 42–45° and diluted 1 : 1 with doubly distilled H_2O at the same temperature under a safe light. The slides should be dipped into the emulsion and allowed to drain and cool vertically for approximately 15 min. To avoid edge artifact due to stretching of the emulsion along the tissue, the slides should be allowed to dry slowly in a humidified chamber. This can be achieved by drying the slides vertically in a slide rack on top of a wet paper towel in an enclosed, light-safe box or container for 1 hr and then allowing the slides to dry further in the humid chamber for another 4 hr without the wet paper towel. The slides should then be stored with desiccant at 4° in a light-safe slide box lined with lead if the risk of exposure to other sources of radioactivity exists. It is also wise to produce two sets of *in situ* hybridization slides so that they can be developed after different incubation times to better assess the exposure time with the emulsion.

Before development, the slides should be allowed to reach room temperature. Under light-safe conditions and safe light, the slides can be developed for 3 min with D-19 developer (Kodak), stopped with 1% (v/v) acetic acid, and fixed with a 1 : 1 (v/v) dilution of Kodak rapid fixer. The developer, 1% acetic acid, and fixative solution all need to be at the same temperature to avoid wrinkling and stretching of the emulsion. Usually, 15° is used. After fixation, the slides are washed for 20 min with running distilled water. At this point, the slides may be counterstained with a variety of histological stains such as hematoxylin and eosin.

Cell Adhesion to Collagen under Static Conditions

Introduction

Essentially all assays of cell adhesion to collagen carried out under static conditions involve plating known numbers of cells on collagen or other substrate and after a set time washing off the nonadherent cells. The major differences between the various assays reported in the literature involve how the remaining, adherent cells are detected. The simplest but most labor-intensive method is simply to count the cells under the microscope. Cells can either be fixed and stained with crystal violet or toluidine blue for counting at a later time, or live cells may be counted by phase-contrast optics. Alternatively, if the assays are carried out in 96-well enzyme-linked immunosorbent assay (ELISA) plates, the stained, adherent cells may be lysed and the absorbance of the resulting solution read in an ELISA plate reader.[17] The elimination of the cumbersome cell-counting step makes it possible to examine cell adhesion under many conditions simultaneously. However, this method lacks the sensitivity needed to detect low numbers of adherent cells accurately, especially when the total mass of the adherent cellular material is small, as is the case with platelets. Metabolic labeling with [^{3}H]thymidine[18] or with [^{35}S]methionine[19] has also been used with actively growing cells in culture, but these labels are of little use in studies using quiescent cells or platelets. To overcome these problems, Haverstick et al.[20] labeled platelets with $Na_2^{51}CrO_4$ and then allowed them to adhere to various substrates. After washing to remove nonadherent platelets, the adherent platelets were detected by measuring the ^{51}Cr released into the supernatant on lysing the platelets with SDS. One advantage of this method is that, to some extent, its sensitivity can be increased by increasing both the specific activity of the ^{51}Cr during labeling and the length of time the cells are permitted to take up the isotope. Because of this, the method has been adapted for use with a variety of cultured cells.[21] The major disadvantage of radioisotopic detection is the use of radioactivity itself.

An alternate approach to detecting adherent cells, also increasing the sensitivity above that achieved by cell staining, involves the use of colorimetric assays to detect enzyme activities within adherent cells. These assays can easily be carried out in 96-well plates and their end points determined by reading the absorbance in an ELISA plate reader. Several

[17] E. Ruoslahti, E. G. Hayman, M. D. Pierschbacher, and E. Engvall, this series, Vol. 82, p. 803.
[18] T. J. Wickham, P. Mathias, D. A. Chcrcsh;, and G. R. Ncmcrow, Cell (Cambridge, Mass.) 73, 309 (1993).
[19] R. G. LeBaron, J. D. Esko, A. Woods, S. Johansson, and M. Höök, J. Cell Biol. 106, 945 (1988).
[20] D. M. Haverstick, J. F. Cowan, K. M. Yamada, and S. A. Santoro, Blood 66, 946 (1985).
[21] W. D. Staatz, K. F. Fok, M. M. Zutter, S. P. Adams, B. A. Rodriguez, and S. A. Santoro, J. Biol. Chem. 266, 7363 (1991).

enzymes have been used for this purpose, including the mitochondrial dehydrogenases[22] and the lysosomal enzyme hexosaminidase.[23] Because these detection systems are both enzymatic and proportional to cell mass, they can easily be optimized for a wide variety of different cell types.[24] We have optimized the hexosaminidase assay for use with platelets and find that it is as sensitive as the ^{51}Cr detection system, but without the inherent disadvantages associated with using radioisotopes. The reader is also referred to earlier treatments of cell adhesion to collagen in this series.[25-30]

Preparation of Substrates

Monomeric Collagen. Monomeric collagens can be isolated by methods described earlier in this series,[31] or may be purchased from commercial sources. In general, the various collagens are readily soluble in 3% acetic acid and stock solutions at collagen concentrations of 1 mg/ml can be made and stored at 4° for several weeks or even months if they are handled aseptically. To prepare substrates for adhesion assays, the stock solutions of collagen are diluted to 10–25 μg/ml with distilled water and aliquots are added to polystyrene dishes or multiwell plates. Coating is carried out for 1 to 3 hr at room temperature or overnight at 4°, using 100 μl/well for 96-well plates, 300 μl/well for 24-well plates, or 1 ml/dish for 35-mm petri dishes. Only untreated polystyrene plates or those such as the Immulon II products (Dynatech, Chantilly, VA), which are modified to enhance protein binding, should be used, because plastics designed for tissue culture use are specifically treated to reduce protein binding. After coating, the collagen solutions are removed and the remaining protein-binding sites on the plates are blocked for 1–2 hr at room temperature with 0.5% (w/v) bovine serum albumin (BSA) in divalent cation-free Hanks' balanced salt solution (adhesion buffer). The volumes used for blocking should be twice that used for coating.

[22] L. M. Jost, J. M. Kirkwood, and T. L. Whiteside, *J. Immunol. Methods* **147,** 153 (1992).
[23] U. Landegren, *J. Immunol. Methods* **67,** 379 (1984).
[24] P. K. Haugen, J. B. McCarthy, A. P. N. Skubitz, L. T. Furcht, and P. C. Letourneau, *J. Cell Biol.* **111,** 2733 (1990).
[25] H. K. Kleinman, this series, Vol. 82, p. 503.
[26] S. A. Santoro and L. W. Cunningham, this series, Vol. 82, p. 509.
[27] B. Öbrink, this series, Vol. 82, p. 513.
[28] C. G. Hellerqvist, this series, Vol. 82, p. 530.
[29] F. Grimell and M. H. Bennett, this series, Vol. 82, p. 535.
[30] S. C. Strom and G. Michalopoulous, this series, Vol. 82, p. 544.
[31] E. J. Miller and R. K. Rhodes, this series, Vol. 82, p. 33.

Denatured Collagen. Denatured collagen substrates are easily produced by diluting stock collagen solutions to 10–25 μg/ml, as noted above, and then heating them to 95° for 10–15 min. After the solutions have cooled to room temperature, they may be coated onto plates as described above. It is important to distinguish between denatured collagen and gelatin when making adhesion substrates because cells often behave differently on these two substrates. In denatured collagen the α chains remain intact but assume random conformations, whereas in commercially obtained gelatin the peptide bonds of the collagen chains have been randomly hydrolyzed by prolonged boiling to yield a mixture of many short peptides.

Isolated Collagen α Chains

Collagen α chains can be purified by the methods described in detail by Miller and Rhodes.[31] Briefly, non-cross-linked collagen α chains can be separated from the cross-linked β and γ chains of collagen by gel-filtration chromatography on Sepharose CB-4 (Pharmacia, Piscataway, NJ) or Bio-Gel A-5 (Bio-Rad, Richmond, CA) in the presence of 2 M guanidine hydrochloride. The individual α chains can then be separated by ion-exchange chromatography on carboxymethyl cellulose. Smaller quantities of these chains can also be purified by reversed-phase high-performance liquid chromatography (HPLC) using a Vydac C_{18} column (Separations Group, Hesperia, CA).[32] The purified collagen α chains may be dissolved in 0.03% acetic acid or distilled water and used to prepare adhesion substrates as described above for native collagen.

Collagen Peptides. The most commonly used collagen peptides are those generated by cyanogen bromide (CNBr) cleavage of collagen at methionine residues. To generate these peptides purified collagen α chains are dissolved in 70% formic acid at 5–10 mg/ml and digested with CNBr at a final concentration of 10–20 mg/ml. Samples are digested for 4 hr (or overnight) at 30° in tubes gassed with N_2. After digestion, the samples are diluted 20-fold with distilled water, frozen, and lyophilized for at least 24 hr to remove the CNBr and formic acid. The resultant peptides can be fractionated by HPLC on a 300-Å pore C_8 reversed-phase column (Aquapore RP 300; Brownlee Laboratories, Santa Clara, CA). The digest is bound to the column in aqueous 0.5% (v/v) trifluoroacetic acid and eluted with a gradient of 0 to 22% (v/v *n*-propanol over 176 min. After lyophilization, these fractions can be dissolved in small volumes of either distilled water or 0.01% (v/v) acetic acid and used to make adhesion substrates as described above. Alternatively, the cyanogen bromide

[32] C. Guidry, E. J. Miller, and M. Höök, *J. Biol. Chem.* **265,** 19230 (1990).

digests can be fractionated by ion-exchange chromatography on CM-cellulose at 45°, as described by Butler *et al.*[33] Substrates composed of proteolytic fragments of individual collagen α chains or synthetic peptides containing collagen seuqences are generally coated at peptide concentrations of 25–50 μg/ml overnight at 4°, then blocked and used as described above. In our experience, peptides with fewer than 20 amino acid residues do not consistently bind to polystyrene dishes. To study the adhesive properties of such short peptides it may be necessary to couple them either to a carrier protein such as BSA or ovalbumin and then to prepare substrates with the peptide conjugates. Coupling is commonly done using a commercially available heterobifunctional reagent such as *N*-succinimidyl 3-(2-pyridyldithio)propionate (SPDP) according to manufacturer instructions (Pierce Chemical, Rockford, IL). This strategy has been described in detail elsewhere in this series.[9] Alternatively, peptides may be coupled directly to chemically activated plates, such as the Covalent-1 microtiter plates available from Covalent Immunology (Medfield, MA), which covalently bind primary amino groups in the peptides.[34]

Cell and Platelet Adhesion Using Hexosaminidase Detection

To assess the adhesive capacity of cells or platelets for collagen, it is essential that adhesion assays be carried out with uniform suspensions of individual cells. It is equally important, however, that treatments used to prepare these cell suspensions do not destroy or inactivate collagen receptors on the cell surfaces. Thus, the preparation and handling of the cells and platelets to be used in adhesion assays must be done with care.

Preparation of Cultured Cells. Cell monolayers should, if at all possible, be released from their flasks by treatment with Versene solution [0.02% (v/v) EDTA in PBS] for a few minutes at 37°. Many cell types can be released within 10–15 min of this treatment and, with gentle pipetting, will yield uniform suspensions of single cells. However, prolonged treatment with EDTA is toxic to some types of cells, and therefore it is advisable to test this method on new cell lines before beginning adhesion assays.

If cells will not release from their flasks with EDTA, brief trypsinization may be used if the cells are removed from the trypsin as quickly as possible and are permitted to restore their receptor complement before being used in an adhesion assay. Culture flasks should be rinsed briefly with 5–10 ml of Versene solution to remove serum proteins that inhibit trypsin and

[33] W. T. Butler, K. A. Piez, and P. Bornstein, *Biochemistry* **6**, 3771 (1967).
[34] G. C. Blanchard, C. G. Taylor, B. R. Busey, and M. L. Williamson, *J. Immunol Methods* **130**, 263 (1990).

then incubated at 37° with 1–2 ml of trypsin at 0.005% (v/v) in Versene for 5–10 min. As soon as the cells are released from the substrate, they are resuspended in 10 ml of adhesion buffer containing 0.1% (v/v) soy trypsin inhibitor (Sigma) and gently centrifuged. The cell pellets may then be resuspended in 5–10 ml of the same buffer and allowed to recover for 2–4 hr at room temperature. Immediately prior to use in adhesion assays the cells are enumerated using either a Coulter counter (Coulter, Hialeah, FL) or a hemocytometer and the cells are adjusted to a density of 10^5 cells/ml using adhesion buffer.

Preparation of Platelets. Draw venous blood into a syringe fitted with a 19-gauge needle and containing a volume of 39 mM citric acid, 75 mM trisodium citrate, 135 mM glucose (ACD) anticoagulant. The blood is then centrifuged for 20 min (160 g, 20°) and the platelet-rich plasma (PRP) is removed and supplemented with an additional 1/10 vol of ACD and prostaglandin E_1 (PGE$_1$) to a final concentration of 10 μM. The platelets are then pelleted by centrifugation for 10 min (1300 g, 20°) and then are washed twice by resuspension in 25 ml of adhesion buffer containing 2.5 ml of ACD and 10 μM PGE$_1$ and recentrifugation for 10 min (1300 g, 20°). The final platelet pellet is resuspended in adhesion buffer containing 10 μM PGE$_1$, the platelet count is determined using a Coulter counter or hemocytometer, and the platelet number adjusted to between 120,000 and 200,000 platelets/μl with adhesion buffer. For most experimental conditions adhesion buffer may be replaced with 0.05 M Tris-HCl, 0.15 M NaCl, 0.5% (w/v) bovine serum albumin, and 0.09% (w/v) glucose. If the adhesive properties of activated platelets are being studied, the PGE$_1$ should be omitted from the final wash and resuspension steps and an aliquot of the desired activating agent can be added to the blocked wells just prior to adding the platelets.

Plating the Cells and Platelets. Although both the $\alpha_1\beta_1$ integrin- and the $\alpha_2\beta_1$ integrin-mediated adhesion of cells to collagen require the presence of either Mg^{2+} or Mn^{2+} in the medium for ligand-binding activity,[1,35] the harvesting and washing procedures described above are carried out in the absence of divalent cations. The concentrations of divalent cations must, therefore, be adjusted before the cells are added to the substrates. Separate 200 mM stock solutions of MgCl$_2$, CaCl$_2$, and EDTA and 10 μM MnCl$_2$ can be used to adjust these divalent cation concentrations. The blocking solution is then decanted from the substrate wells and aliquots of the cells and/or platelets are added to the blocked substrates. Volumes of cell and platelet suspension commonly used are 100 μl/well for 96-well plates, 300 μl/well for 24-well plates, and 1-ml/dish for 35-mm petri dishes. Aliquots

[35] D. C. Turner, L. A. Flier, and S. Carbonetto, *J. Neurosci.* **9**, 3287 (1989).

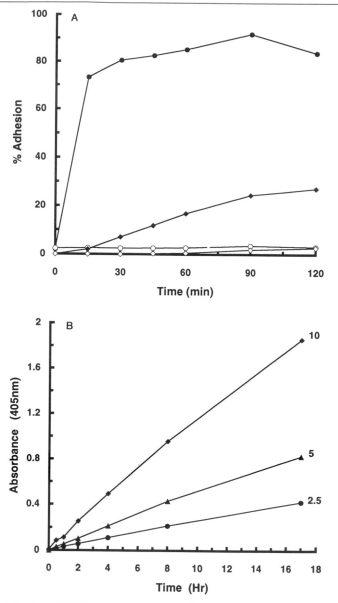

FIG. 3. Adhesion of T47D breast carcinoma cells and platelets adhering to collagen (A) The time required for optimal adhesion of cells to collagen depends in part on cell type. Platelets and T47D breast carcinoma cells were permitted to adhere to collagen substrates for the indicated times. The nonadherent cells were then removed by washing and the number of adherent cells was estimated using the hexosadminidase detection system described in text. Note that the Mg^{2+}-dependent adhesion of the larger carcinoma cells (●) was almost

containing known numbers of cells should be set aside at the beginning of each assay to construct a calibration curve for the number of cells that are found to adhere to collagen under the various conditions studied.

Both the time and the temperature at which cell adhesion is carried out must be optimized for the cell line being studied. Many cell lines such as T47D, a breast ductal carcinoma cell line, adhere quickly to collagen at 37° and, as shown in Fig. 3A, adhesion is almost maximal within 15 min. T47D cells, however, adhere poorly at room temperature. In contrast, platelets, endothelial cells, and other cell lines readily adhere to collagen at room temperature, although they do so more slowly than at 37°. Figure 3A illustrates the differences in adhesion of T47D cells and platelets.

At the end of the adhesion period, the initial cell suspension is removed by aspiration from the edge of each well. The nonadherent cells are then washed from the plates by a process of rinsing and aspiration. Each well is washed three times with adhesion buffer supplemented with the same complement of divalent cations used during the adhesion phase of the assay and each wash is 1.5–2× the volume used for adhesion. Thus, we commonly use 200 μl/well for 96-well plates, 500 μl/well for 24-well plates, and 1.5 ml/dish for 35-mm petri dishes.

After the third wash the cells are lysed by the addition of the hexosaminidase substrate p-nitrophenol-N-acetyl-β-D-glucosaminide in 0.25% (v/v) Triton X-100. To prepare the substrate, p-nitrophenol-N-acetyl-β-D-glucosaminide is dissolved at 7.5 mM in 0.1 M citrate buffer, pH 5, and this solution is then mixed with an equal volume of 0.5% Triton X-100. The reagent can be aliquoted and stored at −20° for several months. The assay is carried out by adding an aliquot of substrate to each well and incubating at 37° in a humidified environment. After a predetermined interval (usually 1–5 hr, depending on cell type) the assay is stopped and the color developed by the addition of 1.5 vol of 50 mM glycine–5 mM EDTA (pH 10.4). The absorbances are then read at 405 nm. The hexosaminidase activity in a given cell type is proportional to the lysosome content of the cells. Thus, the incubation time for this enzymatic step must be optimized for each type of cell being studied. In optimizing the

---

maximal by 15 min, whereas maximal adhesion for platelets (◆) was not reached until 90 to 100 min. Neither T47D cells (○) nor platelets (◇) adhered to collagen in the presence of EDTA. (B) Hexosaminidase activity as a function of time and platelet count. Aliquots containing known numbers of platelets in 50 μl of HBSS–0.5% BSA were incubated with 60 μl of the hexosaminidase substrate described in text for the indicated times, from 5 min to 17 hr. The reaction was stopped and the color developed by the addition of 90 μl of 50 mM glycine–5 mM EDTA (pH 10.4) and the absorbance read at 405 nm. The numbers to the right of each line indicate the number of platelets ($\times 10^{-5}$) in each aliquot.

assay for platelets we incubated aliquots of known platelet densities with hexosaminidase substrate for times ranging from 30 min to 18 hr. The reactions were then stopped and the absorbancies read at 405 nm. Under these conditions (Fig. 3B) the rate of color development remained linear for at least 18 hr. On the basis of these and comparable data we find that optimal color development times for many types of cultured cells are between 1 and 3 hr, whereas for platelets it is about 4–6 hr. When adhesion assays are carried out in 96-well plates, 60 μl of substrate can be followed by 90 μl of glycine and the absorbancies can be read with an ELISA plate reader. If 24-well plates are used, 120 μl of substrate and 180 μl of glycine can be used, whereas with 1-ml plates 0.5 ml of substrate and 0.75 ml of glycine provide sufficient volume to read the absorbance in a spectrophotometer equipped with a 1-ml cuvette.

Pretreatment with Antibodies and Peptides. To determine the effects of antibodies or soluble peptides on the adhesion of cells or platelets to collagen, aliquots of cells are incubated with the antibody or peptide of interest for 15–20 min before being plated onto an adhesion substrate. This incubation should be carried out in the presence of those divalent cations required for normal functioning of the receptor being studied.

Cell and Platelet Adhesion Using ^{51}Cr Detection

Cells in culture are labeled by adding 500 μCi of $Na_2{}^{51}CrO_4$ in 3 ml of fresh culture medium to a 25-cm^2 flask of cells and incubating overnight at 37°. Attached, monolayer cells are washed three times with 5 ml of divalent cation-free Hanks' balanced salt solution (HBSS) to remove any label not taken up by the cells and the cells are detached as described above. Nonattached cells that grow in suspension are gently pelleted by centrifugation and resuspended in the ^{51}Cr-containing medium, labeled overnight, and then washed three times with divalent cation-free HBSS.

Platelets to be labeled with ^{51}Cr are pelleted from platelet-rich plasma, as described above, and gently resuspended in 0.5 ml of adhesion buffer containing 10 μM PGE$_1$ and 300 μCi of $Na_2{}^{51}CrO_4$. After incubating at room temperature for 1 hr the platelets are washed twice by resuspension in 25 ml of adhesion buffer containing a 1/10 vol of ACD and 10 μM PGE$_1$, followed by centrifugation for 10 min (1300 g, 20°).

The labeled cells and platelets are then resuspended in adhesion buffer, and divalent cation concentrations are adjusted and they are allowed to adhere to substrates in the same manner as described above for nonlabeled cells.

At the end of the adhesion period, the nonadherent cells are aspirated from the wells and the wells are washed three times as described for hexosaminidase detection. After the third wash is aspirated from the wells, the cells are lysed and the ^{51}Cr solubilized by the addition of 3% (w/v) SDS to each well. Essentially complete recovery of the label can be obtained if two successive 100-μl aliquots are used for wells of 96-well plates, 300 μl/well for 24-well plates, and 1 ml/dish in 35-mm petri dishes. The two SDS washes are then pooled and their ^{51}Cr content measured in a γ counter. To control for the integrity of the labeled cells, an aliquot of the labeled cell suspension should be centrifuged and the isotope in both the supernatant and the pelleted cells determined. Healthy, thoroughly washed cells retain about 95% of the ^{51}Cr. If more than 5–10% of the label is found in the supernatant, the cells may be insufficiently washed. If further washing does not reduce the fraction of the label in the supernatant to less than 8%, the cells may not be intact.

Platelet Adhesion to Collagen under Flow Conditions

Introduction

The standard "static" adhesion assays described above have been employed in most studies of cell adhesion to collagen. However, some cellular interactions with collagen, such as those of platelets with collagen, occur at the interface of flowing blood with the extracellular matrix and are subject to shear forces. Special assay systems have been designed to examine adhesion under flow conditions. The system of Sakariassen et al.[36] described below is an adaptation of the system described by Muggli et al.[37] and has been tested extensively, both in rheological and adhesion studies. The reader is referred elsewhere for additional background.[36–45]

[36] K. S. Sakariassen, P. A. M. M. Aarts, P. G. de Groot, W. P. M. Houdijk, and J. J. Sixma, J. Lab Clin. Med. 102, 522 (1983).
[37] R. Muggli, H. R. Baumgartner, T. B. Tschopp, and H. Keller, J. Lab. Clin. Med. 95, 195 (1980).
[38] H. R. Baumgartner, Microvasc. Res. 5, 167 (1973).
[39] H. L. Goldsmith and V. T. Turitto, Thromb. Haemostasis 55, 415 (1986).
[40] T. Karino and H. L. Goldsmith, Microvasc. Res. 17, 217 (1979).
[41] J. J. Sixma, P. F. E. M. Nievelstein, W. P. M. Houdijk, J. H. F. I. van Breugel, G. Hindriks, and P. G. de Groot, Ann. N.Y. Acad. Sci. 509, 103 (1987).
[42] J. J. Sixma, A. Pronk, P. F. E. M. Nievelstein, J. J. Zwaginga, G. Hindriks, P. Tijburg, J. D. Banga, and P. G. de Groot, Ann. N.Y. Acad. Sci. 614, 181 (1991).
[43] V. T. Turitto, R. Muggli, and H. R. Baumgartner, Ann. N.Y. Acad. Sci. 283, 284 (1977).
[44] V. T. Turitto and H. J. Weiss, Science 203, 541 (1980).
[45] V. T. Turitto, Prog. Hemostasis Thromb. 6, 139 (1982).

Surface

Collagen. Collagens type I–VII can be obtained commercially from several companies [e.g., collagen types I, III, IV, and V (Sigma); type VII (Chemicon Int., Inc., Temucula, CA); and type VI (Heyl GmbH & Co., Berlin, Germany)]. Before using commercially available collagens, they should be checked for integrity and cross-contamination with other major collagens (types I, III, IV, and V), proteoglycans, or plasma proteins. Collagen preparations can be assayed by ELISA, inhibition-ELISA, and Western blotting.[46]

Fibrillar Collagen. Collagen fibrils are prepared by dissolving collagen in 50 mM acetic acid at 1 mg/ml concentration and dialyzing this solution against 20 mM Na$_2$HPO$_4$, pH 7.0, at 4° for 48 hr.[47] Fibril formation can be followed spectrophotometrically at 313 nm.[48] Subsequently, collagen fibrils are collected by centrifugation at 30,000 g for 30 min at 4° and resuspended in 20 mM Na$_2$HPO$_4$, pH 7.0. If necessary, the collagen concentration in this preparation can be determined by hydroxyproline analysis.

Coverslips. Glass coverslips of the highest quality must be used as a carrier for collagens. We use 18 × 18 mm glass coverslips (Menzel Glaser, Braunschweig, Germany) that have been soaked in acid dichromate for at least 24 hr; to ensure a clean (fat-free) glass surface. Before a coverslip is used for spraying, itis rinsed in deionized water to remove traces of chromosulfuric acid. Subsequently, it is rinsed in 100% ethanol and deionized water and allowed to dry at room temperature. When certain preparations such as cyanogen bromide-derived collagen fragments are not well retained on a coverslip, Denhardt-precoated coverslips can be used. Denhardt-coated coverslips can be prepared as described.[49] Thermanox coverslips (Miles, Naperville, IL) can be used as a carrier for a collagen surface when it is necessary to obtain (cross-section) information about aggregate size distribution, and the ratio of contact and spread platelets. For further information on the procedure for cross-section evaluation of platelet adhesion, see Zwaginga *et al.*[50]

Spraying of Collagen. Purified collagens are solubilized in 50 mM acetic acid, or fibrillar collagens are suspended in 20 mM Na$_2$HPO$_4$, pH

[46] C. Schröter-Kermani, I. Oechsner-Welpelo, and R. Kittelberger, *Immunol. Invest.* **19,** 475 (1990).

[47] R. R. Bruns and J. Gross, *Biopolymers* **13,** 931 (1977).

[48] B. R. Williams, R. A. Gelman, D. C. Poppke, and K. A. Piez, *J. Biol. Chem.* **253,** 6570 (1978).

[49] E. U. M. Saelman, L. F. Horton, M. J. Barnes, H. R. Gralnick, K. M. Hese, H. K. Nieuwenhuis, P. G. de Groot, and J. J. Sixma, *Blood* **82,** 3029 (1993).

[50] J. J. Zwaginga, J. J. Sixma, and P. G. de Groot, *Arteriosclerosis (Dallas)* **10,** 49 (1990).

7.0, and sprayed on glass coverslips (18 × 18 mm; Menzel Glaser) with a retouching airbrush (model 100; Badger Brush Co., Franklin Park, IL), connected to a nitrogen cylinder and operating at a pressure of 1 atm.[37,51] To obtain a uniform reactive collagen surface, fine collagen droplets are sprayed at a rate that allows the instantaneous drying of droplets on the glass surface at room temperature. About 85% of the applied collagen will be deposited on the glass surface, as can be determined by weighing the coverslips with a microbalance before and after spraying.[51] As a control on the spraying procedure, a coverslip can be stained with Coomassie Brilliant Blue to check for a uniform collagen deposition. After spraying the coverslips are blocked with 1% (w/v) human albumin solution in PBS for 30 min at room temperature. Electron microscopic evaluation has revealed that spraying acid-solubilized collagen on the coverslip apparently allows fiber formation on the glass surface. No essential change in adherent platelet morphology is observed when the collagen density on the coverslip is varied from 3 to 30 $\mu g/cm^2$. At the standard collagen density of 30 $\mu g/cm^2$, homogeneous platelet aggregate formation and highly reproducible platelet coverage of the surface are obtained.

Collagen-Containing Matrices. Endothelial cell matrix (ECM) can be used as a physiologically relevant collagen-containing matrix. Human umbilical vein endothelial cells are isolated and grown to confluence on a Thermanox coverslip as described.[50] Matrices are isolated by exposing endothelial cells to 0.1 M NH$_4$OH for 15 min at room temperature. The coverslips are subsequently rinsed three times with phosphate-buffered saline (10 mM sodium phosphate, 150 mM NaCl, pH 7.4).

Flow System

Perfusion Setup. A schematic representation of a perfusion setup is shown in Fig. 4. Steady flow is produced by a roller pump (P) that draws blood from a prewarmed (37°) reservoir through silastic tubing and delivers it to a funnel (F). The flow rate must be checked at 37° and is adjusted according to the empirically derived formula $Q = [(1/6)h^2L]j(o)$. In this formula flow Q is in microliters per second, h is height of the inlet of the perfusion chamber (the slit, in millimeters), L is the slit width (in millimeters), and $j(o)$ is the shear rate in inverse seconds (1/sec). The shear rate can be directly related to the forces acting in the vascular system. Under the influence of gravity, blood flows through the perfusion chamber (C) back into the reservoir. The chamber, the reservoir containing the perfusate, and part of the tubing are maintained in a water bath at

[51] H. J. Weiss, V. T. Turitto, and H. R. Baumgartner, *Thromb. Haemostasis* **65**, 202 (1991).

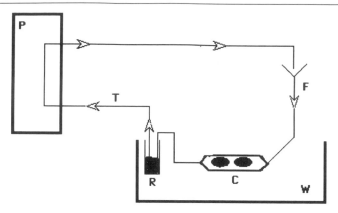

FIG. 4. Schematic representation of a perfusion system. The roller pump (P) draws blood from the perfusate reservoir (R) through silastic tubing (T). The blood is delivered to a funnel (F) and flows under gravitational force through the perfusion chamber (C) back into the reservoir. Both the perfusate reservoir and the chamber are kept in a 37° water bath (W).

37°. In this way, a recirculating system that produces a steady flow can be used to study adhesion under well-defined flow conditions.

Perfusion Chamber. The perfusion chamber used in this system is essentially that described by Sakariassen *et al.*[36] A parallel plate rectangular perfusion chamber, accommodating two coverslips, can be purchased from Ph. G. de Groot (Department of Haematology, University Hospital, Utrecht, the Netherlands).

Operation. Before the blood is circulated through the system, duplicate coverslips are inserted in the chamber and 15 ml of prewarmed Hepes-buffered saline (HBS: 10 mM HEPES, 150 mM NaCl, pH 7.35) must be drawn through the system to wash and warm up the system. Then 15 ml of whole blood, prewarmed at 37°, is recirculated through the chamber for 5 min at a calibrated shear rate. The pressure that the rollers from the pump exert on the tubing must not be excessive because this will result in platelet activation and lysis of erythrocytes. The setting must be such that blood is just drawn from the chamber, that is, a low occlusive setting. To minimize cell damage, it is also important to use low pumping rates by decreasing the number of pump cycles. Low pumping rates and high shear rates can be achieved by using perfusion chambers with small slit heights (<1 mm). The 5-min time interval of blood recirculation is chosen because after this time interval no additional platelet adhesion to collagens has been observed.[51] Shorter exposure times result in less adhesion but can be used to investigate early adhesive mechanisms as proposed by Weiss *et al.*[51] After 5 min of recirculation, blood is removed from the system and 15 ml of prewarmed HBS is drawn through the system to wash the coverslips. It is important to exclude air bubbles from the chamber during the transition from blood to HBS because these air bubbles can damage and/or nonspecifically activate the adherent platelets and may even detach them from the surface.

Quantitation/Evaluation

At the end of the static adhesion assay or perfusion, the coverslips are removed, rinsed with HBS, fixed with 0.5% (v/v) glutardialdehyde, dehydrated in methanol, and stained with May-Grunwald/Giemsa.[50] Platelet adhesion and morphology are evaluated with a light microscope at ×1000 magnification, connected to an image analyzer (i.e., AMS 40-10; Saffron Walden, East Sussex, United Kingdom). Evaluation is performed on 30 fields, perpendicular to the flow axis, in the center of each glass coverslip, and on 60 fields in the center of each Thermanox coverslip. In this way, reliable surface coverage values can be determined. Alternatively, platelets can be labeled with ^{111}In and perfusates can be reconstituted with labeled platelets, plasma, and packed erythrocytes (see below). Platelet coverage is subsequently determined by counting coverslips in a γ counter.

Single-Platelet Disappearance Controls

Before and after perfusion, samples of the perfusate are routinely taken to measure single-platelet disappearance (SPD) as a check on platelet clumping due to handling of blood, damage to platelets due to the roller pump, addition of reagents, and plasma proteins from exogenous sources.[52] Samples of blood (100 μl) are taken before and after an experiment and added to 900 μl of 0.5% (v/v) glutardialdehyde in PBS and 900 μl of 0.16% (w/v) K_2EDTA/PBS. After mixing well, sample platelets can be stored at room temperature but must be counted within 24 hr, using a Coulter counter. The percentage SPD is determined by the number of platelets in glutaraldehyde (A_G) subtracted from the number of platelets in EDTA (A_E) taken after the experiment divided by the number of platelets in EDTA (A_E) taken after the experiment: $(A_E - A_G)/A_E$. SPD values must be smaller than 20% and drop to zero when aggregation blocking reagents such as RGD peptides are used. They may be markedly increased when cross-linking antibodies are included.

---

[52] A. J. M. Verhoeven, M. E. Mommersteeg, and J. W. N. Akkerman, *Biochim. Biophys. Acta* **800,** 242 (1986).

Blood

Blood and Donors. Whole blood must be obtained from healthy volunteer donors who have not taken aspirin or other platelet function inhibitors in the preceding week. Blood can be anticoagulated with a 1/10 vol of low molecular weight heparin (200 U/ml) in 0.15 MNaCl [LMWH (Fragmin); Kabi Pharmacia, Stockholm, Sweden] when platelet adhesion to collagens underdivalent cation-dependent conditions is investigated. Alternatively, blood can be anticoagulated with 3.1% (w/v) trisodium citrate, a weak chelator of divalent cations, which still allows the formation of aggregates on several collagenous surfaces although a shear rate-dependent decrease in adhesion is usually observed because of depletion of divalent cations ($Ca^{2+} \leq 40$ μM, $Mg^{2+} \leq 80$ μM). Blood from patients with platelets deficient in membrane glycoproteins is a valuable tool for investigation of platelet function under flow conditions.[51,53,54]

Status of Blood. It is important to check a sample for preactivation of platelets and to perform a platelet count. If platelets are preactivated, reflected by an apparent increase in platelet volume, as determined with a Coulter counter, the blood should not be used because a considerable portion of the platelets will have formed small aggregates (doublets and triplets). The marked decrease of single platelets will result in low coverage of the adhesive surface. Blood with platelet counts less than 100,000/μl should not be used. Patterns of antibody inhibition may be altered when uncontrolled, activated platelets are used.

Reconstitution of Perfusates. To investigate the role of plasma proteins, blood used in perfusates may be reconstituted with plasma containing known concentrations of plasma proteins. Reconstituted perfusates are prepared as follows. Platelet-rich plasma (PRP) is obtained from whole blood by centrifugation (10 min at 200 G, 20°). One volume of Krebs–Ringer buffer (4 mM KCl, 107 mM NaCl, 20 mM NaHCO$_3$, and 2 mM Na$_2$SO$_4$, containing 19 mM trisodium citrate, pH 5.0) is added to 1 vol of PRP. The final pH is about pH 6. A platelet pellet is obtained by centrifugation (10 min at 500 g, 20°). The platelet pellet is resuspended in Krebs–Ringer buffer, pH 6.0, and washed twice by centrifugation (10 min at 500 g, 20°). After the second wash, platelets are resuspended to a concentration of 200,000 platelets/μl in dialyzed plasma containing unfractionated heparin (5 U/ml). Erythrocytes are washed three times with PBS containing 5 mM D-glucose (2000 g at 20°, twice for 5 min, the last time

[53] H. J. Weiss, V. T. Turitto, and H. R. Baumgartner, *J. Lab. Clin. Med.* **92,** 750 (1978).
[54] H. K. Nieuwenhuis, K. J. Sakariassen, W. P. M. Houdijk, P. F. E. M. Nievelstein, and J. J. Sixma, *Blood* **68,** 692 (1986).

for 15 min). Washed erythrocytes are added to the perfusate to obtain a hematocrit of 0.40, 15 min before perfusion.

Preparation of Components Used in Reconstitution of Blood. The isolation of plasma proteins, von Willebrand factor, and fibronectin has been described elsewhere.[55,56] When the participation of plasma proteins in platelet adhesion to collagen is investigated, collagen-sprayed coverslips are preincubated with von Willebrand factor (1 ristocetin cofactor unit/ml) or with fibronectin (300 μg/ml) for 1 h at room temperature, rinsed with HBS, and used directly. Plasma can be depleted of von Willebrand factor or fibronectin by immunodepletion and/or by affinity chromatography.

Inhibition by Antibodies/Peptide. Before reagents are used in a flow studies, it is advisable to test their inhibitory capacity in small-scale adhesion experiments under static conditions as well as in collagen-induced platelet aggregation assays. The concentration of reagent needed to inhibit adhesion/aggregation completely under flow conditions is often several times higher than that required in static adhesion or aggregation assays. When using inhibitory reagents in a flow experiment, it is important to check for single platelet disappearance as discussed above. A peptide or antibody is usually added to the perfusate and incubated at 37° just before the experiment is performed. If possible, incubation times should be kept as short as possible because these reagents can have deleterious effects on the platelets. In addition, peptides are prone to degradation by plasma proteases. When antibodies are used, an incubation time of 30–45 min has been found to be satisfactory in most cases. For protease-resistant peptides such as dRGDW, a 5-min incubation time was found optimal.[57]

Use of Synthetic Peptides

The use of synthetic peptides to prove integrin function has been considered in detail elsewhere in this series.[9] The approach is also applicable to the study of collagen receptors. Although some simple linear peptides derived from regions of collagen shown to promote adhesive activity have weak inhibitory activity in adhesion assays,[58,59] the sequences appear to be more active when incorporated into a triple-helical structure typical of collagenous molecules. The strategy for generation of triple-helical

[55] E. Ruoslahti, E. G. Hayman, M. D. Pierschbacher, and E. Engvall, this series, Vol. 82, p. 803.
[56] J. A. van Mourik and I. A. Mochtar, *Biochim. Biophys. Acta* **221,** 677 (1970).
[57] E. U. M. Saelman, K. M. Hese, H. K. Nieuwenhuis, A. Uzan, I. Cavero, G. Marguerie, J. J. Sixma, and P. G. de Groot, *Arteriosclerosis Thromb.* **13,** 1164 (1993).
[58] W. D. Staatz, J. J. Walsh, P. Pexton, and S. A. Santoro, *J. Biol. Chem.* **265,** 4778 (1990).
[59] W. D. Staatz, K. F. Fok, M. M. Zutter, S. P. Adams, B. A. Rodriguez, and S. A. Santoro, *J. Biol. Chem.* **266,** 7363 (1991).

structures is based on the observation of Sakakibara *et al.*[60] that a peptide consisting of 5 to 10 repeating units of Gly-Pro-Hyp will adopt a stable triple-helical conformation. Two examples of this application are described. Dedhar *et al.*[61] examined the properties of an RGDT (Arg-Gly-Asp-Thr) sequence incorporated into a triple helix in the peptide (Pro-Hyp-Gly)$_4$-Ala-Pro-Gly-Leu-Arg-Gly-Asp-Thr-Gly-(Pro-Hyp-Gly)$_4$ and we have examined the properties of a triple-helical peptide containing the DGEA (Asp-Gly-Glu-Ala) sequence. The peptide (Gly-Pro-Hyp)$_5$-Gly-Ala-Asp-Gly-Glu-Ala-(Gly-Pro-Hyp)$_5$ is 10- to 30-fold more effective as an inhibitor of platelet adhesion to collagen than related linear DGEA-containing peptides (Fig. 5).

The two sequences illustrate the key features in the design of triple-helical peptides. Each end of the peptide should be composed of a stretch of four to five repeats of the Gly-Pro-Hyp prototypical collagen sequence. The sequence of interest should be positioned to maintain glycine residues at every third position throughout the entire peptide sequence. This feature is essential for maintenance of the triple-helical structure. In addition to the usual chemical analysis, the presence of triple-helical structure should be established by spectroscopic techniques (circular dichroism spectroscopy is one useful approach). The "melting curve" should be established by studies at increasing temperatures, typically over the range of 4–50°.[62] This information is critical for the use of these peptides in adhesion studies because the experiments using the peptides need to be carried out at temperatures sufficiently low to ensure triple-helical conformation. This may often necessitate temperatures of 17–20°. The required length of the peptides gives rise to several drawbacks, namely low synthetic yields and high cost.

Use of Three-Dimensional Collagen Gels

Three-dimensional matrices may be useful models for the study of tissue morphogenesis and remodeling, and of cell invasion and metastasis. For example, endothelial cells and various epithelial cells grow in an organized manner and show characteristics of morphological differentiation when cultured in three-dimensional collagen gel matrices.[63–65] In the

[60] S. Sakakibara, K. Inouze, K. Shudo, Y. Kishida, Y. Kobayashi, and D. J. Prockop, *Biochim. Biophys. Acta* **303,** 198 (1973).

[61] S. Dedhar, E. Ruoslahti, and M. D. Pierschbacher, *J. Cell Biol.* **104,** 585 (1987).

[62] R. K. Rhodes and E. J. Miller, *Biochemistry* **17,** 3442 (1978).

[63] M. L. Li, J. Aggeler, D. A. Farson, C. Hatier, J. Hassell, and M. J. Bissell, *Proc. Natl. Acad. Sci. U.S.A.* **84,** 136 (1987).

[64] M. Pignatelli and W. F. Bodner, *Proc. Natl. Acad. Sci. U.S.A.* **85,** 5561 (1988).

[65] J. R. Gamble, L. J. Matthias, G. Meyer, P. Kaur, G. Russ, R. Faull, M. C. Berndt, and M. A. Vadas, *J. Cell Biol.* **121,** 931 (1993).

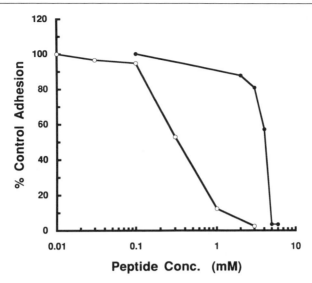

FIG. 5. Enhanced inhibitory activity of the DGEA sequence when constrained in a triple-helical conformation (●). Linear DGEA peptide; (○) triple-helical peptide.

best studied example, growth in collagen gels is required to maintain cell-specific gene expression of primary cultures of mammary epithelial cells.[63] Several studies using inhibitory antibodies implicate the $\alpha_2\beta_1$ integrin in various *in vitro* responses of cells to collagen gels.[66–68]

Collagen Gel Composition. Collagen gels are frequently made from one of two sources: relatively pure collagen preparations or the more heterogeneous Engelbreth-Holm–Swarm (EHS) extracellular matrix. Gels composed of stromal collagen, usually derived from rat tail collagen, are composed of 95–98% collagen I and the remainder of collagen III.[64] These gels have the advantage of being simple and of more defined composition, but may not entirely mimic the complex environment cells encounter *in vivo*. Gels derived from extracellular matrix of the EHS tumor, such as Matrigel,[69] contain (at a minimum) collagen IV, laminin, proteoglycans, and entactin. Although EHS tumor matrix more closely resembles a base-

[66] C. E. Klein, D. Dressel, T. Steinmayer, C. Mauch, B. Eckes, T. Krieg, R. B. Bankert, and L. Weber, *J. Cell Biol.* **115,** 1427 (1991).
[67] J. A. Shiro, B. M. C. Chan, W. T. Roswit, P. D. Kassner, A. P. Pentland, M. E. Hemler, A. Z. Eisen, and T. S. Kupper, *Cell (Cambridge, Mass.)* **67,** 403 (1991).
[68] F. Berdichevsky, C. Gilbert, M. Shearer, and J. Taylor-Papadimitriou, *J. Cell Sci.* **102,** 437 (1992).
[69] H. K. Kleinman, M. L. McGarvey, J. R. Hassell, V. L. Star, F. B. Cannon, G. W. Laurie, and G. R. Martin, *Biochemistry* **25,** 312 (1986).

ment membrane, its composition is less well defined and more subject to variation than that of collagen I gels.

Preparation of Collagen Gels. The collagen used for collagen I gels is commercially available in solution form (Collaborative Biomedical Products, Bedford, MA) or lyophilized (Sigma). The collagen is dissolved in 0.1% acetic acid at a concentration of 3–5 mg/ml. This solution can be sterilized by adding chloroform and dialyzing into a buffered salt solution in the cold.[70] The collagen solution and all other components of the gel are stored at 4° and kept cold until gelation is desired. To form a gel, the solution is neutralized by adding an equal volume of neutralizing buffer (2× Hanks' balanced salt solution plus 50 mM HEPES) and enough culture medium to make a final gel concentration of 0.7–1.5 mg/ml. Alternatively, the collagen solution can be dialyzed against three changes of water at 4° and diluted in medium prior to use.[71] Cells are added to the solution at this point (see below), transferred into petri dishes or wells, and then warmed to 37° for 5–30 min until gelation occurs.

EHS matrix, or Matrigel, is also commercially available (Collaborative Biomedical Products) or can be isolated as described,[69] and is stored at −20°. The commercial gel solution is thawed on ice, and to it is added a 30–50% volume of ice-cold culture medium. The amount of medium required varies with each lot of Matrigel and must be empirically determined. Gelation occurs after warming to 37°.

Cells can be incorporated into the gel in several ways. Commonly, they are seeded into the gel by including them, usually at a concentration around 1–3 × 10^5/ml, in the gel mixture prior to gelation. Alternatively, cells are sandwiched between two layers of collagen by seeding cells onto a previously made collagen gel, allowing them to attach for 2–3 hr, and layering a second collagen gel on top of the cell layer. Cells can also be cultured two-dimensionally by seeding them on top of a collagen gel. Once cells are incorporated into the gel, culture medium is layered over the gels at twice the volume. Medium is changed every 2–4 days by gently replacing part of the volume. To create floating collagen gels, which frequently result in better cell responses than attached collagen gels, the edge of the gel is rimmed with a sterile Pasteur pipette, and the dish is shaken gently until the gel is free floating.

To study the role of collagen receptors in cellular responses to collagen, it is usually necessary to add inhibitory antibodies or peptides to the gels. After cells are cast into the gels, and the gels have polymerized, an equal volume of solution containing antibodies or peptides at twice the desired

[70] B. Obrink, this series, Vol. 82, p. 516.
[71] C. H. Streuli and M. J. Bissel, *J. Cell Biol.* **110,** 1405 (1990).

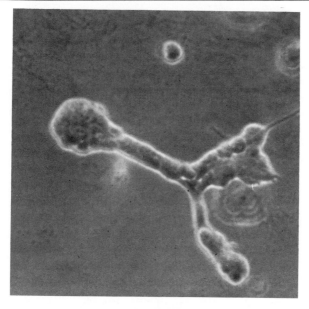

FIG. 6. Morphological organization of well-differentiated human breast carcinoma cells in a collagen gel. T47D cells were cultured for 10 days in a three-dimensional matrix composed of collagen I (1.3 mg/ml) (Collaborative Biomedical Products).

final concentration is added to the gel. The solution should be changed as needed every few days.

Culture of Mammary Cells in Collagen I Gels. In addition to the well-documented effects of three-dimensional collagen matrices on primary cells, certain established cell lines differentiate morphologically when cultured in collagen gels.[64,72] We have found that the well-differentiated human breast carcinoma cell line T47D can undergo morphological differentiation when cultured in gels composed of 1.3 mg of collagen I per milliliter (Fig. 6). Cells were incorporated into gels at a concentration of 2×10^5 cells/ml. It was necessary to culture cells for 8–10 days before morphological changes were noted. Increasing the density of cells seeded into the gels did not shorten this interval, suggesting culture in collagen gels initiates a time-dependent program of cellular differentiation.

Collagen Gel Contraction. Several cell types, such as fibroblasts and melanoma cells, contract collagen gels, a process believed to relate to wound contraction and tissue remodeling. The collagen receptor $\alpha_2\beta_1$ integrin has been implicated in this process by the use of inhibitory antibod-

---

[72] R.Montesano, G. Schaller, and L. Orci, *Cell (Cambridge, Mass.)* **66,** 697 (1991).

ies.[66,67] Cells at a density of $1-3 \times 10^5$ cells/ml are cast into collagen gels (0.65–1 mg/ml) as described above. The gels are rimmed with a sterile pipette and allowed to float in the medium. Gel diameter (in millimeters) is monitored over several hours (platelets) to 2–4 days (fibroblasts and melanoma cells).

Invasion Assays Using Collagen Gels. Many tumor cells migrate through collagen gels, providing an *in vitro* assessment of invasive potential.[73] Several permutations of invasion assays employing collagen gels have been developed. For example, gels composed of either collagen I or Matrigel (50 μg/ml) can be cast on top of a polycarbonate membrane containing pores that are 8–12 μm in diameter. This membrane is suspended in a transwell or Boyden chamber, and frequently a chemoattractant, such as fibronectin (10 μg/ml) or 3T3 fibroblast-conditioned medium, is added to the lower compartment.[73] Cells in serum-free medium containing 0.1% (w/v) BSA are seeded on top of the gel at 10^5/ml and incubated for 2–24 hr. Noninvasive cells are scraped off, and cells invading through to the membrane are fixed. Filters can be effectively and quickly stained using Diff-quick (Baxter, McGraw, IL), mounted on slides with Permount (Sigma), and counted. A variation of this assay is to cast collagen I gels, which are sometimes supplemented with various stimuli such as fibronectin or collagen IV, into multiwells. Cells are cultured at a concentration of 50,000/ml on top of the gels for 1–6 days, and invasion is monitored on a daily basis. Cells migrating into the gel are counted at successive 20-μm levels using an inverted phase-contrast microscope.[74,75]

An additional assay is the Matrigel outgrowth assay, which takes advantage of the fact that metastatic cells tend to form invasive colonies when embedded in three-dimensional gels of Matrigel. Cells are cultured in Matrigel for 2–10 days, and scored qualitatively for colonies showing outwardly migrating groups of cells. For examples of high and low invasive phenotypes, see Bae *et al.*[76]

Antisense Approaches to Study of Collagen Receptors

Although blocking antibodies and peptides have been useful to determine the role of collagen receptors in a number of *in vitro* studies, their

[73] A. Albini, Y. Iwamoto, H. K. Kleinman, G. R. Martin, S. A. Aaronson, J. M. Kozlowski, and R. N. McEwan, *Cancer Res.* **47**, 3239 (1987).

[74] A. E. Faassen, J. A. Schrager, D. J. Klein, T. R. Oegema, J. R. Couchman, and J. B. McCarthy, *J. Cell Biol.* **116**, 521 (1992).

[75] A. E. Faassen, D. L. Mooradian, R. T. Tranquillo, R. B. Dickinson, P. C. Letourneau, T. R. Oegema, and J. B. McCarthy, *J. Cell Sci.* **105**, 501 (1993).

[76] S.-N. Bae, G. Arand, H. Azzam, P. Pavasant, J. Torri, T. L. Frandsen, and E. W. Thompson, *Breast Cancer Res. Treat.* **24**, 241 (1993).

use is limited in long-term or *in vivo* investigations, or in systems for which such antibodies do not exist. For example, it is difficult to study the role of integrins in the multiple steps involved in invasion and metastasis *in vivo* by using inhibitory peptides or antibodies, which have short half-lives *in vivo*. Additionally, examples of cell-signaling events being triggered by blocking antibodies suggest that interpretation of the role of integrins in *in vitro* studies in which blocking antibodies have been used may not always be straightforward.[77,78]

It would thus be of benefit in the study of collagen receptors to be able to decrease or eliminate expression of the receptor on the cell surface and thereby determine its function in cell adhesion. To this end, there has been interest in antisense approaches to decrease expression of integrin subunits. For example, Lallier and Bronner-Fraser[79] were able to inhibit partially neural crest cell adhesion and integrin subunit expression by using synthetic antisense oligonucleotides. The effects were temporary and reversible owing to the short half-life of oligonucleotides. Using a different approach, Hayashi *et al.*[80] expressed antisense RNA for chick β_1 integrin and were able to decrease partially β_1 integrin expression and adhesion to fibronectin. The paucity of studies using antisense techniques to decrease integrin expression attests to the difficulty of this approach.

Considerations in Choosing Antisense Approach

Three strategies are currently available for generating antisense reagents: oligodeoxynucleotides, antisense RNA, or ribozymes (which are not discussed here). For a more complete discussion, see Colman[81] or Neckers and Whitesell.[82] The use of synthetic oligodeoxynucleotides, which are added at high molar ratios directly to the medium, differs from generation of antisense RNA in that one does not need to generate an antisense expression construct or transfect cells. However, oligodeoxynucleotides are degraded by nucleases in culture and are thus short lived. In contrast, antisense RNA is generated by introducing an appropriate cDNA into cells, either transiently or stably. Transient transfections are more rapid and may result in greater levels of expression of the introduced DNA than stable transfections, but may also result in a heterogeneous

[77] A. S. Menko and D. Boettiger, *Cell (Cambridge, Mass.)* **51,** 57 (1987).
[78] Z. Werb, P. M. Tremble, O. Behrendtsen, E. Crowley, and C. H. Damsky, *J. Cell Biol.* **109,** 877 (1989).
[79] T. Lallier and M. Bronner-Fraser, *Science* **259,**692 (1993).
[80] Y. Hayashi, T. Iguchi, T. Kawashima, Z. Z. Bao, C. Yacky, D. Boettiger, and A. F. Horwitz, *Cell Struct. Funct.* **16,** 241 (1991).
[81] A. Colman, *J. Cell Sci.* **97,** 399 (1990).
[82] L. Neckers and L. Whitesell, *Am. J. Physiol.* **265,** L1 (1993).

population of transfected cells expressing different levels of antisense RNA. Stable transfectants may take a few months to generate, but result in clonal populations with defined and quantifiable levels of antisense RNA expression and reduction of target protein levels. The stable and defined phenotype may be important in interpreting data involving cell–matrix interactions, because it is not clear how a mixed population of cells with variable changes in receptor levels may affect cell adhesion data.

Devising an Antisense Expression Construct: Use of Epstein–Barr Virus-Based Vector

It is not entirely clear how best to choose the complementary target sequence that will serve as the template for antisense mRNA production. Sequences complementary to the 5' end that cover the translation start site, or to the 3' end, or those complementary to intron–exon boundaries, have all been successful under certain circumstances.[81,82]

Success with antisense RNA techniques requires a high ratio of antisense : sense mRNA expression. To achieve this, it is important to consider, for each given system, the strength of the promoter being used to drive the antisense mRNA, and the copy number per transfected cell of the construct that will be used. An effective way to achieve multiple copies per cell is to use the self-replicating Epstein–Barr virus (EBV) episome as a vector to express antisense RNA. When transfected into cells expressing the EBV nuclear antigen (EBNA), these episomes, which contain the EBV origin of replication (oriP), are stably maintained in cells under selection conditions. Additionally, the expression of the antisense RNA can be reversed by removing the transfected cells from selection conditions. Using this approach, Hambor et al.[83] effectively inhibited CD8 expression in T cells. Similar episomal systems based on the bovine papillomavirus or the SV40 (simian virus 40) large T antigen are also available.

Use of Antisense RNA to Decrease α_2 Integrin Expression on Mammary Epithelial Cells

We have had success in decreasing α_2 integrin levels on the well-differentiated breast carcinoma cell line T47D, using an antisense RNA approach. A 1.3-kb fragment of the α_2 integrin cDNA, representing the first 353 amino acids and the translation start site, was cloned into the multiple cloning site of the pREP 4 expression vector (InVitrogen, San Diego, CA) to generate pREP4a2' (Fig. 7). The pREP 4 vector contains

---

[83] J. E. Hambor, C. A. Hauer, H.-K. Shu, R. K. Groger, D. R. Kaplan, and M. L. Tykocinski, *Proc. Natl. Acad. Sci. U.S.A.* **85**, 4010 (1988).

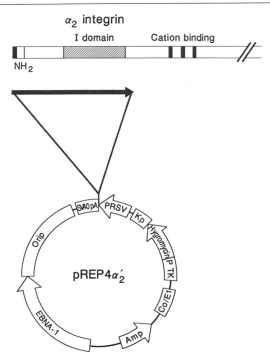

FIG. 7. DNA construct, pREP4α2', used to generate antisense RNA expressing cells. A 1.3-kb fragment of the α_2 integrin cDNA, comprising bp 1 to 1296, was cloned in the antisense orientation relative to the RSV promoter in the pREP4 vector (InVitrogen). The cDNA fragment covers the first 353 amino acids of the coding sequence, as indicated by the map of the α_2 integrin protein.

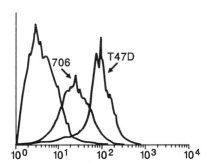

FIG. 8. Flow cytometric analysis of T47D cells expressing α_2 integrin antisense RNA. Cells were incubated with anti-α_2 integrin antibody, P1E6 (GIBCO), followed by goat–anti-mouse–fluorescein isothiocyanate (FITC) secondary antibody, and analyzed on a Beckman FACScan instrument. A representative clone (706) is shown in comparison to the original cell line (T47D) and staining with secondary antibody only (unlabeled profile).

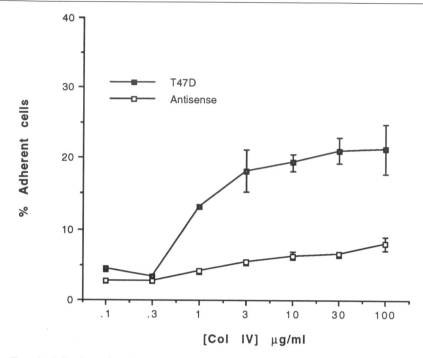

FIG. 9. Adhesion of cells expressing α_2 integrin RNA to collagen. A representative antisense (□) clone and T47D (■) cells were allowed to adhere to increasing concentrations of type IV collagen.

an EBV oriP, a hygromycin resistance gene for selection in mammalian cells, and the Rous sarcoma virus (RSV) promoter to drive mRNA expression. This vector was transfected into an EBNA-expressing T47D cell line. To generate the EBNA⁺ T47D cell line, cells were cotransfected with pCMVEBNA, which contains the EBNA gene driven by the cytomegalovirus (CMV) promoter, and pSV2neo, to provide selection by neomycin resistance (both from Clontech, Palo Alto, CA) at a 10:1 ratio of CMV-EBNA : pSV2neo. Transfectants were selected in 300 μg of G418 (Geneticin; GIBCO-BRL, Gaithersburg, MD) per milliliter. For all transfections, a 75-cm³ flask of cells at 70% confluency was transfected with 20 μg of plasmid DNA, using the Lipofectin reagent (GIBCO-BRL), for 8 hr. After transfection of pREP4a2′ into EBNA⁺ T47D cells, transfectants were selected in hygromycin B (125 μg/ml; Calbiochem, San Diego, CA), a concentration determined to kill 100% of T47D cells in the absence of the hygromycin reistance gene. After about 1 month, small colonies of cells were scraped off the flask and expanded as clones. Cells were maintained in hygromycin B to avoid loss of the pREP4a2' episome.

Analysis of Transfected Cells

As an initial screen, clones were analyzed for decreased α_2 integrin expression by flow cytometry (Fig. 8), using the P1E6 antibody (GIBCO-BRL) against α_2 integrin. Clones showing decreased α_2 integrin expression in this manner were then analyzed for their ability to attach to collagens I and IV in an adhesion assay. It had previously been determined that $\alpha_2\beta_1$ integrin was the major collagen receptor on T47D cells, because blocking anti-α_2 integrin antibodies markedly reduced T47D cell adhesion to collagen. Clones exhibiting decreased α_2 integrin expression also showed decreased adhesion to both collagens I and IV (Fig. 9), and diminished ability to spread on collagen (not shown). The analysis of such antisense expressing cells should include analysis of the expression of other integrin subunits to establish specificity. The adhesion onto control substrates not mediated by the targeted integrin should be examined and shown to be unaffected. Southern blotting procedures should be used to establish the presence of vector within the cells. An attempt should be made to establish the presence of the antisense mRNA, although experience with several systems suggests that this latter goal is not always readily achievable.[81,82]

Using this approach, we have been able to demonstrate that antisense techniques can be used to decrease integrin expression to a point that has consequences for cellular function. This approach should represent an additional tool with which to investigate further the role of collagen receptors in cell–matrix interactions.

[9] Receptors on Platelets

By RANDALL J. FAULL, XIAOPING DU, and MARK H. GINSBERG

Introduction

Hemostasis and thrombosis in health and disease depend on platelet adhesive interactions, as sequential platelet attachment, spreading, secretion, and aggregation generate a hemostatic plug. Platelets are normally nonadhesive within the circulation, but in the presence of the appropriate stimulus will attach to each other, or to other cells, or to extracellular

Copyright © 1994 by Academic Press, Inc.
All rights of reproduction in any form reserved.

matrix components. For example, injury to the endothelial cell monolayer exposes the subendothelial matrix, permitting access to proteins (such as fibronectin, vitronectin, and von Willebrand factor) that can support platelet adhesion. Subsequent cell spreading stabilizes the platelet–matrix interaction to resist the shear stress of the overlying flowing blood. The adherent platelets also secrete a number of substances. Some of these stimulate circulating platelets to acquire new adhesive properties, with subsequent high-affinity fibrinogen binding, and platelet aggregation and attachment to the matrix-bound platelets.

Specific receptor–ligand interactions are the basis of these crucial adhesive events. Platelets express a number of cell surface adhesive proteins, of which five are members of the integrin family ($\alpha_{IIb}\beta_3$, $\alpha_v\beta_3$, $\alpha_2\beta_1$, $\alpha_5\beta_1$, and $\alpha_6\beta_1$).[1] This chapter describes assays used in our laboratory for the functional assessment of these receptors. Particular emphasis is placed on $\alpha_{IIb}\beta_3$ (also known as GPIIb-IIIa), which is the most prominent of the platelet adhesive receptors ($\sim$40,000 receptors/cell) and the most abundant cell surface protein (accounting for $\sim$15% of the protein mass of the platelet membrane). It is also found in platelet α granules,[2] and this internal pool is surface expressed following platelet activation.[3] $\alpha_{IIb}\beta_3$ is the platelet receptor for fibrinogen,[4–8] although platelet activation[4,9] and the presence of divalent cations[10] are necessary for ligand-binding competence with a dissociation constant (K_d) of approximately 0.3 μM.[9,11] The available evidence supports the conclusion that the alteration in ligand-binding affinity is due to a conformational change in the receptor.[12,13] The biochemical basis for this change is uncertain, but it does

[1] M. H. Ginsberg, J. C. Loftus, and E. F. Plow, *Thromb. Haemostasis* **59**, 1 (1988).
[2] J. D. Wencel-Drake, E. F. Plow, T. J. Kunicki, V. L. Woods, D. M.Keller, and M. H. Ginsberg, *Am. J. Pathol.* **124**, 324 (1986).
[3] K. Niiya, E. Hodson, R. Bader, V. Byers-Ward, J. A. Koziol, E. F. Plow, and Z. M. Ruggeri, *Blood* **70**, 475 (1987).
[4] J. S. Bennett and G. Vilaire, *J. Clin. Invest.* **64**, 1393 (1979).
[5] J. S. Bennett, J. A. Hoxie, S. F. Leitman, G. Vilaire, and D. B. Cines, *Proc. Natl. Acad. Sci. U.S.A.* **80**, 2417 (1983).
[6] J. S. Bennett, G. Vilaire, and D. B. Cines, *J. Biol. Chem.* **257**, 8049 (1982).
[7] L. V. Parise and D. R. Phillips, *J. Biol. Chem.* **260**, 10698 (1985).
[8] B. S. Coller, E. I. Peerschke, L. E. Scudder, and C. A. Sullivan, *J. Clin. Invest.* **72**, 325 (1983).
[9] G. A. Marguerie, E. F. Plow, and T. S. Edgington, *J. Biol. Chem.* **254**, 5357 (1979).
[10] D. R. Phillips and A. K. Baughan, *J. Biol. Chem.* **258**, 10240 (1983).
[11] G. A. Marguerie, N. Ardaillou, G. Cherel, and E. F. Plow, *J. Biol. Chem.* **257**, 11872 (1982).
[12] T. E. O'Toole, J. C. Loftus, X. Du, A. A. Glass, Z. M. Ruggeri, S. J. Shattil, E. F. Plow, and M. H. Ginsberg, *Cell Regul.* **1**, 883 (1990).
[13] P. J. Sims, M. H. Ginsberg, E. F. Plow, and S. J. Shattil, *J. Biol. Chem.* **266**, 7345 (1991).

allow macromolecular ligands better access to the receptors.[14] There is sufficient fibrinogen in the plasma ($10\ \mu M$) to ensure saturation of these receptors following platelet activation. Functional $\alpha_{IIb}\beta_3$ is absolutely required for platelet aggregation, and the bleeding diathesis in Glanzmann's thrombasthenia is due to either absence of or expression of a dysfunctional form of $\alpha_{IIb}\beta_3$.[15–18] $\alpha_{IIb}\beta_3$ is also a receptor for fibronectin, vitronectin, and von Willebrand factor,[19–22] although fibrinogen would occupy most receptors on activated platelets in the circulation owing to its higher plasma concentration. However, fibronectin, vitronectin, and von Willebrand factor, but not fibrinogen, are present in the subendothelial matrix, and these ligands may be more important in that microenvironment.

Platelets express only low numbers of the vitronectin receptor $\alpha_v\beta_3$,[23,24] and its functional significance is uncertain. They also express receptors for collagen ($\alpha_2\beta_1$), fibronectin ($\alpha_5\beta_1$), and laminin ($\alpha_6\beta_1$), and each of these (unlike $\alpha_{IIb}\beta_3$) is capable of supporting the adhesion of resting platelets to their respective ligands.[25–28]

Two important nonintegrin platelet adhesive receptors should be mentioned, although they are not discussed in detail in this chapter. The first

[14] B. S. Coller, *J. Cell Biol.* **103**, 451 (1986).

[15] A. T. Nurden and J. P. Caen, *Br. J. Haematol,* **28**, 253 (1974).

[16] D. R. Phillips and P. P. Agin, *J. Clin. Invest.* **60**, 535 (1977).

[17] M. H. Ginsberg, A. Lightsey, T. J. Kunicki, A. Kaufmann, G. A. Marguerie, and E. F. Plow, *J. Clin. Invest.* **78**, 1103 (1986).

[18] J. C. Loftus, T. E. O'Toole, E. F. Plow, A. A. Glass, A. L. Frelinger, and M. H. Ginsberg, *Science* **249**, 915 (1990).

[19] E. F. Plow and M. H. Ginsberg, *J. Biol. Chem.* **256**, 9477 (1981).

[20] L. V. Parise and D. R. Phillips, *J. Biol. Chem.* **261**, 14011 (1986).

[21] R. Pytela, M. D. Pierschbacher, M. H. Ginsberg, E. F. Plow, and E. Ruoslahti, *Science* **231**, 1559 (1986).

[22] E. F. Plow, A. H. Srouji, D. Meyer, G. A. Marguerie, and M. H. Ginsberg, *J. Biol. Chem.* **259**, 5388 (1984).

[23] S. C.-T. Lam, E. F. Plow, S. E. D'Souza, D. A. Cheresh, A. L. Frelinger, and M. H. Ginsberg, *J. Biol. Chem.* **264**, 3742 (1989).

[24] B. S. Coller, U. Seligsohn, S. M. West, L. E. Scudder, and K. J. Norton, *Blood* **78**, 2603 (1991).

[25] W. D. Staatz, S. M. Rajpara, E. A. Wayner, W. G. Carter, and S. M. Santoro, *J. Cell Biol.* **108**, 1917 (1989).

[26] T. J. Kunicki, D. J. Nugent, S. J. Staats, R. P. Orchekowski, E. A. Wayner, and W. G. Carter, *J. Biol. Chem.* **263**, 4516 (1988).

[27] R. S. Piotrowicz, R. P. Orchekowski, D. J. Nugent, K. Y. Yamada, and T. J. Kunicki, *J. Cell Biol.* **106**, 1359 (1988).

[28] A. Sonnenberg, P. W. Modderman, and F. Hogervorst, *Nature (London)* **336**, 487 (1988).

is GPIb-IX, which is a receptor for von Willebrand factor.[29,30] Platelets from patients with the Bernard-Soulier syndrome lack GPIb-IX, and have a marked bleeding diathesis.[31] The second is P-selectin (or GMP-140, GPIIa, PADGEM, CD62), a member of the selectin family of carbohydrate-binding adhesive proteins.[32,33] P-Selectin is contained within α granules and not on the surface of resting platelets, but is surface expressed following platelet activation. It mediates platelet interactions with monocytes and neutrophils.[34]

This chapter describes the following assays for investigating the function of $\alpha_{IIb}\beta_3$: (1) soluble ligand-binding assays using ^{125}I-labeled fibrinogen and fibronectin, in order to measure changes directly in ligand-binding affinity following platelet stimulation, (2) flow cytometric analysis of platelet $\alpha_{IIb}\beta_3$, using monoclonal antibodies that recognize conformation-sensitive epitopes on the receptor, and (3) estimation of fibrinogen binding to purified $\alpha_{IIb}\beta_3$ "captured" onto plastic by a receptor-specific monoclonal antibody.

Binding of Radiolabeled Soluble Ligand (Fibrinogen or Fibronectin) to $\alpha_{IIb}\beta_3$ on Platelets

The binding of both soluble fibrinogen and soluble fibronectin to resting platelets is negligible, but they bind to $\alpha_{IIb}\beta_3$ with moderate affinity following platelet activation by agonists such as thrombin, ADP, and epinephrine.[4-9,19] This property, as well as the availability of reliable methods for purification of both proteins, means they are convenient soluble ligands for direct measurements of changes in the binding affinity (i.e., functional state) of $\alpha_{IIb}\beta_3$.

Preparation of Human Platelets

Platelets are isolated from fresh ACD (acid–citrate–dextrose)-anticoagulated human blood by differential centrifugation and gel filtration on Sepharose 2B (Pharmacia, Piscataway, NJ), as previously described (this series, Vol. 169, p. 11, and Ginsberg et al.[35]).

[29] Z. M. Ruggeri, L. De Marco, L. Gatti, R. Bader, and R. R. Montgomery, *J. Clin. Invest.* **72,** 1 (1983).

[30] K.-J. Kao, S. V. Pizzo, and P. A. McKee, *J. Clin. Invest.* **63,** 656 (1979).

[31] A. T. Nurden, D. Didry, and J. P. Rosa, *Blood Cells* **9,** 333 (1983).

[32] R. P. McEver, *Thromb. Haemostasis* **65,** 223 (1991).

[33] E. Larsen, A. Cell, G. E. Gilbert, B. C. Furie, J. K. Erban, R. Bonfanti, D. D. Wagner, and B. Furie, *Cell (Cambridge, Mass.)* **59,** 305 (1989).

[34] G. I. Johnston, R. G. Cook, and R. P. McEver, *Cell (Cambridge, Mass.)* **56,** 1033 (1989).

[35] M. H. Ginsberg, L. Taylor, and R. G. Painter, *Blood* **55,** 661 (1980).

Preparation of Human Fibrinogen and Fibronectin

Fibrinogen is purified from human plasma anticoagulated with a one-sixth volume of ACD solution [0.065 M citric acid, 0.085 M sodium citrate, 2% (w/v) dextrose].[9,36]

1. Centrifuge 1 liter of plasma for 20 min at 2500 rpm at 4°, to remove any residual cells (discard pellet).

2. Cool 1 liter of plasma to 0°, and add 215 ml of a 1 : 1 (by volume) mixture of 95% ethanol and 0.055 M sodium citrate, pH 7.0, at 4°. Stir while cooling to −3° in an ethanol bath, and let stand for 15 min (fine white precipitate forms).

3. Centrifuge for 20 min at 2500 rpm at −3°.

4. To the precipitate add 550 ml of a 1 : 11.5 (by volume) mixture of 95% ethanol and 0.055 M sodium citrate, pH 6.4, at 4°, and stir for 30 min at −3°.

5. Centrifuge for 30 min at 2500 rpm at −3°.

6. Dissolve the precipitate in 275 ml of 0.055 M sodium citrate, 0.2 M ε-aminocaproic acid, pH 6.4, at room temperature (takes ∼20 min). It is important to stir carefully so that foam does not form, but also to do this step as quickly as possible as prolonged exposure to room temperature is harmful.

7. Cool to 0° with slow constant stirring, and while stirring for 10 to 15 min add 49 ml of a mixture of 9.6 ml of absolute ethanol and 63.2 ml of 0.055 M sodium citrate, pH 6.4, at 4°. Let stand for 15 min at 0°, and a gummy precipitate will form.

8. Centrifuge for 20 min at 2500 rpm at −3°, and save the supernatant (contains the fibrinogen).

9. Slowly add 140 ml of a mixture of 22.5 ml of absolute ethanol and 90.0 ml of 0.055 M sodium citrate, pH 6.4, at 4°, stirring while cooling to −3°. Let stand for 15 min and a fine white precipitate will form.

10. Centrifuge for 20 min at 2500 rpm at −3°.

11. Dissolve the precipitate in 1000 ml of 0.055 M trisodium citrate, pH 7.0, at 4°, with occasional stirring; this takes about 60 min.

12. While stirring, add dropwise 351.35 ml of saturated ammonium sulfate, pH 7.0, at 4° (26% saturation). Stir for 60 min at 4°, and a coarse white precipitate will form.

13. Centrifuge for 30 min at 2500 rpm at 4°.

14. Dissolve the precipitate overnight in the smallest volume possible (∼50 ml) of phosphate-buffered saline (PBS: 0.15 M NaCl, 0.01 M sodium

[36] R. A. Kekwick, M. E. MacKay, M. H. Nance, and B. R. Record, *Biochem. J.* **60,** 671 (1955).

phosphate) containing 1 mM benzamidine and 0.2 M ε-aminocaproic acid, pH 7.0, then dialyze against PBS for 2–3 days (change the PBS twice daily).

15. Check the OD_{280} ($E_{1\,cm}$ = 15.5), aliquot, and store at −70°.

Fibronectin is also prepared from ACD-anticoagulated fresh human plasma[37] (less than 2 days old, stored at room temperature). The entire procedure is performed at room temperature.

1. Centrifuge the plasma at 5000 rpm for 30 min.

2. Filter through Whatman (Clifton, NJ) filter paper (#2) by gravity, then add 0.05% (w/v) sodium azide, 5 mM ethylenediaminetetraacetic acid (EDTA), and 1 mM benzamidine.

3. Preequilibrate gelatin–Sepharose in a siliconized glass column by washing with two to four bed volumes of PBS–5 mM EDTA–0.05% (w/v) sodium azide–1 mM benzamidine, pH 7.0 (column stored in PBS–5 mM EDTA–0.05% sodium azide when not in use).

4. Load plasma onto the column (up to five plasma volumes per column volume), and wash with PBS–5 mM EDTA–0.05% sodium azide–1 mM benzamidine until the OD_{280} is less than 0.05 (two to three column volumes).

5. Wash successively with the same volume of 1 M NaCl–1 mM benzamidine, pH 7.0; 1 M urea–1 mM benzamidine; and PBS–5 mM EDTA–0.05% sodium azide.

6. Elute the column with 1 M NaBr–0.02 M acetic acid, pH 5.0, using a fraction collector; fibronectin comes off as viscous strands.

7. Pool the fractions with peak absorption at 280 nm ($E_{1\,cm}$ = 1.28), and immediately dialyze against PBS, pH 7.0, with three changes of buffer. The fibronectin typically has a concentration of 1–2mg/ml, and the yield is about 50%.

8. Freeze in aliquots and store at −70°. Thaw at 37°, and avoid repeated freeze-thawing.

Labeling of Fibrinogen and Fibronectin

Fibrinogen is labeled with iodine-125 using a modified chloramine-T method.

1. To a tube containing 1 mg of fibrinogen in 50–200 μl of 0.05 M phosphate buffer (pH 7.2) add 1 mCi of ^{125}I and 10 μl of 0.1% (w/v) chloramine-T (in phosphate buffer). Label larger amounts of fibrinogen in batches of 1 mg or less.

2. Mix and incubate for 1 min at room temperature.

[37] J. Forsyth, E. F. Plow, and M. H. Ginsberg, this series, Vol. 215, p. 311.

3. Add 10 μl of 0.1% (w/v) sodium metabisulfite and 25 μl of 1% (w/v) potassium iodide (both in phosphate buffer) to stop the reaction.

4. Dialyze against 0.3 M NaCl–5 mM HEPES, pH 7.2 [pretreated with Chelex 100 (Bio-Rad, Richmond, CA) to remove divalent cations] for 24 to 48 hr, with several changes of buffer.

5. To determine the concentration, read the OD_{280} of a 1:10 dilution of labeled fibrinogen.

6. Store aliquots at −70°.

Fibronectin is labeled with ^{125}I according to the following modified chloramine-T method.

1. Add 1 mCi of ^{125}I and 10 μl of 0.1% (w/v) chloramine-T (in phosphate-buffered saline, pH 7.2) to a tube containing 1 mg of fibronectin (approximate fibronectin concentration, 1 mg/ml in PBS). Label the fibronectin in batches of 1 mg or less.

2. Mix and incubate for 5 min at room temperature.

3. Stop the reaction by adding 10 μl of 0.1% (w/v) sodium metabisulfite and 25 μl of 1% (w/v) potassium iodide (both in PBS).

4. Pool aliquots, and transfer 10 μl to 990 μl of PBS plus 1% (w/v) bovine serum albumin (BSA).

5. To the remainder add a one-tenth volume of 10% (w/v) BSA (heat inactivated, 56° for 60 min) plus 2 mM phenylmethylsulfonyl fluoride (PMSF) in PBS. Dialyze overnight at 4° against PBS to remove free iodine.

6. Determine the concentration of labeled fibronectin by trichloroacetic acid (TCA) precipitation of pre- and postdialysis samples.

7. Store aliquots at −70°.

Soluble Fibrinogen- and Fibronectin-Binding Assays

The basic assay for measuring the binding of either ^{125}I-labeled fibrinogen or fibronectin to platelets is described below. Depending on the needs of the experiment, a number of variations can be introduced, provided that the core method (i.e., separation of bound from unbound ligand by centrifugation through a sucrose gradient) is left intact. These can include variations in platelet or ligand concentration, incubation time, temperature, divalent cation composition or concentration, agonists, inhibitors, or controls. Platelets should be used within 2 hr of preparation, although they are stable for at least 12 hr in platelet-rich plasma. The incubations are performed in 1.5-ml polypropylene Eppendorf centrifuge tubes, and the centrifugation through a sucrose gradient is performed in 400-μl microcentrifuge tubes (e.g., West Coast Scientific, Emeryville, CA). The latter should be pretested to check that they are sufficiently robust for microcentrifugation, yet soft enough for the tips to be comfortably amputated with a razor blade. The microcentrifugation steps are performed at ~12,000 rpm in the horizontal position—a recommended microcentrifuge is the Beckman 11 (Beckman, Palo Alto, CA). All dilutions (including platelets) are made into a modified Tyrode's buffer [NaCl (137.5 mmol/liter), KCl (2.6 mmol/liter), NaHCO$_3$ (12 mmol/liter), glucose (5.5 mmol/liter), 0.1% (w/v) bovine serum albumin, pH 7.4, plus divalent cations as required (e.g., 2 mmol of MgCl$_2$ per liter)].

1. Microcentrifuge the stock solution of ^{125}I-labeled fibrinogen or fibronectin for 5 min to pellet microaggregates, and dilute as required into the modified Tyrode's buffer.

2. To a 1.5-ml tube add 40 μl of ^{125}I-labeled ligand of the desired concentration, 40 μl of buffer or competing ligand, and 80 μl of the platelet suspension (final concentration of platelets should be 0.5–1.0 × 10^8 cells/ml).

3. Add 40 μl of the platelet stimulus (e.g., human α-thrombin) to make a total volume of 200 μl), and incubate for the desired time at room temperature.

4. Following the incubation, layer triplicate 50-μl aliquots of the mixed suspension onto 300 μl of 20% (w/v) sucrose (in the same modified Tyrode's buffer) in the microcentrifuge tubes.

5. Microcentrifuge for 3 min at room temperature.

6. Amputate the tube tips with a razor blade, taking care to cut as close as practical to the cell pellet. Change the razor blade frequently to keep it sharp.

7. Count the tips in a γ scintillation spectrometer, as well as an aliquot of the labeled ligand, in order to quantify both the input and bound ligand.

Comments

The nonsaturable binding of the labeled ligand can be estimated by the inclusion of both a platelet-free tube and another containing a 50× excess of unlabeled ligand. Alternative methods for estimating specific binding of the ligand include the inclusion of tubes containing either (1) a molar excess of EDTA or ethylene glycol-bis(β-aminoethyl ether)-N,N,N',N'-tetraacetic acid (EGTA) (to chelate divalent cations), (2) specific blocking monoclonal antibodies directed against either $\alpha_{IIb}\beta_3$ or the ligand, or (3) peptides containing the arginine-glycine-aspartic acid (RGD) sequence.

The assay is readily utilized to study the effects on $\alpha_{IIb}\beta_3$ function of agonists such as α-thrombin, ADP, or phorbolmyristate acetate (PMA) (see above), or monoclonal antibodies that inhibit or activate integrin function.[12,38-40]

The data can be expressed as either counts per minute (cpm) bound, picograms bound per specified number of platelets (calculated from the specific activity of the ligand), or molecules of ligand bound per platelet. The latter is calculated assuming that the molecular mass of fibrinogen is 3.4×10^5 Da and that of fibronectin is 4.4×10^5 Da. Binding isotherms that are generated can be analyzed by Scatchard plots or by utilizing nonlinear least-squares curve fitting (the LIGAND program[41]). The latter assumes simple mass action, binding to a finite number of independent binding sites, and that the nonsaturable binding component is a constant fraction of the free ligand.

Analyzing Function of $\alpha_{IIb}\beta_3$ Using Flow Cytometry

The availability of monoclonal antibodies that recognize conformationally sensitive epitopes on $\alpha_{IIb}\beta_3$ now permits analysis, using flow cytometry, of the functional state of the receptor on platelets.[42,43] These antibodies include PAC-1,[44] which, like fibrinogen, binds to $\alpha_{IIb}\beta_3$ following platelet activation in an RGD-inhibitable manner; and a series of antibodies that bind with higher affinity to the ligand-occupied form of $\alpha_{IIb}\beta_3$ (the "LIBS" antibodies[38-40]). The advantages of this method are its simplicity and the lack of need for radioactive iodine. A disadvantage is that the total number of receptors on the cell surface is more difficult to quantitate directly. However, the percentage occupied can easily be estimated by comparison with the fluorescence signals of antibodies that recognize the nonactivated form of the $\alpha_{IIb}\beta_3$ complex. This method has been used to distinguish platelet aggregation disorders due to defective ligand-binding capacity (neither PAC-1 binding nor increased LIBS antibody binding following treatment with ADP, and no increased LIBS antibody binding in the presence of RGD-containing peptide) from those due to defective platelet activation (neither PAC-1 binding nor increased LIBS antibody binding

[38] A. L. Frelinger, S. C.-T. Lam, E. F. Plow, M. A. Smith, J. C. Loftus, and M. H. Ginsberg, *J. Biol. Chem.* **263**, 12397 (1988).

[39] A. L. Frelinger, I. Cohen, E. F. Plow, M. A. Smith, J. Roberts, S. C.-T. Lam, and M. H. Ginsberg, *J. Biol. Chem.* **265**, 6346 (1990).

[40] A. L. Frelinger, X. Du, E. F. Plow, and M. H. Ginsberg, *J. Biol. Chem.* **266**, 17106 (1991).

[41] P. J. Munson and D. Rodbard, *Anal. Biochem.* **107**, 220 (1980).

[42] S. J. Shattil, M. Cunningham, and J. A. Hoxie, *Blood* **70**, 307 (1987).

[43] M. H. Ginsberg, A. L. Frelinger, S. C.-T. Lam, J. Forsyth, R. McMillan, E. F. Plow, and S. J. Shattil, *Blood* **76**, 2017 (1990).

[44] S. J. Shattil, J. A. Hoxie, M. Cunningham, and L. F. Brass, *J. Biol. Chem.* **260**, 11107 (1985).

following ADP, but increased LIBS antibody binding in the presence of RGD-containing peptide.[43]

The same agonists, inhibitors, or activating antibodies as described above can be studied using this assay system.

1. The monoclonal antibodies are directly conjugated to fluorescein isothiocyanate according to standard methods to achieve a fluorescein/protein molar ratio of 3 to 6[42].

2. Platelet-rich plasma is obtained from fresh venous blood anticoagulated with a one-tenth volume of 3.8% (w/v) sodium citrate. The erythrocytes and leukocytes are sedimented at 180 g for 20 min at room temperature.

3. Five-microliter aliquots of the platelet-rich plasma are added to polypropylene tubes containing 25 μl of a fluoresceinisothiocyanate-conjugated monoclonal antibody (10^{-9} to 10^{-6} mol/liter) in Tyrode's buffer [NaCl (137.5 mmol/liter), KCl (2.6 mmol/liter), NaHCO$_3$ (12 mmol/liter), MgCl$_2$ (2 mmol/liter), 1% (w/v) bovine serum albumin, pH 7.4].

4. The volume is made up to 50 μl with agonists and/or peptides diluted in the same buffer, and the mixture incubated without stirring for 15 min at room temperature.

5. Dilute samples to 0.5 ml with Tyrode's buffer for analysis on a flow cytometer. Acquire the light scatter and fluorescence signals at logarithmic gain, and analyze 10,000 platelets/sample.

Comments

The results are conveniently expressed as mean platelet fluorescence intensity in arbitrary fluorescence units, or as histograms of log platelet fluorescence intensity in arbitrary units on the abscissa and platelet number on the ordinate.

Fibrinogen Binding to Purified $\alpha_{IIb}\beta_3$ Captured onto Plastic by Specific Monoclonal Antibody

Most of the available evidence supports the model that both platelet activation and ligand binding induce a conformational change in $\alpha_{IIb}\beta_3$.[13,38,45] That this is an intrinsic property of the receptor rather than due to changes in the adjacent microenvironment is supported by the observations that activating anti-$\alpha_{IIb}\beta_3$ monoclonal antibodies can induce high-affinity fibrinogen binding to both paraformaldehyde-fixed platelets

[45] L. V. Parise, S. L. Helgerson, B. Steiner, L. Nannizzi, and D. R. Phillips, *J. Biol. Chem.* **262**, 12597 (1987).

and to isolated purified $\alpha_{IIb}\beta_3$.[12,40] The following assay was designed to study the functional properties of purified $\alpha_{IIb}\beta_3$ receptor "captured" onto plastic microtiter wells by an anti-β_3 monoclonal antibody. This led to the observation that RGD-containing ligand-mimetic peptides can also induce high-affinity fibrinogen binding to isolated $\alpha_{IIb}\beta_3$ receptors.[46] The capturing monoclonal antibody (MAb) generally used in the assay (MAb 15[46]) is directed against the β_3 chain and has no known effect on platelet function. Other nonfunctional antibodies should also be suitable, and they can be directed against either chain of the receptor.

Purification of $\alpha_{IIb}\beta_3$

The $\alpha_{IIb}\beta_3$ can be purified by either RGD-affinity chromatography[21] or in a "resting" form by extraction from platelet membranes and sequential chromatography. Moreover, this assay can also be applied to $\alpha_{IIb}\beta_3$ in a crude platelet lysate.[46] The lysate is prepared from platelets isolated from fresh ACD-anticoagulated human blood as described and referenced above. The platelets are washed three times in modified Tyrode's buffer, pH 6.5 [N-2-hydroxyethylpiperazine-N'-2-ethanesulfonic acid (HEPES, 5 mmol/liter), NaCl (150 mmol/liter), KCl (2.5 mmol/liter), NaHCO$_3$ (12 mmol/liter), glucose (5.5 mmol/liter), 0.1% (w/v) bovine serum albumin], containing prostaglandin E$_1$ (250 ng/ml) and $\sim 1 \times 10^9$ platelets/ml, solubilized in ice-cold lysing buffer [HEPES (10 mmol/liter), NaCl (150 mmol/liter), octyl-β-D-glucopyranoside (SO mmol/liter), CaCl$_2$ (1 mmol/liter), MgCl$_2$ (1 mmol/liter), PMSF (1 mmol/liter), N-ethylmaleimide (8 mmol/liter), and prostaglandin E$_1$ (PGE$_1$, 250 ng/ml, pH 7.4]. Insoluble particles are removed by centrifuging the lysate at 100,000 g for 30 min.

Fibrinogen Binding to Solubilized Captured $\alpha_{IIb}\beta_3$

1. Coat MAb 15 onto microtiter wells (Immulon 2 removawell strips; Dynatech Laboratories, Chantilly, VA) by addition of 50 μl/well of a 15-μg/ml solution in sodium bicarbonate buffer (pH 9.5) and incubation at 4° for 48 hr.

2. Drain the wells and block with 150 μl/well of 2.5% (w/v) bovine serum albumin in 10 mM HEPES–saline, pH 7.4, for 2 hr at room temperature or overnight at 4°. The plates can be stored at $-20°$ for at least 1 month.

3. Incubate platelet lysates (or purified $\alpha_{IIb}\beta_3$) with or without synthetic peptides (e.g., 1 mM RGDS, GRGDSP, GRGESP, HHLGGAKQAGDV) or activating monoclonal antibody for 30 min at 4°.

[46] X. Du, E. F. Plow, A. L. Frelinger, T. E. O'Toole, J. C. Loftus, and M. H. Ginsberg, *Cell (Cambridge, Mass.)* **65,** 409 (1991).

4. Wash the wells three times with modified Tyrode's buffer, pH 7.4 [HEPES (2.5 mmol/liter), NaCl (150 mmol/liter), KCl (2.5 mmol/liter), $CaCl_2$ (1 mmol/liter), $MgCl_2$ (1 mmol/liter), $NaHCO_3$ (12 mmol/liter), glucose (5.5 mmol/liter), 0.1% (w/v) bovine serum albumin], and drain.

5. Add platelet lysate (or purified $\alpha_{IIb}\beta_3$), 50 μl/well, and incubate overnight at 4°.

6. Wash the wells twice with modified Tyrode's buffer, pH 7.4, containing octyl-β-D-glucopyranoside (50 mmol/liter), then twice with the modified Tyrode's buffer, pH 7.4, only.

7. Add 2% (w/v) bovine serum albumin in modified Tyrode's buffer, pH 7.4 (150 μl/well) for 1 hr at room temperature, then wash once with modified Tyrode's buffer, pH 7.4.

8. Add 50 μl per well of ^{125}I-labeled fibrinogen at the desired concentration (e.g., 100 nM final concentration), diluted in modified Tyrode's buffer containing 2% (w/v) bovine serum albumin, and incubate for 4 hr at room temperature.

9. Wash the wells three times with modified Tyrode's buffer, and count the bound radioactivity with a γ scintillation spectrometer.

Comments

The amount of $\alpha_{IIb}\beta_3$ captured by MAb 15 should be quantified in parallel each time by the binding of a ^{125}I-labeled monoclonal antibody gainst $\alpha_{IIb}\beta_3$.

Nonspecific binding of the ^{125}I-labeled fibrinogen can be estimated in several ways: (1) by addition of the labeled fibrinogen in the presence of either RGD-containing peptide, blocking monoclonal antibody, or excess unlabeled fibrinogen; (2) by performing the assay using microtiter wells coated with an irrelevant antibody; or (3) by adding lysis buffer alone to the MAb 15-coated wells.

Acknowledgments

This work was supported by NIH Grants HL28235, HL48728, and AR27214. X.D. was supported by a research fellowship from the American Heart Association, California Affiliate. This is Publication No. 8516-VB from The Scripps Research Institute.

[10] Hyaluronic Acid Receptors

By IVAN STAMENKOVIC and ALEJANDRO ARUFFO

Hyaluronan (hyaluronic acid, HA) is an abundant connective tissue polysaccharide belonging to the glycosaminoglycan class of molecules. Unlike most other glycosaminoglycans, hyaluronan is not covalently associated with a protein core. Chemical analysis has shown that hyaluronan is a linear polymer composed of repeating disaccharide units with the structure $[\beta$-D-glucuronic acid-$(1 \rightarrow 3)$-β-N-acetyl-D-glucosamine-$(1 \rightarrow 4)]_n$.[1] Because of its structural properties, hyaluronan was long thought to be an inert substance whose chief function was to regulate the hydration of extracellular matrix (ECM). The discovery of the existence of a specific hyaluronate receptor on the cell surface stimulated interest in the molecule and its relationship with cells. Hyaluronan is synthesized in the plasma membrane of fibroblasts by addition of carbohydrates to the reducing end of the polysaccharide, while the nonreducing extremity penetrates the extracellular milieu. Production of hyaluronan is increased at sites of limb development, wound healing, and tumor invasion.[2] Although there is still much speculation as to its principal physiological function, hyaluronan has been proposed to fulfill diverse roles that can be subdivided into two broad categories. The first includes functions resulting from interaction with other components of the extracellular matrix. The second comprises events related to interactions with cells.

Within the ECM, hyaluronan is bound to several proteoglycans. This is perhaps best illustrated in cartilage, where hyaluronan plays an important structural role, and where it is bound to the cartilage proteoglycan aggrecan.[3,4] The bond is stabilized by link proteins[5,6] that participate in the generation of aggregates with a molecular mass in the vicinity of 10^8 Da, that are deposited within the collagen network. The physical properties

[1] K. Meyer and J. W. Palmer, *J. Biol. Chem.* **107**, 629 (1934).

[2] B. P. Toole, *in* "Cell Biology of Extracellular Matrix" (E. D. Hay, ed.), p. 259. Plenum, New York, 1981.

[3] R. G. LeBaron, D. R. Zimmermann, and E. Ruoslahti, *J. Biol. Chem.* **267**, 10003 (1992).

[4] T. E. Hardingham and A. J. Fosang, *FASEB J.* **6**, 861 (1992).

[5] F. Deak, I. Kiss, K. J. Sparks, W. S. Argrave, G. Hampikian, and P. F. Goetinck, *Proc. Natl. Acad. Sci. U.S.A.* **83**, 3766 (1986).

[6] K. J. Doege, J. R. Hassell, B. Caterson, and Y. Yamada, *Proc.Natl. Acad. Sci. U.S.A.* **83**, 3761 (1986).

Copyright © 1994 by Academic Press, Inc.
All rights of reproduction in any form reserved.

of hyaluronan allow it to behave as an efficient osmotic buffer that helps regulate water homeostasis and plasma protein distribution in tissues.[7,8]

Hyaluronan is also implicated in a variety of cellular functions that include cell proliferation and activation,[9,10] cell–cell interactions,[11] and migration.[11] These functions are dependent on interactions between hyaluronan and specific receptors on the cell surface.[11]

Hyaluronan-binding sites were first identified in cartilage link protein, with the use of monoclonal antibodies (MAb) and synthetic peptides.[12] The sites were found to be located in tandemly repeated sequences containing clusters of positively charged amino acids. Highly homologous sequences are found at the amino terminus of aggrecan,[3,4] the fibroblast proteoglycan versican,[3] and a tumor necrosis factor-inducible protein, TSG-6.[13] In aggrecan, this domain has been confirmed to be the hyaluronan-binding region.[12]

Although several distinct cell surface hyaluronan-binding molecules are likely to exist, so far only two have been characterized at the molecular level. The first is CD44, a broadly distributed cell surface glycoprotein[14,15] with multiple isoforms generated by alternate splicing of at least 10 exons encoding the membrane-proximal portion of the extracellular domain.[16] The amino terminus of CD44 displays 30% homology with link protein tandem repeats[14] and contains the hyaluronan-binding domain.[17] The second cell surface hyaluronan receptor is a fibroblast protein known as RHAMM, an acronym for the receptor for hyaluronan-mediated motility.[18] Interestingly, RHAMM does not display significant homology to the link protein hyaluronan-binding domain.[18] Studies on CD44 have shown that

[7] T. C. Laurent, in "Chemistry and Molecular Biology of the Intracellular Matrix" (E. A. Balaz, ed.), p. 703. Academic Press, London, 1970.

[8] T. C. Laurent and J. R. E. Fraser, *FASEB J.* **6**, 2398 (1992).

[9] M. Brecht, U. Mayer, E. Schlosser, and P. Prehm, *Biochem. J.* **239**, 445 (1986).

[10] P. W. Noble, F. R. Lake, P. H. Henson, and D. W. H. Riches, *J. Clin. Invest.* **91**, 2368 (1993).

[11] C. B. Underhill, *Ciba Found. Symp.* **143**, 87 (1989).

[12] P. F. Goetinck, N. S. Stripe, P. A. Tsonis, and D. Carlone, *J. Cell Biol.* **105**, 2403 (1987).

[13] T. H. Lee, H.-G. Wisniewski, and J. Vilcek, *J. Cell Biol.* **116**, 545 (1992).

[14] I. Stamenkovic, M. Amiot, J. M. Pesando, and B. Seed, *Cell (Cambridge, Mass.)* **56**, 1057 (1989).

[15] L. A. Goldstein, D. F. Zhou, L. J. Picker, C. N. Minty, R. F. Bargatze, J. F. Ding, and E. C. Butcher, *Cell (Cambridge, Mass.)* **56**, 1063 (1989).

[16] G. R. Screaton, M. V. Bell, D. G. Jackson, F. B. Cornelis, U. Gerth, and J. I. Bell, *Proc. Natl. Acad. Sci. U.S.A.* **89**, 12160 (1992).

[17] R. J. Peach, D. Hollenbaugh, I. Stamenkovic, and A. Aruffo, *J. Cell Biol.* **122**, 257 (1993).

[18] C. Hardwick, K. Hoare, R. Owens, H. P. Holn, M. Hook, D. Moore, V. Cripps, L. Austen, D. M. Nance, and E. A. Turley, *J Cell Biol.* **117**, 1343 (1992).

the CD44 molecule may participate in lymphocyte activation,[19-22] cell–cell adhesion,[23] and cell–ECM interaction.[24,25]

We have developed new and adapted conventional approaches for the study of hyaluronan receptors and their functional role in physiological and pathological conditions. Our work has focused on the CD44 molecule and we detail some of the lines of investigation that have contributed to our understanding of the function of CD44. It is hoped that this will provide a framework for the study of the role of other cell surface molecules that bind hyaluronan and other glycosaminoglycans.

Development of Soluble Recombinant CD44–Immunoglobulin Fusion Proteins

Developments in recombinant DNA technology, such as the polymerase chain reaction (PCR) methodology, have allowed the facile preparation of recombinant genes encoding soluble forms of cell surface receptors. Using this methodology recombinant genes encoding the extracellular domain of CD44 have been generated.[26] To facilitate protein purification and the use of the soluble receptor in structure–function studies the recombinant genes encoding soluble forms of the CD44 receptor consisted of a cDNA fragment encoding the complete extracellular domain of CD44 genetically fused onto a DNA fragment encoding the hinge, CH2, and CH3 domains of human immunoglobulin G (IgG). This chimeric gene encodes a fusion protein containing two functional domains: the amino-terminal domain corresponds to the extracellular domain of CD44, whereas the carboxy-terminal domain consists of the Fc portion of human IgG$_1$ (Fig. 1). Using this approach, chimeric genes encoding soluble forms of various wild-type isoforms of CD44, as well as truncated and mutant forms of CD44, have been prepared. Fusion proteins derived from these

[19] S. Huet, H. Groux, B. Caillou, H. Valentin, A.-M. Prieur, and A. Bernard, *J. Immunol.* **143**, 798 (1989).

[20] Y. Shimizu, G. A. Van Seventer, R. Siraganian, L. Wahl, and S. Shaw, *J. Immunol.* **143**, 2457 (1989).

[21] S. M. Denning, P. T. Le, K. H. Singer, and B. F. Haynes, *J. Immunol.* **144**, 7 (1990).

[22] D. S. A. Webb, Y. Shimizu, G. A. Van Seventer, S. Shaw, and T. L. Gerrard, *Science* **249**, 1295 (1990).

[23] T. St. John, J. Meyer, R. Idzerda, and W. M. Gallatin, *Cell (Cambridge, Mass.)* **60**, 45 (1990).

[24] S. Jalkanen, R. F. Bargatze, L. R. Heron, and E. C. Butcher, *Eur. J. Immunol.* **16**, 1195 (1986).

[25] W. G. Carter and E. A. Wayner, *J. Biol. Chem.* **263**, 4193 (1988).

[26] A. Aruffo, I. Stamenkovic, M. Melnick, C. B. Underhill, and B. Seed, *Cell (Cambridge, Mass.)* **61**, 1303 (1990).

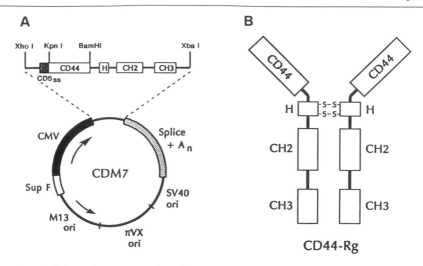

FIG. 1. Schematic representation of the Ig expression vector (A), including the CD44–Ig chimera containing nucleotide sequences encoding the CD5 secretory signal peptide, and the CD44–Rg dimer (B).

constructs have been used to establish the function of CD44 as a hyaluronan receptor,[26] assess the different ability of various CD44 isoforms to bind hyaluronan,[27,28] examine the tissue distribution of HA during development,[29] determine the role of CD44-mediated cell adhesion in tumor growth and metastasis,[30,31] and identify domains and individual residues within these domains that are responsible for hyaluronan binding.[17]

Chimeric constructs encoding the CD44–Rg fusion proteins are typically prepared by PCR. Synthetic oligonucleotide primers flanking the cDNA fragment encoding the complete extracellular domain of the different CD44 isoforms or fragments thereof are used to amplify the CD44 domains of interest. The primers are designed to contain endonuclease restriction sites so as to facilitate ligation of PCR-amplified sequences to the Ig expression vector. The Ig vector was developed in the CDM8 plasmid[32] and contains elements of the CDM8 polylinker. However, the presence of polylinker endonuclease restriction sites in the genomic DNA of human IgG allows only the *Xho*I site to be used at the 5′ end. In-frame

[27] I. Stamenkovic, A. Aruffo, M. Amiot, and B. Seed, *EMBO J.* **10,** 343 (1991).
[28] A. Bartolazzi and I. Stamenkovic, unpublished observations, 1994.
[29] B. A. Fenderson, I. Stamenkovic, and A. Aruffo, *Differentiation (Berlin)* **54,** 85 (1993).
[30] M.-S. Sy, Y.-J. Guo, and I. Stamenkovic, *J. Exp. Med.* **174,** 859 (1991).
[31] M.-S. Sy, Y.-J. Guo, and I. Stamenkovic, *J. Exp. Med.* **176,** 623 (1992).
[32] A. Aruffo and B. Seed, *Proc. Natl. Acad. Sci. U.S.A.* **84,** 8573 (1987).

ligation to the IgG Fc portion is achieved using a *Bam*HI site. Thus, forward and reverse oligonucleotide primers for amplification of sequences encoding the extracellular domain of CD44 require the presence of an *Xho*I site and a *Bam*HI, or *Bam*HI-compatible, site, respectively. Because the nucleotide sequence corresponding to the extracellular domain of CD44 contains a *Bam*HI restriction site, the *Bam*HI-compatible *Bgl*II site was used. To ensure that the ligation to Ig Fc sequences is in frame, the reverse primer must be designed so that the GAT trinucleotide of the *Bam*HI and *Bgl*II restriction sites (GGA TCC and AGA TCT, respectively) forms a codon. For amplification of the entire extracellular domain of CD44, the following oligonucleotide primers would be used:

Forward primer: 5′ CAC GGG **CTC GAG** ATG GAC AAG TTT TGG TGG CAC GCA 3′
Reverse primer: 5′ CAC G**AG ATC T**TC TGG AAT TTG GGG TGT CCT TAT AGG 3′

Restriction sites are shown in boldface type. The ATC trinucleotide of the *Bgl*II site that forms a codon is the reverse complement of GAT. An overhang of four to six nucleotides 5′ to the restriction site appears necessary to stabilize restriction endonuclease interaction with DNA substrate.

The PCR reactions are performed as follows: 2 μg of template (full-length CD44 cDNA) is mixed with 1 μg of each primer, a 0.5 mM concentration of each dNTP, and a buffer containing 500 mM KCl, 100 mM Tris-HCl (pH 8.4), 40 mM MgCl$_2$, and bovine serum albumin (BSA, 0.6 mg/ml) (for a 10× solution) in a total volume of 100 μl. Because the primers have a long (24–27 nucleotide) uninterrupted complementarity to the template, high-stringency conditions can be used for amplification. We typically use the following: 94° for 1 min, 60° for 2 min, 72° for 3 min for a total of 30 cycles. Ten microliters of amplified fragments is tested on a 1% (w/v) agarose gel. The remaining 90 μl is phenol extracted once, ethanol precipitated, and resuspended in a 30- to 50-μl volume of water, depending on the abundance of the amplified material. Appropriate restriction enzymes and buffer are added and digestion allowed to proceed overnight at 37°. The vector is digested simultaneously with corresponding endonucleases and a fraction of digested amplicon and vector electrophoresed on a 1.5% (w/v) low-melt (NUSieve) agarose minigel (American Bioanalytical, Natick, MA), slabs of agarose containing appropriate fragments are excised and melted at 70° for 5 min, and 5 μl of each fragment is added to a ligation mix containing 5 μl of 10× ligase buffer (New England BioLabs, Beverly, MA), 35 μl of water, and 0.1 μl of T4 DNA ligase (New England BioLabs). Ligation is allowed to proceed for 4 hr at 22° or overnight at 15°. A fraction of the ligation mix is used to transform MC1061/

P3 *Escherichia coli* cells and single colonies are picked and episomal DNA tested for content of appropriate inserts.

The plasmid constructs containing inserts of appropriate size are amplified, purified on a cesium gradient, and transfected into appropriate mammalian host cells.We typically use COS cells because the Ig expression vector contains a simian virus 40 (SV40) origin of replication, faciliating the overproduction of recombinant proteins. In addition, being simian fibroblasts, COS cells possess the machinery necessary for correct processing of most mammalian glycoproteins. Exceptions might include proteins whose specific glycosylation might be essential for function. Because glycosylation is often cell type specific, COS cells may be deficient in certain glycosyltransferases. One example is α-2,6-sialyltransferase, whose expression in COS cells is insufficient to promote cellular interactions that depend on α-2,6-linked sialic acid.[33,34] For the most part, however, when the polypeptide structure is itself critical, recombinant proteins overproduced in COS cells conserve their native function.

COS cells are transfected by the DEAE-dextran method, which in our hands provides an average transfection efficiency of about 50%. The transfection medium consists of DEAE-dextran (final concentration, 400 μg/ml), 100 μM chloroquine in Dulbecco's Modified Eagle's Medium (DME) supplemented with 10% (v/v) NUserum (Collaborative Research, Waltham, MA) (a semisynthetic serum that reduces the likelihood of DNA precipitation). Transfection medium and 2 μg or less of recombinant plasmid DNA per milliliter are added to plates of semiconfluent COS cells, and the incubation allowed to proceed for 4 hr at 37°. Following transfection the medium is removed, cells are shocked with a 10% (v/v) solution of dimethyl sulfoxide (DMSO) in phosphate-buffered saline (PBS) for 2 min, the DMSO–PBS is aspirated, and the cells are overlaid with 8–10 ml of DME–10% fetal bovine serum (FBS) overnight. On the following day, the medium is aspirated and replaced by 10 ml of serum-free DME. Three to 5 ml of serum-free medium is added every 2 days and the cells cultured for a total of 8–10 days. Use of serum-free medium eliminates contamination of purified Rg molecules by bovine immunoglobulins and has no detectable adverse effect on COS cells during the culture period. Eight to 10 days following transfection, supernatants are harvested and Rg proteins purified.

The purification protocol exploits the presence of human IgG$_1$ and murine IgG$_{2b}$ carboxy-terminal domains of the Rg fusion proteins. Super-

[33] D. Sgroi, A. Varki, S. Braesch-Andersen, and I. Stamenkovic, *J. Biol. Chem.* **268,** 7011 (1993).
[34] L. D. Powell, D. Sgroi, E. R. Sjöberg, I. Stamenkovic, and A. Varki, *J. Biol. Chem.* **268,** 7019 (1993).

natants are absorbed onto protein A–Sepharose columns. Typically a 1.5-ml column of packed protein A–Sepharose beads (Repligen, Cambridge, MA) is prepared per liter of supernatant. The supernatant is allowed to flow by gravity over the column at 1 ml/min for at least two passages (usually with the aid of a pump, as several liters of supernatant are purified at a time). Protein A beads are then eluted with 3 ml of 0.1 M citric acid, pH 3.3. The eluate is immediately neutralized with Tris base to a pH of 8.0 and dialyzed overnight against PBS in a Spectra/Por dialysis bag (Houston, TX). Recovered protein is then quantified using an Amersham (Arlington Heights, IL) protein assay kit. For most of the Rgs we have produced, yields of 2–6 μg of Rg/ml of supernatant may be expected.

Although the procedures outlined above represent a standard for Rg production, some problems have been encountered in the preparation of CD44–Rg. For example, it was observed that a chimeric gene encoding the complete extracellular domain of the 90-kDa isoform of CD44 fused to human IgG directed the expression of a CD44–Rg protein that was inefficiently secreted by COS cells. When the natural CD44 amino-terminal secretory signal sequence of this protein was replaced with the signal sequence from another lymphocyte cell surface protein, CD5, CD44–Rg was expressed and secreted.[26] This is in contrast to the chimeric genes encoding other CD44 isoforms that direct the expression and secretion of their respective CD44–Rg fusion proteins even though they have the same amino-terminal signal sequence as the 90-kDa isoform of CD44.[17] Similar PCR technology has been used to prepare recombinant immunoglobulin fusions of both truncation and point mutants of CD44.[17,26]

The ability of CD44 to function as an HA receptor appears to be regulated *in vivo*.[35,36] Isoforms that bind to HA as soluble recombinant proteins or when expressed in certain tumor cell lines have been reported not to bind to hyaluronan on normal cells.[35] In particular, the 90-kDa isoform of CD44 predominantly expressed by lymphocytes appears not to bind HA when the cells are in a resting state.[35] Activating the cells or treating them with the appropriate anti-CD44 MAb has been shown to drive the protein into a conformation that allows hyaluronan binding.[35] On the other hand, the Ig fusion of this CD44 isoform appears to be locked in an HA-binding conformation. These observations suggest that CD44–Rg may not invariably reflect the behavior of cell surface CD44, and therefore data obtained using recombinant soluble forms of CD44 must be interpreted with caution.

[35] R. Hyman, J. Lesley, and R. Schulte, *Immunogenetics* **33**, 392 (1991).
[36] Q. He, J. Lesley, R. Hyman, K. Ishihara, and P. W. Kincade, *J. Cell Biol.* **119**, 1711 (1992).

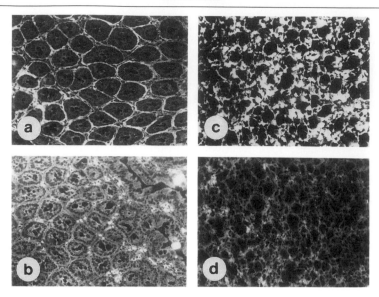

Fig. 2. Immunofluorescence histology of intestine (a and b) and renal papilla (c and d) using CD44–Rg. Frozen tissues, mounted onto microscope slides, were incubated with CD44–Rg (a and c) and an unrelated Rg molecule (b and d), washed, incubated with a goat anti-human affinity-purified fluorescein-conjugated antibody, washed again, and examined by fluorescence microscopy as described in text.

Use of CD44–Rg in Tissue Staining

One of the advantageous applications of RG molecules in general, and of CD44–Rg in particular, is the possibility of performing tissue staining in a manner similar to that applied to the use of monoclonal antibodies (MAbs). Frozen tissue sections are mounted onto slides, air dried for 10 min, and incubated with CD44–Rg (5 μg/ml in PBS or serum-free DME) for 1 hr at room temperature. Slides are washed three times in PBS and incubated with fluorescein- or rhodamine-labeled, affinity-purified goat anti-human or goat anti-mouse antibody (Cappel, Malvern, PA) at a 1 : 100 dilution for 30 min at room temperature. Following a second series of washings in PBS, coverslips are mounted with mounting fluid and the tissues examined under an epifluorescence microscope at the appropriate wavelength. This approach has allowed the identification of CD44 as a hyaluronan receptor[26] and evaluation of hyaluronan tissue localization in mammalian organs (Fig. 2) at different stages of mammalian development[29] and under pathological conditions.[37] It is being applied to the identification of putative distinct ligands for some of the other CD44 isoforms.

[37] K. Nishikawa, G. Andres, A. K. Bhan, R. T. McCluskey, A. B. Collins, J. L. Stow, and I. Stamenkovic, *Lab Invest.* **68,** 146 (1993).

Alternative ways to examine interactions between hyaluronan and its receptors are provided by using labeled hyaluronan. Two simple and effective hyaluronan labeling approaches include radiolabeling and fluorescence.

Use of Radiolabeled Hyaluronan

Radiolabeled hyaluronan has been useful in the identification of cell surface HA-binding sites and their affinity for hyaluronan.[38,39] Cells that produce hyaluronan, such as lymph node stromal cells (described below) are grown at subconfluence in the presence of [³H]acetate (200 μCi/ml) for 24–36 hr. Supernatants are harvested and subjected to pronase digestion [1 mg/ml, in the presence of 0.02% (w/v) sodium azide] at 37° overnight. Radiolabeled hyaluronan is precipitated with cetylpyridinium chloride in 0.03 M NaCl and hyaluronan extracted from the precipitate with 0.4 M NaCl, 0.1% (v/v) cetylpyridinium chloride, and reprecipitated with 3 vol of 1.3% (v/v) potassium acetate in 95% ethanol. The precipitate is washed in potassium acetate–ethanol, dissolved in water, and centrifuged at 30,000 g for 1 hr. Alternatively, radiolabeled hyaluronan can be obtained as above, using [³H]glucosamine.

Radiolabeled hyaluronan-binding assays are performed on whole cells or on whole or solubilized cell membranes according to methods developed by Underhill et al.[39] Subconfluent, HA receptor-expressing adherent cells (roughly 1–2 × 10⁸ total) are detached with 0.5 mM ethylenediaminetetraacetic acid (EDTA), washed in PBS, and resuspended in fresh PBS. Following centrifugation at 90 g for 10 min, the cell pellet is resuspended in 5 ml of 0.01 M Na$_2$HPO$_4$ containing protease inhibitors [1 mM phenylmethylsulfonyl fluoride (PMSF)]. The cell suspension is maintained on ice for 15–20 min following which the cells are lysed with a Polytron (Brinkman, Westbury, NY). Two volumes of 0.5 M sucrose in 0.02 M Tris, pH 7.3, is added and nuclei removed by centrifuging the resulting mixture at 2000 g for 15 min at 4°. The supernatant is then subjected to high-speed centrifugation (30,000 g for 20 min at 4°) and the pelleted membrane fractions stored at -20° or immediately used in binding assays. Although samples obtained by this approach are crude and contain a substantial amount of subcellular components, the pellet obtained from high-speed centrifugation contains most of the hyaluronan-binding activity.[39]

Hyaluronan-binding protein was found to be optimally extractable from the crude membrane preparation by digitonin and sodium deoxycholate

[38] R. L. Goldberg, C. B. Underhill, and B. P. Toole, *Anal Biochem.* **125,** 59 (1982).
[39] C. B. Underhill, G. Chi-Rosso, and B. P. Toole, *J. Biol. Chem.* **258,** 8086 (1983).

because neither reagent interfered with the HA-binding assay. Deoxycholate was found to be superior to digitonin, but had to be maintained at a pH of about 8.0, because at lower pH its solubility significantly decreased. Membrane preparations are solubilized by agitation in a 20 mM Tris (pH 8.0) solution containing 0.5% (v/v) sodium deoxycholate for 2–3 hr. The solution is then diluted 1 : 4 (v/v) in 20 mM Tris (pH 8.0) and the protein concentration determined.

To perform the HA-binding assays, crude membrane suspensions in 20 mM Tris, pH 8.0, and sodium deoxycholate extracts are mixed with [^{3}H]hyaluronan in microfuge tubes. The volume is adjusted to 0.5 ml and contains 2–5 μg of [^{3}H]hyaluronan, 0.04% (v/v) sodium deoxycholate, 0.5 M NaCl, and 20 mM Tris, pH 8.0. Samples are shaken at room temperature for 10 min and 0.5 ml of saturated ammonium sulfate, pH 7.5, is added and the samples mixed vigorously. Then 50 μl of nonfat milk is added to each sample, and the samples are mixed briefly and centrifuged for 5 min at 14,000 g at 4°. Supernatants are removed and the pellets rinsed with 50% saturated ammonium sulfate and dissolved in 1 ml of water. Samples are then processed for scintillation counting.

This biochemical approach, used by Underhill and collaborators, helped define CD44 HA-binding activity well before the genetic characterization of the molecule. It remains a valid method for identifying new cell surface HA receptors.

Use of Fluoresceinated Hyaluronan

Fluorescein-conjugated hyaluronan has been used to examine the microcirculation, tissue permeability, tissue distribution, and receptor binding of hyaluronan. Two procedures have been developed by De Belder and Wik to prepare fluorescein-labeled hyaluronan.[40] In the first and most commonly used protocol, fluorescein isothiocyanate (FITC, 0.03 g) is conjugated to sodium hyaluronate (0.2 g) dissolved in 40 ml of formamide (with an agitator for 24 hr at room temperature) and further diluted with 50 ml of methyl sulfoxide. A small amount of sodium hydrogen carbonate (0.1 g) is added to prevent acidity and dibutyltin dilaurate (0.1 g) is added as a catalyst. The mixture is then stirred over steam for 30 min. The fluorescein-conjugated product is diluted in 50 ml of water and precipitated in 2 liters of ethanol containing a few drops of saturated, aqueous sodium chloride. One hour later, the sediment is collected and spun for 10 min at 3000 rpm at 4°. The pellet is resuspended in 10 ml of water and reprecipitated with ethanol two more times (or as many additional times as are

[40] A. N. De Belder and K. O. Wik, *Carbohydr. Res.* **44,** 251 (1975).

needed to remove the unconjugated FITC). The material can be checked for unconjugated FITC by running the product on a thin-layer chromatography (TLC) plate developed with chloroform : methanol (3 : 1). Unbound FITC can be detected by ultraviolet (UV) light as a fast-moving, free fluorescent substance. The FITC-conjugated material is dried and stored at 4°. A yield of about 150 mg is expected from this procedure.

In the second protocol hyaluronan (0.05 g) in 40 ml of water is further diluted in 20 ml of methyl sulfoxide and a solution containing acetaldehyde (25 μl), cyclohexyl isocyanide (25 μl), and fluoresceinamine (25 mg) in methyl sulfoxide is added. The pH of the reaction is monitored and kept in the range of pH 5–7. The reaction is allowed to proceed for 5 hr at room temperature. The product of the reaction was precipitated with ethanol (800 ml). Fluorescein-conjugated hyaluronan is then precipitated two more times with ethanol following resuspension in water, dried, and stored at 4°.

Using viscosimetry, gel chromatography, and sedimentation and diffusion coefficient measurements De Belder and Wik showed that only a small portion of the hyaluronan is degraded following conjugation.[40] Also, they found that both procedures yielded a similar degree of substitution on the carbohydrate (0.5–0.01) and that the labeling was uniform.

Binding of CD44–Rg to Immobilized Hyaluronan

The binding of CD44 to hyaluronan can be conveniently assayed by enzyme-linked immunosorbent assay (ELISA). Hyaluronan is immobilized in the wells of a 96-well plastic dish. Typically, the polystyrene ELISA plates are coated with 50 μl/well of a stock 5-mg/ml solution of HA diluted to 5 mg/ml in 50 mM sodium carbonate, pH 9.6, for 16 hr at 22°. Following washing with PBS–Tween the binding of the CD44–Rg fusion proteins to the immobilized hyaluronan is assayed using a goat anti-human F(ab')$_2$ horseradish peroxidase (HRP)-conjugated antibody, washed, incubated with EIA chromogen reagent, and the optical density measured in a dual-wavelength (450 and 630 nm) ELISA plate reader.

This assay system has been used to assess the binding of soluble isoforms of CD44 to hyaluronan and to other glycosaminoglycans including chondroitin A, B, and C sulfate, heparan sulfate, and keratan sulfate. These studies have shown that the 90-kDa isoform of CD44 preferentially binds hyaluronan and weakly binds chrondroitin A and C sulfate (Fig. 3). Similarly, this assay has been used to assess the binding of different CD44 isoforms and mutants of CD44, which has allowed the identification of the hyaluronan-binding site of CD44.[17]

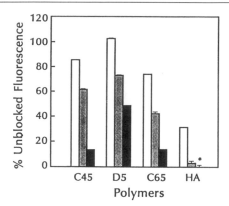

FIG. 3. Blocking of CD44–Rg reactivity with cultured lymph node stromal cells by glycosaminoglycans. Indicated glycosaminoglycans were added to cells at 5 (open bars), 50 (hatched bars), or 500 (solid bars) $\mu g/ml$, followed by CD44–Rg (5 $\mu g/ml$). Inhibition of binding was assessed by indirect immunofluorescence using flow cytomery.

Identification of Hyaluronan-Binding Site in 90-kDa Isoform of CD44

All CD44 isoforms share the same amino-terminal domain (first 5 exons, 221 amino acids).[14,16,27] This domain contains a region that bears 30% amino acid identity with the HA-binding domain of chicken and rat cartilage link proteins (CLP, initial 131 amino acids).[5,6,14] These observations suggested that the hyaluronan-binding domain of CD44 is contained within this region of the protein. Truncation mutagenesis was used to identify the minimal CD44 domain that can mediate hyaluronan binding. This domain was found to consist of the initial 186 amino acids of CD44 and to contain the region of the protein that is highly homologous to CLP as well as a region of 55 amino acids that is highly conserved among species.[41–44] The truncated CD44 proteins were prepared as immunoglobulin fusion proteins, using PCR-based methodology. In particular, PCR primers flanking the region of the CD44 to be amplified, and encoding restriction sites that would enable the subcloning of the amplified PCR fragment (as described above), were used to carry out 25 cycles of amplification (1 min at 94°, 2 min at 55°, and 3 min at 72°). The PCR fragments

[41] D. F. H. Zhou, J. F. Ding, L. J. Picker, R. F. Bargatze, E. C. Butcher, and D. V. Goeddel, J. Immunol. **143,** 3390 (1989).
[42] R. L.Idzerda, W. G. Carter, C. Nottenburg, E. A. Wayner, W. M. Gallatin, and T. St. John, Proc. Natl. Acad. Sci. U.S.A. **86,** 4659 (1989).
[43] C. Nottenburg, G. Rees, and T. St. John, Proc. Natl. Acad. Sci. U.S.A. **86,** 8521 (1989).
[44] B. T. Bosworth, T. St. John, W. M. Gallatin, and J. A. Harp, Mol. Immunol. **28,** 1131 (1991).

were then subcloned into a mammalian expression vector containing sequences encoding the constant region of human IgG_1. These soluble CD44 mutants were then prepared by transient expression in COS cells and their ability to bind to hyaluronan examined using the ELISA assay described above.

Inspection of the sequences in this minimal hyaluronan-binding domain showed that it contains two clusters of positively charged residues.[17] Because glycosaminoglycans have been shown to interact predominantly with basic peptides on their receptors,[45] these clusters of basic residues were targeted for site-directed mutagenesis. The site-directed mutagenesis was carried out using the PCR strategy outlined above. The only difference was that for each point mutant a PCR primer containing the base substitution that would result in the desired amino acid mutation was used.

It is critical that each mutant generated by this PCR approach be sequenced prior to use. In many instances undesired mutations have been found in the PCR product even though a low number of amplification cycles is used to generate these PCR fragments.

Using this approach it was observed that the CD44 interaction with hyaluronan required both clusters of positive charge and that replacing Arg-41, which is located in the most amino-terminal cluster of basic residues, with alanine completely abolished the CD44 hyaluronan-binding interaction.[17] Studies designed to identify the hyaluronan-binding site of RHAMM and CLP have shown that in the former case the interaction is mediated by two clusters of positively charged residues in the membrane-proximal extracellular domain,[46] whereas in the latter it is mediated by a set of tandemly repeated sequences that are also rich in basic amino acids. In the study of the hyaluronan-binding site of RHAMM, truncation mutagenesis, synthetic peptides corresponding to sequences in the protein shown to have the desired binding activity, and fusion proteins in which the hyaluronan-binding domains of the protein were grafted on other portions of the molecule known not to have hyaluronan-binding activity were used to identify the hyaluronan-binding site in this protein.[46] In the case of CLP antibody blocking studies and peptide fragments were used to map the hyaluronan-binding site.[12]

Taken together, studies on the hyaluronan-binding sites of a number of different receptors that interact with this glycosaminoglycan have shown that there is no consensus sequence for hyaluronan binding. Rather, it appears that this protein–carbohydrate interaction is mediated by two clusters of positively charged residues. In some cases, these two clusters

[45] R. L. Jackson, S. J. Busch, and A. D. Cardin, *Physiol. Rev.* **71**, 481 (1991).
[46] B. Yang, L. Zhang, and E. A. Turleys, *J. Biol. Chem.* **268**, 1 (1993).

of basic residues are closely spaced (10-amino acid spacing, RHAMM)[18] whereas in others (CD44 and CLP) the two clusters of basic residues are distal to each other (100 amino acids apart).[5,6,14,15]

Development of Stable CD44 Transfectants

A useful means to study the role of CD44 is provided by stable expression of the CD44 molecule in adherent and nonadherent cells that lack constitutive CD44 expression. Among lymphoid cells, Burkitt lymphoma-derived cell lines typically lack CD44-specific RNA transcripts.[27] Numerous adherent cell lines derived from various types of human malignancies are deficient in CD44 expression[14,27,47,48] and provide suitable candidates for the study of CD44-induced alterations in interactions with other cells and with substrate. Expression vectors containing CD44 cDNAs are introduced into cells by calcium phosphate transfection or by electroporation and selected for resistance to appropriate antibiotics. Expression of CD44 in resistant clones is tested by immunofluorescence and/or immunoprecipitation, using anti-CD44-specific MAb. Stable transfectants expressing CD44 isoforms have been useful in determining the role of CD44 in mediating cell–cell and cell–ECM interactions, and in demonstrating that different isoforms of CD44 confer different adhesion properties to cells.[27,30] A variety of different cell types have been used to express stably human,[27,30,31,48] rat,[49] and murine[36] CD44 isoforms. Various isoforms of human,[27,48] rat,[49] and murine[36] CD44 have thus been tested for adhesion properties. In addition, truncated forms of the receptors[36,48] have been examined in an attempt to determine which intra- and extracellular motifs are required for mediating adhesion.

To introduce cDNA clones into human lymphoid and epithelial tumor cell lines, we use electroporation, which in our hands has provided a superior yield in transfectants compared to other methods. Typically, we use a Bio-Rad (Richmond, CA) Gene Pulser and adjust the electroporation parameters to 250 V/960 μF. Although these parameters have given favorable results for melanoma and several lymphoma cell lines, Namalwa cells tolerate 400 V/960 μF well, with a superior yield in numbers of transfectants. Each new candidate cell line may require different conditions, but the most common range of voltage is 200–350 V, and the capacitance is commonly 960 or 500 μF. Typically, 0.4 ml of a 10^7-cells/ml

[47] M. Hofmann, W. Rudy, M. Zoller, C. Tolg, H. Ponta, P. Herlick, and U. Gunthert, *Cancer Res.* **51,** 5292 (1991).

[48] L. Thomas, H. R. Byers, J. Vink, and I. Stamenkovic, *J. Cell Biol.* **11,** 971 (1992).

[49] U. Gunthert, M. Hofmann, W. Rudy, S. Reber, M. Zoller, I. Haussmann, S. Matzku, A. Wenzel, H. Ponta, and P. Herrlick, *Cell (Cambridge, Mass.)* **65,** 13 (1991).

suspension is aliquoted into 0.4-cm electroporation cuvettes. DNA is added at a test vector : selection vector ratio of 15 : 1, which corresponds to 15 μg of the cDNA clone to be expressed and 1 μg of the plasmid containing the antibiotic resistance gene. An alternative method that has proved successful in stable expression of CD44 in Epstein–Barr virus (EBV)-transformed Burkitt lymphoma lines was the use of vectors containing the *EBNA-1* gene, which allows episomal maintenance of transfected DNA. The EBV vector p205[50] and the CDM7 vector containing CD44 cDNA inserts were, respectively, linearized with *Xba*I and *Spe*I (*Xba*I compatible) and ligated together. The pCDM7CD44p205 plasmid thus created contains CD44 cDNA, whose expression is driven by the CDM7 cytometalovirus (CMV) promoter, the *EBNA-1* gene, allowing stable episomal expression, and the hygromycin B phosphotransferase gene (hygromycin resistance selection gene) driven by a herpes simplex virus (P_{tk}) promoter. The large size of this combined vector did not appear to be an impediment on efficient stable expression of CD44 in Namalwa cells.[27]

Data obtained so far suggest that different isoforms of CD44 might confer different properties in different species. Human lymphoma and melanoma cells expressing CD44H adhere to hyaluronate-coated plates and to hyaluronate-producing cells.[27,30,48] However, the same cells transfected with higher molecular weight isoforms containing variable exons 8–10 (CD44E), 7–10, and 6–10 adhere to hyaluronate-coated surfaces only slightly more than CD44-negative parental cells[27,30,48] (also unpublished data[28]). The reason for this difference is not clear. It would appear, however, that shedding of the higher molecular weight forms might play a role. Murine homologs of the above human isoforms are reported to bind soluble hyaluronate.[36] Thus the discrepancy between the function of murine and human CD44 isoforms may be due to shedding or, possibly, to differences in posttranslational modifications and folding.

Use of CD44 Transfectants in Adhesion Assays

A variety of different approaches can be used to assess the role of CD44 in mediating cell adhesion to HA. One of the simplest consists of coating plastic tissue culture dishes or glass coverslips with HA. Best results are usually obtained when plastic surfaces that have not been treated for tissue culture are used, because this greatly reduces the amount of background adhesion, particularly when adherent cells are assayed. The amount of HA used to coat plates is important because HA sticking to plates is not efficient. In our experience, best results were obtained when the coating

---

[50] J. L. Yates, N. Warren, and B. Sugden, *Nature (London)* **313**, 812 (1985).

concentration of HA was 5 mg/ml. Typically, plates are coated with a solution of HA in PBS, pH 8.5–9.0, at 4° overnight, washed two or three times with PBS, and incubated with 0.5% (w/v) BSA in PBS for an additional 2–4 hr at 4° to block nonspecific binding sites. Plates are either used immediately or stored frozen. It has been our experience that best results are obtained when freshly prepared plates are used.

Nonadherent cells are metabolically radiolabeled, typically with [^{3}H]thymidine, and total incorporation of radiolabel in the number of cells that are to be tested for attachment determined. That same number of cells is then seeded onto the HA-coated plates in PBS or RPMI without serum, and allowed to settle for 30 to 40 min at 4°. This approach usually yields the lowest background. Adhesion assays may also be conducted at 37° for 15–20 min with comparable results, although the background adhesion is usually more prominent. Removal of nonadherent cells can be performed by gently swirling the plates and removing the supernatant with a Pasteur pipette. Usually three or four washes suffice. An alternative approach is to fill the wells with PBS, following attachment, invert the plates, and centrifuge them gently at 500 rpm in a Sorvall (Newtown, CT) 6000 centrifuge for 5 min at 4°. The cover is removed, the supernatant decanted, and cell attachment determined. Typically the cells are lysed in a 0.1% (v/v) Nonidet P-40 (NP-40) or sodium dodecyl sulfate (SDS) solution and the released radioactive counts determined in a β counter. Key to obtaining reproducible results in this type of assay is the homogeneity of HA coating. The use of a positively charged surface can facilitate even coating of wells.

For adherent cells, essentially the same type of protocol can be used. In this case, it is all the more important to use wells that have not been treated for tissue culture and to perform adhesion at 4°. Prior to initiating the assays, cells are detached from plates with EDTA, washed, and resuspended in PBS or RPMI with or without serum.

In addition to coating plates with HA, an alternative attachment assay may be performed using adherent cells that produce HA as substrate. A variety of different fibroblast lines produce or can be induced to produce HA. However, murine and rat lymph node stromal cells readily produce large amounts of HA and can be conveniently used in these adhesion assays.[26] These cells can be prepared according to procedures developed by Ise et al.[51] Rat cervical lymph nodes are excised, connective tissue is removed, and the nodes are squashed to extrude the majority of lymphocytes and minced in DME. Tissue fragments are allowed to settle in 15-

[51] Y. Ise, K. Yamaguchi, K. Sato, Y. Yamamura, F. Kitamura, T. Tamatani, and M. Miyasaka, Eur. J. Immunol. **18**, 1235 (1988).

ml polystyrene tubes, the medium is removed, and the remaining tissue is washed several times with medium to remove any nonadherent hematopoietic cells. The tissue is then resuspended in DME containing 0.5% (w/v) collagenase type II from *clostridium histolyticum* (Sigma Co., St. Louis, MO) and incubated at 37° for 30 min. The tissue is then mechanically disaggregated and the released stromal cells filtered through a 100-μm nylon mesh, washed several times in DME, resuspended in DME supplemented with 20% (v/v) FBS, and seeded onto culture dishes. Cells are routinely subcultured at confluence by trypsin–EDTA treatment and a 1 : 5–1 : 10 dilution into fresh medium. Morphologically, these cells appear as plump fibroblasts. They can be distinguished by a high degree of sulfate incorporation, promotion of pseudoemperipolesis (crawling of lymphocytes underneath the stromal cells), and abundant production of HA.

For adhesion assays, stromal cells are allowed to adhere and form a confluent monolayer. Supernatant is removed and transfectants seeded onto the monolayer in RPMI. Thirty-minute incubations at 4° are performed. Gentle washing or centrifugation as described above can be performed to remove nonadherent cells.

Use of CD44 Transfectants in Motility/Migration Assays

Stable expression of CD44 in adherent cells has facilitated motility studies. Various approaches can be used for assessment of CD44-induced migration on HA surfaces. We have adapted a method based on procedures described by Goodman *et al.*[52] Sterile glass coverslips are incubated for 6 hr at 4° in 35-mm petri dishes containing 1 ml of a solution of hyaluronic acid at varying concentrations in PBS. The coverslips are washed several times in PBS and incubated for 12 hr at 4° in heat-denatured BSA to block nonspecific binding sites. Following washing in PBS, the coverslips are ready for use. Although it is difficult to assess exactly how much of the HA has bound to the glass surface, our experience suggests that coating concentrations of 5 mg/ml give the best results.[48]

To perform migration assays, adherent CD44 transfectants, or CD44-expressing cell lines, are detached from tissue culture plates using 0.5 mM EDTA and washed extensively in PBS. A single drop of a 10^6-cells/ml suspension is placed onto the center of the HA-coated coverslip, which is immersed in medium supplemented with 10% (v/v) serum, so as to obtain a cell density ranging between 0.5 and 1 cell/10^4 μm^2. Cells are allowed to attach to the surface for 2 to 4 hr at 37°. At this time migration studies are begun. Migration is typically studied for a period of 3 hr,

[52] S. L. Goodman, G. Risse, and K. von der Mark, *J. Cell Biol.* **109,** 799 (1989).

although shorter and longer time periods can be used depending on the type of cell studied and its inherent migration rate on a given substrate. Cell migration is observed under an inverted microscope (Nikon Diaphot) with a $\times 10$ phase-contrast objective in an attached, hermetically sealed Plexiglas Nikon NP-2 incubator at $37°$ in a 5% CO_2 environment. The microscope is connected to a video camera (Dage-MTI 65DX) and a time-lapse video cassette recorder. Image analysis can then be performed by playing back the video images saved at specific times during the experiment, and tracing the migration paths of individual cells on the video monitor. Migration rate is defined as the algebraic sum of the two-dimensional migration distances divided by the number of hours during which the experiment was performed, and expressed as micrometers per hour. Migration of several individual cells needs to be evaluated (we have typically assessed migration rates of 20–50 individual cells). The integrated migration path of each cell is represented as a "spider" diagram.[48] CD44H-expressing cells and transfectants revealed significantly increased migration on hyaluronate-coated substrate with respect to CD44-negative counterparts. In addition, the increase in migration rate was observed on HA only, and not on other glycosaminoglycans.[48]

This approach provides a convenient way to determine HA receptor-dependent cell motility and to study the effect of different concentrations of HA as well as possible mechanisms of interference with migration, using monoclonal antibodies and soluble HA receptor fusion proteins.

Soluble CD44–Rg can be used to block CD44-expressing cell migration. In this approach, HA-coated coverslips are preincubated with CD44–Rg at 10 μg/ml at $37°$ for 30 min. Coverslips are washed in PBS and seeded with cells to be tested as above. We have also performed experiments in which cells and soluble CD44–Rg (25 μg/ml) were added simultaneously to the coverslips. Under both conditions, soluble CD44–Rg was found to specifically block migration promoted by expression of CD44H.[48] Thus CD44–Rg provides a suitable reagent for competitive inhibition of CD44–HA interaction.

Transfection of cytoplasmic deletion mutants of CD44H has shown that the cytoplasmic domain is required to enable appropriate interactions with HA to occur. Murine "tailless" CD44H has been observed to lose the capacity to bind soluble HA, but both human and murine CD44H cytoplasmic deletion mutants can promote significant transfectant attachment to HA-coated surfaces. This discrepancy is most likely due to the fact that surface-bound HA might cross-link CD44 molecules, causing enough aggregation of cell surface CD44 for adhesion to occur. However, the cytoplasmic domain of CD44 is required for migration, because cells expressing cytoplasmic deletion mutants of CD44H revealed migration patterns on HA similar to those of untransfected parental cells.[48]

Role of Hyaluronic Acid Receptors in Haptotactic Invasiveness through Hyaluronic Acid-Coated Membranes

Invasiveness of tumors and the capacity of normal cells to penetrate tissues depend on the capability of the cells to transmigrate tissue barriers and consequently on mechanisms that facilitate such transmigration. A convenient means to test whether expression of HA receptors facilitates haptotactic invasiveness is provided by an adaptation of procedures described by Aznavoorian *et al.*[53] Haptotaxis is defined as the migration of cells in a concentration-dependent manner on insoluble step gradients of substratum-bound attractants.[53] Modified Boyden chambers (Transwell 24, 8-μm pore; Costar, Cambridge, MA), with 8-μm pore polycarbonate membranes, were coated on the inferior surface of the membrane by overnight flotation of the upper chamber units on solutions of varying concentration of HA and various other glycosaminoglycans at 37°. Control units are floated on PBS or a solution of 0.5% (w/v) BSA. Following coating, both sides of each membrane are washed with PBS and 10^5 cells are seeded onto each upper chamber unit and incubated for 24 hr at 37°. Membranes are then washed several times in PBS and the entire unit immersed for 10 min in 10% (v/v) formalin and stained with Gill's hematoxylin. Fixed membranes are removed, mounted on glass slides, and the cells on the side of the filter opposite the side of seeding are counted under an optical microscope. We have found that constitutive expression of CD44H dramatically increases haptotactic invasiveness of human melanoma cells *in vitro*[54] and that introduction of CD44H into cells that are CD44-deficient induces haptotaxis (unpublished observations[55]).

Role of Hyaluronic Acid Receptors in Tumor Growth and Dissemination *in Vivo*

Experiments performed *in vitro* suggest that expression of HA receptors augments cellular migration in a two-dimensional migration assay as well as cellular transmigration capability in a haptotactic invasion assay system. How do these properties influence tumor cell behavior *in vivo*?

[53] S. Aznavoorian, M. L. Stracke, H. Krutzsch, E. Schiffmann, and L. A. Liotta, *J. Cell Biol.* **110,** 1427 (1990).
[54] L. Thomas, T. Etoh, I. Stamenkovic, M. C. Mihm, Jr., and H. R. Byers, *J. Invest. Dermatol.* **100,** 115 (1993).
[55] L. Thomas and I. Stamenkovic, unpublished observations, 1992.

The first obvious issue is to determine whether expression of HA receptors modifies the rate of tumor growth *in vivo*. The most straightforward way to answer this question is to inject HA receptor-positive and -negative tumor cells of the same origin subcutaneously into nude (*nu/nu*) or SCID (severe combined immunodeficiency) mice. Tumor growth can be readily assessed over a period of 2–8 weeks, depending on the cell type and the impact of HA receptor expression on the rate of growth readily determined. Typically 5×10^6 to 1×10^7 cells are injected in the retroscapular region, so that any tumor mass can be visualized without requiring surgery. Sufficient numbers of cells should be injected so that the experiment can be performed over a reasonable span of time, typically a few weeks. Tumor growth can thus be easily calibrated and expressed in weight using the formula $(d^2D)/2$, where d represents the smallest and D the largest diameter of the tumor. We have observed that expression of CD44H in human lymphomas and melanomas augments the rate of tumor development in nude and SCID mice following injection of transfected cells (Fig. 4).[30]

The second issue is to determine whether expression of HA receptors

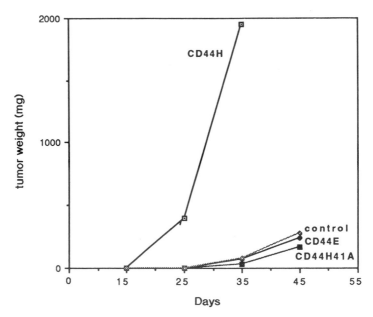

FIG. 4. Promotion of tumor growth *in vivo* by CD44H expression is dependent on CD44H–HA interaction. SCID mice were injected subcutaneously with a human melanoma cell line (MC) transfected with cDNAs encoding the indicated CD44 molecules or an unrelated cell surface receptor (control). Tumor growth was determined as described in text.

might modify the pattern of dissemination of tumors. Tumor cells expressing HA receptors are injected intravenously into the tail vein of nude and
SCID mice and animals monitored for signs of metastasis. The Burkitt
lymphoma Namalwa provides a suitable means to monitor *in vivo* tumor
growth because the cells secrete human IgM. An ELISA assay was therefore developed and the serum level of human IgM monitored over a period
of 6–12 weeks. Ninety-six-well ELISA plates were coated with purified
goat anti-human IgM antibodies (0.5 mg/ml; Southern Biotechnology Associates, Birmingham, AL) and murine serum samples serially diluted. The
second antibody was an affinity-purified alkaline phosphatase-conjugated
goat-anti-human chain-specific IgG (Southern Biotechnology Associates).
Detectable serum IgM invariably correlated with development of tumor
masses at autopsy.[30] These studies showed that Namalwa cells transfected
with CD44H developed tumors more rapidly than parental cell counterparts. Although most transfectant- and parental cell-derived tumors developed at the same sites following intravenous injection, two significant
exceptions were observed. The first was the adrenal medulla and the
second the renal papilla and medulla.[30] These sections of both organs are
rich in HA, and in the kidney, tumor cell infiltration was confined to areas
that contain HA.[30] Thus expression of functional HA receptors in tumor
cells augments the rate of local and disseminated tumor development.

Although the data obtained from the above experiments are highly
reproducible, it still remains to be definitively demonstrated that the enhancement of tumor development associated with expression of CD44H
is due to CD44–HA interaction. We have attempted to resolve this issue
using a recombinant CD44H mutant that does not bind HA. The approach
was based on subcutaneous injections of a human melanoma transfected
with cDNAs encoding CD44H and CD44H in which the arginine residue
at position 41 has been mutated to an alanine (CD44H41R-A), into nude
and SCID mice. These experiments showed that CD44H41R-A transfectants developed tumors at a rate comparable to that of parental, CD44-
negative melanoma cells (Fig. 4). This model provides the most direct *in
vivo* evidence that CD44–HA interaction plays a role in determining the
rate of tumor growth and dissemination.

Blocking of Hematogenous Dissemination of CD44-Positive Tumors Using Soluble CD44–Rg

The observation that expression of CD44H augments the rate of metastatic growth of intravenously injected CD44-transfectant tumor cells[30]
raises the possibility that administration of CD44–Rg might provide a
means to inhibit metastasis, or at least reduce its rapidity. To test this

possibility, CD44H-transfectants cells were injected intravenously into nude mice together with 100, 300, or 500 μg CD44–Rg per milliliter and control Rg molecules and followed for signs of tumor development. Rg molecules were injected at the initial dose three additional times at 2-day intervals.[31] High doses of CD44–Rg were observed not only to reduce the rate of metastasis, but to block tumor dissemination altogether.[31] This unexpected result may have several explanations. The first is that soluble CD44–Rg may block CD44H-binding sites. However, considering the abundance of hyaluronan in connective tissue, saturation of all CD44H-binding sites by the dose of CD44–Rg injected seems unlikely, and therefore this mechanism may only provide part of the explanation. An alternative possibility might be that CD44 transfectants bind hyaluronan in the serum, which then serves as a molecular bridge for CD44–Rg. The result would be a form of "opsonization" of the tumor cells, which might undergo complement or natural killer cell-mediated lysis and elimination in the reticuloendothelial system. In either case, exploitation of CD44H–HA interaction may provide novel adjunct therapeutic avenues for the control of tumor growth and dissemination.

Acknowledgments

We thank Robert Peach for advice. This work was supported by National Institutes of Health Grant CA55735 to I.S. and the Bristol Myers Squibb Pharmaceutical Research Institute. I.S. is a Scholar of the Leukemia Society of America.

Section III

Extracellular Matrix Components That Do Not Self-Assemble

[11] Regulation by Heparan Sulfate in Fibroblast Growth Factor Signaling

By ALAN C. RAPRAEGER, SCOTT GUIMOND, ALISON KRUFKA, and BRADLEY B. OLWIN

Biosynthesis of Proteoglycans: Overview

As proteins destined for the cell surface are sorted within the Golgi apparatus, they are decorated with carbohydrate and arrive at the plasma membrane as membrane-anchored glycoproteins, or are secreted in the extracellular milieu. Information inherent in the polypeptide structure not only ensures delivery of the protein to its ultimate destination, but is also deciphered by batteries of enzymes within the Golgi that modify the protein by the addition of carbohydrate. In the case of proteoglycans, recognition of a serine, often within an SGXG sequence flanked by acidic residues, leads to addition of xylose to the serine by ester linkage.[1-3] Galactosyltransferase adds galactose residues and thus completes the linkage regions for the ensuing synthesis of a specific type of glycosaminoglycan chain, namely, chondroitin sulfate(s), keratan sulfate, or heparan sulfate. The biochemical nature of the glycosaminoglycan chain that is synthesized has profound influence on the biological activity of the protein.

Chondroitin Sulfates

Proteins bearing chondroitin sulfate bind to a variety of extracellular proteins, including collagens and fibronectin, and are postulated to occupy highly hydrated conformations in the extracellular space. An example of this is *aggrecan,* an abundant and essential structural component of cartilage.[1-3] Chondroitin sulfate is synthesized as a repeating disaccharide of N-acetylgalactosamine and glucuronic acid. The polymer is then modified at each disaccharide by sulfotransferases, which transfer sulfate from the high-energy sulfate donor phosphoadenosine phosphosulfate (PAPS) to the hydroxyl group at C-4 or C-6 on the hexosamine, thus modifying the entire polymer to contain either N-acetylgalactosaminyl 6-O-sulfate,

[1] M. A. Bourdon, T. Krusius, S. Campbell, N. B. Schwartz, and E. Ruoslahti, *Proc. Natl. Acad. Sci. U.S.A.* **84**, 3194 (1987).

[2] J. D. Esko, *Curr. Opin. Cell Biol.* **3**, 805 (1991).

[3] D. R. Zimmerman and E. Ruoslahti, *EMBO J.* **8**, 2975 (1989).

Copyright © 1994 by Academic Press, Inc.
All rights of reproduction in any form reserved.

or N-acetylgalactosaminyl 4-O-sulfate.[4,5] Some regions within the chains are converted to dermatan sulfate by a specialized enzyme that epimerizes glucuronic acid at C-6, generating iduronic acid. The enhanced rotational movement of the chain around the iduronic acid residue allows the dermatan sulfate regions greater opportunity to interact with binding domains within protein ligands.

Heparan Sulfate

The synthesis of heparan sulfate is characterized by a high degree of heterogeneity in the sugar backbone and by its sulfation pattern compared to the chondroitin sulfates. Initially synthesized as a repeating disaccharide of glucuronic acid and N-acetylglucosamine (Fig. 1), the sugar polymer is then acted on by a series of concerted enzymatic steps that can lead to dramatic modification of the chain. These modifications have three important features: (1) the activity of each enzyme in the reaction sequence is dependent on the preceding enzyme completing its chore; (2) the battery of enzymes often acts only at discrete sites within the chain rather than throughout the entire polymer; and (3) the extent, and possibly the type, of these modifications varies with cell type.[6] The net result is heparan sulfate proteoglycans that emerge from the *trans*-Golgi network and arrive at the cell surface with varying potentials for binding biological ligands. The extracellular ligands for these chains are numerous and range from cell surface lipoproteins to extracellular matrix ligands and growth factors such as the fibroblast growth factors that contain heparin-binding domains.

Heparin and Heparin-Binding Domains

Heparin is a close but specialized cousin of heparan sulfate. Heparin is the product of mast cells, which synthesize copious amounts of the polyanion on a core protein backbone that is almost entirely Ser-Gly repeats. Once synthesis is complete, the core protein is destroyed and the soluble heparin chains are packed within storage granules with other mast cell secretory products. The synthesis of these chains uses largely the same machinery as heparan sulfate synthesis, but goes further to completion. Thus, heparin can be viewed as a form of heparan sulfate in which the modification reactions have proceeded nearly to completion. Until recently, a defining characteristic of heparin was its anticoagulant

---

[4] U. Lindahl and M. Höök, *Annu. Rev. Biochem.* **47**, 385 (1978).
[5] J. T. Gallagher, *Curr. Opin. Cell Biol.* **1**, 1201 (1989).
[6] U. Lindahl, M. Kusche, K. Lidholt, and L.-G. Oscarsson, *Ann. N. Y. Acad. Sci.* **556**, 36 (1989).

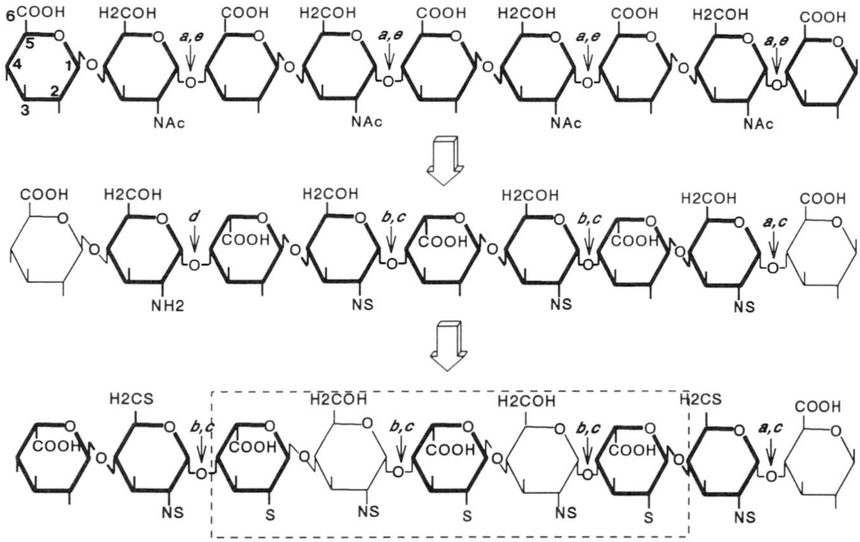

Fig. 1. Structural modifications during the synthesis of heparin and/or heparan sulfate glycosaminoglycan. The initial heparin or heparan sulfate chain is synthesized as a repeating disaccharide of α-1,4-glucuronate-β-1,4-N-acetylglucosamine. A series of concerted modification reactions generate regional variation within the chain, involving deacetylation, N-sulfation, epimerization of glucuronic acid to iduronic acid, and 6-O-sulfation/2-O-sulfation, as shown in the subsequent steps below. At each step, the sugars are set in boldface if they have been modified. The oligosaccharide shown is an example of the types of modification that generally take place. Oligosaccharides with many different sulfation sequences are possible, particularly in heparan sulfate. In heparin, these reactions go largely to completion whereas heparan sulfate is less modified. The modified regions in heparan sulfate occur in block regions, possibly due to the local action of complexes of multiple enzymes. The glycosaminoglycan can be cleaved by enzymatic and chemical steps (small arrows) that recognize regions of the glycosaminoglycan, largely on the basis of its degree of modification. Thus, heparinase III (or heparitinase) cleaves many sites within heparan sulfate because it recognizes sites that are poorly modified or not modified at all. Conversely, heparinase I (heparinase) recognizes highly modified sites that are prevalent in heparin. The boxed residues represent a pentasaccharide of heparin that is capable of binding FGF-2, apparently not requiring sulfates at C-6 of the glucosamine although heparin would be predominantly sulfated at this site. Interestingly, the entire 9-sugar oligosaccharide shown in this example would be insufficient to promote FGF-2 signaling on Swiss 3T3 cells, where a 12-sugar length is required. Thus, binding and activity of the fragments may not directly correlate. NAc, Acetylated amine; NS, sulfated amine; CS, sulfated carbon; S, sulfate bound via ester linkage. Cleavage reactions: (a) heparinase III (heparitinase); (b) heparinase I (heparinase); (c) low-pH (1.5) nitrous acid; (d) high-pH (3.9) nitrous acid; and (e) deacetylation (hydrazinolysis), followed by high-pH (3.9) nitrous acid.

activity, resulting from a specialized glucosaminyl 3-*O*-sulfate within a pentasaccharide sequence that blocks the blood coagulation cascade by binding antithrombin and inhibiting its activation of thrombin.[6] However, this same anticoagulant activity has now been described on heparan sulfate present in proteoglycans at the surface of endothelial cells, suggesting a role for this specialized carbohydrate domain in generating a nonthrombogenic lining of blood vessels.[7,8]

The ready availability of soluble heparin from pharmaceutical stores has promoted its use in describing domains within biological ligands that interact with heparan sulfate, leading to the description of "heparin-binding" domains within a large number of proteins.[9] These are typically ligands that bind specifically to heparan sulfate proteoglycans at cell surfaces or in the extracellular matrix. The heparan sulfate-binding domains within these proteins contain the basic amino acids lysine and arginine, which interact with sulfated sequences within the heparan sulfate chain. Accumulating evidence suggests that these interactions are more than adventitious binding of charged polymers. This suggests that the conformation of the protein domain presents charged amino acids that coincide with negatively charged residues of the heparan sulfate chain. Furthermore, it suggests that the sulfation "sequence" of the heparan sulfate and the conformational flexibility of the chain, as determined in part by the epimerization reaction generating iduronate, have profound influence on the ability of a cell to recognize distinct ligands. As discussed below, this becomes apparent when the binding of discrete domains within heparin or heparan sulfate to fibroblast growth factor (FGF) family members is examined.

Heparan Sulfate Proteoglycans

Proteins bearing heparan sulfate chains are constituents of the cell surface and the extracellular matrix.[10-15] Among the most ubiquitous is

[7] J. A. Marcum, D. H. Atha, L. M. S. Fritze, P. Nawroth, D. Stern, and R. D. Rosenberg, *J. Biol. Chem.* **261,** 7507 (1986).
[8] T. Kojima, C. W. Leone, G. A. Marchildon, J. A. Marcum, and R. D. Rosenberg, *J. Biol. Chem.* **267,** 4859 (1992).
[9] R. L. Jackson, S. J. Busch, and A. D. Cardin, *Physiol. Rev.* **71,** 481 (1991).
[10] L. Kjéllen and U. Lindahl, *Annu. Rev. Biochem.* **60,** 443 (1991).
[11] M. Jalkanen, K. Elenius, and A. Rapraeger, *Trends in Glycosci. Glycotechnol.* **5,** 107 (1993).
[12] M. Bernfield, R. Kokenyesi, M. Kato, M. T. Hinkes, J. Spring, R. L. Gallo, and E. J. Lose, *Annu. Rev. Cell Biol.* **8,** 365 (1992).
[13] J. T. Gallagher, J. E. Turnbull, and M. Lyon, *Biochem. Soc. Trans.* **18,** 207 (1990).
[14] A. C. Rapraeger, *Curr. Opin. Cell Biol.* **5,** 844 (1993).
[15] A. C. Rapraeger, in "Molecular and Cellular Aspects of Basement Membranes" (D. H. Rohrbach and R. Timpl, eds.), p. 267. Academic Press, San Diego, 1993.

perlecan, a major constituent of the basal lamina that underlies all epithelial cells. At the cell surface, heparan sulfate is present on transmembrane proteins (e.g., the *syndecans*[12]), as well as on a class of heparan sulfate-bearing proteins linked via glycosylphosphatidylinositol to lipid tails embedded in the outer leaflet of the plasma membrane (e.g., *glypican*[16,17]). In each of these three examples, the core proteins clearly have major roles in controlling the ultimate destination of the proteoglycan, namely, secretion to the extracellular matrix, or localization to specialized domains of the cell surface. At these sites, cell type-specific domains inherent in the heparan sulfate may influence the ability of the cell to respond to exogenous ligands.

Fibroblast Growth Factors

Many growth factors bind heparin at greater than physiological ionic strength, suggesting these may be ligands for heparan sulfate *in vivo*. However, the fibroblast growth factors described here not only bind heparin or heparan sulfate, but demonstrate enhanced activity when bound to the glycosaminoglycan.

The fibroblast growth factors were the first growth factors shown to bind heparin with high affinity. They currently are composed of a family of nine related polypeptides (FGFs 1–9) with diverse roles in cell growth and differentiation.[18-21] This family of related polypeptides ranges from 16 to 25 kDa and contains two conserved cysteines. FGF-1 (acidic FGF) and FGF-2 (basic FGF) are considered the prototypes of the family.[22,23] Fibroblast growth factors are perhaps the most notable ligands for heparan sulfate, as binding to heparan sulfate can be shown to be an obligate prerequisite for growth factor activity. Whereas heparan sulfate has other important ligands, the direct consequences of their binding cannot be demonstrated with such clarity and ease.

The FGF-binding site for heparan sulfate is not structurally defined and is likely to be formed from multiple regions brought together by

[16] G. David, V. Lories, B. Decock, P. Marynen, J. J. Cassiman, and H. Van den Berghe, *J. Cell Biol.* **111**, 3165 (1990).
[17] G. David, *Adv. Exp. Med. Biol.* **313**, 69 (1992).
[18] C. Basilico and D. Moscatelli, *Adv. Cancer Res.* **59**, 115 (1992).
[19] W. H. Burgess and T. Maciag, *Annu. Rev. Biochem.* **58**, 575 (1989).
[20] M. Miyamoto, K. Naruo, C. Seko, S. Matsumoto, T. Kondo, and T. Kurokawa, *Mol. Cell. Biol.* **13**, 4251 (1993).
[21] A. Tanaka, K. Miyamoto, N. Minamino, M. Takeda, B. Sato, H. Matsuo, and K. Matsumoto, *Proc. Natl. Acad. Sci. U.S.A.* **89**, 8928 (1992).
[22] D. Gospodarowicz, *J. Biol. Chem.* **250**, 2515 (1975).
[23] D. Gospodarowicz, J. Weseman, and J. Moran, *Nature (London)* **256**, 216 (1975).

protein folding.[24] The crystal structures of FGF-1 and FGF-2 reveal that a large number of positively charged amino acid side chains reside near the carboxyl terminus; however, deletion mutagenesis suggests that additional regions play critical roles in binding heparan sulfate.[25,26]

Requirement of Heparan Sulfate for Fibroblast Growth Factor Signaling

Purification and Iodination of Fibroblast Growth Factors

Purification. Purification of FGF-1 (acidic FGF) and FGF-2 (basic FGF) was initially accomplished without using heparin columns; however, the discovery by Lobb and Fett[27] that both growth factors could be readily purified by heparin affinity chromatography led to an explosion of FGF research, including the rapid discovery of five additional FGF family members.

Bovine FGF-1 and FGF-2 are readily purified from bovine brain by a two-step chromatographic procedure. First, 5 kg of bovine brain is homogenized in 5 liters of 0.15 M ammonium sulfate containing 1 mM phenylmethylsulfonyl fluoride (PMSF), 1 mM ethylenediaminetetraacetic acid (EDTA), leupeptin (1 μg/ml), and aprotinin (20 KIU/ml), then acidified to pH 4.5. The homogenate is centrifuged at 6000 g for 30 min at 4°C and the FGFs partially purified by a differential ammonium sulfate fractionation. The homogenate is made 40% (w/v) in ammonium sulfate and stirred for 1 hr, centrifuged at 10,000 g for 30 min at 4°C and the supernatant concentrated by bringing the ammonium sulfate to a final concentration of 75% (w/v). The pellet is resuspended in approximately 500 ml of 20 mM Tris (pH 7.4) containing 1 mM EDTA, leupeptin (1 μg/ml), and aprotinin (20 KIU/ml), then dialyzed against 20 mM Tris, pH 7.4, containing 1 mM EDTA. The conductivity of the sample is determined and NaCl crystals are added to bring the conductivity to 0.6 M NaCl. The material is batch applied to 25 ml of heparin–agarose [note that Bio-Rad (Richmond, CA) heparin–agarose yields a more consistent elution profile and less release of heparin than other commercially available heparin–Sepharose columns], washed successively with five column volumes of 0.6 M NaCl, and eluted with a gradient of 0.6 to 3.0 M NaCl. This procedure separates

[24] A. Baird, D. Schubert, N. Ling, and R. Guillemin, *Proc. Natl. Acad. Sci. U.S.A.* **85,** 2324 (1988).
[25] J. Zhang, L. S. Cousens, P. J. Barr, and S. R. Sprang, *Proc. Natl. Acad. Sci. U.S.A.* **88,** 3446 (1991).
[26] X. Zhu, H. Komiya, A. Chirino, S. Faham, G. M. Fox, T. Arakawa, B. T. Hsu, and D. C. Rees, *Science* **251,** 90 (1991).
[27] R. R. Lobb and J. W. Fett, *Biochemistry* **23,** 6295 (1984).

the majority of the FGF-1 from the FGF-2, which elute at ~1.0 and 2.5 M NaCl, respectively. The individual peaks of FGF-1 and FGF-2 are pooled and reapplied to 5.0-ml heparin–agarose columns that have not been used previously. The FGF-1 pool is diluted 5-fold and batch applied to 3 ml of heparin–agarose, washed with 10 column volumes each of 0.6 and 0.8 M NaCl, and eluted with 1.5 M NaCl. The FGF-1 pool is then diluted 10-fold and applied to a Mono S (Pharmacia, Piscataway, NJ) column, and eluted with a gradient from 0 to 1.0 M NaCl. The FGF-2 pool from the first heparin–agarose column is diluted 3-fold and batch applied to a 3.0-ml heparin–agarose column and washed successively with 10 column volumes of 0.6, 1.0, and 1.5 M NaCl. The FGF-2 is eluted with 2.5 M NaCl, diluted 10-fold and applied to a Mono S column, washed with dilution buffer, and eluted with a gradient from 0 to 1.5 M NaCl. Both proteins are greater than 95% pure and have no cross-contamination of FGF-1 or FGF-2. Again, it must be stressed that the heparin–agarose columns be new and not cross-contaminated by the two FGFs. The yields from 5 kg of bovine brain averages 1–2 and 0.1–0.2 mg, respectively, for FGF-1 and FGF-2.

Recombinant FGF-2 is purified from yeast (Zymogenetics, Seattle, WA) that express an artificial human FGF-2 cDNA. A 6-liter culture of yeast is pelleted and resuspended in lysis buffer [100 ml of 1.5 M NH$_4$SO$_4$, 10 mM EDTA (pH 7.0), leupeptin (10 μg/ml), and 1 mM PMSF]. The lysed sample is passed twice through a French press at 16,000 psi, and the lysate is adjusted to pH 4.0 with 5 N HCl and stirred for 1 hr on ice. The lysate is then clarified by centrifugation at 15,000 g and filtered twice through Millipore (Bedford, MA) Millex-GV syringe filters (0.22-μm pore size). The lysate is brought to 20 mM N-2-hydroxyethylpiperazine-N'-2-ethanesulfonic acid (HEPES), buffered at pH 7.4, and 0.6 M NaCl. The pH is then adjusted to pH 7.4 and the lysate batch applied to a heparin–agarose column preequilibrated in 20 mM HEPES, pH 7.4, containing 0.6 M NaCl that has previously been used only for recombinant FGF-2 purification. The column is washed with the equilibration buffer until the protein concentration of the effluent is zero. The FGF-2 is then eluted with 3 M NaCl, yielding approximately 40–60 mg of FGF-2 that is greater than 98% pure.

Iodination. FGF-1 and FGF-2 either purified from bovine brain or produced as recombinant molecules are radiolabeled with ^{125}I, using a procedure that involves chloramine-T. For FGF-2, 1 μg of FGF-2 is incubated with 3 μg of chloramine-T in 1 M HEPES (pH 7.4) and 1 mCi of Na[125]I in a final volume of 70 μl for 2 min at 22°. For FGF-1, the procedure is identical except that 1 μg of chloramine-T is used. Both reactions are terminated by the addition of 100 μl of 0.05 M dithiothreitol (DTT). The mixture is then added to a 0.1-ml (packed volume) heparin–agarose column prepared in a 1-ml Pipetman tip with a small amount of glass wool in the tip to prevent the resin from flowing through. The columns are then washed five times, each time with 1.0 ml of 20 mM HEPES, pH 7.4, containing 0.6 M NaCl and 0.2% (w/v) RIA-grade bovine serum albumin (BSA) (Sigma Chem. Co., St. Louis, MO). Both columns (one containing ^{125}I-labeled FGF-1 and the other containing ^{125}I-labeled FGF-2) are eluted by addition of 1.0 ml of the wash buffer containing an additional 2.4 M NaCl and five fractions containing four drops each are collected (~100 μl/fraction). The fractions are then assayed for ^{125}I counts per minute (cpm) in a γ counter and the fractions containing the highest ^{125}I counts per minute are subjected to an MM14 cell cycle exit assay or 3T3 fibroblast [3H]thymidine incorporation assay to determine the concentration and specific activity in comparison to a known standard. Typically, 100 μl of ^{125}I-labeled FGF, ranging from 1000 to 4000 disintegrations per minute (dpm)/fmol, is purified by this procedure. The ^{125}I-labeled FGF-1 and FGF-2 are stored at 4°, as freezing will shorten the biological half-life owing to the concentration of radiolabel in the frozen sample. Typically, both ^{125}I-labeled FGF-1 and FGF-2 retain 80% of their biological activity for 2 weeks and 50% for 4 weeks at 4°.

Binding of Fibroblast Growth Factor to Cell Surfaces

Binding to Low- versus High-Affinity Sites. Fibroblast growth factor binding to the cell surface is to both low- and high-affinity sites.[28,29] The low-affinity sites (K_d of 1–5 nM) represent binding to heparan sulfate at the cell surface or the surrounding extracellular matrix. The high-affinity sites (K_d of 10–100 pM) represent binding to protein receptors,[30–32] likely to be in a complex with the heparan sulfate.

Several different procedures have been developed to measure binding of FGF to intact cells. A number of procedures use a high ionic strength wash at neutral pH to remove FGF bound to low-affinity sites. These sites are presumed to be heparan sulfate, but may include other low-affinity sites as well. Although the high ionic strength washes are effective, loss of high-affinity binding sites may also occur. The extent of loss of high-affinity binding sites may depend on the cell type assayed. The assay described below has been used for a number of different cell types including Swiss, NIH, and BALB/c 3T3 cells, C3H10T1/2 cells, skeletal muscle

[28] D. Moscatelli, *J. Cell. Physiol.* **131,** 123 (1987).
[29] B. B. Olwin and S. D. Hauschka, *Biochemistry* **25,** 3487 (1986).
[30] L. W. Burrus, M. E. Zuber, B. A. Lueddeckey, and B. B. Olwin, *Mol. Cell. Biol.* **12,** 5600 (1992).
[31] M. Jaye, J. Schlessinger, and C. A. Dionne, *Biochim. Biophys. Acta* **1135,** 185 (1992).
[32] D. E. Johnson and L. T. Williams, *Adv. Cancer Res.* **60,** 1 (1993).

myoblasts, NMuMG (normal murine mammary gland) cells, and primary cultures of myoblasts and fibroblasts.

Cells grown to high density are rinsed three times in binding buffer [cell growth medium containing 0.2% (w/v) Pentex-grade BSA in place of serum and buffered with 25 mM HEPES, pH 7.4] preequilibrated at 4°. The desired concentrations of ^{125}I-labeled FGF and unlabeled FGF are added to the binding buffer and then added to the cells. The cells are then incubated with gentle agitation at 4° for 3 hr to reach equilibrium. All washing and elution of bound ^{125}I-labeled FGF must be performed as rapidly as possible. All solutions are kept on ice. Cells are washed three times with a one-third growth volume of binding buffer (i.e., if the cells were grown in 10 ml of medium, three 3-ml washes are performed). All of the subsequent washes are retained for determining bound ^{125}I-labeled FGF. ^{125}I-Labeled FGF bound to low-affinity sites is removed by washing twice with 20 mM HEPES, pH 7.4, containing 2 M NaCl and 0.2% (w/v) BSA. The remaining specifically bound ^{125}I-labeled FGF is removed by washing two times with sodium citrate (pH 4.0) containing 2 M NaCl and 0.2% (w/v) BSA, and all remaining bound ^{125}I-labeled FGF is removed by washing with 1% (w/v) sodium dodecyl sulfate (SDS)–0.5 M NaOH solution. All washes are with one-third growth volumes as described above.

Cross-Linking of Fibroblast Growth Factor to Cell Surface Receptors. As an alternative to ligand binding, a nonquantitative approach involves covalently cross-linking the ^{125}I-labeled FGF to protein receptors. This permits visualization of the cross-linked products by autoradiography following separation by SDS–polyacrylamide gel electrophoresis (SDS–PAGE). This procedure is identical to the intact cellular binding protocols through the wash steps and can be performed with or without the high ionic strength washes. Cross-linking will not occur to the carbohydrate side chains of the heparan sulfate proteoglycan as no nucleophiles are present for interaction with the cross-linking reagent. Once the cells have been washed, disuccinimidyl suberate (DSS) is added to a final volume of 0.015 M from a stock solution of 0.5 M DSS dissolved in dimethyl sulfoxide (DMSO). The cells are incubated on ice with the reagent for 30 min and the reaction terminated by washing with 50 mM Tris, pH 7.4, containing 100 mM NaCl. The cells are then scraped in this buffer and centrifuged at 400 g. The pellet is resuspended, at 50 μl/10^7 cells, in 20 mM HEPES (pH 7.4), 50 mM NaCl, 2.5 mM EDTA, leupeptin (10 μg/ml), aprotinin (20 KIU/ml), 1 mM PMSF, and 1% (v/v) Triton X-100 and incubated on ice for 10 min. The lysate is then centrifuged at 16,000 g and the supernatant made 1× in SDS–PAGE sample buffer and boiled for 5 min. Usually 10–20 μl is loaded per lane of an SDS–PAGE gel.

Abolishing Fibroblast Growth Factor Binding to Heparan Sulfate

Degradation of Heparan Sulfate with Heparinases. To date, all cells that express endogenous FGF receptors also display heparan sulfate at their surfaces or are in contact with extracellular matrix rich in heparan sulfate proteoglycans. Thus, critical demonstration of heparan sulfate participation in the signaling event requires that the endogenous heparan sulfate be abolished. In the past, attempts to do this have relied largely on the addition of heparin as a competitor, although it is now recognized that this has the obvious pitfall that soluble heparin can supplant the activity of the heparan sulfate coreceptor. Alternatively, heparinases have been employed to digest the endogenous heparan sulfate. Although this approach can prove fruitful it is also fraught with difficulty and the potential for misinterpretation. First, heparinases are endoglycosidases that cleave at specific sites within the heparan sulfate chain and may release highly sulfated fragments of the heparan sulfate glycosaminoglycan; these fragments are capable of binding FGF and influencing its activity. Thus, these enzymes may generate potent signaling partners rather than abolishing heparan sulfate activity. Second, these enzymes are fragile; they have pH optima in the range of pH 7.0–7.5, but are relatively unstable at this pH. Thus, prolonged treatment necessary to assess cellular effects is difficult without repeated addition of high enzyme concentrations. Third, the enzymes can act only on heparan sulfate that has arrived at the cell surface. As the cell is continually synthesizing new proteoglycan, a competition is set up at the cell surface between the heparinase and the heparan sulfate-binding ligand. As the affinity of FGF for heparan sulfate is high ($\sim$1–5 nM), the likelihood that the ligand binds and protects a fragment of heparan sulfate is high. A means of nullifying the heparan sulfate prior to its encounter with the ligand is more advantageous.

Metabolic Inhibitors of Heparan Sulfate Synthesis. An alternative approach is to disrupt the synthesis of the heparan sulfate itself. An approach that has been used for over a decade is to attempt to disrupt the synthesis of the proteoglycan itself. This approach employs xylosides, which are competitive inhibitors of the xylosyltransferase required to initiate glycosaminoglycan synthesis on the core protein[33]; an excess of xyloside, such as methylumbelliferyl-β-D-xyloside or p-nitrophenyl-β-D-xyloside, acts as substrate for the enzyme and swamps out synthesis on the core protein. Drawbacks to this method are severalfold: (1) they are relatively inefficient at blocking heparan sulfate attachment, compared to greater effectiveness at blocking chondroitin sulfate attachment; (2) even

---

[33] T. A. Fritz, F. N. Lugemwa, A. K. Sarkar, and J. D. Esko, *J. Biol. Chem.* **269**, 300 (1994).

successful blockage of proteoglycan formation is accompanied by the secretion of copious amounts of glycosaminoglycan chains synthesized on the competitive acceptor, thus failing to abolish glycosaminoglycan synthesis; and (3) the concentrations of xyloside needed to affect heparan sulfate are often toxic to the cell.

An alternative approach is to disrupt the sulfation of the glycosamino-glycan chain. The sulfate residues are generally required for binding of heparan sulfate glycosaminoglycan to its ligands; indeed, the sulfation pattern may regulate ligand recognition. An effective approach used to disrupt sulfation is to use sodium chlorate.[34,35] Sodium chlorate is a competitive inhibitor of the ATP-sulfurylase (sulfate adenylyltransferase) that participates in the formation of phosphoadenosine phosphosulfate (PAPS), the high-energy sulfate donor required by the sulfotransferases adding sulfate groups to newly synthesized heparan sulfate chains in the Golgi. In the presence of 1–30 mM concentrations of chlorate ion, the ATP-sulfurylase is inhibited and the heparan sulfate proteoglycan arrives at the cell surface bearing nonsulfated heparan sulfate or chondroitin sulfate chains. The potential drawback to the use of chlorate is that it reduces all sulfation and is not limited to heparan sulfate chains. Thus, other glycosaminoglycans, glycoproteins, glycolipids, and so on, that bear sulfate are affected. It remains the obligation of the investigator, therefore, to demonstrate that the noted effect is specific to the sulfated component in question. In the case of FGFs, this can be done by adding back heparin or heparan sulfate and restoring the blocked effect (see below). Indeed, this provides the opportunity to question whether or not specific types or modifications of the exogenous heparan sulfate affect activity.

We have used the chlorate inhibition approach on several types of cells (Table I). In the initial work exploring its effect on FGF-2 signaling, it was used to inhibit mitogenesis of Swiss 3T3 fibroblasts and the differentiation of mouse MM14 myoblasts.[36,37] These two cell types represent different approaches to the use of the inhibitor and are explained here in detail. The Swiss 3T3 cell response is complicated by their response to a variety of mitogens and by their relative insensitivity to the sulfation inhibitor, whereas the myoblasts are easily affected and have a biological response that is absolutely dependent on FGF.

Swiss 3T3 cells treated with 30 mM sodium chlorate in complete medium continue to respond to FGF-2. This suggests that the sulfate concen-

[34] J. R. Farley, G. Nakayama, D. Cryns, and I. H. Segel, *Arch. Biochem. Biophys.* **185**, 376 (1978).

[35] K. M. Keller, P. R. Brauer, and J. M. Keller, *Biochemistry* **28**, 8100 (1989).

[36] B. B. Olwin and A. Rapraeger, *J. Cell Biol.* **118**, 631 (1992).

[37] A. C. Rapraeger, A. Krufka, and B. B. Olwin, *Science* **252**, 1705 (1991).

TABLE I

INHIBITION OF SULFATION AND FIBROBLAST GROWTH FACTOR RESPONSE

Cell type	Effective chlorate concentration (mM)	Assay	Sulfate deprivation	Ref.
Swiss mouse 3T3	30	Binding of FGF-1, -2, -4, -5; mitogenesis	Yes	a
MM14 myoblasts	10	Binding of FGF-1, -2, -4	No	a, b
Mouse brain endothelial	30	Mitogenesis	Yes	c
Bovine aortic endothelial	30	Mitogenesis	Yes	c
PC-12	30	Neurite outgrowth	Yes	c
Adrenal corticoendothelial	5	Binding of FGF-2; mitogenesis	Yes	d
NMuMG-R1 epithelial	30	Binding of FGF-2; mitogenesis	Yes	e

[a] A. C. Rapraeger, A. Krufka, and B. B. Olwin, *Science* **252**, 1705 (1991).
[b] B. B. Olwin and A. C. Rapraeger, *J. Cell Biol.* **118**, 631 (1992).
[c] A. Krufka and A. C. Rapraeger, unpublished (1994).
[d] M. Ishihara, D. J. Tyrrell, G. B. Stauber, S. Brown, L. S. Cousens, and R. J. Stack, *J. Biol. Chem.* **268**, 4675 (1993).
[e] J. Reiland and A. C. Rapraeger, *J. Cell Sci.* **105**, 1085 (1993).

tration in complete medium outcompetes the chlorate for the ATP-sulfurylase, allowing sulfated proteoglycan to reach the cell surface and bind FGF. This is confirmed by measuring the specific binding of iodinated FGF-2 to monolayers of 3T3 cells. To enhance the efficacy of the chlorate treatment, therefore, sources of endogenous sulfate must be removed from these cells. Sulfate can be removed from the medium by (1) curtailing use of streptomycin sulfate as a bacteriostatic agent, (2) formulating culture medium that is devoid of sulfate salts (e.g., substituting magnesium chloride for magnesium sulfate), and (3) dialyzing the serum against sulfate-free DMEM. In addition, as the 3T3 cells can metabolize the sulfur in cysteine and methionine, generating sulfate that is used in these reactions, the cysteine is removed from the medium.[35] The medium formulation used for the 3T3 cells is shown in Table II.

The mitogenic response of the Swiss 3T3 cells to FGFs has been assessed by [³H]thymidine incorporation and confirmed by counting of increased cell numbers. For these assays, the cells are starved in serum-free medium containing sodium chlorate prior to stimulation. To remove exogenous proteoglycan synthesized prior to chlorate treatment, the cells

TABLE II
Culture of Swiss Mouse 3T3 Cells in Sodium Chlorate

I. Formulation for 500 ml of low-sulfate, no-cysteine DMEM, low-salt DMEM (LS, NC, LS-DMEM)

  A. Combine the following:

      10 ml   50× L-tryptophan (80 mg of L-tryptophan in 100 ml of doubly distilled H_2O; add a few drops of NaOH to solubilize)

      20 ml   25× L-tyrosine (360 mg of L-tyrosine in 200 ml of doubly distilled H_2O, pH to 10.4; may have to warm to solubilize)

      10 ml   50× "no-cysteine" amino acids (420 mg of L-arginine hydrochloride, 210 mg of L-histidine hydrochloride hydrate, 525 mg of L-isoleucine, 525 mg of L-leucine, 725 mg of L-lysine hydrochloride, 150 mg of L-methionine, 330 mg of L-phenylalanine, 432 mg of L-threonine, and 470 mg of L-valine in 100 ml of doubly distilled H_2O)

      50 ml   10× glycine-serine (150 mg of glycine plus 210 mg of serine in 500 ml of doubly distilled H_2O)

      5 ml   100× sulfate-free calcium, magnesium (200 mM $CaCl_2$ plus 145 mM $MgCl_2$)

      50 ml   10× DMEM salts, calcium-free, magnesium-free, NaCl-free [10 ml of 100× KCl (40 g/liter), 10 ml of 100× Na_2PO_4 (12.5 g/liter), 10 ml of 100× $Fe(NO_3)_3$ (10 mg/liter), 4.5 g of dextrose, and 3.0 ml of phenol red; bring to 100 ml]

      5 ml   100× nonessential amino acids (GIBCO-BRL), Gaithersburg, MD)

      20 ml   25× MEM vitamins (GIBCO-BRL)

      1.875 g of sodium bicarbonate

  B. Bring to 500 ml

  C. Adjust pH to 7.2, filter sterilize

II. Calf serum

  A. Heat inactivate at 56° for 30 min, dialyze 500 ml vs 5 liters of buffer:

      1×: 1 M NaCl, 20 mM HEPES, pH 7.4

      2×: 20 mM HEPES, pH 7.4

      1×: 0.15 M NaCl, 20 mM HEPES, pH 7.4

  B. Filter sterilize, aliquot, freeze

III. Formulation for complete media

	LS, NC, LS-DMEM (ml)	NaCl (g)	NaClO$_3$ (g)	Na$_2$SO$_4$ (g)	Serum (ml)	Glutamine (mM)
No treatment	100	0.640	0	0	10	4
30 mM chlorate	100	0.466	0.318	0	10	4
30 mM chlorate/ 10 mM sulfate	100	0.408	0.318	0.142	10	4

are cultured in 30 mM chlorate for 48 hr in DMEM containing 10% (v/v) calf serum and then suspended with trypsin. The cells are then replated into assay plates for an additional 24 hr in DMEM containing 0.1% (w/v) BSA and 30 mM chlorate. As controls, chlorate is either removed from selected plates at this time or is continued in the presence of 10 mM sodium sulfate, which overcomes the effect of the competitive inhibitor and restores sulfation of the proteoglycan. Following the 24-hr starvation period, these treatments are continued in the presence of FGF-2 for 24 hr. The Swiss 3T3 cells typically respond half-maximally to 3–10 pM FGF-2. Thus, routine proliferation studies are conducted in the range of 1–100 pM. During the last 6 hr of the stimulation period, [^{3}H]thymidine (2 μCi/ml) is added and incorporation is assessed by removal of the labeling medium followed by three washes with 10% (w/v) trichloroacetic acid (TCA) and dissolution of the cell layers in 0.1 N NaOH [or 10% (w/v) SDS] and quantification by scintillation counting. The timing of the [^{3}H]thymidine addition is experimentally determined to coincide with the first S phase as cells escape from the starvation block.

Potential nonspecific effects of chlorate on the cells are questioned in several ways. The first is to restore sulfation by the addition of exogenous sulfate, thus outcompeting the inhibitor for the ATP-sulfurylase. On Swiss 3T3 cells, 5–10 mM sodium sulfate restores sulfation of the proteoglycans in 30 mM chlorate, restores FGF-2 binding to heparan sulfate, and restores mitogenic signaling. A second means is to assess the impact of the chlorate treatment on the activity of other growth factors. Such factors need to be chosen with some caution, as several growth factors can be affected by removal of heparan sulfate. Epidermal growth factor (EGF) is an acceptable candidate as it has no clear requirement for heparan sulfate binding. This sets it apart, however, from heparin-binding EGF.[38] On Swiss 3T3 cells, chlorate has no inhibitory effect on EGF activity or on the ability of the cells to grow in serum. Finally, a critical assessment is the ability of exogenous heparin, heparan sulfate, or proteoglycan to restore growth factor signaling. On the Swiss 3T3 cells, about 10–100 ng of exogenous heparin per milliliter restores full mitogenic signaling. Assuming an average molecular mass of 10 kDa for the heparin and a K_d of 1 nM, this amount of heparin would occupy about 90% of the FGF monomers when present at 100–300 pM, confirming that occupancy of the growth factor by the glycosaminoglycan is a prerequisite for binding and signaling. Other growth factors may be affected by chlorate treatment in other ways.

[38] S. Higashiyama, K. Lau, G. E. Besner, J. A. Abraham, and M. Klagsburn, *J. Biol. Chem.* **267**, 6205 (1992).

For example, Rifkin and co-workers[39] have demonstrated that binding to heparan sulfate can limit the diffusion of heparin-binding growth factors, thus maintaining effective concentrations at local sites. Treatment with chlorate would abrogate this effect and potentially lead to reduced signaling. Such effects might come into play for other growth factors that bind heparan sulfate, even weakly. However, the addition of exogenous heparin would be predicted not to restore signaling as it does with FGF-2, as the soluble heparin would increase diffusion by competing with the endogenous heparan sulfate immobilized at cell surfaces. For this reason, the ability of exogenous heparin or heparan sulfate to restore signaling in the absence of endogenous heparan sulfate becomes an important criterion for a primary role in signaling.

The mouse MM14 myoblasts provide a different approach to assessing FGF-2 signaling.[29,36,40] In the presence of FGF-2, this cell line is maintained as a proliferating and nondifferentiated cell population. However, if the cells are deprived of FGF during the G_1 phase of the cell cycle, they exit the cell cycle at G_1 and embark on a terminal differentiation pathway culminating in fusion into myotubes and the expression of muscle-specific genes. If FGF-2 is withheld from MM14 myoblast cultures for as little as 12 hr, the entire population of cells will quantitatively differentiate, as measured by failure to incorporate [^{3}H]thymidine into newly synthesized DNA or by the immunodetection of muscle-specific myosin. In the absence of serum the cells will become quiescent but remain undifferentiated in the presence of FGF. Thus, FGF-2 acts as a repressor of terminal differentiation in these cells. The quantitative differentiation of the cells is dependent on culture conditions and requires prescreening of the cells for specific lots of horse serum. The cells can be grown only at low densities (about $1 \times 10^6/100$-mm tissue culture dish), must be fed with FGF and serum at 12 hr intervals, and are passaged every 48 hr. Detailed information on cell growth conditions are available from B. B. Olwin (Department of Biochemistry, Purdue University, West Lafayette, IN).

Heparan Sulfate-Deficient Cell Lines

Essentially all adherent cells express heparan sulfate at their cell surfaces and in the neighboring extracellular matrix. Thus, inhibitors must be used on these cells to deprive them of endogenous heparan sulfate. Two categories of cells that fail to express heparan sulfate, however, are (1) mutants defective in heparan sulfate synthesis, and (2) circulating

[39] R. Flaumenhaft, D. Moscatelli, and D. B. Rifkin, *J. Cell Biol.* **111**, 1651 (1990).
[40] B. B. Olwin and S. D. Hauschka, *J. Cell Biol.* **110**, 503 (1990).

lymphoid cells, which are heparan sulfate deficient although they apparently contain the machinery for glycosaminoglycan synthesis and express heparan sulfate when enmeshed in the matrix of a tissue, such as the bone marrow or lymph nodes.

Mutants. Heparan sulfate mutants were first derived by Esko,[2] screening for Chinese hamster ovary (CHO) cells that showed reduced radiosulfate incorporation following mutagenesis procedures. This approach has identified a number of CHO cell mutants defective in the expression of chondroitin sulfate, heparan sulfate, or both. In particular, the heparan sulfate-deficient cell line has been used to examine the requirement of heparan sulfate for FGF-2 binding to tyrosine kinase receptor 1.[41] These cells, which fail to express FGF receptors, were transfected with FGF receptor 1 (FGFR-1), but fail to bind FGF-2. This failure is overcome by supplying the cells with exogenous heparin or heparan sulfate at concentrations of 1–10 ng/ml, which promotes binding to the receptor. The CHO mutants expressing FGF receptor fail to respond to FGF, but together with the activity studies on Swiss 3T3 and MM14 cells this demonstrated for the first time the requirement for a receptor–FGF–heparan sulfate complex for FGF-2 signaling.

Lymphoid Cells. Lymphoid cells have proved useful for extending these studies. These cells are also negative for FGF receptors but, in contrast to the CHO cells, do display a response to FGFs when receptor and heparan sulfate or heparin are supplied. Circulating B lymphocytes do not express cell surface heparan sulfate, although they display syndecan 1 at their surface in the bone marrow as pre-B cells and when in immune tissues as plasma cells.[42,43] Ornitz *et al.*[44] have used the BaF3 lymphocytic cell line, which is (1) heparan sulfate negative, (2) FGF receptor negative, and (3) interleukin 3 (IL-3) dependent; these cells require IL-3 for proliferation and cell survival. When transfected with FGFR-1 and furnished with exogenous heparin, these cells (F32 cell clone) proliferate and survive in the absence of IL-3. Thus, FGFR-1, exogenous heparin, and FGF-2 are required for signaling. As no two of these components alone will relieve the cells of their IL-3 dependence, the direct participation of all three (namely, receptor tyrosine kinase, heparan sulfate proteoglycan, and FGF) in FGF signaling is likely. Additional experiments utilizing transfected

[41] A. Yayon, M. Klagsbrun, J. D. Esko, P. Leder, and D. M. Ornitz, *Cell (Cambridge, Mass.)* **64,** 841 (1991).

[42] R. D. Sanderson, P. Lalor, and M. Bernfield, *Cell Regul.* **1,** 27 (1989).

[43] R. C. Ridley, H. Xiao, H. Hata, J. Woodliff, J. Epstein, and R. D. Sanderson, *Blood* **81,** 767 (1993).

[44] D. M. Ornitz, A. Yayon, J. G. Flanagan, C. M. Svahn, E. Levi, and P. Leder, *Mol. Cell. Biol.* **12,** 240 (1992).

lymphoid cells are likely to reveal specific proteoglycans and receptors compatible for signaling and, eventually, domains present within the FGFs and their receptors that interact with heparan sulfate proteoglycans.

Cleavage/Fractionation of Heparan Sulfate–Heparin

The participation of heparan sulfate in FGF signaling suggests that specific sulfation sequences inherent in these glycosaminoglycan chains may recognize certain FGF family members and/or receptors. The most direct way to question this specificity has been to cleave heparin or heparan sulfate into fragments of defined length or of defined sulfation patterns (Fig. 1) and to test these fragments for activity in FGF signaling.

Enzymatic Digestion. The cleavage of these glycosaminoglycans into fragments can be accomplished either by enzymatic or chemical means. Enzymes of bacterial origin that recognize specific sulfation sites within the glycosaminoglycan chain[45] have been isolated, purified, character-ized, and termed heparinase I, II, and III. These enzymes essentially recognize the degree of modification that has occurred during the epimeri-zation and sulfation of the nascent glycosaminoglycan chain in the Golgi. Heparinase I (referred to as heparinase or heparin lyase) recognizes N-sulfated regions and specifically cleaves disaccharides that contain 2-O-sulfated iduronate (N-sulfoglucosaminyl-α-1,4-induronosyl 2-O-sulfate or N-sulfoglucosaminyl 6-O-sulfate α-1,4-iduronosyl 2-O-sulfate). This site occurs only where chain modification has proceeded nearly to completion; hence, this site is prevalent in heparin and is more rare in heparan sulfate. Heparinase III, however, recognizes a sequence that bears little or no modification beyond the initial synthesis of the sugar backbone, namely, disaccharides that contain glucuronic acid (e.g., N-acetylglucosaminyl-α-1,4-glucuronate). These sites are much more prevalent in heparan sulfate and this enzyme is commonly referred to as heparitinase.

Cleavage of heparan sulfate or heparin chains with these enzymes will generate fragments of variable length depending on their sulfation pattern. If starting with isolated heparan sulfate proteoglycans, the heparan sulfate chains are obtained by complete digestion of the core protein with pronase (5 mg/ml) at 37°. Alternatively, the glycosaminoglycans are released from the core protein by mild alkaline borohydride treatment (0.05 N NaOH in 1 M sodium borohydride, 15 hr, 4°), although this risks removing sulfate groups at the C-2 position of the iduronate. Borohydride is present in the alkali to reduce the "reducing end" of the glycosaminoglycan chain, which prevents depolymerization of the chain by the alkali treatment. The re-

[45] R. J. Linhardt, J. E. Turnbull, H. M. Wang, D. Loganathan, and J. T. Gallagher, *Biochemis-try* **29**, 2611 (1990).

leased chains are isolated by DEAE-Sephacel (Pharmacia) ion-exchange chromatography; the chains would be expected to elute at about 0.4–0.8 M NaCl in a linear salt gradient in 50 mM sodium acetate, pH 5.0.

To generate heparan sulfate or heparin fragments, the chains are cleaved with heparinases for variable amounts of time (e.g., from 1 hr to overnight) and the resulting fragments of varying size are chromatographed on Sephadex G-50 (or Bio-Gel P-10) in 1 M NaCl, by which oligosaccharides ranging from disaccharides to at least dodecasaccharides can be resolved. An excellent description of the conditions for cleavage and enzymatic units required is in Linkhardt et al.[45] and Gallagher's group has used these methods effectively to isolate bioactive heparan sulfate fragments.[46–50] As the heparinases are lyases, rather than hydrolases, they produce a 4,5-unsaturated hexuronic acid residue at the nonreducing terminal of the oligosaccharide. This unsaturated bond absorbs in the ultraviolet (UV) at 232 nm; absorbance at this wavelength is used to measure the production of these fragments and to follow their behavior in subsequent chromatographic steps. If necessary, this residue can be eliminated by treatment with 10 mM mercuric acetate in 130 mM sodium acetate (pH 5.0) for 30 min at room temperature.[51] Removal of the mercuric acetate is accomplished by several passes through a PD-10 (Pharmacia Fine Chemicals) desalting column equilibrated with 1 M NaCl in water. The fragments are precipitated overnight by 70% (v/v) cold ethanol containing 1.3% (w/v) potassium acetate to help nullify charges on the glycosaminoglycan, dried, and stored as a lyophilized powder.

Chemical Cleavage. Chemical cleavage of the heparan sulfate or heparin chains employs limited deaminative cleavage with nitrous acid. A low-pH (pH 1.5) procedure cleaves at N-sulfated glucosamine (GlcNS) units[52] and the resulting 2,5-anhydro-D-mannose residues are reduced with NaBH$_4$ to 2,5-anhydro-D-mannitol. If desired, NaB^{3}H$_4$ can be used to introduce tritium label into the anhydromannitol, thus labeling the fragment. For partial depolymerization, 1 g of heparin in 20 ml of ice-cold water is combined with 42 mg of NaNO$_2$ and acidified to pH 1.5 with dilute

[46] J. E. Turnbull and J. T. Gallagher, *Biochem. J.* **265,** 715 (1990).
[47] J. E. Turnbull and J. T. Gallagher, *Biochem. J.* **277,** 297 (1991).
[48] J. E. Turnbull and J. T. Gallagher, *Biochem. J.* **273,** 553 (1991).
[49] J. E. Turnbull, D. G. Fernig, Y. Ke, M. C. Wilkinson, and J. T. Gallagher, *J. Biol. Chem.* **267,** 10337 (1992).
[50] J. E. Turnbull and J. T. Gallagher, *Biochem. Soc. Trans.* **21,** 477 (1993).
[51] U. Ludwigs, A. Elgavish, J. D. Esko, E. Meezan, and L. Rodén, *Biochem. J.* **245,** 795 (1987).
[52] J. E. Shively and H. E. Conrad, *Biochemistry* **15,** 3932 (1976).

sulfuric acid.[53] As the N-sulfaminylglucosamine is prevalent in heparin, limited cleavage will produce heparin fragments of varying length, whereas exhaustive cleavage will produce largely disaccharides. A similar cleavage of heparan sulfate, in which these groups are more rare, will produce fragments that are likely to contain regions of low modification.

An alternative nitrous acid procedure can be used to recognize sites within the chain that are poorly modified. A two-step procedure can be used to cleave N-acetylated GlcN, which is rare in heparin but more prevalent in heparan sulfate. The first step is to deacetylate the amine; a sample of 5–10 mg of heparan sulfate is treated at 100° for 4 hr in 1 ml of hydrazine, containing 30% (v/v) water and 1% (w/v) hydrazine sulfate.[52] After repeated evaporation in the presence of toluene, the sample is passed through Sephadex G-15, lyophilized, and subjected to the second step of the procedure, a deaminative cleavage with nitrous acid at pH 3.9,[52] which recognizes N-unsubstituted GlcN units.

Selective Desulfation. A combination of these cleavage procedures can be used on heparin or heparan sulfate to derive fragments of varying length and sulfation, relying on the variation inherent in the glycosaminoglycans. This variation can be modified, however, by preferentially removing specific sulfate groups. These sulfates occur primarily as glycosaminyl N-sulfate, glucosaminyl 6-O sulfate, and iduronosyl 2-O-sulfate. These sulfates can all be removed by solvolysis of the heparin pyridinium salt by DMSO containing water or methanol.[54] Limited solvolysis, however, can be used to remove the N-sulfates and 6-O-sulfates preferentially. The glycosaminoglycan is applied in water to Dowex 50W-X 8, 50–100 mesh (H$^+$ form) and eluted in water. The passthrough fraction is neutralized with pyridine to pH 7 and lyophilized. N-Desulfation occurs readily in DMSO containing 5% (v/v) water at 50° for 90 min.[54–56] Preferential 6-O-desulfation can be accomplished by treatment with 90% (v/v) DMSO–10% (v/v) water at 110° for 5 hr,[56] which removes the 6-O-sulfates as well as the N-sulfates. Re-N-sulfation is then carried out by reaction with trimethylamine–sulfur trioxide. Glycosaminoglycan (50 mg) is dissolved in 2 ml of 1.0 M sodium carbonate and added to 50 mg of trimethylamine–sulfur trioxide under nitrogen, then heated at 50° for 24 hr with occasional agitation.[57] Resulfated material is dialyzed against running deionized water for 24 hr, then against distilled water and lyophilized. As

[53] M. Maccarana, B. Casu, and U. Lindahl, *J. Biol. Chem.* **268**, 23898 (1993).
[54] Y. Inoue and K. Nagasawa, *Carbohydr. Res.* **46**, 87 (1976).
[55] K. Nagasawa and Y. Inoue, *Metab. Carbohydr. Chem.* **8**, 291 (1980).
[56] K. Nagasawa and Y. Inoue, *Metab. Carbohydr. Chem.* **8**, 287 (1980).
[57] T. Irimura, M. Nakajima, and G. L. Nicolson, *Biochemistry* **25**, 5332 (1986).

prolonged solvolysis will remove 2-O-sulfate groups as well as N- and 6-O-sulfates, the final disaccharide composition of the material should be confirmed by analytical procedures described below. If removal of the 2-O-sulfates from iduronate is desired, selective 2-O-desulfation is performed by lyophilization in the presence of base. The sodium salt of 40 mg of heparin in 10 ml of water is brought to pH 12.5–12.8 by addition of sodium hydroxide, followed by lyophilization.[58]

Molar amounts of the oligosaccharides can be determined by weighing, if sufficient quantities have been produced and are purified, or by measuring total hexuronic acid content, using the carbazole reaction,[59] and correcting for total sugar content (e.g., hexosamine and hexuronic acid).

Analysis of the disaccharide composition of the starting material or isolated fragments can be conducted by high-voltage paper electrophoresis or strong anion-exchange high-performance liquid chromatography (HPLC) and comparison with known standards.[52,53,60] Disaccharides are generated chemically by successive treatment with the pH 3.9 nitrous acid procedure (together with deacetylation of the glucosamine) and the pH 1.5 nitrous acid procedure. Alternatively, exhaustive digestion with heparinase I (heparinase) and heparinase III (heparitinase) will cleave the chains to disaccharides.

FGF Signaling with Exogenous Heparin or Heparan Sulfates in Heparan Sulfate-Deficient Systems

Binding Studies. The finding that exogenous heparin or heparan sulfate restores FGF signaling on heparan sulfate-deficient cells provides the opportunity to search for specific sulfation sequences within the glycosaminoglycan chain that may be specific for regulating either (1) the activity of specific FGF family members or (2) the signal that is generated on binding to the receptor tyrosine kinase. Sequences of heparin or heparan sulfate that bind to FGF-2 have been defined using three approaches. One approach has been to identify sequences that bind most strongly to FGF-2. This approach has utilized heparinase III to cleave native heparan sulfate at undersulfated regions, generating fragments bearing a range of sulfation patterns.[49] When these fragments are bound to an affinity column of FGF-2 and eluted in a salt gradient, a fragment consisting of 14 sugars is eluted. Within the fragment is a decasaccharide composed of a fivefold repeated disaccharide [(iduronosyl 2-O-sulfate)-α-1,4-(N-sulfoglucosamine)]$_5$. This repeating disaccharide represents a highly modified region of the heparan

[58] M. Jaseja, R. N. Rej, F. Sauriol, and A. S. Perlin, *Can. J. Chem.* **67**, 1449 (1989).
[59] T. Bitter and H. M. Muir, *Anal. Biochem.* **4**, 330 (1962).
[60] M. J. Bienkowski and H. E. Conrad, *J. Biol. Chem.* **260**, 356 (1985).

sulfate chain, but one that is not fully sulfated, as it lacks the glucosaminyl 6-O-sulfate. This suggests that whereas sulfates are required for binding FGF-2, the 6-O-sulfate is not necessary.

A second approach has attempted to identify the minimal sequence that binds FGF-2.[53] This approach has fragmented heparin using the low-pH (1.5) nitrous acid procedure to generate fragments bearing a range of sulfation patterns. The fragments or native chains are labeled at their reducing termini by treatment with sodium [³H]borohydride, generating [³H]anhydromannitol. These labeled fragments are incubated with FGF-2 and passed through nitrocellulose, which retains the protein ligand but does not bind heparin or its fragments. Thus, specific glycosaminoglycan fragments that are retained must be bound to the FGF-2 and represent specific FGF-binding moieties. In addition, the fragments identified by this procedure can be examined for their ability to compete with labeled heparin or heparan sulfate binding to FGF-2, allowing assessment of their relative affinity for the growth factor. These studies have identified a pentasaccharide sequence (hexuronosyl-N-sulfoglucosaminyl)$_2$-iduronosyl 2-O-sulfate as the minimal binding sequence for FGF-2. It is important to note (1) that this sequence again implicates iduronate 2-O-sulfate as a recognition site on the glycosaminoglycan chain and (2) it again fails to implicate 6-O-sulfate on the glucosamine as a binding requirement.

Activity Studies. The third approach, which we have used, has examined the activity of fragments generated by nitrous acid cleavage.[61] This assay is based on the finding that exogenous heparin restores FGF-2 signaling on Swiss 3T3 cells treated with chlorate. Chlorate-treated, starved cells are incubated with heparin fragments or modified heparins ranging from 0.1 to 100 nM in the presence of 30–100 pM FGF-2. For native heparin, restoration of FGF-2 activity is seen with as little as 0.1 nM, with full activity noted at 10 nM. Modified heparins, such as (1) heparin deprived of the 2-O-sulfate on iduronic acid, (2) heparin deprived of the 6-O-sulfate on the N-sulfoglucosamine, or (3) heparin fragments shorter than a dodecasaccharide, fail to show bioactivity even at 1000-fold higher concentrations. This suggests that a minimal length of 12 sugars is necessary on these cells to achieve signaling; this length is more than the minimal length required to bind FGF-2. This implies that two heparan sulfate-binding sites may be present on the growth factor or that binding may occur on the growth factor and the receptor simultaneously. Alternatively, only one binding site may be present but it may require a particular conformation of the fragment for which the 12-sugar length is necessary.

[61] S. Guimond, M. Maccarana, B. B. Olwin, U. Lindahl, and A. C. Rapraeger, *J. Biol. Chem.* **268**, 23906 (1993).

These types of experiments also demonstrate the importance of the 2-*O*-sulfate and 6-*O*-sulfate residues. A caveat is that removal of the 6-*O*-sulfates also leads to the removal of about one-third of the 2-*O*-sulfates and this reduction in 2-O-sulfation, rather than the removal of 6-*O*-sulfates per se, may have impaired the biological activity. Regardless of the explanation, this contrasts with the requirements for FGF binding, as the 6-O-desulfated binds FGF-2 in the solid-phase assay described above and competes fully with native heparin binding to the growth factor. Indeed, this predicts that the 6-O-desulfated heparin would act as a competitive inhibitor of FGF-2 action on cells, either in the presence of exogenous heparin or as a competitor with cell surface heparan sulfate. This can be demonstrated by examining the activity of FGF-2 on chlorate-treated 3T3 cells in the presence of native heparin combined with a 100-fold excess of 6-O-desulfated heparin; the 6-O-desulfated heparin effectively competes for binding to the FGF-2 and attenuates the activity of the native heparin. Similarly, the 6-O-desulfated heparin competes with FGF-2 binding to cell surface heparan sulfate and, at concentrations of about 100 μg/ml, effectively blocks FGF-2 signaling on mouse MM14 myoblasts as shown by their differentiation despite the continued presence of FGF-2.

In contrast to these effects with FGF-2, FGF-1 and FGF-4 have different requirements. Neither the 2-O-desulfated nor the 6-O-desulfated heparin enhance or inhibit FGF-1 signaling. In contrast, FGF-4 can use either the 6-O-desulfated or the 2-O-sulfated heparins as cofactors for stimulation of mitogenesis in Swiss 3T3 cells. Thus, the sulfation sequences that regulate diverse members of the FGF family appear to be different.

Summary

The integral role of heparan sulfate proteoglycans in FGF signaling provides a potential means of regulating FGF activity. This regulation may be used by the cell, where the modification of heparan sulfate glycosaminoglycans during their synthesis in the Golgi can produce cell type- and potentially ligand-specific sulfation sequences. The description of these sequences will not only provide information on how this regulation is achieved, perhaps lending insight into other heparan sulfate–ligand interactions, but may also discern sulfated mimetics that can be used to disrupt or alter FGF signaling. These mimetics may be useful in the treatment of disease, or in understanding how FGF signaling via discrete pathways within the cell leads to specific cellular responses, such as activation of mitogenic signaling pathways, calcium fluxes, and cellular differentiation.

[12] Role of Transforming Growth Factor β and Decorin in Controlling Fibrosis

By J. R. HARPER, R. C. SPIRO, W. A. GAARDE, R. N. TAMURA,
M. D. PIERSCHBACHER, N. A. NOBLE, K. K. STECKER, and
W. A. BORDER

Introduction

The processes of wound healing and tissue regeneration involve a complex interplay between growth factors, cytokines, extracellular matrix (ECM) components, and adhesion receptors on a variety of cell types. Both beneficial and detrimental consequences comprise the attempt by the body to maintain an infection-free, homeostatic condition. The extracellular matrix and its many components clearly play a central role in the rebuilding of damaged tissue, by providing a substrate to support the migration and growth of wound-healing cells at the site of tissue injury and signaling an end to the process of tissue remodeling. The identification of critical functional domains of the extracellular matrix and its molecular components and the ability to synthesize the active peptide sequences have enabled biotechnology to develop derivatives of these components to accelerate and improve the quality of the healing process. On the other hand, the negative aspect, or "dark side,"[1] of wound healing is believed to involve a variety of growth factors and cytokines commonly associated with the inflammatory process.[2] One such cytokine shown to play a key role in the development of fibrosis is transforming growth factor β (TGF-β). Transforming growth factor β is a pluripotent growth factor that has effects on cell prolieration,[3,4] alters the expression of the integrin class of cell adhesion receptors,[5,6] and increases the expression of ECM components, such as collagen, biglycan, and fibronectin.[7,8] In addition to

[1] W. A. Border and E. Ruoslahti, *J. Clin. Invest.* **90**, 1 (1992).

[2] A. B. Roberts, M. E. Joyce, M. E. Bolander, and M. B. Sporn, *in* "Growth Factors in Health and Disease" (B. Westermark, C. Betsholtz, and B. Hökfelt, eds.), p. 89. Elsevier, Amsterdam, 1990.

[3] C. C. Bascom, N. J. Sipes, R. J. Coffey, and H. L. Moses, *J. Cell. Biochem.* **39**, 25 (1989).

[4] H. L. Moses, E. Y. Yang, and J. A. Pietenpol, *Cell (Cambridge, Mass.)* **63**, 245 (1990).

[5] J. Heino, R. A. Ignotz, M. E. Hemler, C. Crouse, and J. Massague, *J. Biol. Chem.* **264**, 380 (1989).

[6] R. A. Ignotz, J. Heino, and J. Massague, *J. Biol. Chem.* **264**, 389 (1989).

[7] A. Bassols and J. Massague, *J. Biol. Chem.* **263**, 3039 (1988).

[8] V.-M. Kahari, H. Larjava, and J. Uitto, *J. Biol. Chem.* **266**, 10608 (1991).

Copyright © 1994 by Academic Press, Inc.
All rights of reproduction in any form reserved.

its stimulatory effect on matrix expression and deposition, TGF-β may affect the natural turnover and metabolism of the ECM by inducing specific inhibitors of matrix-degrading proteases, such as plasminogen-activator inhibitor,[9] and by directly inhibiting the expression of certain proteases themselves.[10] Collectively, the multiple actions of TGF-β on ECM-related macromolecules lead to an accumulation of matrix, resulting in fibrosis.[11] Decorin, a small proteoglycan found as a component of collagen fibrils in most extracellular matrices, has been shown to bind and influence certain activities of TGF-β,[12] and thus has potential as a therapeutic agent for the treatment of fibrotic diseases.

Studies in several animal models of fibrosis have implicated TGF-β as a key causative agent in fibrosis, or scar formation. The first report demonstrating a key role for TGF-β used a model of experimental rat glomerulonephritis induced by injecting an anti-thymocyte serum (ATS), leading to mesangial cell lysis and acute inflammation. In this model, treatment of diseased rats with a TGF-β-neutralizing antibody significantly reduced the accumulation of pathologic matrix in glomeruli.[13] Subsequent studies demonstrated that the systemic administration of the small proteoglycan, decorin, was also capable of preventing matrix accumulation or acclerating the clearance of fibrotic matrix in glomeruli of ATS-treated animals.[14] In this chapter we present an example of the efficacy of decorin in preventing fibrosis in this model. Decorin targeting to fibrotic tissues and its affinity for mesenchymal collagens (types I and III) will be illustrated by intravenous decorin injection and an *in vitro* overlay technique.

Studies are underway to test the efficacy and specific matrix targeting of decorin in a variety of models of fibrotic disease in which TGF-β has been implicated by the use of neutralizing antibodies. For example, intratracheal instillation of the chemotherapeutic agent, bleomycin, induces a pulmonary fibrotic lesion akin to idiopathic pulmonary interstitial fibrosis. Neutralizing antibodies against TGF-β are effective in preventing or ameliorating the fibrotic response in this disease model, demonstrating a causal role for TGF-β in this condition.[15] Moreover, incisional wounds

[9] M. Laiho, O. Säkselä, A. P. Andreasen, and J. Keski-Oja, *J. Cell Biol.* **103,** 2403 (1986).
[10] S. M. Wahl, J. B. Allen, B. S. Weeks, H. L. Wong, and P. E. Klotman, *Proc. Natl. Acad. Sci. U.S.A.* **90,** 4577 (1993).
[11] S. Tomooka, W. A. Border, B. C. Marshall, and N. A. Noble, *Kidney Int.* **42,** 1462 (1992).
[12] Y. Yamaguchi, D. M. Mann, and E. Ruoslahti, *Nature (London)* **346,** 281 (1990).
[13] W. A. Border, S. Okuda, L. R. Languino, M. B. Sporn, and E. Ruoslahti, *Nature (London)* **346,** 371 (1990).
[14] W. A. Border, N. A. Noble, T. Yamamoto, J. R. Harper, Y. Yamaguchi, M. D. Pierschbacher, and E. Ruoslahti, *Nature (London)* **360,** 361 (1992).
[15] S. N. Giri, D. M. Hyde, and M. A. Hollinger, *Thorax* **48,** 959 (1993).

to the brain result in collagen/glial scar formation in the wound bed. Transforming growth factor β_1 expression at the mRNA and protein levels has been shown to be directly involved in this extracellular matrix accumulation by similar studies using neutralizing antibodies.[16]

To develop an extracellular matrix proteoglycan, such as decorin, as a therapeutic agent for the treatment of fibrosis or scarring, it is necessary and useful to have an *in vitro* assay for predicting the *in vivo* activity of decorin. Throughout the purification process and following formulation of a biological agent that has potential therapeutic value, such as decorin, it is necessary to monitor the functional activity in order to assess the stability of the compound. This stability is usually expressed as a specific activity with respect to either *in vivo* or *in vitro* activity units per milligram of protein. Because animal studies are expensive, time consuming, and often generate variable results, *in vitro* assays of biological potency or efficacy are preferable. To develop a reproducible bioassay that can be properly validated for use in drug development, it is useful to understand the mechanism of action of the compound, so that an attempt can be made to recreate this activity in the laboratory.

The mechanism by which decorin reduces the fibrotic response to inflammation and tissue injury may relate to several characteristics of its core protein. One such characteristic is the binding of decorin to various collagens.[17-19] Originally described as a component of the type I and II collagen fibrils, decorin has been localized to the d and e bands of the D-period.[20,21] Decorin molecules integrated into collagen fibrils formed *in vitro* have a dramatic effect on the optical properties of the fibrils, and possibly other physical characteristics. We have adapted the fibrillogenesis assay originally described by Vogel and co-workers[22,23] into a microtiter assay format that can be used to determine the functional integrity of the decorin core protein.

Related to its collagen-binding activity, decorin has been shown to bind specifically to the complement component C1q.[24] We have established

[16] A. Logan, M. Berry, A. M. Gonzalez, S. A. Frautschy, M. B. Sporn, and A. Baird, *Eur. J. Neurosci.* **6**, 355–363 (1994).
[17] D. J. Bidanset, C. Guidry, L. C. Rosenberg, H. U. Choi, R. Timpl, and M. Hook, *J. Biol. Chem.* **267**, 5250 (1992).
[18] E. Hedbom, D. Heinegard, *J. Biol. Chem.* **264**, 6898 (1989).
[19] G. Pogany and K. G. Vogel, *Biochem. Biophys. Res. Commun.* **189**, 165 (1992).
[20] B. Obrink, *Eur. J. Biochem.* **33**, 387 (1973).
[21] G. A. Pringle and C. M. Dodd, *J. Histochem. Cytochem.* **38**, 1405 (1990).
[22] K. G. Vogel and J. A. Trotter, *Collagen Relat. Res.* **7**, 105 (1987).
[23] K. G. Vogel, M. Paulsson, and D. Heinegard, *Biochem. J.* **223**, 587 (1984).
[24] R. Krumdieck, M. Hook, L. C. Rosenberg, and J. E. Volanakis, *J. Immunol.* **149**, 3695 (1992).

an assay that assesses the integrity of the collagen and C1q-binding sites on the decorin core protein. The binding of decorin to collagens and C1q is quantitated by a modification of the affinity coelectrophoresis method of Lee and Lander.[25] This method may be useful to determine the binding affinity and specificity of most proteoglycans to biologically relevant ligands.

Finally, in a gel-contraction assay that may relate to wound contraction during healing and scar formation, TGF-β induces fibroblasts to increase collagen gel contraction in a dose-dependent manner.[26] We have established a collagen gel assay for monitoring the ability of decorin to inhibit the TGF-β-induced contraction.

These bioassays exemplify how basic understanding of proteoglycan functional properties can be adapted to serve as methods used in the development of this class of molecules as biological therapeutic compounds. As they are described, these assay methods may well be useful in predicting the *in vivo* efficacy of decorin produced, purified, or preserved by different methods.

Collagen Fibrillogenesis

The original *in vitro* collagen fibrillogenesis protocol[23] has been adapted to a 96-well microtiter plate format using a Molecular Devices microtiter plate reader, equipped with the SoftMax software package (Molecular Devices Corp., Menlo Park, CA) for monitoring the kinetics of fibril formation. The protocol provides a relatively simple means of collecting and analyzing data from 96 samples simultaneously, which is particularly useful for analyzing replicate samples. The new protocol also reduces the quantity of reagents required for each sample, compared to previously published protocols. In addition, the SoftMax software is capable of determining maximum rates of change and the time required to reach this rate, two important collagen fibril formation parameters that are influenced by decorin. It is also possible to obtain maximum absorbance values and kinetic curves, although this requires the use of a spreadsheet program and a graphics program.

The assay is initiated by distributing appropriately diluted decorin samples into triplicate wells of a 96-well microtiter plate (Titertek, McLean, VA). The diluting buffer is composed of 0.01 M sodium phosphate, 0.14 M sodium chloride, pH 7.4 (PBS). Serial dilutions of decorin can easily and accurately be prepared in the microtiter plate, using a

[25] M. K. Lee and A. D. Lander, *Proc. Natl. Acad. Sci. U.S.A.* **88,** 2768 (1991).
[26] R. Montesano and L. Orci, *Proc. Natl. Acad. Sci. U.S.A.* **85,** 4894 (1988).

multichannel pipette. To each well containing 75 μl of diluted decorin sample, 150 μl of phosphate buffer (0.06 M sodium phosphate, 0.28 M sodium chloride pH 7.4) is added. The reaction is initiated by the addition of 75 μl containing 800 μg of rat tail type I collagen (Collaborative Biomedical Products, Bedford, MA) per milliliter in H_2O. Following the addition of collagen with a multichannel pipette, the mixture is triturated 10× to mix the contents of each well, and pipette tips must be changed between each addition of collagen to avoid cross-contamination. The plate is then placed in a Molecular Devices plate reader that has been programmed to read absorbance at 405 nm at appropriate time intervals (e.g., 5 min) for an appropriate length of time (e.g., 12 hr). Owing to the automated nature of this data collection, additional assays may be performed overnight, to obtain maximal use of the plate reader.

The SoftMax software facilitates the calculation of maximum rate of change in A_{405} and the time to reach the maximum rate, and is capable of generating plots of the maximum rate or time to reach maximum rate versus decorin concentration. Curve fitting using several algorithms is possible using this software. To obtain kinetic curves and maximum A_{405} values, the raw data can be accessed and downloaded to a spreadsheet program for further analysis. After the maximum A_{405} values have been determined for each sample, the values can be returned to SoftMax to take advantage of its analytical capabilities to generate plots of mean maximum A_{405} versus decorin concentration (Fig. 1). Using this assay, we have found that several conditions causing denaturation of the core protein also alter this activity of decorin.

Affinity Coelectrophoresis Analysis of Decorin–Collagen Binding

The affinity coelectrophoresis (ACE) method of Lee and Lander[25] has been adapted for analyzing specific proteoglycan–protein interactions. As originally described, the ACE gel method takes advantage of the fact that the electrophoretic mobility of most proteins at neutral pH is significantly less than that of highly charged glycosaminoglycans (GAGs) or intact proteoglycans. Consequently, it was demonstrated that electrophoretic migrations of radiolabeled GAGs or proteoglycans through zones containing binding proteins is retarded in a dose-dependent fashion, relative to the unbound charged molecule.[25,27] The major advantage of ACE gels over solid-phase binding-type assays is that binding can be measured at physiological pH, ionic strength, and under conditions in which the

---

[27] J. D. San Antonio, J. Slover, J. Lawler, M. J. Karnovsky, and A. D. Lander, *Biochemistry* **32**, 4746 (1993).

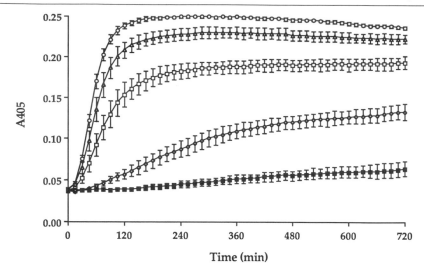

FIG. 1. Change in absorbance at 405 nm during *in vitro* fibrillogenesis of rat tail type I collagen in the presence of increasing concentrations of recombinant human (rHu) decorin. Serial dilutions of rHu decorin (○, 0.00; △, 0.082; □, 0.741; ◇, 6.667; ■, 60.00 μg/ml) were prepared in a 96-well microtiter plate (75 μl/well) followed by addition of 60 mM sodium phosphate–0.28 M sodium chloride (pH 7.4) (150 μl/well). Rat tail type I collagen at 800 μg/ml was added (75 μl/well) and triturated to mix the contents of each well. Absorbance at 405 nm was measured every 15 min for 12 hr using a Molecular Devices plate reader.

respective molecules are freely mobile. In addition, the technique requires only small amounts of purified material and can yield binding affinity values that agree well with those determined by more conventional techniques. The ACE gel technique is employed here for the analysis of decorin binding to collagen and collagen-like proteins, such as C1q (Fig. 2).

The affinity coelectrophoresis method utilizes a 4-mm thick, 1% (w/v) low melting point agarose (SeaPlaque; FMC Corp., Philadelphia, PA) gel in 50 mM sodium 3-(N-morpholino)-2-hydroxypropane sulfonic acid (MOPSO, pH 7.0)–125 mM sodium acetate–0.5% (w/v) 3-[(cholamido-propyl)-dimethyl-ammonic]-1-propanesulfonate (CHAPS), poured onto a GelBond film (FMC) fitted to a Plexiglas casting tray (75 × 100 mm). A specialized Teflon comb is used to divide the gel into nine parallel rectangular well chambers, each 45 × 4 × 4 mm, held 3 mm apart. A rough comb that is 66 × 37 × 1 mm is used to create a slot running horizontally and adjacent (≈4 mm) to the nine vertical, rectangular wells. Molten (~70°) 1% (w/v) agarose (20–30 ml) is poured into the cast and allowed to solidify. Protein samples are prepared in the above gel buffer (350–500 μl) at twice the desired concentrations, mixed with an equal volume of molten 2%

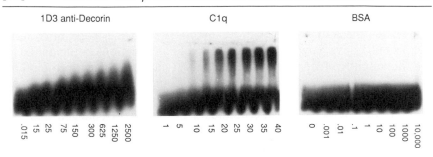

FIG. 2. Analysis of decorin by affinity coelectrophoresis-C1q binding. Electrophoresis was performed on ^{125}I-labeled decorin through zones containing increasing concentrations (nanomolar) of an anti-decorin monoclonal antibody (1D3, left), human C1q protein (middle), and bovine serum albumin (BSA, right). A concentration-dependent reduction in the electrophoretic mobility of the labeled decorin is apparent in both the anti-decorin and C1q panels.

(w/v) agarose at 37°, and applied to appropriate rectangular wells, using a transfer pipette. For collagen samples, a stock solution is serially diluted in 0.5 N acetic acid and neutralized with an equal volume of 0.5 N NaOH. This is followed by mixing with an equal volume of 100 mM MOPSO, pH 7.0, and then with the molten 2% agarose solution prior to loading into the rectangular ACE gel wells. After allowing the agarose to solidify for approximately 20 min, the protein-loaded ACE gel is removed from the casting tray and submerged in electrophoresis buffer (gel buffer minus CHAPS) in a horizontal electrophoresis chamber (Hoefer (SuperSub); Hoefer, San Francisco, CA). Approximately 0.2 ng of purified human recombinant decorin that has been iodinated (Iodo-Beads; Pierce, Rockford, IL) to a specific activity of ~50,000 cpm/ng is then mixed with electrophoresis buffer containing 0.5% (v/v) bromphenol blue and 6% (w/v) sucrose and added to the 66 × 1 mm slot well. Electrophoresis is conducted at 70–80 V for 2.5–3 hr with a constant recirculation of buffer and cooling of the apparatus to maintain buffer temperature at 20–25°. Gels are then air dried overnight and autoradiographed at −80°. A representative decorin ACE gel (Fig. 2) demonstrates the concentration-dependent retardation of mobility of ^{125}I-labeled decorin by an anti-decorin monoclonal antibody (1D3) and by the C1q component of complement that contains a collagen-like triple-helical domain. No retardation of mobility is observed with equal concentrations of bovine serum albumin. Similar results are obtained with various collagen types. An estimation of the dissociation constant (K_D) can be obtained from this type of autoradiogram by using the protein concentration at which the decorin is half-shifted from being fully mobile to maximally retarded.[27] This yields a K_D value between 5 and 10 nM with C1q, which is in agreement with previously published values obtained from solid-phase binding assays.[24] A more accurate assessment of the K_D can be obtained from the slope of the plot $R/[\text{protein}]_{\text{total}}$ vs R, where R represents the retardation coefficient based on the mobility of free decorin and decorin moving through protein-containing zones (see Lee and Lander[25] and San Antonio et al.[27] for a detailed explanation).

The specificity of the interaction of decorin with collagen and Clq in ACE gels is easily demonstrated by competition experiments using unlabeled decorin. The inclusion of excess unlabeled decorin in the labeled sample competes directly for the binding observed in ACE gels in a concentration-dependent manner. Competition for binding has also been observed by mixing decorin core protein (minus GAG chain after chondroitinase ABC treatment) with the protein samples prior to loading the rectangular wells. This method is particularly useful for testing decorin fragments and peptides that do not have the GAG side chain. These types of analyses may be employed to localize the collagen-binding domain(s) on decorin.

Fibroblast–Collagen Gel Contraction Assay of Decorin Effects on Cell Responses to Transforming Growth Factor β

Fibroblasts cultured within matrices of type I collagen will spread and reorganize the collagen fibers, resulting in contraction of the gels.[26] The rate and degree of contraction depends on several factors such as the concentration of collagen, the number of cells, the serum concentration, and the presence of modulators of cellular activity. Transforming growth factor β and platelet-derived growth factor (PDGF) have been shown to stimulate contraction by fibroblasts, whereas endothelins have been shown to have a similar effect on gel contraction by lipocytes.[28] This assay has been adapted to assess the ability of decorin to influence TGF-β-induced contraction of collagen gels, as it may relate to the efficacy of decorin in preventing fibrosis in vivo.

Murine 3T3 cells are incorporated into collagen gels, because they have been reported to contract such gels in response to TGF-β stimulation. Under the proper conditions, maximal TGF-β-induced contraction occurs within 48 hr, providing a relatively rapid and convenient readout of a cellular response to TGF-β. It is easy to modify either gels or medium in order to study the effects of such changes on contraction. Decorin, or any other test article, can be tested either as an integrated component of the collagen gel, or by addition together with TGF-β to gels of collagen alone. This system is also suitable for using immunohistochemistry to

[28] D. C. Rockey, C. N. Housset, and S. L. Friedman, J. Clin. Invest. 92, 1795 (1993).

examine matrix composition and organization, as well as to examine the expression of specific mRNAs and proteins resulting from treatment.

Wells are blocked with 2% (w/v) bovine serum albumin (BSA) in PBS for 1 hr at 37° and rinsed three times with sterile PBS before use. Prior to use in the assay, NIH 3T3 cells are cultured in Dulbecco's modified Eagle's medium (DMEM) supplemented with glutamine, penicillin, streptomycin, and 10% (v/v) fetal calf serum (FCS). Cells are harvested from log-phase cultures with trypsin–ethylenediaminetetraacetic acid (EDTA) and split at a ratio of 1:3, 1 to 2 days prior to the assay. Following the addition of an equal volume of culture medium to neutralize the trypsin–EDTA, the cell pellet is washed three times with serum-free DMEM. The number of viable cells is determined, using trypan blue exclusion and a hemocytometer, and the cell pellet is then resuspended in serum-free DMEM to a viable cell concentration of 3×10^6 cells/ml.

On ice, the collagen gel mixture is prepared by mixing $5 \times$ DMEM (one-fifth of 90% of the final volume), 1.0 M N-2-hydroxyethylpiperazine-N'-2-ethanesulfonic acid (HEPES, pH 7.5) to 50 mM, 0.5 M NaOH to twice the molar concentration of the acetic acid from the collagen solution, rat tail collagen type I (Collaborative Research, Waltham, MA) to 1.5 mg/ml, and water to 90% of the final volume. When decorin is added to this mixture, the final decorin concentration is 0.2 mg/ml. Control gel mixtures receive an equivalent amount of either PBS or ovalbumin. Cells are added to the collagen solution at a final concentration of 3×10^5 cells/ml, then the collagen–cell suspension is transferred to previously coated 24-well plates (0.7 ml/well), where the polymerization occurs by incubating at 37° for 1 hr. After polymerization, the gels are dislodged from the wells by vigorously introducing 1 ml of serum-free DMEM into the wells, and triplicate gels are then transferred to each well of a six-well plate with broad-tipped forceps. Transferring the gels is a delicate procedure and care should be taken to minimize damage to the gels. The six-well plates contain 4 ml of assay medium [DMEM, 0.5% (v/v) FCS, 50 mM HEPES (pH 7.5), 2 mM glutamine, penicillin (100 units/ml), and streptomycin (100 μg/ml)] per well. Transforming growth factor β is added to a final concentration of 1 ng/ml. We have chosen to eliminate serum from the gel mixtures and to reduce the serum content in the medium to 0.5% to minimize the effects serum may have on TGF-β and/or decorin. Although serum is required for 3T3 fibroblasts, WI-38 fibroblasts (human lung) do not appear to require serum for contraction. The plates are placed on a rocker platform (Bellco, Vineland, NJ) with the speed set at 4 in a 37° tissue culture incubator.

Photographs of the gels are taken using a Sony CCD-Iris color video camera connected to a Sony mavigraph color video printer (UP-5200MD) at the start of the experiment and periodically for the next 48 hr. The outline of each gel is then traced and measured using a Summagraphics Bit Pad Plus and the Digitize version 1.15 software (RockWare Scientific Software, Wheat Ridge, CO). The gel areas are averaged and plotted against time to show the rate and extent of contraction.

We routinely observe an inhibition of the TGF-β-mediated contraction when fibroblasts are interacting with collagen/decorin fibrils (Fig. 3). This inhibition may be due to a mechanical strengthening of the gels, resulting in the gels being more resistant to the contractile forces of the cells. Results from preliminary centrifugal compression studies suggest that decorin-containing collagen gels are more rigid than collagen gels formed in the absence of decorin. Decorin may also be binding TGF-β to neutralize its activity or decorin may be acting directly on the cells to alter their ability to respond to TGF-β. We are considering the potential role of cell metabolism and the expression of specific TGF-β responsive genes as a way to address these issues.

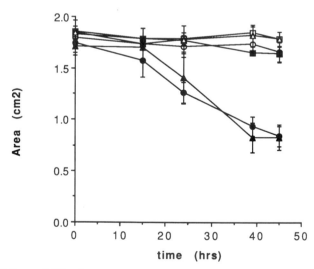

Fig. 3. Collagen (CN) gel contraction assay. Comparison of decorin and ovalbumin for their effect on TGF-β-induced gel contractions by NIH 3T3 cells. Gels containing fibroblasts were polymerized in the presence of decorin (DC) or ovalbumin (Ov) and induced to contract by the addition of TGF-β (B). Gels containing decorin did not exhibit the usual time-dependent contraction over a 48-hr period. Gel (media): (○) CN (B−); (●) CN (B+); (□) CN/DC (B−); (■) CN/DC (B+); (△) CN/Ov (B−); (▲) CN/Ov (B+).

Localization of Decorin to Fibrotic Matrix in Experimental Rat Glomerulonephritis

Acute inflammation resulting from a single injection of ATS into rats leads to increased TGF-β expression, followed by accumulation of fibrotic matrix in the mesangium. Intravenous administration of recombinant human decorin prevents this extracellular matrix buildup. The molecular profiles of normal and pathological extracellular matrix components in the kidney serve as the end points in evaluating this experimental fibrotic disease.[14] Types I and III collagen, as well as extra domain A-containing (EDA⁺) fibronectin, are not usually found in significant quantities in the normal rat glomerulus by routine immunohistochemical methods. However, these components are elevated in the glomeruli of diseased rats in the absence of decorin treatment and are significantly reduced following intravenous (iv) decorin treatment (Fig. 4). The collagen-binding activity of decorin is implicated as an important aspect of this efficacy, by the

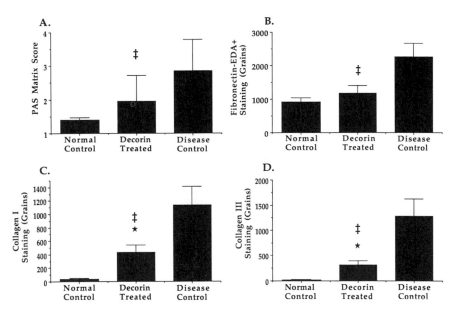

FIG. 4. Histological and immunohistochemical analyses of kidneys from control rats or rats having ATS-induced glomerulonephritis. Decorin treatment had a significant effect ($p < 0.05$) (‡) on all four parameters examined. Only PAS staining for carbohydrate in the mesangial matrix (A) and EDA⁺ fibronectin (B) were indistinguishable from normal control animals (∗). Types I and III collagen, although significantly reduced by decorin treatment ($p < 0.05$), remained elevated above control levels (C and D).

apparent targeting of decorin administered intravenously as well as by decorin exposed to the fibrotic matrix *in vitro*.

The binding of decorin to collagen, demonstrated above using the ACE gel system, may be responsible for its targeting to the site of tissue injury, because mesenchymal collagens are often exposed to the circulation following such tissue damage. To illustrate this tissue localization, decorin was shown to bind preferentially to fibrotic glomeruli in frozen sections, using anti-decorin monoclonal antibodies specific to human decorin. Unlike normal mesangial matrix, fibrotic mesangial matrix in human and experimental rat disease contains both types I and III collagen,[29-31] most likely serving as the anchor for decorin within the scar tissue. Frozen sections of kidney from either normal or ATS-treated rats were overlaid with a solution of human decorin and binding of this decorin to fibrotic glomeruli was analyzed as outlined below. In the overlay assay, decorin appeared to bind a component(s) of the tubulointerstitial matrix of both normal and diseased kidneys, whereas significant binding to glomeruli was observed only in cases of glomerular fibrosis in ATS-treated animals (Figs. 5B and D). Preferential localization of decor into glomeruli exhibiting accumulated fibrotic matrix, but not normal glomeruli, illustrates the apparent specificity of binding to matrix components found only in fibrotic glomeruli.

The detection of decorin binding to mesenchymal collagens *in vitro* by the ACE gel and to fibrotic matrix in the overlay method appears to be relevant to the tissue distribution of decorin in glomeruli following iv injection. As soon as 30 min and up to 24 hr following the last decorin injection, human decorin was localized to fibrotic glomeruli using standard immunohistochemical methods (Figs. 5A and C). The fact that frozen sections of normal rat kidneys were negative for human decorin staining in tubulointerstitial stroma and glomeruli (Fig. 5A) suggests that the matrix components responsible for decorin binding are either absent or inaccessible to injected decorin in these tissues in uninjured animals.

Method for Detecting Human Decorin following Intravenous Administration or Overlaid on Fibrotic Rat Glomeruli *in Vitro*

Fresh kidneys are embedded in frozen tissue embedding medium (Tissue Prep; Fisher Scientific, Pittsburgh, PA) and snap frozen in

[29] A. D. Glick, H. R. Jacobson, and M. A. Haralson, *Hum. Pathol.* **23,** 1373 (1992).
[30] A. Oomura, T. Nakamura, M. Arakawa *et al.*, *Virchows Arch. A: Pathol. Anat.* **415,** 151 (1989).
[31] S. Adler, L. J. Striker, G. E. Striker, D. T. Perkinson, J. Hibbert, and W. G. Couser, *Am. J. Pathol.* **123,** 553 (1986).

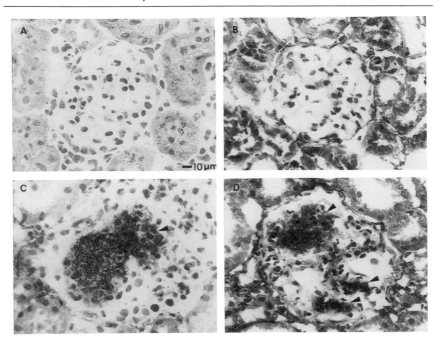

FIG. 5. Localization of human decorin bound to ECM components in rat kidney. Decorin binding to normal rat glomeruli (A and B) or fibrotic glomerular matrix (C and D) was examined. Analysis was performed either on frozen tissue sections incubated with decorin in an overlay assay (B and D), or for 30 min following intravenous injection *in vivo* (A and C). Decorin was localized by immunohistochemical methods described in text. Magnification: ×280.

2-methylbutane that has been cooled in liquid nitrogen. Cryostat sections are cut 4 to 5 μm in thickness and placed onto poly(L-lysine)-coated glass slides. Sections are air dried overnight at room temperature. Following a rehydration step performed in PBS, the slides are blocked with 5% (v/v) normal horse serum diluted in PBS for 10 min at room temperature. Excess blocking buffer is removed by draining and blotting on a paper towel, and then a solution of human decorin is overlaid and incubated for 30 min at 37°. After removing excess decorin solution, slides are washed three times in PBS–Tween. The primary antibody step consists of adding mouse anti-human decorin (1C6; Telios Pharmaceuticals, San Diego, CA) that has been prediluted in PBS containing 5% (v/v) normal horse serum and 7.5% (v/v) normal rat serum, followed by a 1-hr incubation at 37° in a humid chamber. Three 10-min washes in PBS containing 0.05% Tween-20 precede the addition of the secondary antibody. Peroxidase-conjugated, affinity-purified goat anti-mouse IgG (Cat. No. 115-035-100; Jackson Immuno-

Research, Inc., Westgrove, PA) is diluted in PBS containing 5% (v/v) normal horse serum and 7.5% (v/v) normal rat serum, then added to sections and incubated for 1 hr at 37°. Prior to color development, slides are washed three times in PBS–Tween for 3 min each. Color development is achieved by applying freshly prepared 0.06% (w/v) 3,3'-diaminobenzidine tetrahydrochloride (Sigma, St. Louis, MO) in PBS, containing 0.01% (v/v) hydrogen peroxide. Development of color is allowed to proceed for 5 min at room temperature. The three times-washed slides are finally counterstained lightly with hematoxylin (1–2 min) and washed with tap water prior to preparation for microscopy. Stained sections are sequentially dehydrated in a series of 70, 80, 95, and 100% ethanol (two to three changes each) and cleared for 2 min with xylene. Coverslips are applied using Permount or other permanent, nonaqueous mounting media, and then the slides are ready for examination by light microscopy.

Conclusion

The beneficial role of growth factors, such as TGF-β, in the repair and healing of wounded tissue is still uncertain. On the other hand, current research is available that implicates TGF-β as central to the process of scar formation and fibrotic disease in several organ systems. As the understanding of natural growth factor regulation increases, so does the potential of developing biological therapeutics to control diseases such as acute and chronic renal disease, liver and pulmonary fibrosis, and scarring due to traumatic injury to the skin and possibly the central nervous system. This chapter represents an example of how basic research methods and functional attributes of an extracellular matrix component have been extended into the area of biotechnology and used to control the functional quality and potency of a biological therapeutic.

Section IV

Extracellular Matrix in Invertebrates

[13] Extracellular Calcium-Binding Protein SPARC/Osteonectin in *Caenorhabditis elegans*

By JEAN E. SCHWARZBAUER, FREDERIQUE MUSSET-BILAL, and CAROL S. RYAN

The nematode *Caenorhabditis elegans* is an excellent genetic system for dissecting protein function. Beginning with the pioneering work of Brenner,[1] numerous mutations have been generated and characterized phenotypically.[2] Ease of culture, transparency, and small size, (fewer than 1000 nongonadal nuclei), have allowed the determination of a complete cell lineage map by direct observation of living nematodes.[3] Colocalization of genetic and physical loci is made possible by an extensive *C. elegans* genome map.[4,5] The ability to identify genes corresponding to particular mutations has advanced significantly with the development of methods for transformation of mutants with wild-type genes.[6] The ability to introduce mutations into specific genes is now becoming possible by Tc1 transposon insertion or excision.[7] A comprehensive volume describing all aspects of nematode biology[2] is an excellent resource for anyone studying *C. elegans,* from novice to expert. In addition, *The Worm Breeder's Gazette,* published quarterly by the *Caenorhabditis* Genetics Center (CGC, University of Minnesota, St. Paul, MN), contains short research articles and technical notes contributed by members of the nematode community and represents a unique mechanism for keeping abreast of the latest techniques and the most recent results from other laboratories. The CGC, supported by the NIH National Center for Research Resources, also maintains a large collection of normal and mutant strains for distribution on request.

Because nematodes have a relatively simple anatomy, they are an attractive system for analyzing specific cellular structures common to all organisms. We are dissecting the extracellular matrix (ECM) and basement membrane structures in *C. elegans* in order to characterize the functions

[1] S. Brenner, *Genetics* **77,** 71 (1974).

[2] W. Wood, ed., "The Nematode *Caenorhabditis elegans.*" Cold Spring Harbor Lab., Cold Spring Harbor, NY, 1988.

[3] J. E. Sulston and H. R. Horvitz, *Dev. Biol.* **56,** 110 (1977).

[4] S. J. O'Brien, ed., *Genet. Maps* **5,** 3.111 (1990).

[5] A. Coulson, J. Sulston, S. Brenner, and J. Karn, *Proc. Natl. Acad. Sci. U.S.A.* **83,** 7821 (1986).

[6] A. Fire, *EMBO J.* **5,** 2673 (1986).

[7] R. H. A. Plasterk, *BioEssays* **14,** 629 (1992).

METHODS IN ENZYMOLOGY, VOL. 245
Copyright © 1994 by Academic Press, Inc.
All rights of reproduction in any form reserved.

and interactions of known matrix proteins as well as to identify novel matrix molecules. We have begun with the extracellular matrix-associated protein SPARC/osteonectin. This protein has been well characterized biochemically in mammalian systems and its sequence and expression patterns have been determined.[8-10] SPARC is a small secreted glycoprotein that has calcium-binding activity.[11-13] It is found at particularly high levels in bone and in Reichert's membrane during embryonic development. Addition of purified SPARC to attached and spread endothelial cells results in reduced cell attachment and rounded cell morphology.[14] These data taken together suggest that SPARC is involved in modulating cell interactions with extracellular matrix directly or by affecting the structure or interactions within the matrix itself. SPARC characterization in *C. elegans* should shed some light on the *in vivo* functions of this abundant protein.

Caenorhabditis elegans SPARC was initially cloned by polymerase chain reaction (PCR) amplification using degenerate oligonucleotide primers.[15] Full-length cDNA and genomic DNA clones were then isolated by standard hybridization screening of λ phage libraries using the PCR-derived clone as a probe. The sequences of the cDNA and of all of the exon–intron boundaries within the single gene were determined. These analyses showed that gene organization and protein structure are conserved from nematodes through mammals. The protein sequence is 38% identical between mammalian and nematode SPARC and greater than 60% similar.

Sequence analyses of several basement membrane proteins show that this structure is highly conserved from nematodes to humans. Collagen type IV,[16] perlecan,[17] laminin, and SPARC[15] have already been identified and other matrix proteins such as nidogen/entactin are likely to be present

[8] J. D. Termine, H. K. Kleinman, S. W. Whitson, K. M. Conn, M. L. McGarvey, and G. R. Martin, *Cell (Cambridge, Mass.)* **26,** 99 (1981).

[9] I. J. Mason, A. Taylor, J. G. Williams, H. Sage, and B. L. M. Hogan, *EMBO J.* **5,** 1465 (1986).

[10] P. W. H. Holland, S. J. Harper, J. H. McVey, and B. L. M. Hogan, *J. Cell Biol.* **105,** 473 (1987).

[11] J. D. Termine, A. B. Belcourt, K. M. Conn, and H. K. Kleinman, *J. Biol. Chem.* **256,** 10403 (1981).

[12] J. Engel, W. Taylor, M. Paulsson, H. Sage, and B. Hogan, *Biochemistry* **26,** 6958 (1987).

[13] P. Maurer, U. Mayer, M. Bruch, P. Jeno, K. Mann, R. Landwehr, J. Engel, and R. Timpl, *Eur. J. Biochem.* **205,** 233 (1992).

[14] E. H. Sage, R. B. Vernon, S. E. Funk, E. A. Everitt, and J. Angello, *J. Cell Biol.* **109,** 341 (1989).

[15] J. E. Schwarzbauer and C. S. Spencer, *Mol. Biol. Cell* **4,** 941 (1993).

[16] M. H. Sibley, J. J. Johnson, C. C. Mello, and J. M. Kramer, *J. Cell Biol.* **123,** 255 (1993).

[17] T. M. Rogalski, B. D. Williams, G. P. Mullen, and D. G. Moerman, *Genes Dev.* **7,** 1471 (1993).

as well. Along with basement membranes, the collagenous cuticle represents the other major extracellular matrix structure in nematodes. Connections from the outer cuticle through the hypodermis and basement membrane to the muscle cells form an integral part of the nematode "skeleton." Therefore, it is not surprising that defects in basement membrane proteins and cuticle collagens affect nematode body shape and mobility. Nonlethal mutations within matrix proteins result in mutant phenotypes such as uncoordinated (Unc), roller (Rol), and dumpy (Dpy). For example, the *unc-52* gene encodes perlecan,[17] whereas some of the *rol* and *dpy* genes encode cuticular collagens.[18,19] Undoubtedly, genetic analyses of other matrix proteins will also provide insights into their roles in tissue structure and function.

Our approach to *C. elegans* SPARC function has been largely a biochemical and cell biological one. We have used expression of bacterial fusion proteins to look at domain function. Transgenic nematodes have been established to characterize the expression and distribution of this protein in *C. elegans*. Antibodies have been raised against SPARC and used to identify and characterize the protein in nematode extracts. Protocols for fixing, permeabilizing, and sectioning nematodes have been tested for use in the localization of SPARC by immunofluorescence. Finally, the expression of recombinant *C. elegans* SPARC proteins is being characterized using a mammalian expression system. In this chapter we describe these approaches to the analysis of SPARC structure and function.

Domain Functions Analyzed Using Bacterial Fusion Proteins

The conservation in structure suggests that SPARC performs many of the same functions in mammals and nematodes. We have tested one of these functions *in vitro*. Mammalian SPARC has two calcium-binding domains, amino-terminal acidic domain I and carboxy-terminal domain IV containing a classic EF-hand calcium-binding motif. To test whether amino- and carboxy-terminal domains of nematode SPARC also bind calcium, we express these domains as bacterial fusion proteins with maltose-binding protein (MBP) (New England BioLabs, Inc., Beverly, MA). The MBP portion of the fusion protein is 42 kDa in molecular mass and the cDNA-encoded segment is attached at the carboxy terminus. In the absence of a heterologous insert, the MBP vector encodes an MBP-*lacZα* fusion protein of 52 kDa. The gene is under the control of a P_{tac} promoter, which is inducible by isopropyl-β-D-thiogalactoside (IPTG).

---

[18] J. M. Kramer, R. P. French, E.-C. Park, and J. J. Johnson, *Mol. Cell. Biol.* **10**, 2081 (1990).
[19] A. D. Levy, J. Yang, and J. M. Kramer, *Mol. Biol. Cell* **4**, 803 (1993).

Engel *et al.*[12] have predicted the disulfide bonding pattern for mouse SPARC and it includes an intradomain disulfide bond formed between the cysteine residues that flank the EF-hand motif in domain IV. These residues are conserved in nematode SPARC. We therefore included these residues in our domain IV fusion protein construction.[15] Convenient *Sfa*NI restriction sites at positions 710 and 794 are used and *Xba*I linkers are attached to make the reading frame compatible with MBP and to introduce a termination codon at the 3′ end. The resulting fragment is cloned into the *Xba*I site of the vector pMALcRI, the plasmid is transfected into *Escherichia coli* strain HB 101, and colonies are screened for expression of the correct-sized fusion protein. To prepare the fusion protein, a slightly modified version of the New England BioLabs protocol is used. A bacterial culture at saturation is diluted 1 : 10 (v/v) into fresh LB and grown for 30 min at 37° with shaking. IPTG is added from a stock solution to a final concentration of 0.3 mM and bacteria are grown for an additional 3 hr. Bacteria are collected by centrifugation, resuspended in lysis buffer [10 mM sodium phosphate (pH 7.0), 30 mM NaCl, 0.25% (v/v) Tween 20, 10 mM 2-mercaptoethanol (2-ME), 10 mM ethylenediaminetetraacetic acid (EDTA), 10 mM EGTA], and frozen at $-20°$ overnight. After thawing slowly in cold water, the bacteria are lysed by sonication, the soluble fraction is diluted with column buffer plus Tween 20 [10 mM sodium phosphate (pH 7.0), 500 mM NaCl, 1 mM sodium azide, 0.25% (v/v) Tween 20], and applied to amylose resin. 2-Mercaptoethanol is omitted from all steps after cell lysis. Bound proteins are eluted with column buffer plus 10 mM maltose, dialyzed into 20 mM Tris-HCl (pH 8), 50 mM NaCl, and stored at $-80°$. MBP-domain IV is expressed at high levels and milligram quantities are usually obtained from a 100-ml culture of bacteria. To remove any traces of 2-ME remaining from the lysis buffer and to allow formation of the intradomain disulfide bond, the fusion protein is dialyzed extensively against Tris-NaCl buffer.

To determine whether domain IV from *C. elegans* SPARC is able to bind calcium, we use an *in vitro* $^{45}Ca^{2+}$-binding assay.[20] One microgram of MBP-domain IV and MBP-*lacZα* is electrophoresed through a 12% (w/v) polyacrylamide-sodium dodecyl sulfate (SDS) gel and transferred to a nitrocellulose filter (Sartorius, Bohemia, NY) using the Mini-PROTEAN II apparatus (Bio-Rad, Richmond, CA). The filter is soaked in 25 mM Tris-HCl (pH 7.5), 100 mM NaCl overnight at 4°. Three 20-min, room temperature washes with 10 mM imidazole hydrochloride (pH 6.8), 60 mM KCl, 5 mM MgCl$_2$ are followed by a 10-min incubation in the same buffer plus 1 μCi of $^{45}CaCl_2$/ml (Du Pont-New England Nuclear, Wilming-

[20] S. Chakravarti, M. F. Tam, and A. E. Chung, *J. Biol. Chem.* **265**, 10597 (1990).

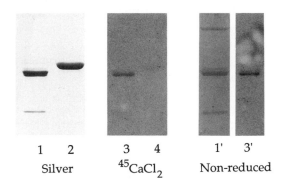

1   2   3   4   1'   3'

Silver ^{45}CaCl$_2$ Non-reduced

FIG. 1. Calcium-binding activity of *C. elegans* SPARC domain IV. One-microgram aliquots of purified MBP-domain IV (lanes 1, 3, 1', and 3') and MBP-*lacZα* (lanes 2 and 4) fusion proteins were electrophoresed through 12% (w/v) polyacrylamide-SDS gels with (lanes 1–4) and without (lanes 1' and 3') reduction. Parallel gels were either stained with silver (lanes 1, 2, and 1') or transferred to nitrocellulose (lanes 3, 4, and 3'). Proteins immobilized on nitrocellulose were incubated with ^{45}CaCl$_2$ and autoradiograms of bound calcium are shown here. Note that MBP-domain IV binds calcium, but MBP-*lacZα* does not. Reduced (lanes 1 and 2) and nonreduced (lane 1') silver-stained samples are different exposures of the same gel. Lower levels of nonreduced proteins migrated in the position of monomer due to the presence of dimers and larger aggregates.

ton, DE). After two 10-min washes in distilled water, the filter is air dried and exposed to Kodak (Rochester, NY) XAR-5 film or to a storage phosphor screen for quantitating bound calcium.

Calcium-binding activity of MBP-domain IV is determined after electrophoresis under reducing and nonreducing conditions. A parallel gel is prepared in an identical manner and the proteins are visualized by silver staining.[21] MBP-*lacZα* is used as a negative control. Figure 1 shows the results of silver staining and calcium binding. Reduced samples show equivalent amounts of MBP-*lacZα* and MBP-domain IV protein by silver stain, but only MBP-domain IV (lane 3) and a calmodulin-positive control (not shown) bind ^{45}Ca^{2+}. No binding by MBP-*lacZα* is observed. Nonreduced MBP-domain IV also binds calcium (Fig. 1, lane 3'). However, nonreduced fusion protein bands are always present at lower levels owing to formation of dimers and other aggregates that do not migrate into the running gel (compare lanes 1 and 1'). Quantitation of bound calcium is carried out using a Molecular Dynamics PhosphorImager (Sunnyvale, CA). Relative protein levels are determined by scanning densitometry of dried, silver-stained gels. Calcium counts bound are divided by the amount of protein to yield normalized calcium binding per unit protein. This value

[21] C. R. Merril, D. Goldman, and M. L. VanKeuren, this series, Vol. 104, p. 441.

is five times higher for nonreduced than for reduced MBP-domain IV. Apparently, the intradomain disulfide bond between cysteines flanking the EF-hand motif stabilizes this domain and allows a higher level of calcium binding. In a similar experiment, an MBP-domain I fusion protein also shows calcium-binding activity.[15]

The calcium-binding activity of MBP-domain IV provides a simple one-step purification procedure. Fusion protein expression is induced and bacteria are grown and collected as before. Bacteria are lysed by sonication in 10 mM Tris-HCl (pH 8.8), 30 mM NaCl, 0.25% (v/v) Tween 20, 10 mM EDTA, and debris is removed by centrifugation. CaCl$_2$ is then added to a final concentration of 40 mM. In parallel lysates containing MBP-$lacZ\alpha$, most of the bacterial proteins, including the fusion protein, are precipitated at this calcium concentration and can be removed by centrifugation at 10,000 rpm for 15 min. However, MBP-domain IV remains in solution with only a few other minor contaminants. This differential solubility yields a protein that is greater than 90% pure.

Sites of Expression of SPARC in Vivo

SPARC is likely to be expressed by cells that contact a basement membrane or other types of extracellular matrix structure. To identify the cells that express SPARC, we took advantage of technology for transforming nematodes with fusion genes that contain the lacZ gene under the control of a heterologous promoter. Some useful lacZ cassettes for expression have been designed and tested by Fire et al.[22] and form the basis for our construction. We use the nuclear localization signal (NLS)/lacZ/unc-54 cassette from the plasmid pPD22.04. In this cassette, the β-galactosidase coding sequence is preceded by an NLS and followed by the unc-54 3' untranslated region and polyadenylation signal. The nuclear localization signal directs the β-galactosidase into the nucleus to aid in identifying expressing cells and the unc-54 3' end provides efficient transcription termination and polyadenylation.

The entire SPARC gene is contained within a 3.6-kb region of an 8-kb EcoRI-to-BamHI genomic fragment. The gene plus 4 kb of upstream sequence and 0.8 kb of downstream sequence is used intact to retain any regulatory elements flanking the gene or within introns. The lacZ gene is inserted into the first exon of the SPARC gene in frame with the initiator ATG. This position disrupts the SPARC gene, thus preventing production of SPARC protein from the transgene. Two restriction sites, Fnu4HI and BsmAI, are used to insert the NLS/lacZ/unc-54 cassette, thus deleting

[22] A. Fire, S. W. Harrison, and D. Dixon, *Gene* **93**, 189 (1990).

amino acids 7 through 10 from the signal sequence. The resulting plasmid, pWSP8-*lacZ*, is injected into the gonads of L4 larvae along with the pRF4 plasmid carrying the dominant marker *rol-6 (su1006)*.[23] *rol-6 (su1006)* is a mutant collagen gene that serves as a dominant marker for DNA transformation. Nematodes transformed with this gene have helically twisted cuticles and bodies, which causes the animals to roll over and to move in circles. This phenotype is readily detected with a dissecting microscope, beginning at the L3 larval stage. Therefore, other abnormalities appearing at earlier stages should not be masked by the presence of this dominant Rol phenotype.

Individual F_1 animals expressing the Rol phenotype are picked and used to establish stable lines. More than 10 lines were established and 6 of these were further characterized.[12] Wild-type and transgenic nematodes are maintained at either 15 or 20° on NGM-agar plates with a lawn of *Escherichia coli* OP50, using established procedures.[1,24] Several animals expressing the Rol phenotype are transferred to fresh plates weekly. For long-term storage, animals are frozen in a phosphate-buffered glycerol solution at −85°.[24]

To monitor expression of β-galactosidase from the SPARC promoter, nematodes are fixed and stained, using the procedure of Fire.[6] Animals are collected from plates and washed once with distilled water. The pellet is resuspended in 50 μl of water and the nematodes are transferred to a glass depression slide. The slide is placed in a desiccator jar and the animals are lyophilized under vacuum. A vacuum pump that pulls a strong vacuum is important at this step. Completely freeze-dried nematodes are then permeabilized by overlaying with acetone for 2–4 min. After evaporation of the acetone, the animals are incubated in freshly prepared staining solution [200 mM sodium phosphate (pH 7.5), 1 mM MgCl$_2$, 5 mM K$_3$Fe(CN)$_6$, 5 mM K$_4$Fe(CN)$_6$, 0.004% (w/v) SDS, and 0.012% (v/v) X-gal added from a 3% (w/v) stock dissolved in dimethylformamide]. For most genes, staining is usually allowed to develop overnight. However, β-galactosidase expression under control of the SPARC promoter is high and stained cells can be seen after only about 10 min of staining. To prevent overstaining, the staining reaction is stopped by pelleting the nematodes and resuspending in distilled water. Figure 2 shows a composite photograph of an adult hermaphrodite stained for SPARC–β-galactosidase expression. Body wall muscle cells are stained along the length of the animal. Unhatched and hatched larvae are also stained, indicating that

[23] C. C. Mello, J. M. Kramer, D. Stinchcomb, and V. Ambros, *EMBO J.* **10,** 3959 (1991).
[24] J. Sulston and J. Hodgkin, *in* "The Nematode *Caenorhabditis elegans*" (W. Wood, ed.), p. 587. Cold Spring Harbor Lab., Cold Spring Harbor, NY, 1988.

FIG. 2. Expression of SPARC-*lacZ* in transgenic nematodes. Transgenic nematodes carrying a SPARC-*lacZ* fusion gene were stained for β-galactosidase expression. A composite photograph of an adult hermaphrodite is shown here. Body wall muscle cells are stained along the length of the animal. The head is to the left.

expression begins early during development and continues throughout the lifetime of the animal.[15]

To confirm the presence of the transgene in these and other transgenic nematodes, we use single-animal PCR amplification. Primers are designed to amplify the ends of the genomic insert in the pWSP8 plasmid. About 200 bp of sequence is determined across the *Eco*RI site at the 5' end and the *Bam*HI site at 3' end of the genomic insert. Segments with at least 50% GC content are used to make oligonucleotide primers. These SPARC-specific primers are paired with primers from the flanking vector sequences: positions 2665–2685 and 117–136 in pGEM2, and positions 156–175 in pSP73. Polymerase chain reaction-amplified products are designed to be between 200 and 500 bp in length. Individual nematodes are frozen in a 2.5-μl drop containing proteinase K (60 μg/ml), 10 mM Tris-HCl (pH 8.2), 50 mM KCl, 2.5 mM MgCl$_2$, 0.45% (v/v) Tween 20, and 0.05% (w/v) gelatin and lysed by incubating at 60° for 1 hr and at 95° for 15 min. Samples are subjected to PCR amplification using *Taq* DNA polymerase (Promega-Biotech, Madison, WI) exactly as described by Williams *et al.*[25] Amplification conditions consist of the following for 30 cycles: 30 sec (94°), 1 min (58°), 30 sec (72°). Ten-microliter aliquots of the PCR reactions are electrophoresed through an 8% (w/v) polyacrylamide gel in 90 mM Tris base, 90 mM boric acid, 2 mM EDTA (pH 8.3), and bands are visualized with ethidium bromide. This protocol yields readily visible ethidium bromide-stained bands with few nonspecific bands.

Production and Characterization of Domain-Specific Antibodies

The MBP-domain IV fusion protein is used to raise SPARC-specific antibodies. The molecular context of domain IV in this fusion protein is similar to native SPARC in that it lies at the carboxy terminus of the molecule and it forms an intradomain disulfide bond. In addition, we have

[25] B. D. Williams, B. Schrank, C. Huynh, R. Shownkeen, and R. H. Waterston, *Genetics* **131**, 609 (1992).

found that this fusion protein is functional and can bind calcium. The protein is purified by amylose affinity chromatography and dialyzed into 20 mM Tris-HCl (pH 8), 50 mM NaCl. MBP-domain IV (0.2 mg in Freund's complete adjuvant) is injected subcutaneously into a rabbit. The rabbit is boosted 2 weeks later with the same amount of protein in Freund's incomplete adjuvant. The first immune bleed is collected 2 weeks after the boost and has significant anti-MBP-domain IV reactivity by immunoblotting.

The anti-domain IV antiserum reacts with both MBP-$lacZ\alpha$ and MBP-domain IV on Western blots. To determine whether anti-domain IV antibodies are present in the antiserum, we use several approaches. The MBP fusion proteins contain a factor X proteolytic cleavage site between the MBP and the SPARC portions. We first attempted to show SPARC-specific recognition of the fusion protein by immunoblotting after cleavage with factor X. The small size of the domain IV segment (28 amino acids) prohibits visualization by silver staining or detection by antibodies. We next used preabsorption to determine reactivity. A bacterial lysate is made from cells expressing the vector-derived MBP-$lacZ\alpha$ protein after induction with IPTG. Bacteria are disrupted by sonication in lysis buffer, insoluble debris is removed, and the supernatant containing most of the bacterial proteins including MBP-$lacZ\alpha$ is dialyzed into 0.1 M NaHCO$_3$ (pH 8.3), 0.5 M NaCl. Total bacterial proteins are coupled to CNBr-activated Sepharose following the procedure recommended by Pharmacia (Piscataway, NJ) to give a final concentration of 6 mg of protein/ml Sepharose. Anti-MBP-domain IV antiserum is then incubated with bacterial protein-Sepharose at a ratio of 10 to 1 at 4° overnight with mixing. This preabsorbed antiserum no longer recognizes MBP-$lacZ\alpha$ by immunoblotting but retains the ability to detect MBP-domain IV. Partial cleavage of MBP-domain IV with factor X yields the intact fusion protein and the MBP portion; only the intact protein is recognized by the preabsorbed antiserum. These results confirm that antibodies against SPARC epitopes are present in the antiserum.

Identification of SPARC Protein in Nematode Lysates

SPARC-specific antisera should be useful for analyzing SPARC expression and distribution *in vivo*. The level of SPARC mRNA by Northern analysis and the amount of β-galactosidase produced under the control of the SPARC promoter are relatively high. Therefore, SPARC protein should be detectable by immunoblotting of whole nematode extracts.

The quality of the total nematode lysates is critical to identifying SPARC protein. Nematodes are collected from plates just as the bacteria are exhausted and with all ages represented. For a typical preparation, animals are washed from two 15-cm plates using M9 buffer (22 mM KH$_2$PO$_4$, 42 mM Na$_2$HPO$_4$, 86 mM NaCl, 1 mM MgSO$_4$). Bacteria are removed by pelleting the nematodes at 1500 rpm for 5 min through a 10-ml cushion of 5% (w/v) sucrose in M9 buffer. Pelleted nematodes are washed with M9 and, after resuspension in M9, are incubated for 30 min to allow them to digest any remaining bacteria. Animals are again collected by centrifugation, all of the M9 is removed, and the pellet is resuspended in an equal volume of fresh M9 buffer. Sodium dodecylsulfate is added to a final concentration of 2% (w/v) and the suspension is boiled for 5 min to lyse all animals and to solubilize nematode proteins. Immediately after boiling, the sample is cooled on ice and protease inhibitors are added. We routinely use 10 mM EDTA, 2 mM phenylmethylsulfonyl fluoride, 1 mM iodoacetic acid, and 1 mM N-ethylmaleimide. Two 15-cm plates yield about 500 μl of lysate and 2 to 5 μl gives reasonable amounts of proteins by SDS–polyacrylamide gel electrophoresis (PAGE) and silver staining. Lysates are compared to a bacterial extract by silver staining to estimate the approximate level of bacterial protein contamination. Usually, the nematode extracts contain few if any identifiable bacterial proteins. When contamination is obvious, the extracts are discarded.

SPARC is predicted to be between 27 and 30 kDa, depending on whether the conserved N-linked glycosylation site is used.[15] On the basis of homology to mammalian SPARC, the nematode protein should have as many as nine intrachain disulfide bonds. This extensive folding will yield a relatively compact protein with different mobilities under reducing and nonreducing conditions. Extracts are electrophoresed using a 12% (w/v) polyacrylamide-SDS gel and reduced and nonreduced lysates are compared. After transfer to nitrocellulose, the filters are incubated in a 1 : 500 (v/v) dilution of anti-MBP-domain IV antiserum in buffer A [25 mM Tris-HCl (pH 7.5), 150 mM NaCl, 0.1% (v/v) Tween 20]. This is followed by incubations with 1 : 10,000 (v/v) dilutions of biotinylated goat anti-rabbit IgG and streptavidin-horseradish peroxidase (HRP) (GIBCO-BRL, Gaithersburg, MD) in buffer A. Proteins are then visualized using chemiluminescence reagents (Renaissance; Du Pont-New England Nuclear) and exposure to Kodak XAR-5 film. As illustrated in Fig. 3, the anti-MBP-domain IV antiserum detects a protein band that migrates at approximately 30 kDa in the reduced nematode lysate. In a parallel nonreduced sample, this protein band is absent but a new band is present, migrating ahead of the 26-kDa molecular mass marker. Therefore, nematode SPARC shows significant levels of intrachain disulfide bonding, similar to mammalian SPARC.

For some experiments, it is advantageous to separate nematode proteins into different detergent-soluble fractions. A large-scale fractionation

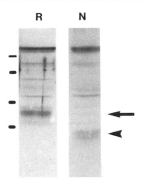

FIG. 3. Immunoblotting with anti-SPARC antiserum. Total nematode lysates were prepared and reduced (R) and nonreduced (N) extracts were separated by SDS–PAGE in a 12% (w/v) gel. After transfer to nitrocellulose, SPARC protein was detected using the anti-MBP-domain IV antiserum and chemiluminescence. Reduced SPARC migrates at about 30 kDa (arrow), whereas the nonreduced protein migrates significantly faster, ahead of the 26-kDa marker (arrowhead). Several background bands are seen in both lanes and are also detected by the preimmune serum (not shown). Molecular mass markers of 58, 48, 36, and 26 kDa are indicated at the left.

procedure, in which nematodes are lysed with a French press, has previously been reported by Francis and Waterston.[26] For fractionating small numbers of animals, we have adapted a procedure for lysing animals by vortexing them in the presence of glass beads. Nematodes are collected and washed as described above. Packed nematodes (25 μl) are then resuspended in 25 μl of 10 mM Tris-HCl (pH 8.0), 2 mM EDTA. Protease inhibitors are added along with 50 μl of packed glass beads (0.3–0.4 mm diameter; Sigma Chemical Co., St. Louis, MO) followed by vortexing for 1 min. Detergent is then added to give a final concentration of 1% (v/v) Nonidet P-40 (NP-40) or 2% (w/v) deoxycholate (DOC). After vortexing for 5–10 min, samples are spun for 10 min at 4° and supernatants containing detergent soluble proteins are analyzed by SDS–PAGE and silver staining. Not surprisingly, DOC-soluble protein fractions look similar to SDS lysates. However, NP-40 extracts contain only a subset of nematode proteins. This approach should be useful for making a variety of nematode protein fractions by changing the concentration or type of detergent, modifying the buffer conditions, or using various denaturants.

Microscopy of Nematodes

One of our goals is to localize SPARC protein within the body of the nematode by using immunofluorescence microscopy. Many different

[26] G. R. Francis and R. H. Waterston, *J. Cell Biol.* **101,** 1532 (1985).

procedures for fixing, permeabilizing, and staining embryos and adult animals have been tried. Acceptable staining with rhodamine-phalloidin has been obtained using many different published approaches. However, inefficient permeabilization can yield unacceptably high background staining with rhodamine- or fluorescein-labeled second antibody. In our hands, the protocol of Francis and Waterston[26] gives the cleanest and most reproducible indirect immunofluorescence staining of muscle and basement membrane structures. In this procedure, a French press is used to fracture nematodes followed by fixation with paraformaldehyde, permeabilization with either acetone or NP-40, and staining.

As an alternative to whole animal staining, we have developed a protocol for preparing paraffin sections of nematodes by adapting a procedure described by Sulston and Hodgkin.[24] Washed nematodes are fixed in 3.7% (v/v) formaldehyde in M9 buffer for several hours, oriented in a groove made on an agar pad, and immobilized by overlaying with a drop of 60° agar. Using a dissecting microscope, a small agar block containing several animals in the same orientation is cut out and dehydrated by 10-min incubations in increasing concentrations of alcohol (50, 70, 90, and 95%, v/v), three incubations in absolute ethanol, and three changes of xylene. After an overnight incubation at 65° in a 1 : 1 (v/v) mixture of xylene : paraffin, samples are incubated in two changes of straight paraffin, oriented, and embedded in paraffin. This method yields paraffin blocks with animals in a known orientation, so that sections can be cut cross-wise or lengthwise. Sections of 5 μm are cut, deparaffinized with Americlear (Baxter Scientific, McGaw Park, IL), rehydrated with decreasing alcohol concentrations, and stained with hematoxylin and eosin. Figure 4 shows cross-sections through the pharynx and the body of a nematode where gonad, intestine, and body wall are clearly visible.

Characterization of SPARC Constructs for Expression in Transgenic Nematodes

The SPARC gene, *ost-1,* maps to the left end of the long arm of chromosome IV on the *C. elegans* physical map. No obvious candidate mutations have been identified in the vicinity. Therefore, we are using a reverse genetic approach to study SPARC function in *C. elegans.* This approach requires that we microinject animals with plasmids carrying different versions of the SPARC gene and establish lines expressing these SPARC proteins. Although this approach yields valuable results regarding function and localization, negative results are difficult to interpret. It is essential that a mechanism be in place to demonstrate that recombinant

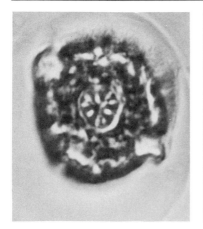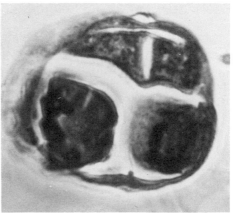

FIG. 4. Nematode cross-sections. Fixed animals were oriented and embedded in paraffin. Sections were stained with hematoxylin and eosin. *Left:* A cross-section through the pharynx. *Right:* A cross-section through the body. Intestine (top) and gonad are surrounded by the body wall.

SPARC polypeptides are expressed and secreted in transgenic animals. To this end, we have taken the approach of testing the expression of SPARC constructions in a mammalian system.

Two different eukaryotic expression vectors have been used for transient expression of cDNAs in COS cells. The retroviral vector pLJ works well for stable expression of recombinant proteins.[27–29] Because pLJ contains the SV40 origin of replication, transfection of pLJ constructs into COS cells also yields reasonable levels of protein by transient expression from the amplified plasmid. To make the cDNA compatible with the pLJ *Bam*HI cloning site, an essentially complete *C. elegans* SPARC cDNA was engineered to have a *Bam*HI site in the 5' untranslated region and a *Bgl*II site was generated at a *Dra*I site downstream of the termination codon. The other vector, pcDNAI (Invitrogen, San Diego, CA), was designed specifically for COS cell expression. This vector uses the cytomegalovirus (CMV) promoter to drive expression followed by a multiple cloning site, simian virus 40 (SV40) IVS, and a polyadenylation site. The multiple cloning site makes insertion of different cDNAs straightforward. Using the *Eco*RI and *Eco*RV sites, an *Eco*RI-to-*Dra*I fragment containing the entire SPARC coding region was inserted into this vector.

[27] J. E. Schwarzbauer, C. S. Spencer, and C. L. Wilson, *J. Cell Biol.* **109,** 3445 (1989).
[28] F. J. B. Barkalow and J. E. Schwarzbauer, *J. Biol. Chem.* **264,** 7812 (1991).
[29] J. E. Schwarzbauer, *J. Cell Biol.* **113,** 1463 (1991).

Plasmids are transfected into COS cells using the standard DEAE-dextran protocol.[30] Twenty-four hours after glycerol shock, the medium is removed and replaced with labeling medium containing 25 μCi of [^{35}S]methionine per milliliter and 0.1 times the normal level of unlabeled methionine. Labeled cell-conditioned medium is collected 24 hr later and used for immunoprecipitation with the anti-MBP-domain IV antiserum. After separation in SDS–polyacrylamide gels, the dried gels are exposed to a storage phosphor screen and protein bands are detected using a Molecular Dynamics PhosphorImager. We routinely use this approach to monitor the expression of both intact proteins and recombinant protein fragments. High levels of *C. elegans* SPARC were secreted in a COS cell transient expression assay, indicating that the protein folds properly, forms the appropriate disulfide bonds, and is otherwise able to transit the secretory pathway.

Concluding Remarks

Although *C. elegans* is usually considered as a system primarily for applying a genetic approach, we are finding that this organism can also be used successfully for biochemical and cell biological experiments. The basement membrane and its associated proteins, such as SPARC/osteo-nectin,[15] type IV collagen,[16] and perlecan,[17] are highly conserved. Therefore, the results of functional analyses obtained using *C. elegans* can be interpreted in the context of relevant data from mammalian systems. The resilience of the nematode and its simple anatomy are other attributes that make this an attractive system for dissecting extracellular matrix function.

Acknowledgments

This work is supported by grants from the NIH (CA44627), the ACS (CB-79), and the March of Dimes Birth Defects Foundation. J.E.S. was an Established Investigator of the American Heart Association.

---

[30] J. Sambrook, E. F. Fritsch, and T. Maniatis, eds., "Molecular Cloning: A Laboratory Manual," 2nd ed. Cold Spring Harbor Lab., Cold Spring Harbor, NY, 1989.

[14] *Drosophila* Extracellular Matrix

By L. I. FESSLER, R. E. NELSON, and J. H. FESSLER

Introduction

Extracellular matrices (ECM) occur as major skeletal components, as the immediate environment of cells (pericellular matrix), and as a fluid, mobile phase, (i.e., the interstitial fluid and blood). Research on *Drosophila* ECM has concentrated on the role of ECM in relation to cells, especially during development.[1-4] Evidence suggests that at this level *Drosophila* and vertebrates share fundamental problems and molecular mechanisms.

The exoskeleton of *Drosophila* differs radically from vertebrate endoskeleton, is based on chitin, and is not further considered here. *Drosophila* has an open circulatory system with hemolymph having the joint functions of interstitial fluid and blood, and ECM components, which are primarily concentrated next to cells, are also detectable in hemolymph. There may be some hemolymph components that serve functions analogous to some vertebrate blood proteins. In this chapter we focus on wandering *Drosophila* blood cells, called hemocytes, which are major producers of the pericellular ECM components.

The combined power of classic genetic analysis and recombinant DNA technology is most readily applied to the metazoans *Caenorhabditis elegans, Drosophila,* and mice. Each of these systems has particular advantages for investigating the roles of ECM during development. Some advantages of the *Drosophila* system are a relatively small genome with mostly small and relatively few introns, an extensive set of mutants that are readily mapped cytologically on the few, polytene chromosomes, and a 9-day life cycle that is accessible to optical and biochemical analyses at all stages of development, using synchronized mass cultures. In addition, transposon systems have been developed both for finding mutants associated with specific cells and tissues (enhancer trap searches and mutants) and for inserting specific DNA constructs into the germ line, thereby creating lines of transgenic animals. Unfortunately, genetic homologous recombination has so far not been possible in *Drosophila,* which is a disadvantage relative to mice. As a partial substitute for "gene knockout"

---

[1] J. H. Fessler and L. I. Fessler, *Annu. Rev. Cell Biol.* **5,** 309 (1989).
[2] M. Hortsch and C. S. Goodman, *Annu. Rev. Cell Biol.* **7,** 505 (1991).
[3] T. A. Bunch and D. L. Brower, *Curr. Top. Dev. Biol.* **28,** 81 (1993).
[4] N. H. Brown, *BioEssays* **15,** 383 (1993).

Copyright © 1994 by Academic Press, Inc.
All rights of reproduction in any form reserved.

experiments, numerous deletions of relatively small regions of chromosomes are known, and new chromosomal combinations can be set up fairly quickly by breeding. Bate and Martinez Arias,[5] Ashburner,[6] and Campos-Ortega and Hartenstein[7] provide a general introduction to this invertebrate system.

Three types of cell culture are particularly useful for *Drosophila* ECM research. Several immortal cell lines have been established from *Drosophila* embryos and can be grown in mass culture[8–10] (for a detailed description, see Sang[11]). They secrete ECM components into their culture medium, and this chapter describes how biochemical amounts of some individual ECM proteins are isolated from the conditioned medium. Although the specific cell types that gave rise to these cell lines are not known, it is likely that at least some of the progenitor cells were related to the hemocytes that make several ECM components *in vivo*.

Short-term primary cultures are readily set up from dissociated *Drosophila* embryos.[11] Differentiation of several cell types proceeds *in vitro* in these mixed cultures, over approximately the same time period as this differentiation occurs in whole embryos. These primary cultures can be initiated on substrate coatings of specific, individual *Drosophila* or vertebrate ECM glycoproteins, in the absence of fetal calf serum.[12,13] This allows study of the effects of a given ECM component on cell adhesion, spreading, and differentiation. Additional ECM components are subsequently synthesized by differentiated cells that arise in the culture.

A third type of cell culture utilizes *Drosophila* S2 cell lines that have been permanently transformed with one or more DNA constructs that either code for an ECM receptor, such as an integrin,[14,15] or for a part of an ECM molecule, such as a portion of a laminin chain (L. I. Fessler, unpublished observations, 1994). Cells that specifically express high concentrations of an ECM receptor are then tested for spreading on coatings

[5] M. Bate and A. Martinez Arias, "The Development of *Drosophila Melanogaster*." Cold Spring Harbor Lab., Cold Spring Harbor, NY, 1993.

[6] M. Ashburner, "Drosophila." Cold Spring Harbor Lab., Cold Spring Harbor, NY, 1989.

[7] J. A. Campos-Ortega and V. Hartenstein, "The Embryonic Development of *Drosophila melanogaster*." Springer-Verlag, Berlin and New York, 1985.

[8] G. Echalier and A. Ohanessian, *C.R. Hebd. Seances Acad. Sci., Ser. D* **268**, 1771 (1969).

[9] V. T. Kakpakov, V. A. Gvozdev, T. P. Platova, and L. G. Polukarova, *Genetika* **5**, 67 (1969).

[10] I. Schneider, *J. Embryol. Exp. Morphol.* **27**, 353 (1972).

[11] J. H. Sang, *in* "Advances in Cell Culture" (K. Maramorosch, ed.), p. 125. Academic Press, New York, 1981.

[12] T. Volk, L. I. Fessler, and J. H. Fessler, *Cell (Cambridge, Mass.)* **63**, 525 (1990).

[13] D. Gullberg, L. I. Fessler, and J. H. Fessler, *Dev. Dyn.* **119**, 116 (1994).

[14] T. A. Bunch and D. L. Brower, *Development (Cambridge, UK)* **116**, 239 (1992).

[15] M. Zavortink, T. A. Bunch, and D. L. Brower, *Cell Adhesion and Communication* **1**, 251 (1993).

of isolated ECM components. The manufacture of parts of an ECM poly-peptide chain by *Drosophila* cells makes it likely that *Drosophila*-specific posttranslational modifications, such as glycosylation, are also carried out correctly.

Removal of ECM is important during development and wound healing. Targeted, hormone-controlled breakdown of *Drosophila* ECM occurs, especially as part of the changes of metamorphosis.[16] Several gelatinases occur in *Drosophila,* and one of them corresponds to human gelatinase A.[17]

This chapter primarily concerns the preparation and properties of individual *Drosophila* ECM proteins. For the major *Drosophila* ECM components, basement membrane collagen IV[18-23] and laminin,[24-32] the homologies with the corresponding vertebrate proteins are notable. The most highly conserved portions are the sites of molecular interactions with other ECM components. These amino acid sequence homologies have been useful in the design of probes to search for additional human protein chains of a given protein family, such as the various α chains of the type IV collagens. Two *Drosophila* genes belonging to the vertebrate tenascin family, *ten*[a] [33] and *ten*[m],[34] have been identified by molecular cloning based

[16] L. I. Fessler, M. L. Condic, R. E. Nelson, J. H. Fessler, and J. W. Friström, *Development (Cambridge, UK)* **117,** 1061 (1993).

[17] E. Woodhouse, E. Herstburger, W. Stetler-Stevenson, L. Liotta, and A. Shearn, *Cell Growth Dev.* **5,** 151 (1994).

[18] J. E. Natzle, J. M. Monson, and B. J. McCarthy, *Nature (London)* **296,** 368 (1982).

[19] J. M. Monson, J. Natzle, J. Friedman, and B. J. McCarthy, *Proc. Natl. Acad. Sci. U.S.A.* **79,** 1761 (1982).

[20] J. P. Cecchini, B. Knibiehler, C. Mirre, and Y. Le Parco, *Eur. J. Biochem.* **165,** 587 (1987).

[21] B. Blumberg, A. J. MacKrell, P. F. Olson, M. Kurkinen, J. M. Monson, J. E. Natzle, and J. H. Fessler, *J. Biol. Chem.* **262,** 5947 (1987).

[22] B. Blumberg, A. J. MacKrell, and J. H. Fessler, *J. Biol. Chem.* **263,** 18328 (1988).

[23] G. P. Lunstrum, H. P. Bachinger, L. I. Fessler, K. G. Duncan, R. E. Nelson, and J. H. Fessler, *J. Biol. Chem.* **263,** 18318 (1988).

[24] L. I. Fessler, A. G. Campbell, K. G. Duncan, and J. H. Fessler, *J. Cell Biol.* **105,** 2383 (1987).

[25] D. J. Montell and C. S. Goodman, *Cell (Cambridge, Mass.)* **53,** 463 (1988).

[26] D. J. Montell and C. S. Goodman, *J. Cell Biol.* **109,** 2441 (1989).

[27] H. C. Chi and C. F. Hui, *Nucleic Acids Res.* **16,** 7205 (1988).

[28] H. C. Chi and C. F. Hui, *J. Biol. Chem.* **264,** 1543 (1989).

[29] K. Garrison, A. J. MacKrell, and J. H. Fessler, *J. Biol. Chem.* **266,** 22899 (1991).

[30] M. Kusche-Gullberg, K. Garrison, A. J. MacKrell, L. I. Fessler, and J. H. Fessler, *EMBO J.* **11,** 4519 (1992).

[31] A. J. MacKrell, M. Kusche-Gullberg, K. Garrison, and J. H. Fessler, *FASEB J.* **7,** 375 (1993).

[32] C. Henchcliffe, L. Garcia-Alonso, J. Tang, and C. S. Goodman, *Development (Cambridge, UK)* **118,** 325 (1993).

[33] S. Baumgartner and R. Chiquet-Ehrismann, *Mech. Dev.* **40,** 165 (1993).

[34] S. Baumgartner and R. Chiquet-Ehrismann, submitted for publication (1994).

on homologies. We speculate that some ECM proteins that currently seem to be unique to *Drosophila* may eventually be found to have human homologs.

The secreted proteins, unique to *Drosophila*, that have been identified are glutactin,[35] tiggrin,[36] peroxidasin,[37] and the slit protein.[38-40] Their domain structure and occurrence during development have been documented. Partial characterization of the following materials has been published. The large proteoglycan-like protein, papilin,[41] is synthesized fairly early in embryogenesis and then becomes a part of basement membranes. More restricted localization of a proteoglycan in developing imaginal disks has been visualized by immunolocalization with a monoclonal antibody.[42] A sulfated proteoglycan isolated from larvae has been identified as a heparan sulfate proteoglycan.[43] Immunolocalization with antibodies recognizing a 240-kDa protein is seen during the formation of cell membranes following the initial syncytial blastoderm stage, and during subsequent cell sheet organization, and this protein may be an ECM component.[44] A 350-kDa protein, recognized by a monoclonal antibody, has a restricted distribution in the embryonic and pupal neuropil of the central nervous system (CNS), and it is suggested that this protein is a special component of the CNS ECM.[45] It is not clear whether the secreted amalgam protein,[46] found in the nervous system, is to be considered an ECM protein. Other secreted proteins, with specialized functions, might also be included as ECM proteins, but these are not discussed here. Early studies of *Drosophila* ECM suggested that there is a homolog of vertebrate fibronectin.[47] However, extensive investigations failed to extend these observations.

[35] P. F. Olson, L. I. Fessler, R. E. Nelson, R. E. Sterne, A. G. Campbell, and J. H. Fessler, *EMBO J.* **9,** 1219 (1990).

[36] F. J. Fogerty, L. I. Fessler, T. A. Bunch, Y. Yaron, C. G. Parker, R. E. Nelson, D. L. Brower, D. Gullberg, and J. H. Fessler, *Development (Cambridge, UK)* **120,** 1747 (1994).

[37] R. E. Nelson, L. I. Fessler, Y. Takagi, B. Blumberg, P. F. Olson, C. G. Parker, D. Keene, and J. H. Fessler, *EMBO J.* **13,** 3438 (1994).

[38] J. M. Rothberg, D. A. Hartley, Z. Walther, and S. Artavanis-Tsakonas, *Cell (Cambridge, Mass.)* **55,** 1047 (1988).

[39] J. M. Rothberg, J. R. Jacobs, C. S. Goodman, and S. Artavanis-Tsakonas, *Genes Dev.* **4,** 2169 (1990).

[40] J. M. Rothberg and S. Artavanis-Tsakonas, *J. Mol. Biol.* **227,** 367 (1992).

[41] A. G. Campbell, L. I. Fessler, T. Salo, and J. H. Fessler, *J. Biol. Chem.* **262,** 17605 (1987).

[42] D. L. Brower, M. Piovant, R. Salatino, J. Brailey, and M. J. Hendrix, *Dev. Biol.* **119,** 373 (1987).

[43] V. Cambiazo and N. C. Inestrosa, *Comp. Biochem. Physiol. B* **97B,** 307 (1990).

[44] V. Garzino, H. Berenger, and J. Pradel, *Development (Cambridge, UK)* **106,** 17 (1989).

[45] M. J. Go and Y. Hotta, *J. Neurobiol.* **23,** 890 (1992).

[46] M. A. Seeger, L. Haffley, and T. C. Kaufman, *Cell (Cambridge, Mass.)* **55,** 589 (1988).

[47] D. Gratecos, C. Naidet, M. Astier, J. P. Thiery, and M. Semeriva, *EMBO J.* **7,** 215 (1988).

Some receptors of ECM molecules, the *Drosophila* integrins, were first identified in a monoclonal antibody screen.[48,49] The ubiquitous $\alpha_{PS2}\beta_{PS}$ and $\alpha_{PS1}\beta_{PS}$ integrins and their splice variants have extensive homologies to vertebrate integrin α and β chains.[50–55] Additional integrins are present at more restricted sites.[56] *Drosophila* cell adhesion and receptors have been reviewed by Hortsch and Goodman,[2] Bunch and Brower,[3] and Brown[4]. As more ECM components are identified, more receptor–ligand interactions will be defined.

Isolation of Extracellular Matrix Proteins from Cell Cultures

Extracellular matrix proteins are secreted into the culture medium by a number of established cell lines, originally derived from embryos.[56a] Kc cells[8,9] have been especially advantageous, because appreciable quantities of ECM proteins are secreted into the medium. Different sublines of Kc cells secrete somewhat different mixtures of ECM proteins. Several Kc cell lines, for example, the Kc 7E10 cell line,[57] produce the well-characterized proteins[1] collagen IV, laminin, papilin, glutactin, and peroxidasin.[57a] These cells are nonadherent and do not express PS integrins. The Kc 167 cell line is more adherent; it synthesizes integrins and a new ECM protein, tiggrin, which has been identified as a ligand of $\alpha_{PS2}\beta_{PS}$ integrin.[36] The Kc 167 cells also secrete papilin, collagen IV, and laminin. In addition,

[48] M. Wilcox, D. L. Brower, and R. J. Smith, *Cell* (*Cambridge, Mass.*) **25**, 159 (1981).

[49] D. L. Brower, M. Wilcox, M. Piovant, R. J. Smith, and L. A. Reger, *Proc. Natl. Acad. Sci. U.S.A.* **81**, 7485 (1984).

[50] M. Leptin, R. Aebersold, and M. Wilcox, EMBO J. **6**, 1037 (1987).

[51] T. Bogaert, N. Brown, and M. Wilcox, *Cell* (*Cambridge, Mass.*) **51**, 929 (1987).

[52] A. J. MacKrell, B. Blumberg, S. R. Haynes, and J. H. Fessler, *Proc. Natl. Acad. Sci. U.S.A.* **85**, 2633 (1988).

[53] N. H. Brown, D. L. King, M. Wilcox, and F. C. Kafatos, *Cell* (*Cambridge, Mass.*) **59**, 185 (1989).

[54] S. Zusman, Y. Grinblat, G. Yee, F. C. Kafatos, and R. O. Hynes, *Development* (*Cambridge, UK*) **118**, 737 (1993).

[55] M. Wehrli, A. Di Antonio, I. M. Fearnley, R. J. Smith, and M. Wilcox, *Mech. Dev.* **43**, 21 (1993).

[56] G. H. Yee and R. O. Hynes, *Development* (*Cambridge, UK*) **118**, 845 (1993).

[56a] Many established *Drosophila* cell lines are kept by P. Cherbas (Department of Biology, University of Indiana, Bloomington, IN) and at the PHLS Center for Applied Microbiology and Research, European Collection of Animal Cell Cultures, Division of Biologics (Porton Down, Salisbury, Wiltshire, United Kingdom).

[57] T. M. Landon, B. A. Sage, B. J. Seeler, and J. D. O'Connor, *J. Biol. Chem.* **263**, 4693 (1988).

[57a] Peroxidasin was previously referred to as protein X. Peroxidasin is described in Nelson *et al.*[37]

many unidentified higher and mostly lower molecular weight proteins are secreted. The Schneider S2 and S3 cell lines[10] produce lower levels of ECM proteins. The slit protein is secreted into the medium and deposited as part of the matrix in cultures of S2 cells.[39]

Stock Kc cells are maintained in tissue culture flasks, but some cells, such as the Kc 7E10 cells, have been adapted for growth in suspension in spinner and roller cultures at room temperature in D22 medium[6] (Sigma, St. Louis, MO) plus penicillin and streptomycin and without fetal calf serum (FCS). The Kc 167 cells attach to tissue culture flasks and can be maintained in D22 medium plus antibiotics with or without 2% (v/v) FCS. These cells also grow in M3 medium (Sigma), which has a defined amino acid composition.[6] For isolation of radiolabeled proteins the amino acid to be added in labeled form is omitted from M3 medium and the amounts of Yeastolate (Difco Lab., Detroit, MI) and Lactalbumin Hydrolysate (Sigma) are decreased to 1/10 of the normal additions. To label sulfated glycoproteins, $H_2^{35}SO_4$ is added to medium containing 1/20 the normal concentration of SO_4^{2-}. Omission of unlabeled SO_4^{2-} yields undersulfated proteins.[41] Bovine ECM components are always potential contaminants of *Drosophila* proteins purified from cell cultures grown in the presence of FCS. The *Drosophila* origin of a newly isolated protein should be proved by obtaining it in labeled form from cell cultures supplied with radioactive amino acids.

Concentration of Proteins from Conditioned Cell Culture Medium

A flow diagram outlines the purification procedure (Fig. 1). Kc 7E10 cells grown at room temperature or at 25° in suspension are maintained in a spinner bottle for several weeks and diluted three- to fourfold with new medium weekly. These cultures can reach a density of about 1–5 × 10^6 cells/ml. Aliquots are then placed into roller bottles, diluted 1 : 1 (v/ v) with new medium, and grown to a cell density of about 10^7 cells/ml, and then diluted 1 : 2 (v/v) and grown to high density two or three more times.[57b] For isolation of proteins from the medium the cells are removed by centrifugation at 1000 g for 10 min and the medium is clarified by centrifugation at 6000 g for 60 min at 4°. For isolation of proteins from adherent cell cultures, for example, Kc 167 cell cultures, the medium is

---

[57b] In these cell cultures collagen IV is folded and secreted normally even though the hydroxylation of proline and lysine residues is incomplete. To obtain sufficient hydroxylation of collagen IV the cultures need to be supplemented daily with ascorbic acid (25 μg/ml). This yields collagen IV with the same degree of hydroxylation as made *in vivo*: pepsin-treated, fully hydroxylated collagen IV has 35% of the total proline and 60% of the total lysine hydroxylated. On reduction the hydroxylated collagen molecule separates into a major α1 IV component and a minor α1′ IV component.[23]

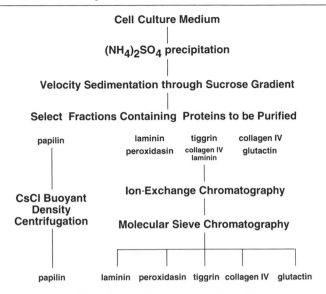

FIG. 1. Flow diagram for ECM protein purification.

decanted and treated as above. Protease inhibitors are added [30 mM ethylenediaminetetraacetic acid (EDTA pH 7.5), 2 mM N-ethylmaleimide (NEM), and 1 mM phenylmethylsulfonyl fluoride (PMSF)] and proteins are precipitated at 4° by addition of solid $(NH_4)_2SO_4$ to 45% saturation. Most of the ECM proteins are precipitated with this concentration of $(NH_4)_2SO_4$. Differential precipitation of proteins with lower $(NH_4)_2SO_4$ concentrations can be useful for partial separation of some ECM molecules. The precipitate is dissolved in 1 ml/100 ml of original culture medium in 0.05 M Tris-HCl (pH 7.5)–0.15 M NaCl (TBS) plus inhibitors (20 mM EDTA, 10 mM NEM, 1 mM PMSF). Following extensive dialysis against TBS any undissolved residue is removed by centrifugation at 8000 g for 30 min at 4°.

Velocity Sedimentation

The protein mixture (1 ml/centrifuge tube) is partially separated by velocity sedimentation at 4° on a 5–20% (w/v) sucrose gradient in TBS [a 0.75-ml, 60% (w/v) sucrose cushion in TBS at the bottom of a tube] in a Beckman SW 41 rotor at 39,000 rpm for 15 to 22 hr, depending on the proteins to be isolated. The sedimentation coefficients given in Table I[57c]

[57c] L. I. Fessler and J. H. Fessler, *J. Biol. Chem.* **249**, 7637 (1974).

TABLE I
SEDIMENTATION COEFFICIENTS OF
EXTRACELLULAR MATRIX PROTEINS[a]

Protein	$s_{20,w}$	Ref.
Collagen IV	4.1S	Lunstrum et al.[23]
Laminin	11.0S	Fessler et al.[24]
Peroxidasin	11.6S	Nelson et al.[37]
Tiggrin	6.45S	Fogerty et al.[36]

[a] Sedimentation coefficients are determined with purified proteins dissolved in TBS plus 0.1% (v/v) Triton X-100. (This detergent is added to avoid losses due to protein adherence to the tube.) The gradient is a 5–20% (w/v) sucrose gradient in TBS plus 0.1% (v/v) Triton, and a 0.1-ml sample is layered on the top of the gradient. Sedimentation is in an SW 60 rotor at 4°. The position of the protein is determined by SDS–PAGE and densitometric analysis. Sedimentation coefficients are calculated, as decribed.[57c]

help in the choice of centrifugation conditions for the isolation of any given protein. Fractions are collected from the bottom of the centrifuge tube and aliquots of individual fractions are analyzed by sodium dodecyl sulfate-polyacrylamide gel electrophoresis (SDS–PAGE) to identify the components in each fraction. An example of the separation of the proteins secreted by Kc 7E10 cells is shown in Fig. 2. Individual fractions, peak fractions or larger pools of material can be chosen for further purification: for example, proteoglycans are in fractions 1–3; laminin and peroxidasin peak in fractions 9–11 (however, appreciable quantities of laminin can be isolated from fractions 5–8 as well); collagen IV is present as monomeric molecules together with glutactin in fractions 13–16, and higher aggregates are found in faster sedimenting fractions; and glutactin peaks in fractions 14–17. To isolate tiggrin the proteins in the medium of Kc 167 cells are partially resolved by velocity sedimentation at 4°, at 39,000 rpm for 19 hr. Tiggrin sediments between laminin and collagen, and is purified from these intermediate fractions.

Ion-Exchange and Molecular Sieve Chromatography

The peak fractions of the proteins to be purified further are pooled and dialyzed against Mono Q buffer: 0.3 M sucrose, 0.05 M Tris-HCl (pH 8.0), 0.005 M EDTA, 1 mM PMSF, and 0.05% (v/v) Triton X-100

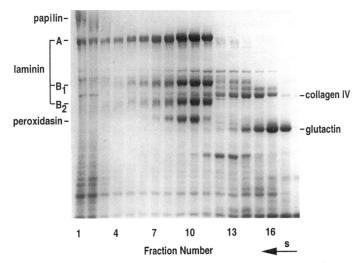

FIG. 2. Partial separation of proteins secreted by Kc 7E10 cells by velocity sedimentation. The Coomassie blue-stained SDS–PAGE electrophoretogram of the reduced proteins is shown for the successive fractions of a 5–20% (w/v) sucrose gradient velocity sedimentation gradient. The proteins in the medium were concentrated by $(NH_4)_2SO_4$ precipitation and a 1-ml sample was sedimented in a Beckman SW 41 rotor at 4°, 39,000 rpm for 15 hr, and then separated into 21 fractions. The electrophoretic band positions of several key polypeptides are indicated. Note that all separations are only partial, and further chromatographic purifications are required. (Reproduced from Campbell *et al.*[41])

containing NaCl at a concentration that permits binding of the proteins at 4° to the Mono Q ion-exchange column in a fast protein liquid chromatography (FPLC) system (Pharmacia-LKB Biotechnology, Piscataway, NJ). A low concentration of NaCl (between 0.05 and 0.15 *M* NaCl) prevents the partial, slow precipitation of collagen IV, if it is present in the mixture. Alternatively, DEAE columns (DE-52; Whatman, Clifton, NJ) give satisfactory separations. The proteins are eluted with a gradient of NaCl, starting with the NaCl concentrations in the sample buffer to 0.5 *M* NaCl. Figure 3 shows the NaCl concentrations required for elution of the identified proteins.

Each protein is purified further by gel filtration on Superose 6 or 12 columns (Pharmacia-LKB Biotechnology) or on an agarose A-50 sizing column. The appropriate fractions containing the proteins to be separated are dialyzed against Mono Q buffer containing 0.15 *M* NaCl. The samples are concentrated either by centrifugation through Centricon 30 microconcentrators (Amicon, Danvers, MA) or by vacuum dialysis. A 2-ml volume is applied to a 1.6 × 50 cm Superose 6 column and 1.0-ml samples

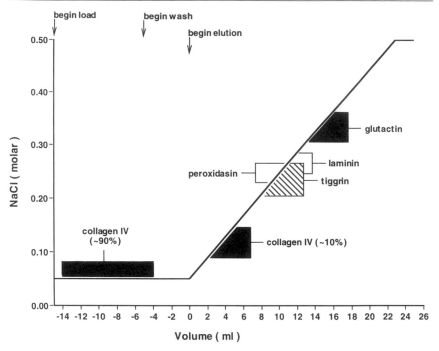

FIG. 3. Elution of ECM proteins from a Mono Q ion-exchange column. This composite diagram shows the ranges of NaCl concentrations required for release of several ECM proteins from a Mono Q column eluted with a linearly increasing gradient of 0.02 M NaCl/ml in Mono Q buffer. Starting samples for this chromatography were separate sets of selected fractions from more than one velocity sedimentation experiment. Each set of pooled sedimentation fractions is indicated by different shading as follows: Kc 167 cell culture media proteins, tiggrin sedimentation pool (▨); Kc 7E10 cell culture media proteins, glutactin and collagen IV sedimentation pool (■); laminin plus peroxidosin sedimentation pool (□). Note that additional proteins, mostly of smaller molecular size, contaminate both the velocity sedimentation and the ion-exchange fractions. They are removed by subsequent molecular sieve chromatography.

are eluted with the above buffer at an elution rate of 0.13 ml/min. Samples are stored frozen in this buffer. Figure 4 shows separations of groups of proteins on a preparative Superose 6 column. Protein separation on a Superose 6 analytical column (1.0 × 30 cm), calibrated with globular molecular weight standards, can provide an estimate of the molecular weights of globular proteins. However, many ECM proteins, such as collagen and laminin, are elongated, asymmetric, and variably flexible molecules and elute from the sizing column in positions that do not correspond to their molecular weights.

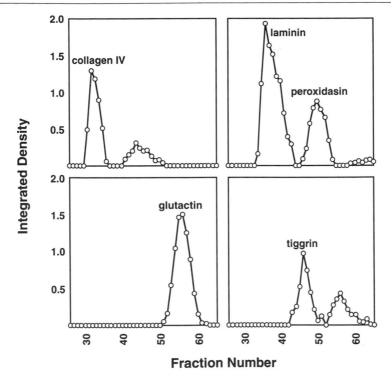

Fraction Number

FIG. 4. Molecular sieve chromatography. Separate sets of selected fractions obtained by the ion-exchange chromatography of Fig. 3 were applied to a preparative Superose 6 column and eluted at 0.13 ml/min with 100 ml of buffer. Aliquots of consecutive 1-ml fractions were reduced and analyzed by SDS–PAGE and stained with Coomassie blue. Each electrophoretogram contained the electrophoresis buffer front and was scanned densitometrically along each complete lane. The integrated, normalized optical densities are plotted. At each named peak only bands corresponding to that protein were seen. Unnamed peaks represent other proteins and contaminants, which also eluted after fraction 63. Totally excluded marker eluted at fraction 29.

Equilibrium CsCl Buoyant Density Centrifugation

Papilin,[41] a proteoglycan-like ECM protein, is synthesized by both Kc cell lines. It is a sulfated protein and is readily detected in the medium of cells labeled with $H_2{}^{35}SO_4$. By velocity sedimentation of total medium proteins it is found together with other ECM aggregates in the fastest sedimenting fractions at the bottom of the tube. To obtain native papilin the fast sedimenting fractions are dialyzed into TBS and resedimented for 10 hr at 39,000 rpm at 4°. This places the monomeric and dimeric papilin molecules within the gradient around fractions 5–10. By ion-exchange chromatography on a Mono Q column it is eluted with a NaCl gradient of 0.25 to 0.45 M. Papilin elutes as a fairly broad peak, following laminin, reflecting heterogeneity in glycosylation and sulfation.

Papilin in denatured form is further purified from the mixture of proteins in the fast sedimenting fractions near the bottom of the gradient (Fig. 2, fractions 1–5) by equilibrium, CsCl buoyant density sedimentation in guanidine hydrochloride (GuCl) solution. GuCl is needed to avoid salting out of proteins when CsCl is added. Samples are dialyzed into TBS followed by dialysis against 4 M GuCl in TBS. CsCl, 1.40 g, is dissolved in 3.56 ml of sample, and this sample is centrifuged in an SW 60 Beckman rotor at 10° for 48 hr at 40,000 rpm. The rotor is not spun at full speed to avoid overstraining with this CsCl load. The gradient is divided into 17 fractions by drop collection from the bottom of the tube. Fractions are dialyzed against TBS and analyzed by SDS–PAGE or on an SDS-agarose gel, as shown in Fig. 5. The peak of the proteoglycans is buoyant at a density of 1.40–1.42 g/ml with some spread of the molecules to lower density fractions. The other proteins are buoyant at or near the top of the gradient.

Characterization of Extracellular Matrix Molecules

Electrophoretic Mobilities of Extracellular Matrix Proteins

The relative electrophoretic mobilities of the nonreduced and reduced proteins, which we have characterized, are shown in Fig. 6. Collagen IV[23] molecules are disulfide-linked trimers, which on reduction separate into a major α1 IV chain and a minor α1' IV chain.[57b] Collagen IV can associate into higher aggregates, as is shown by SDS-agarose gel electrophoresis (Fig. 7) or electron microscopy (Fig. 9). Laminin,[24,25] a disulfide-linked heterotrimer, consists of A, B1, and B2 chains. Each of these chains also contains multiple intramolecular S–S bridges. Peroxidasin[37] is a disulfide-linked homotrimer of 170-kDa subunits. Papilin[41] exists as a disulfide-linked monomer or as higher oligomers (Figs. 5 and 9). The size of the reduced, monomeric core protein is in the range of 400 kDa. The electrophoretic mobility of tiggrin[36] is not altered by reduction. Because the tiggrin subunit (250 kDa) has only one cysteine, intrachain S–S bridges cannot form and most of the molecules isolated from conditioned medium have no intermolecular bonds. Glutactin[35] has intramolecular S–S bridges, and exists as a monomeric chain. The slit protein secreted by S2 cells is about 200 kDa[39] in molecular mass.

Oligomers of ECM proteins such as collagen IV or papilin are too large for separation by SDS–PAGE, but can be separated by SDS-agarose

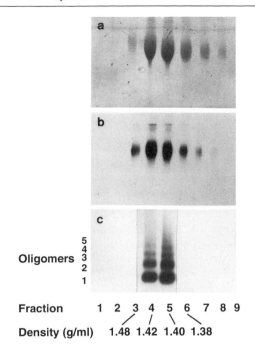

Fraction   1 2 3 4 5 6 7 8 9

Density (g/ml)   1.48 1.42 1.40 1.38

FIG. 5. CsCl buoyant density equilibrium centrifugation. Cell cultures were incubated with $H_2^{35}SO_4$ for 24 hr. The medium proteins were partly separated by velocity sedimentation, and fractions 1–5 of Fig. 1 were pooled and sedimented to buoyant equilibrium in CsCl, 4 M GuCl in TBS buffer. Aliquots of peak fractions were reduced, electrophoresed on SDS–4.5% (w/v) polyacrylamide gels, and visualized by Coomassie blue staining (a) and fluorography (b). Other aliquots were electrophoresed, without reduction, on SDS–1.5% (w/v) agarose, and the fluorograms for fractions 4 and 5 are shown (c). The buoyant densities of the peak fractions are also given. (Reproduced from Campbell *et al.*[41]).

slab gel electrophoresis[58]: 1.5% (w/v) agarose (type I, low EEO; Sigma) is dissolved in 0.19 M Tris-HCl (pH 8.8)–0.25% (w/v) SDS, and a horizontal, 200 × 130 × 4 mm gel is cast. The gel is placed in a horizontal electrophoresis apparatus (as for a gel for nucleic acid separation) and reservoir buffer [0.2% (w/v) SDS, 0.19 M Tris-glycine (pH 8.3)] is added so that the gel is just submerged in the buffer. The samples in TBS–0.2% (w/v) SDS–4% (v/v) glycerol–0.003% bromphenol blue are heated at 100° for 2 min. Electrophoresis is carried out for about 16 hr at 20 mA, constant current. The gels are fixed in 50% (v/v) methanol–7% (v/v) acetic acid, and stained with 0.2% (w/v) Coomassie blue. For radiolabeled proteins

[58] K. G. Duncan, L. I. Fessler, H. P. Bachinger, and J. H. Fessler, *J. Biol. Chem.* **258,** 5869 (1983).

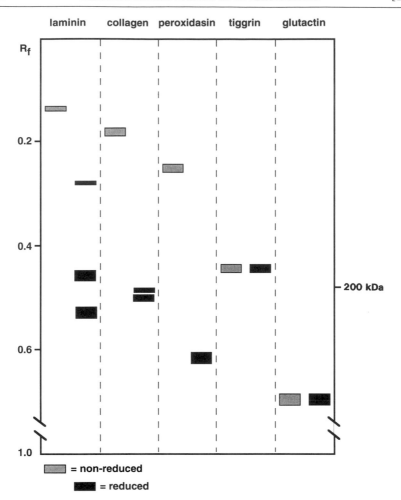

FIG. 6. Relative electrophoretic mobilities of ECM proteins. Diagrammatic representation of the mobilities of purified disulfide-linked and reduced proteins electrophoresed on a 24-cm SDS–5% (w/v) polyacrylamide gel. The molecular weights of the subunits of each protein before posttranslational modification are given in Table II. The 200-kDa marker is myosin.

fluorography is carried out after the gels are soaked in 50% (v/v) methanol–7% (v/v) acetic acid for 1 hr, washed with running water, and soaked in 0.7 M sodium salicylate (adjusted to about pH 6.5) for 30 min and dried. The electrophoretic resolution of disulfide-linked oligomers of *Drosophila* collagen IV is shown in Fig. 7. The semilogarithmic relationship of the

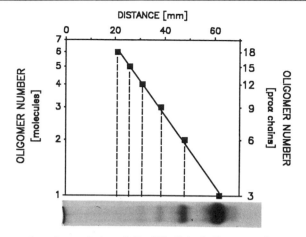

FIG. 7. Electrophoretic resolution of disulfide-linked oligomers of *Drosophila* collagen IV on an SDS-agarose gel. The electrophoretogram stained with Coomassie blue is shown together with a plot of the distance migrated by each band versus the logarithm of the presumed number of collagen IV molecules (left ordinate) or α chains (right ordinate) that are disulfide linked in the complexes. The statistical regression line is shown; its correlation coefficient is 0.85. (Reproduced from Lunstrum *et al.*[23])

mobilities of these bands to a simple integer sequence is consistent with a series of oligomers of internally disulfide-linked, triple-helical molecules. A possible standard for SDS-agarose gel electrophoretograms is oligomers of myosin.

Composition of Extracellular Matrix Proteins

Table II gives the molecular weights of ECM proteins, based on DNA sequence analyses. Posttranslational modifications of these proteins occur. Collagen IV contains hydroxyproline and hydroxylysine; glutactin and papilin, and to a much lesser extent the A chain of laminin, are sulfated. Sulfation of glutactin is on tyrosines and in papilin sulfated glycosaminoglycan (GAG) side chains are present. Most of the proteins are N- or O-glycosylated. However, carbohydrate analyses are incomplete.

From the derived amino acid sequences the locations of possible asparagine N-glycosylation sites can be predicted. Removal of N-linked carbohydrates with peptide: *N*-glycosidase F (PNGase F) (New England Bio-Labs, Beverly, MA) shows to what extent the various proteins isolated from the culture medium are actually *N*-glycosylated. Figure 8 illustrates that peroxidasin, glutactin, and the laminin chains are extensively

TABLE II

MOLECULAR WEIGHTS OF EXTRACELLULAR MATRIX PROTEINS DERIVED
FROM DNA SEQUENCES

Protein	M_r ($\times 10^{-3}$)	Ref.
Collagen IV	172	Blumberg et al.[22]
Laminin		
A chain	409	Kusche-Gullberg et al.[30]; Henchcliffe et al.[32]
B1 chain	196	Montell and Goodman[25]
B2 chain	179	Montell and Goodman[26]; Chi and Hui[27]
Tiggrin	255	Fogerty et al.[36]
Glutactin	117	Olson et al.[35]
Peroxidasin	168	Nelson et al.[37]
Slit protein	~200	Rothberg et al.[39]
Ten[a]	85	Baumgartner and Chiquet-Ehrismann[33]
Ten[m]	281	Baumgartner and Chiquet-Ehrismann[34]

N-glycosylated molecules. Removal of O-linked carbohydrates with O-glyconase (Genzyme, Cambridge, MA) has not been studied extensively with these proteins. Characterization of the carbohydrate side chains of these proteins on Western blots is available by affinity binding of a large

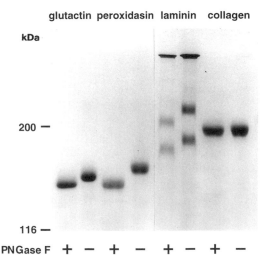

FIG. 8. Comparison of glycosylated and N-deglycosylated ECM proteins. The purified, denatured, reduced proteins incubated without or with PNGase were electrophoresed on a 24-cm SDS–4.5% acrylamide gel. The Coomassie blue-stained gel is shown.

variety of biotinylated lectins and recognition of the bound lectins with the biotin–avidin–peroxidase system (Vector Laboratories, Burlingame, CA).

The GAG side chains of the proteoglycans can be removed by a number of methods. Enzymatic treatment with chondroitinase ABC, heparinase, heparatinase, keratinase, neuraminidase, and PNGase F and H aid in the identification of the GAG side chains. To identify the O-linked carbohydrate chains β elimination in 0.1 N NaOH with 1.0 M NaBH$_4$ at 47° for 48 hr releases the GAG chains, which can then be characterized further. Treatment of the protein with HNO$_2$ decomposes heparan sulfate. Deglycosylation with trifluoromethanesulfonic acid[59] yields intact core protein.

Analysis of Extracellular Matrix Protein Structure by Electron Microscopy

Examples of rotary shadowing of sprayed, or spread, purified proteins are shown in Fig. 9. It is best to complement this approach with the technique of negative contrast with uranyl acetate.[60]

Isolation of Extracellular Matrix Proteins from Embryos, Larvae, Pupae, and Flies

Some ECM proteins (e.g., laminin) can be obtained from embryos by extraction with 1 M NaCl buffer, or from embryo lysates.[25] *Drosophila* embryos are dechorionated with 30% (v/v) clorox, washed well with a 0.7% (w/v) NaCl solution, and either extracted directly or frozen in liquid N$_2$ and powdered in a precooled mortar and pestle. For extraction at 4° all solutions contain protease inhibitors: 1 mM PMSF, 20 mM EDTA, and 1 μg/ml of the following; aprotinin, pepstatin, leupeptin, and possibly other inhibitors [antipain, chymostatin, N-tosyl-L-phenylalanine chloromethyl ketone (TPCK) and Na-p-tosyl-L-lysine chloromethyl ketone (TLCK)]. Embryos are either extracted with 1 M NaCl–50 mM Tris-HCl–20 mM EDTA (pH 7.5) plus protease inhibitors or, for more exhaustive treatment, the embryos are solubilized in 10 mM triethanolamine–1% (v/v) Nonidet P-40 (NP-40)–0.5% (w/v) deoxycholate–0.15 M NaCl (IPB) plus protease inhibitors (pH 8.2) by homogenization (five strokes) in a glass Dounce (Wheaton, Millville, NJ) homogenizer with a loose Teflon pestle (1 g of embryos per 2–5 ml of buffer). The lysate is stirred for 1 hr at 4°. The lysate is sedimented 13,000 g for 30 min and then clarified at 100,000 g for 2 hr at 4°. By affinity chromatography laminin is bound on

---

[59] A. S. Edge, C. R. Faltynek, L. Hof, L. Reichert, Jr., and P. Weber, *Anal. Biochem.* **118**, 131 (1981).

[60] J. Engel and H. Furthmayr, this series. Vol. 145, p. 3.

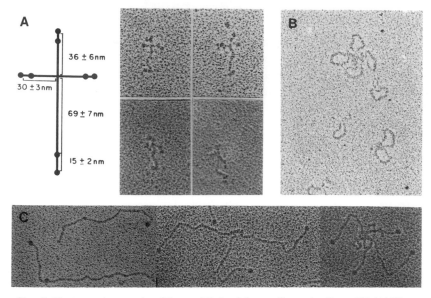

Fig. 9. Electron micrographs of *Drosophila* laminin, papilin, and collagen IV. (A) Diagram of laminin gives the dimensions of the molecules depicted in the electron micrographs of laminin sprayed onto mica and rotary shadowed. (B) Electron micrograph of a papilin monomer, dimer, and tetramer adsorbed onto pentylamine-treated grids, stained with uranyl acetate, and rotary shadowed. The loops are 225 ± 15 nm. (C) Electron micrograph of collagen IV monomers and oligomers. The thread length of a monomeric collagen molecule is 447 ± 6 nm. (Composite of figures from Fessler *et al.*,[24] Campbell *et al.*,[41] and Lunstrum *et al.*[23])

a peanut lectin-agarose column (Vector Laboratories). The column is washed extensively with phosphate-buffered saline (PBS), and equilibrated with IPB buffer. The lysate is passed over the column at 5–10 ml/hr. Unbound proteins, including a number of ECM proteins, are removed by washes with IPB, IPB with 0.5 M NaCl, and IPB. Laminin, proteoglycan, and some other unidentified proteins are eluted with IPB containing 0.5 M D-galactose. These proteins can be visualized by SDS–PAGE and silver staining of the gel or on Western blots with antibodies. Laminin also binds to wheat germ agglutinin-Sepharose (Sigma) and heparin-Sepharose (Pharmacia).[24] Separation of laminin and proteoglycan on a heparin-Sepharose column, followed by separation of the bound proteins on a sizing column, yields pure laminin. Affinity chromatography on lectin columns is a useful tool, which should be explored further when unidentified proteins are to be purified.

Quantitative extraction of matrix proteins, followed by identification of specific proteins on Western immunoblots, is useful for evaluation of the temporal expression during development.[1] Staged dechorionated embryos, larvae, pupae, and adult flies are frozen and powdered in liquid N_2. Equal weights of tissue are homogenized in 10 vol of 2% (w/v) SDS–0.15 M NaCl–0.05 M Tris-HCl (pH 7.5)–0.02 M EDTA– 1 mM PMSF at 100°, heated for 5 min at 100°, and sonicated to fragment the nucleic acid. Complete extraction in most cases requires reduction with 10 mM dithiothreitol (DTT) at 100° for 5 min. The insoluble residue is sedimented and the concentration of soluble protein in each sample is determined with the Bio-Rad (Richmond, CA) protein assay. Known quantities of protein extracted at each stage are then electrophoresed on SDS–PAGE and a Western blot with antibodies to specific ECM proteins is developed. The antigen–antibody complex can then be recognized by binding of ^{125}I-labeled protein A and autoradiography, followed by counting of the radioactive bands. To obtain accurate quantitation increasing volumes of extract from each time period are electrophoresed, and the Western blot is developed with an excess of antibody and ^{125}I-labeled protein A. It is important to determine that the quantity of antigen electrophoresed, and the antigen–antibody–protein A complex formed, are within a linear response range of signal. Other systems for recognition of the antigen–antibody complex, such as the use of alkaline phosphatase-conjugated secondary IgG, or enhanced chemiluminescence systems, are more difficult to quantitate, but are also adequate for general analyses. An example of developmental Western analyses is shown in Fig. 10. Oocytes and early embryos already contain a small quantity of ECM proteins, but after zygotic transcription is initiated much higher levels of proteins are synthesized at the indicated times.

Synthesis, Location, and Interaction of Extracellular Matrix Proteins

Northern Blot Analyses

Analysis of the temporal relative expression of mRNAs encoding the various ECM proteins during *Drosophila* development complements the developmental Western blots. Total RNA is isolated by the guanidinium thiocyanate method[61] from embryos, larvae, pupae, and adults that have been kept at 25° for the specific time intervals. Poly(A)RNA is purified by oligo(dT)-cellulose chromatography. Northern blots are developed by standard methods using antisense RNA probes for specific ECM mRNAs.

[61] J. M. Chirgwin, A. E. Przybyla, R. J. MacDonald, and W. J. Rutter, *Biochemistry* **18**, 5294 (1979).

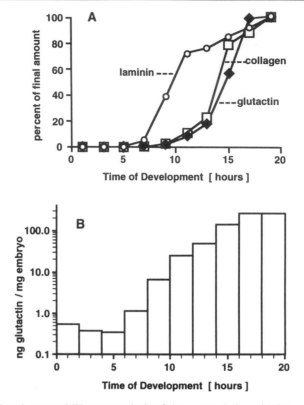

FIG. 10. Developmental Western analysis of the accumulation of ECM proteins during embryogenesis. SDS extracts of embryos, at increasing times (in hours) of development at 25°, were electrophoresed and Western blots with the specific antibodies were developed. The accumulation of each protein was determined, using quantities of each protein that assured a linear response range of signal. (A) Accumulation of the proteins laminin, collagen IV, and glutactin are given as a linear plot. (B) Amount of glutactin found in the embryos at the increasing times of development are plotted on a log scale. This shows that trace amounts of the protein are present in the early embryos before zygotic transcription and translation occur later during embryogenesis. (A) (From Fessler and Fessler.[1])

To detect pairs of mRNA species of different electrophoretic mobilities a single antisense RNA containing segments complementary to both mRNAs is used.[29]

Antibody Preparations

Polyclonal antibodies to ECM proteins are made in rabbits, rats, or mice. Many rabbit preimmune sera react with *Drosophila* tissues, and therefore rabbit sera need to be tested before immunization of rabbits. Antisera prepared in mice and rats have fewer background problems. The purified nonreduced proteins are separated by SDS–PAGE, and the protein bands are identified by Coomassie blue staining, cut out, washed with PBS, fragmented, and injected with adjuvants intradermally and intramuscularly at 2- to 3-week intervals. Alternatively, the purified protein in PBS is mixed with adjuvant and injected. Affinity purification of the antibodies on an antigen-Sepharose column, or by binding to antigen transferred to nitrocellulose after SDS–PAGE, is followed by elution with 0.2 M glycine hydrochloride, pH 2.3, and immediate neutralization with Tris-HCl, pH 9, and dialysis against PBS at 4°.

Monoclonal antibodies have been prepared, using purified antigens (e.g., antibodies to laminin[26]), or by immunizations with whole tissues (e.g., *Drosophila* heads[62] and disks[42,48]) and identification of antibodies localized to specific sites. Brower *et al.*[42,48] generated monoclonal antibodies against imaginal disks and identified an extracellular matrix proteoglycan on the basal surface of imaginal disk epithelia and in a filamentous network between epithelial cells. Fujita *et al.*[62] injected *Drosophila* heads and isolated many monoclonal antibodies, including a number of antibodies that recognize basement membranes. The technique of isolation of proteins with monoclonal antibodies immobilized on Sepharose or agarose has not been used extensively thus far, but may be a useful future approach.

In Situ Hybridization and Immunostaining

A detailed description of the temporal and spatial expression of ECM genes and the deposition of the ECM proteins is beyond the scope of this chapter, but some overall observations are summarized here.

The principal sites of synthesis of ECM components are hemocytes, the cells of the fat body, and some glial cells, as detected both by *in situ* hybridization for RNA and by immunostaining.[23,26,30,35–39,63–65] Other tissues, such as epithelia or muscle, may be able to synthesize these materials at a lower level, but this is difficult to document definitively. Although immunostaining of a hemocyte might be equally likely to show

[62] S. C. Fujita, S. L. Zipursky, S. Benzer, A. Ferrus, and S. L. Shotwell, *Proc. Natl. Acad. Sci. U.S.A.* **79**, 7929 (1982).
[63] Y. Le Parco, B. Knibiehler, J. P. Cecchini, and C. Mirre, *Exp. Cell Res.* **163**, 405 (1986).
[64] C. Mirre, J. P. Cecchini, Y. Le Parco, and B. Knibiehler, *Development (Cambridge, UK)* **102**, 369 (1988).
[65] Y. Le Parco, A. Le Bivic, B. Knibiehler, C. Mirre, and J. P. Cecchini, *Insect Biochem.* **19**, 789 (1989).

a newly synthesized or a phagocytosed protein, the correspondence with *in situ* hybridization is consistent with *de novo* production. Well-staining hemocytes are also prominent components of differentiated primary cell cultures that synthesize radioactively labeled ECM components.[13] Polyclonal antibodies made to reduced laminin chains primarily stains intracellular laminin, suggesting a significant intracellular presence of incompletely assembled molecules.[30] Conversely, relatively poor staining of some cells for glutactin[35] may indicate a short intracellular residence time for completed molecules of this ECM component. Double immunostaining for overlapping pairs of ECM molecules indicates that individual hemocytes synthesize several, separate ECM components, specifically collagen IV, laminin, papilin, tiggrin, and peroxidasin. Different ECM proteins that are all found in a common structure, such as basement membranes, are not necessarily made at the same time. Indeed, all evidence points to a successive deposition of basement membrane ECM components (Fig. 10).[1]

It is difficult to visualize the earliest deposition of some ECM proteins, although these may be among the most important ones developmentally. ECM proteins have been detected during oogenesis[66] and low maternal contributions of some ECM proteins are detected in extracts of very young embryos (see Fig. 10) and by immunostaining.[67] Major subsequent accumulations of ECM are in basement membranes that surround the CNS and muscle, and underlie epithelia.[23,26,30,35,36] ECM proteins are concentrated at muscle insertions[34–36,39] and at locations of mechanical stress. Specialized functions such as axonal guidance are attributed to secreted ECM proteins, such as laminin[26] and the slit protein.[39] In larvae ECM proteins are synthesized by the fat body and hemocytes, in the lymph glands, and by adepithelial cells.[64] Special problems of ECM modification[16] and deposition[42,68] arise in imaginal disks, the larval precursor structures of adult appendages. Immunoelectron microscopy with antibodies to collagen IV and laminin shows localization to excessive layers of basement membranes in the tumor-like imaginal disks of lethal (1) *disks-large-1* mutants.[69]

Extracellular Matrix Ligand–Cell Surface Receptor Interaction

The interaction of cells with ECM proteins involves multiple cell surface receptors. Prominent receptors are *Drosophila* integrins. For identifying which ECM proteins are ligands of *Drosophila* integrins, Bunch and

[66] H. O. Gutzeit, W. Eberhardt, and E. Gratwohl, *J. Cell Sci.* **100**, 781 (1991).
[67] B. Knibiehler, C. Mirre, and Y. Le Parco, *Cell Differ. Dev.* **30**, 147 (1990).
[68] D. Friström, M. Wilcox, and J. Friström, *Development (Cambridge, UK)* **117**, 509 (1993).
[69] L. A. Abbott and J. E. Natzle, *Mech. Dev.* **37**, 43 (1992).

TABLE III
CHROMOSOME LOCATIONS OF GENES ENCODING EXTRACELLULAR
MATRIX PROTEINS

Protein	Chromosome band	Ref.
Collagen IV	25C	Monson et al.[19]; Natzle et al.[18]; Le Parco et al.[71]
Laminin		
A chain	65A 10–11	Montell and Goodman[25]
B1 chain	28D	Montell and Goodman[25]
B2 chain	67C	Montell and Goodman[25]
Glutactin	29D	Olson et al.[35]
Tiggrin	26D 1–2	Fogerty et al.[36]
Peroxidasin	62E/F	Nelson et al.[37]
Ten[a]	11A 6–9	Baumgartner and Chiquet-Ehrismann[33]
Ten[m]	79E1,2	Baumgartner and Chiquet-Ehrismann[34]
Slit protein	52D	Rothberg et al.[38]

Brower used transfected S2 cells.[14,15] For example, the integrin β_{PS} subunit and the two splice variants, $\alpha_{PS2(c)}$ or $\alpha_{PS2(m8)}$ of the α_{PS2} subunit, are expressed in these cells under the control of a heat-shock promoter. Cells stimulated to express integrins at their cell surfaces attach and spread on immobilized ligands, such as vertebrate fibronectin or vitronectin[14,15] and *Drosophila* tiggrin.[36] These ligand–integrin interactions utilize an arginine-glycine-aspartic acid (RGD)-containing sequence and RGD peptides inhibit this interaction. Antibodies recognizing the functional domain of the integrin β chain[70] block this ligand–receptor interaction. Thus far only one *Drosophila* ECM protein, tiggrin, has been demonstrated to be a ligand for the $\alpha_{PS2}\beta_{PS}$ integrins, but other investigations are in progress. Primary cells recognize laminin as a substratum, utilizing non-$\alpha_{PS2}\beta_{PS}$ integrin receptors.[12] What role cell surface proteoglycans play remains to be determined.

DNA Cloning

Antibodies prepared against purified *Drosophila* ECM proteins have been used to isolate cDNA clones from expression libraries (e.g., laminin A, B1, and B2 chains,[25,29] and glutactin[35] and tiggrin[36]). Heterologous antibodies have not been used successfully thus far. With the availability

---

[70] S. Hirano, K. Ui, T. Miyake, T. Uemura, and M. Takeichi, *Development (Cambridge, UK)* **113,** 1007 (1991).

of sufficient quantities of proteins, N-terminal amino acid sequence analyses of intact or fragmented proteins have been possible. This is essential for confirmation of the identity of the cDNA isolated, as described above. Oligonucleotide mixtures were used successfully for the isolation of cDNA clones coding for peroxidasin.[37]

By low-stringency screens with a chicken procollagen I cDNA clone a number of *Drosophila* genes encoding collagen IV (*Dcg1*[18–21] and *Dcg2*[18,63]) and additional collagen-like genes mapping to different locations on the chromosomes were isolated.[71] Some vertebrate proteins are classified as collagens, even though only a minor portion of their amino acid sequence has the canonical $(Gly-X-Y)_n$ sequence required for folding into a triple helix. Correspondingly, there may be *Drosophila* proteins with relatively short $(Gly-X-Y)_n$ repeats.

Vertebrate ECM proteins have highly conserved domains and polymerase chain reaction (PCR) approaches may lead to successful isolations of *Drosophila* cDNAs with homologous domains. Comparisons of *Drosophila* and vertebrate amino acid sequences derived from cDNA analyses have led to the overall conclusion that functional domains of proteins are more conserved than structural regions (laminin, collagen IV, Ten[a], and Ten[m]).Molecular cloning of *ten*[a 33] and *ten*[m 34] genes, homologous to tenascin, was achieved using the PCR with degenerate primers derived from the epidermal growth factor (EGF) portion of chicken tenascin.

Chromosome Location of Genes Encoding Extracellular Matrix Proteins

The availability of genetic approaches makes the *Drosophila* system advantageous. *In situ* hybridization of biotinylated or digoxigenin-labeled DNA probes to the salivary gland polytene chromosomes identifies the gene location.[6] Table III lists the cytological location on chromosomes of the cloned and identified genes encoding ECM proteins.

Acknowledgments

Some of the research described in this chapter was supported by grants from the Muscular Dystrophy Association and USPHS, Grant AG02128, which also supported the cost of writing this chapter. We thank Kathryn Brill for typing of this manuscript.

[71] Y. Le Parco, J. P. Cecchini, B. Knibiehler, and C. Mirre, *Biol. Cell.* **56,** 217 (1986).

Section V

Modern Methods in Extracellular Matrix Research

[15] Expression of Heterologous Integrin Genes in Cultured Eukaryotic Cells

By FILIPPO G. GIANCOTTI, LAURA SPINARDI, FABRIZIO MAINIERO, AND RAYMOND SANDERS

Introduction

The development of methods for the introduction of DNA into cultured eukaryotic cells has made it possible to express cloned genes in cell lines from different species and tissues. Gene transfer generally involves insertion of the cloned sequence of interest in an appropriate expression vector, amplification of the resulting vector in bacteria, and introduction in mammalian cells by transfection. Gene transfer methods have been extensively applied to the study of different classes of adhesion receptors. In the integrin field, these methods have been used to establish structure–function relationships of integrins, in particular with regard to the localization of the ligand-binding and subunit association sites[1-4] and the identification of the cytoplasmic sequences involved in the interaction with the cytoskeleton.[5-10] In addition, these methods have helped to define the role of individual integrins in cell adhesion, migration, and matrix assembly.[11-17] Finally, gene transfer methods are also actively employed

[1] J. C. Loftus, T. E. O'Toole, E. F. Plow, A. Glass, A. L. Frelinger, III, and M. H. Ginsberg, *Science* **249,** 915 (1990).
[2] M. L. Bajt, M. H. Ginsberg, A. L. Frelinger, III, M. C. Berndt, and J. C. Loftus, *J. Biol. Chem.* **267,** 3789 (1992).
[3] R. Briesewitz, M. R. Epstein, and E. E. Marcantonio, *J. Biol. Chem.* **268,** 2989 (1993).
[4] D.-T. Shih, J. M. Edelman, A. F. Horwitz, G. B. Grunwald, and C. A. Buck, *J. Cell Biol.* **122,** 1361 (1993).
[5] J. Solowska, J. L. Guan, E. E. Marcantonio, J. E. Trevithick, C. A. Buck, and R. O. Hynes, *J. Cell Biol.* **109,** 853 (1989).
[6] Y. Hayashi, E. E. Marcantonio, A. Reszka, D. Boettinger, and A. F. Horwitz, *J. Cell Biol.* **110,** 175 (1990).
[7] E. E. Marcantonio, J. L. Guan, J. E. Trevithick, and R. O. Hynes, *Cell Regul.* **1,** 597 (1990).
[8] J. Solowska, J. E. Edelman, S. M. Albelda, and C. A. Buck, *J. Cell Biol.* **114,** 1079 (1991).
[9] A. A. Reszka, Y. Hayashi, and A. F. Horwitz, *J. Cell Biol.* **117,** 1321 (1992).
[10] L. Spinardi, Y.-L. Ren, R. Sanders, and F. G. Giancotti, *Mol. Biol. Cell* **4,** 871 (1993).
[11] F. G. Giancotti and E. Ruoslahti, *Cell (Cambridge, Mass.)* **60,** 849 (1990).
[12] M. J. Elices, L. Osborn, Y. Takada, C. Crouse, S. Luhowskyj, M. E. Hemler, and R. R. Lobb, *Cell (Cambridge, Mass.)* **60,** 577 (1990).
[13] B. M. C. Chan, P. D. Kassner, J. A. Schiro, H. R. Byers, T. S. Kupper, and M. E. Hemler, *Cell (Cambridge, Mass.)* **68,** 1051 (1992).

METHODS IN ENZYMOLOGY, VOL. 245

Copyright © 1994 by Academic Press, Inc.
All rights of reproduction in any form reserved.

to investigate the role of integrins in intracellular signaling[18] and their contribution to tumorigenesis.[11,19]

The integrins are a large family of heterodimeric receptors involved in cell–matrix and cell–cell adhesion.[20–25] Integrins are expressed on virtually all types of cells. Although each cell displays a specific integrin repertoire, many integrins are expressed in several different cell types. It is therefore often difficult to find a suitable negative host cell line to express a certain integrin. In studies aimed at examining the subcellular localization of wild-type and mutant integrin subunits, this problem has been obviated by using heterologous systems in which a cDNA derived from a certain species is introduced in cells of another species and the expressed protein is then detected by means of a species-specific monoclonal antibody.[5–10] The availability of cell lines carrying targeted disruptions of integrin genes[26] will, in the near future, expand the spectrum of functional studies that investigators in the integrin field will be able to perform. In the mean time, the choice of a suitable host cell line remains a critical issue when planning a gene transfer experiment with integrins. A second potential obstacle to the successfull expression of integrins from cDNA is represented by their heterodimeric nature. Both the α and β subunits are transmembrane glycoproteins that combine in the endoplasmic reticulum (ER) soon after biosynthesis.[27] Because only paired subunits undergo processing in the ER and in the Golgi apparatus and are then transported to the cell surface, transfection of an individual integrin subunit leads to

[14] J. S. Bauer, C. L. Schreiner, F. G. Giancotti, E. Ruoslahti, and R. L. Juliano, *J. Cell Biol.* **116**, 477 (1992).
[15] D. I. Leavesley, G. D. Ferguson, E. A. Wayner, and D. A. Cheresh, *J. Cell Biol.* **117**, 1101 (1992).
[16] J. Ylänne, Y. Chen, T. E. O'Toole, J. C. Loftus, Y. Takada, and M. H. Ginsberg, *J. Cell Biol.* **122**, 223 (1993).
[17] Z. Zhang, A. O. Morla, K. Vuori, J. S. Bauer, R. L. Juliano, and E. Ruoslahti, *J. Cell Biol.* **122**, 235 (1993).
[18] J.-L. Guan, J. E. Trevithick, and R. O. Hynes, *Cell Regul.* **2**, 951 (1991).
[19] B. M. C. Chan, N. Matsuura, Y. Takada, B. R. Zetter, and M. E. Hemler, *Science* **251**, 1600 (1991).
[20] C. A. Buck and A. F. Horwitz, *Annu. Rev. Cell Biol.* **3**, 179 (1987).
[21] M. H. Ginsberg, J. C. Loftus, and E. F. Plow, *Thromb. Haemostasis* **59**, 1 (1988).
[22] E. Ruoslahti, *Annu. Rev. Biochem.* **57**, 375 (1988).
[23] M. E. Hemler, *Annu. Rev. Immunol.* **8**, 365 (1990).
[24] T. A. Springer, *Nature (London)* **346**, 425 (1990).
[25] R. O. Hynes, *Cell (Cambridge, Mass.)* **69**, 11 (1992).
[26] E. L. George and R. O. Hynes, this volume [20].
[27] D. A. Cheresh and R. C. Spiro, *J. Biol. Chem.* **262**, 17703 (1987).

successful cell surface expression only if the host cell expresses adequate levels of its partner subunit(s).[28] The β_1 subunit can combine with different α subunits and many cell lines express a variety of $\alpha\beta_1$ integrins. The β_1 subunit is often synthesized in molar excess over the available different α subunits and the fraction of β_1 subunit that does not combine with α subunits undergoes degradation in the ER.[29] Thus, investigators have transfected an individual α subunit in cells that did not express that particular subunit and obtained successful expression at the cell surface of hybrid receptors containing the cDNA-encoded α subunit and endogenous β_1.[12–14] Because the α_v subunit is also often expressed in excess amounts relative to the different β subunits with which it can combine, transfection of an α_v-combining β subunit can lead to successful expression at the cell surface of the corresponding $\alpha_v\beta$ integrin.[15] Finally, the α_6 subunit combines more efficiently with β_4 than with β_1. Thus, introduction of even moderate amounts of the β_4 subunit in cells that express an endogenous $\alpha_6\beta_1$ integrin leads to cell surface expression of the $\alpha_6\beta_4$ integrin and a corresponding decrease of the cell surface levels of $\alpha_6\beta_1$.[30]

In this chapter we focus on gene transfer methods and experimental strategies that appear to be most useful in integrin research. The vectors for gene transfer can be broadly divided in two categories: plasmids and animal viruses. Although the viral vectors have the great advantage that they can introduce foreign DNA in eukaryotic cells efficiently, they have so far had limited application in research on integrins. This is possibly because there is only a limited number of cell types that can be infected by a certain viral vector [e.g., NIH 3T3 and C127 fibroblasts by the bovine papillomavirus, Epstein–Barr virus nuclear antigen (EBNA)-positive primate cells by the Epstein–Barr virus].[31] Retroviruses are perhaps the most promising viral vectors because they can infect a variety of cell types with an efficiency of nearly 100%. Two commonly used retroviral vectors are the LNC vector, carrying a cytomegalovirus (CMV) promotor, and the LSN-2 vector, with a long terminal repeat promotor.[32] In general, the use of retroviral vectors is limited by packaging constraints to DNA inserts smaller than 6–7 kb. In addition, protein expression from retroviral-based

[28] T. E. O'Toole, J. C. Loftus, E. F. Plow, A. A. Glass, J. A Harper, and M. H. Ginsberg, *Blood* **74**, 14 (1989).

[29] J. Heino, R. A. Ignotz, M. E. Hemler, C. Crouse, and J. Massagué, *J. Biol. Chem.* **264**, 380 (1989).

[30] F. G. Giancotti, M. A. Stepp, S. Suzuki, E. Engvall, and E. Ruoslahti, *J. Cell Biol.* **118**, 951 (1992).

[31] N. Muzyczka, ed., *Curr. Top. Microbiol. Immunol.* **158**, (1990).

[32] W. R. A. Osborne and A. D. Miller, *Proc. Natl. Acad. Sci. U.S.A.* **85**, 6851 (1988).

vectors is generally low owing to inefficient processing and translation of the encoded mRNAs. For these reasons, most investigators in the integrin field have preferred plasmid-based vectors.

For the cell biologist, the first and perhaps most important step in eukaryotic expression is the choice of a host cell line suited to the planned experiment. When expressing integrins, one must consider a number of factors. These include the spectrum of integrins already present in the cell, their relative level of expression, and the possible existence of an intracellular pool of pre-β or pre-α subunits in excess of the endogenous partner subunits available for association. If one chooses to use a heterologous sytem, for example, to transfect a chicken or human integrin cDNA in a mouse cell line that already has the integrin in question, then it is necessary to have an antibody to distinguish the cDNA-encoded integrin subunit from the endogenous one. Finally, the ability of a cell to form focal adhesions or hemidesmosomes, its histotype, the level of differentiation, and tumorigenic potential are all factors that must be taken into account.

Constitutive Elements of Mammalian Expression Vectors

The essential elements of all expression vectors are a promotor to mediate transcription of exogenous DNA sequences and signals for efficient termination and polyadenylation of transcripts. Most vectors also contain an enhancer element located upstream of the promotor and a small intron, with a functional splice donor and acceptor site, located between the 3' untranslated region of foreign DNA and the polyadenylation signal. The plasmid vectors used for transient expression generally contain a viral origin of replication to mediate intracellular replication of the transfected plasmid. Some vectors also contain an independent transcription unit encoding a eukaryotic selection marker.

Promoter and Enhancer Elements

The promoter is composed of the TATAA box, located approximately 30 bp upstream of the mRNA transcription initiation site, which functions by designating the start site for polymerase II-mediated transcription, and the CAAT box, approximately 80 bp upstream of the initiation site, which serves as a binding site for cellular factors that facilitate transcription initiation.[33] The enhancers are core sequences that increase the rate of transcription from a certain promoter by 10- to 100-fold and act indepen-

---

[33] W. S. Dynan and R. Tjian, *Nature (London)* **316**, 774 (1985).

dently of their orientation and distance from the promoter.[34] They are the primary target of transcriptional factors involved in the regulation of tissue-specific gene expression and therefore the choice of a given promoter/enhancer combination is most important for efficient expression in a given cell type. Many vectors contain promoter/enhancer combinations that function well in many different cell types such as those derived from the simian virus 40 (SV40) early gene, the adenovirus major late gene, the cytomegalovirus immediate-early region, and the Rous sarcoma virus long terminal repeat (LTR). However, there are still significant differences in potency between these promoter/enhancers in a given cell type. It is thus advisable to conduct preliminary transient transfection experiments with a reporter gene, such as chloramphenicol acetyltransferase (CAT), to determine which is the most effective promoter in a given cell type.

Termination and Polyadenylation Signals

The biogenesis of mRNAs includes a number of steps that immediately follow the termination of transcription. Construction of an expression vector requires an evaluation of the steps that affect the structure of the 3' end of the message, because messages that are not properly processed at their 3' ends are translated inefficiently. Soon after transcription termination, the 3' end of the mRNA is cleaved at a site between two important elements of the 3' untranslated region: the AAUAAA hexamer, located 11–30 nucleotides upstream of the cleavage site, and a GU- or U-rich sequence, located downstream.[35] Both signals are required for efficient cleavage and subsequent polyadenylation. It follows that cDNAs do not contain all the signals necessary for polyadenylation and these elements must be provided in the vector. The most commonly used signals for polyadenylation are derived from SV40 early transcription unit, the mouse β-globin gene, and the bovine growth hormone gene.

Splicing Signals

The presence of a synthetic intron containing at least one splice donor and one acceptor site in the 3' untranslated region upstream of the polyadenylation signals may, in some cases, be beneficial in the expression of cDNAs in in vitro cultured cells.[36] The requirement for a splicing signal

[34] T. Maniatis, S. Goodbourn, and J. A. Fischer, *Science* **236**, 1237 (1987).
[35] N. J. Proudfoot, *Trends Biochem. Sci.* **14**, 105 (1989).
[36] P. Gruss and G. Khouri, *Nature (London)* **286**, 634 (1980).

for expression in cultured cells, however, is not as absolute as it appears to be for expression in transgenic animals.[37,38]

Viral Replicons

A number of animal viruses contain signals, located at the so-called origin of replication, that allow extrachromosomal replication of the viral genome in permissive cells. Plasmid vectors containing such signals will also replicate efficiently in transfected cells in the presence of appropriate *trans*-acting factors. Therefore, one way to increase the number of copies of transcriptional units available for expression is to use plasmid vectors carrying the viral replicon of SV40 or polyomavirus in cells expressing the appropriate T antigen. Because the transfected cells usually die when the plasmid molecules exceed the number of 10^4 per cell, usually at 72–96 hr posttransfection, these system are used for transient, high-level expression. One of the most widely used transient expression system combines vectors carrying the SV40 replicon and the Lg-T antigen-positive COS cells.[39]

Selectable Markers

With the transfection methods available it is possible to introduce a plasmid vector in only a small fraction of the treated cells. In the 48–72 hr following transfection, these cells will transiently express the desired protein even if the plasmid does not possess a viral replicon and therefore cannot replicate. In a minority of transfected cells, tandem arrays of linear plasmid molecules will become stably integrated in the host genome, leading to stable expression of the desired protein.[40] The identification of cells with stably integrated DNA is facilitated by the simultaneous transfection of a gene encoding a selectable marker. The selectable markers used confer resistance to an antibiotic or a drug. Many commercially available eukaryotic expression vectors contain a selectable marker gene. However, because two unlinked plasmids are taken up by the same cell with high frequency (>90%), it is possible to cotransfect the vector encoding the gene of interest and the plasmid containing a selectable marker.[40] The use of a 10 : 1 molar ratio of the two plasmids will ensure that every cell that has resisted the selection conditions has also incorporated the

[37] M.-J. Gething and J. Sambrook, *Nature (London)* **293**, 620 (1981).
[38] R. Brinster, J. Allen, R. Behringer, R. Gelinas, and R. Palmiter, *Proc. Natl. Acad. Sci. U.S.A.* **85**, 836 (1988).
[39] Y. Gluzman, *Cell (Cambridge, Mass.)* **23**, 175 (1981).
[40] M. Wigler, R. Sweet, G. K. Sim, B. Wold, A. Pellicer, E. Lacy, T. Maniatis, S. Silverstein, and R. Axel, *Cell (Cambridge, Mass.)* **16**, 777 (1979).

gene of interest. In addition, the transfection of an excess molar ratio of the gene of interest favors the generation of transfectants that harbor multiple copies of it and therefore express high levels of the desired protein. The selectable markers more commonly employed include the following.

1. Aminoglycoside phosphotransferase (aph): This is the most widely used selectable marker. It is dominant because eukaryotic cells are *aph⁻*. The gene contained in various vectors is of bacterial origin and, when transfected into eukaryotic cells, confers on them the ability to grow in the presence of aminoglycoside antibiotics such as geneticin.[41] Geneticin is toxic to eukaryotic cells because it effectively suppresses protein synthesis.

2. Hygromycin B phosphotransferase (hyg): This is also a dominant selection marker. It confers resistance to the antibiotic hygromycin.[42] It can be used in combination with the geneticin selection, either simultaneously or sequentially, because the resistance to geneticin and to hygromycin can be selected for independently.

3. Thymidine kinase (tk): This enzyme is involved in the salvage pathway for the biosynthesis of thymidine. Cell lines that lack the thymidine kinase gene (*tk⁻*) are not able to grow in medium containing hypoxanthine, aminopterin, and thymidine (HAT medium).[43] Available *tk⁻* cell lines include the mouse Ltk⁻ cells, the human 143 tk⁻ cells, and the Rat-2 fibroblasts.

4. Dihydrofolate reductase (dhfr): This enzyme catalyzes the two steps of the biosynthesis of tetrahydrofolate from folate. Because tetrahydrofolate is a cofactor necessary for the biosynthesis of thymidine, purines, and glycine, *dhfr⁻* cells cannot grow in the absence of these supplements. The only available *dhfr⁻* cells are mutant clones derived from the Chinese hamster ovary (CHO) cell line.[44] Transfection of these cells with an exogenous *dhfr* gene generates clones that can grow in the absence of added nucleotides.[45]

cDNA Sequences

Virtually all full-length cDNAs contain an intact ribosome-binding site, defined by the Kozak consensus sequence.[46] The core elements of this

[41] F. Colbère-Garapin, F. Horodniceanu, P. Kourilsky, and A. C. Garapin, *J. Mol. Biol.* **150**, 1 (1981).
[42] B. Sugden, K. Marsh, and J. Yates, *Mol. Cell. Biol.* **1**, 854 (1985).
[43] E. H. Szybalski and W. Szybalski, *Proc. Natl. Acad. Sci. U.S.A.* **48**, 2026 (1962).
[44] G. Urlaub and L. A. Chasin, *Proc. Natl. Acad. Sci. U.S.A.* **77**, 4216 (1980).
[45] S. Subramani, R. Mulligan, and P. Berg, *Mol. Cell. Biol.* **1**, 854 (1981).
[46] M. Kozak, *J. Cell Biol.* **108**, 229 (1989).

consensus sequence are the AUG codon that functions as translation start site, a G at position $+4$, and an A or a G at position -3. The length of the 5' untranslated sequences of cloned cDNAs is variable and generally does not matter, because the distance between the transcription initiation site and the start codon is not critical in eukaryotes. However, if the 5' untranslated region contains AUG triplets upstream of the start codon, these must be eliminated.[47] It is also advisable, when possible, to eliminate all the ancillary sequences present at the 5' and 3' ends of the cDNA, such as synthetic linkers and homopolymeric tails used for cloning purposes.[48]

Transfection Methods

There are several methods that allow the introduction of DNA into cultured eukaryotic cells. They differ by mechanism of action and effectiveness in different cell types. Some methods, such as that based on the use of DEAE-dextran, can be used only for transient transfection, and others, such as that employing Polybrene, may not be efficient for transient assays. Therefore, if the efficacy of a given method in a certain cell type is untested, it is advisable to test different methods by using a reporter gene such as those encoding CAT or β-galactosidase (for transient assay), or *aph* (for stable transfection). The most commonly used transfection methods are described in the following sections.

Calcium Phosphate Coprecipitation

In this method the DNA to be transfected is mixed with calcium chloride and sodium phosphate under conditions that allow the formation of coprecitates of DNA and calcium phosphate.[40] In this form the DNA can be taken up by the cell by endocytosis. The mechanisms by which the endocytosed DNA is liberated in the cytoplasm and eventually becomes stably integrated in the host cell genome are not known. This method is of general use because it can be used for both transient and stable transfection and generally works well with most of the cell lines commonly used in these experiments. A protocol for calcium phosphate coprecipitation of DNA is subsequently described.

DEAE-Dextran-Mediated Transfection

DEAE-dextran is a polycation that adheres tightly to the surface of DNA. The resulting complex is internalized by the cell by endocytosis.

[47] L. Perez, J. W. Wills, and E. Hunter, *J. Virol.* **6,** 1276 (1987).
[48] C. C. Simonsen, H. M. Shepard, P. W. Gray, D. W. Leung, D. Pennica, E. Yelverto, R. Derynck, P. J. Sherwood, A. D. Levinson, and D. V. Goeddel, *U.C.L.A. Symp. Mol. Cell. Biol.* **25,** 1 (1982).

Thus, DEAE-dextran mediates transfection by a mechanism not dissimilar from calcium phosphate and it is expected to work well in the same cell types that are transfected efficiently by the calcium phosphate method. The two methods, however, differ because the DEAE-dextran protocol does not allow stable integration of the DNA in the host cell genome.[49] Therefore, its use is limited to transient assays.

Electroporation

Electroporation is generally used to introduce DNA in cell types that are resistant to calcium phosphate or DEAE-mediated transfection.[50] It can be used for both transient and stable expression experiments. Electroporation consists of the application of brief (20–100 msec) electric pulses of moderate strength (250–750 V/cm) to suspensions of cells. It causes the formation of small pores in the plasma membrane that allow the penetration of the DNA. In contrast to the calcium phosphate method, electroporation generally leads to the integration of only one copy of the foreign DNA in the host cell genome. Although this feature is advantageous in gene targeting experiments, it limits the level of expression one can obtain in stably transfected cell lines.

There are several other methods of transfection. The encapsulation of DNA in liposomes[51] is a method that generally does not work significantly better than calcium phosphate or DEAE-dextran. However, various synthetic cationic lipids may prove to be more efficient than and as widely applicable as the methods we have described above.[52] Polybrene-mediated transfection[53] is generally not more efficient than other methods for stable introduction of DNA into cells. Protoplast fusion[54] cannot be used for cotransfection of two independent DNAs.

Transient Expression in 293-T Cells

Transient expression systems have two major requirements: the target host cell line should be highly transfectable and constitutive elements of the expression vector should allow high levels of expression in this spe-

[49] J. H. McCutchan and J. S. Pagano, *J. Natl. Cancer Inst. (U.S.)* **41**, 351 (1968).
[50] E. Neumann, M. Schaefer-Ridder, Y. Wang, and P. H. Hofschneider, *EMBO J.* **1**, 841 (1982).
[51] R. J. Mannino and S. Gould-Fogerite, *BioTechniques* **6**, 682 (1988).
[52] P. L. Felgner, T. R. Gadek, M. Holm, R. Roman, H. W. Chan, M. Wenz, J. P. Northrop, G. M. Ringold, and M. Danielsen, *Proc. Natl. Acad. Sci. U.S.A.* **84**, 7413 (1987).
[53] S. Kaway and M. Nishizawa, *Mol. Cell. Biol.* **4**, 1172 (1984).
[54] W. Schaffner, *Proc. Natl. Acad. Sci. U.S.A.* **77**, 2163 (1980).

cific cell line. Most transient transfection experiments employ the COS cell system. We have found the following method to be superior to the COS system. This method is based on the observation that the CMV immediate-early (IE) gene enhancer–promoter complex is transcriptionally "superactive" in adenovirus-transformed human embryonic cell lines.[55] The strength of the CMV promoter complex in these cells is due to the ability of the E1a protein to bind to and *trans*-activate the IE gene enhancer.[56] The human embryonic kidney 293 cells are a suitable host for transient transfections with CMV promotor-based vectors because they are transformed by the adenovirus and they are highly transfectable by the calcium phosphate coprecipitation method described by Chen and Okayama[57] (transfection efficiency is routinely 50 to 70%). Thus, transfection of CMV-based vectors in 293 cells yields high levels of protein expression (~5 μg/100-mm dish of transfected cells).[58] In addition, the 293 cells have been transfected with the large T antigen to produce the 293-T cell line [G. Nolan (Department of Pharmacology, Stanford University, Stanford, CA), personal communication, 1994]. This cell line permits transient replication of vectors containing an SV40 origin of replication. The method we describe employs a vector containing both a CMV promotor and an SV40 origin of replication, such as pRc-CMV (Invitrogen, La Jolla, CA) or pRK-5 (Genentech, San Francisco, CA), and the 293-T cells.

The 293 cells express several integrins, including $\alpha_1\beta_1$, $\alpha_2\beta_1$, $\alpha_3\beta_1$, $\alpha_5\beta_1$, $\alpha_6\beta_1$, and $\alpha_v\beta_1$.[59] Although this may limit the spectrum of functional studies on integrins that one can perform in these cells, it should be noted that in these cells the level of expression of recombinant integrin subunits is severalfold higher than that of endogenous ones. We routinely use this system followed by immunoprecipitation or immunoblotting analysis to verify that the vectors we have constructed encode proteins of the correct size and antigenic reactivity. Figure 1 shows the results of an immunoblotting analysis performed on 293-T cells transiently transfected with wild-type and mutant β_4 cDNAs. In addition, owing to the high level of expression achieved, this system can be used to generate recombinant fragments of integrin subunits for antibody production.

[55] F. L. Graham, J. Smiley, W. C. Russell, and R. Nairu, *J. Gen. Virol.* **36,** 59 (1977).
[56] C. M. Gorman, D. Gies, G. McCray, and M. Huang, *Virology* **171,** 377 (1989).
[57] C. Chen and H. Okayama, *Mol. Cell. Biol.* **7,** 2745 (1987).
[58] L. R. Paborsky, B. M. Fendly, K. L. Fisher, R. M. Lawn, B. J. Marks, G. McCray, K. M. Tate, G. A. Vehar, and C. M. Gorman, *Protein Eng.* 547 (1990).
[59] S. C. Bodary and J. W. McLean, *J. Biol. Chem.* **265,** 5938 (1990).

M Y23 Y20 Y13 I   G   H WT

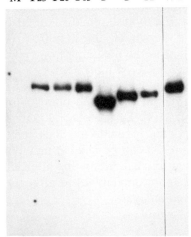

FIG. 1. Transient transfection of 293-T cells with β_4 constructs. The 293-T cells were either mock transfected (M) or transfected with pRc-CMV vectors encoding wild-type (WT) and mutant forms of the integrin β_4 subunit, including single amino acid substitutions (Y23, Y20, and Y13) and internal deletions (I, G, and H). Twenty-four hours after transfection, the cells were extracted and the extracts were subjected to Western blotting with an antibody raised against a 31-mer synthetic peptide reproducing the C terminus of β_4. Bound antibodies were revealed by incubation with ^{125}I-labeled protein A and autoradiography.

Experimental Method

Stock Solutions

BES (2×), pH 6.95: 50 mM N,N-bis(2-hydroxyethyl)-2-aminoethanesulfonic acid, 280 mM NaCl, 1.5 mM Na$_2$HPO$_4$ (filter sterilized)

CaCl$_2$ (0.25 M) (filter sterilized)

Fibronectin (10 μg/ml) in phosphate-buffered saline (PBS), pH 7.4 (filter sterilized)

Glycerol (15%, v/v) in PBS, pH 7.4 (filter sterilized)

Cells. 293-T cells are grown in Dulbecco's modified Eagle's minimum essential medium (DMEM) supplemented with 10% (v/v) calf serum and antibiotics. Because they tend to detach when transfected, they are plated on fibronectin-coated dishes before transfection.

Protocol

1. Coat 100-mm diameter culture dishes with fibronectin (10 μ/ml) for 1 h at room temperature, aspirate solution, and seed 5 × 10^5 293-T cells per dish. Culture the cells for about 10 hr.

2. Precipitate 15–30 μg of vector DNA (the vector needs to be circular when using this coprecipitation method) with ethanol. Resuspend the DNA pellet, which is now sterile, with 0.5 ml of 0.25 M $CaCl_2$. Add 0.5 ml of $2\times$ BES, mix gently, and incubate for 20 min at room temperature. Directly add the suspension to the medium of one plate of cells, swirl, and incubate overnight in a 3% CO_2 atmosphere.

3. To facilitate DNA uptake, subject the cells to a glycerol-shock on the following morning. Gently wash the plates with PBS, and incubate each one with 2 ml of 15% (v/v) glycerol for 3 min. Wash again with PBS and feed with fresh medium. Incubate the cells in a 5% CO_2 atmosphere.

4. If necessary, the cells can be labeled metabolically after an additional 36 hr or subjected to surface labeling after 48 hr. The cells or the medium are harvested approximately 48 hr after the glycerol shock and subjected to the required analyses.

Stable Expression

The generation of cell lines stably expressing integrin subunits from cDNA is the method of choice to establish structure–function relationships. In addition, the strategy of stably expressing integrin cDNAs is often used to explore the effect of a certain integrin on cellular functions, such as adhesion, migration, pericellular matrix assembly, and growth. The analysis of structure–function relationships does not absolutely require that the host cell be negative for the integrin in question because one can resort to species-specific antibodies to distinguish the recombinant molecule from its endogenous counterpart. In contrast, gene transfer experiments examining the effects of a certain integrin on cellular behavior often require either the use of a negative host cell line or the adoption of an overexpression strategy.

Because most mesenchymal and epithelial cell lines synthesize the β_1 subunit in amounts exceeding that of all the α subunits available for association,[29] it is often possible to transfect a certain integrin α subunit that the host cell does not express and obtain expression at the cell surface of hybrid heterodimers containing the recombinant α subunit and endogenous β_1.[12–14] Similarly, one can express an α_v-combining β subunit in cells expressing endogenous α_v[15] or the β_2 subunit in lymphoid cell lines derived from patients affected by leukocyte adhesion deficiency (LAD), a genetic disorder caused by heterogeneous mutations that affect the biosynthesis and association of the β_2 subunit with α subunits.[60] In these approaches

[60] M. L. Hibbs, A. J. Wardlaw, S. A. Stacker, D. C. Anderson, A. Lee, T. M. Roberts, and T. A. Springer, *J. Clin. Invest.* **85,** 674 (1990).

the choice of host cell line is not restricted to the same species of the cDNA to be introduced. In fact, integrin subunits from one vertebrate species generally combine efficiently with partner subunits from another vertebrate species to form functional hybrid heterodimers.[3–10,14,16,17] Figure 2 shows the results of an immunodepletion analysis conducted on CHO cells stably transfected with the human α_5 and β_1 cDNAs. The data indicate that the recombinant α_5 and β_1 subunits combine efficiently with one another but also with endogenous companion subunits. A final approach is to transfect the cDNAs encoding both subunits of a certain integrin if they are both missing in the host cell line. For example, cotransfection of α_{IIb} and β_3 cDNAs (28) or α_L and β_2 cDNAs[61] in COS cells leads to successful expression at the cell surface of fully recombinant hetero-dimers.

We have used the method described below to create clones of CHO and NIH 3T3 cells stably expressing integrins from cDNA. The protocol can be modified for use with other cell lines, but in this case it is advisable to conduct preliminary experiments to determine which enhancer–promotor complex and transfection method works best. In addition, the dose of geneticin (G418) to be used for selection may have to be determined empirically by treating the cells with different doses between 0.1 and 1 mg/ml. The effective dose corresponds to the dose that causes the death of 100% of the cells after 1 week in culture.

Experimental Method

Stock Solutions

G418 (500×): Active G418 (200 mg/ml), 100 mM 4(-2-hydroxyethyl)-1-piperazinoethanesulfonic acid, pH 7.3 (filter sterilize and store at −20°)
BES (2×) pH 6.95: 50 mM N,N-Bis (2-hydroxyethyl)-2-aminoethanesulfonic acid, 280 mM NaCl, 1.5 mM Na$_2$HPO$_4$ (filter sterilized)
CaCl$_2$ (0.25 M) (filter sterilized)
Glycerol (15%, v/v) in PBS, pH 7.4 (filter sterilized)

Cells. CHO cells are grown in Ham's F12 medium supplemented with 10% (v/v) fetal calf serum and antibiotics. NIH 3T3 cells are grown in DMEM plus 10% (v/v) fetal calf serum and antibiotics.

Protocol

1. Seed 5 × 10^5 cells/100-mm diameter dish and culture for about 24 hr in complete medium.

[61] M. L. Hibbs, H. Xu, S. A. Stacker, and T. A. Springer, *Science* **251**, 1611 (1991).

Clone C4   Clone A24

α5   β1   α5   α5   α5   β1   β1   α5   β1   β1   α5   α5   α5
Mab Mab Cyto   Mab Mab Mab Mab Cyto Mab Mab Mab Mab Cyto

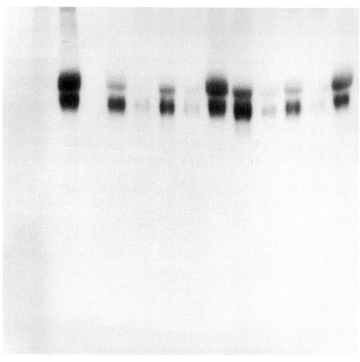

1   2   3   4   5   6   7   8   9   10   11   12   13

FIG. 2. Immunodepletion analysis of CHO cells transfected with human α_5 and β_1 cDNAs. Cells of the control clone C4 and the α_5- and β_1-transfected clone A24[11] were surface labeled with ^{125}I, extracted in detergent buffer, and immunoprecipitated with various antibodies. Clone C4 was immunoprecipitated with the monoclonal antibody (MAb) P1D6 reacting with human α_5 (lane 1), the monoclonal antibody LM534 reacting with human β_1 (lane 2), or the α_5 cyto antiserum reacting with both human and Chinese hamster α_5 (lane 3). Clone A24 was subjected to three sequential rounds of immunoprecipitation with the P1D6 MAb to deplete all heterodimers containing a human α_5 subunit (lane 4, first round; lane 5, third round), followed by three sequential rounds of immunoprecipitation with the LM534 MAb to deplete all heterodimers with a human β_1 subunit (lane 6, first round; lane 7, third round). The sample was finally immunoprecipitated with the α_5 cyto serum (lane 8). Clone A24 was also subjected to sequential immunoprecipitation with the LM534 MAb (lane 9, first round; lane 10, third round), the P1D6 MAb (lane 11, first round; lane 12, third round), and the α_5 cyto serum (lane 13). The immunoprecipitates were separated by SDS–PAGE and analyzed by autoradiography. Note that clone A24 expresses at its surface hybrid heterodimers containing a human β_1 subunit associated with Chinese hamster α_5 (lane 6) and a human α_5 subunit associated with Chinese hamster β_1 (lane 11).

2. Mix 10–15 μg of pRK-5 or pRc-CMV carrying the integrin cDNA insert with 0.5–1 μg of pSV2-neo (or similar vectors) and precipitate the DNA (the vectors need to be circular when using this coprecipitation method) with ethanol. Resuspend the DNA pellet, which is now sterile, with 0.5 ml of 0.25 M CaCl$_2$. Add 0.5 ml of 2x BES, mix gently, and incubate for 20 min at room temperature. Directly add the suspension to the medium of one plate of cells, swirl, and incubate overnight in a 3% CO$_2$ atmosphere. To generate control transfectants follow the same procedure with 10–15 μg of calf thymus DNA and 0.5–1 μg of pSV2-neo.

3. To facilitate DNA uptake, subject the cells to a glycerol shock on the following morning. Gently wash the plates with PBS, and incubate each one with 2 ml of 15% (v/v) glycerol for 3 min. Wash again with PBS, feed with fresh medium, and incubate the plates in a 5% CO$_2$ atmosphere.

4. Culture the plates for about 48 hr until they are subconfluent. Then split in selective medium [complete medium supplemented with G418 (400 μg/ml)]—1 : 6 (one dish), 1 : 9 (two dishes), 1 : 12 (four dishes), and 1 : 18 (five dishes)—and culture for 10–15 days, replacing the medium every 3 days. Plating the cells at different densities will facilitate the subsequent step.

5. Use cloning cylinders to pick at least 20 large, healthy, well-isolated colonies. Subculture each colony in two separate 50-mm diameter dishes. One dish can be used to determine the level of expression at the cell surface by fluorescence-activated cell sorting (FACS) analysis. The other can be expanded for further analyses, such as immunoblotting and immunoprecipitation. Performing a FACS analysis immediately after cloning allows the rapid identification of useful clones. If different integrin cDNAs have been transfected separately, one may want to identify those clones that express similarly high amounts of each transfected recombinant integrin. If a single cDNA has been transfected, then it may be convenient to identify clones expressing different levels of the recombinant integrin. This will allow one to determine if a certain biological effect correlates with the level of expression of the recombinant molecule.

Overexpression Methods

Integrins are abundant components of the cell surface in many cell types. The high level of expression of integrins is likely to reflect their need to function as medium- to low-affinity transmembrane linkers between abundant elements of the extracellular matrix and those equally abundant in the cytoskeleton. It follows that in order to examine some aspects of integrin function by gene transfer methods it may be necessary to achieve high levels of expression. In addition, there is evidence suggesting that integrins must interact with the proper cytoskeletal molecules and activation signals in order to become fully active.[25] Thus, integrins may be fully functional only in the cell types in which they are normally expressed and many gene transfer experiments of integrins may require the use of a host cell that already expresses the integrin to be transfected. In these experiments, therefore, it may be necessary to achieve a level of expression of the recombinant integrin that largely exceeds that of the endogenous one.

In principle, overexpression can be obtained by using a strong promotor, such as the keratin K14 promoter in keratinocytes,[62] or by increasing the number of transcription units integrated in the host cell genome by amplification methods. The most popular amplification method is based on the observation that the gene encoding the enzyme dihydrofolate reductase (dhfr) undergoes amplification in cells exposed to increasing concentrations of methotrexate (MTX).[63,64] The enzyme dhfr catalyzes the biosynthesis of tetrahydrofolate from folate. Folate is a necessary cofactor for the conversion of serine to glycine and the *de novo* biosynthesis of thymidine monophosphate and purines. In addition to glycine, *dhfr*⁻ cells require for survival thymidine and either hypoxanthine or adenosine, because these molecules can be utilized as substrates by the biosynthetic salvage pathway for pyrimidines and purines (Fig. 3). Thus, *dhfr*⁻ cells can be transfected with a *dhfr* gene and the *dhfr*⁺ transfectants can be selected in medium that does not contain nucleotides. Subsequent exposure of the cells to increasing concentrations of MTX, a competitive inhibitor of dhfr, will create the selective pressure needed to isolate those rare cells in which the *dhfr* gene has undergone spontaneous amplification. Genes that have been cotransfected with *dhfr* and have integrated in close proximity with it will also undergo amplification when the cells are treated with MTX. Because clones can be obtained that carry up to 1000 copies of the *dhfr* gene and the transcriptional unit linked to it, this method can be used to obtain high levels of expression.[65]

The dhfr amplification method is commonly used with CHO cells that have been mutated to become *dhfr*⁻.[44] However, this method can be extended to cell types containing an endogenous wild-type *dhfr* gene by using a mutant *dhfr* gene encoding an enzyme that has a lower affinity for MTX and therefore is inhibited by this drug less than the wild-type protein.[66] In addition, there are dominant selective markers such as adeno-

---

[62] A. Leask, M. Rosenberg, R. Vassar, and E. Fuchs, *Genes Dev.* **4**, 1985 (1990).
[63] R. T. Schimke, *Cell (Cambridge, Mass.)* **37**, 705 (1984).
[64] G. R. Stark and G. M. Wahl, *Annu. Rev. Biochem.* **53**, 447 (1984).
[65] R. J. Kaufman, *Genet. Eng.* **9**, 155 (1987).
[66] C. C. Simonsen and A. D. Levinson, *Proc. Natl. Acad. Sci. U.S.A.* **80**, 2495 (1983).

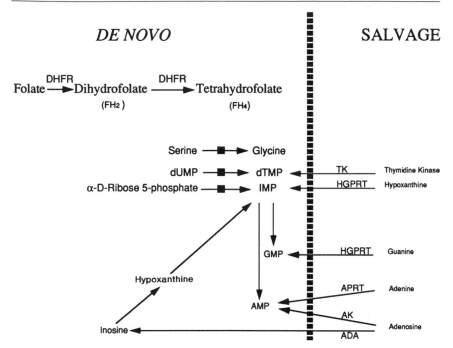

FIG. 3. Essential steps of the biosynthesis of pyrimidines and purines. A number of enzymes involved in the *de novo* biosynthetic pathway (solid boxes) require tetrahydrofolate (FH₄) as a cofactor. This is synthesized from folate by the enzyme dihydrofolate reductase (dhfr). The enzymes involved in the salvage pathway include thymidine kinase (TK), hypoxanthine–guanine phosphoribosyltransferase (HGPRT), adenine Phosphoribosyltransferase (APRT), adenosine kinase (AK), and adenosine deaminase (ADA).

sine deaminase[67] and asparagine synthetase[68] that can also be used for amplification in different cell types.

A protocol that has been used to generate CHO clones overexpressing the $\alpha_5\beta_1$ integrin is detailed below. In this protocol, the *dhfr⁻* CHO cells are transfected simultaneously with a plasmid containing the *dhfr* gene and with plasmids encoding the human α_5 and the β_1 subunits. After selection in nucleotide-free medium, the primary clones that express comparable amounts of the two integrin subunits are identified and subjected to amplification. At each amplification step clones expressing the highest

[67] R. J. Kaufman, P. Murtha, D. E. Ingolia, C.-Y. Yeung, and R. E. Kellems, *Proc. Natl. Acad. Sci. U.S.A.* **83**, 3136 (1986).
[68] I. Andrulis, J. Chen, and P. Ray, *Mol. Cell. Biol.* **7**, 2435 (1987).

levels of both subunits are identified by immunoblotting (Fig. 4) or FACS analysis. Cotrasfection of the two integrin subunits is required if one wishes to attain a high level of expression without perturbing the levels of endogenous integrins in CHO cells.

Experimental Method

Stock Solutions

MTX (5 mM): Methotrexate (2.25 mg/ml) in α^- minimal essential medium (MEM), pH 7.3 (filter sterilize and store at $-20°$)

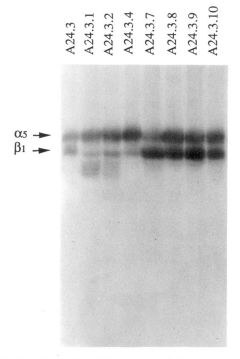

FIG. 4. Coamplification of transfected human α_5 and β_1 sequences in CHO cells. Cells of the indicated clones[11] were extracted in detergent buffer. Aliquots of the extracts containing 200 μg of total protein were separated by SDS–PAGE and subjected to immunoblotting with a mixture of two cytoplasmic peptide antisera: one reacting with the α_5 subunit and the other with the β_1 subunit. Bound antibodies were revealed by incubation with alkaline phosphatase-labeled protein A. Clone A24.3 has undergone one cycle of methatrexate-induced amplification. The other clones are derivatives of A24.3 and were subjected to one additional amplification step. Note that clones A24.3.1, A24.3.2, and A24.3.4 express amplified levels of α_5, but not β_1, and conversely clone A24.3.7 expresses amplified levels of β_1, but not α_5. Clones A24.3.8, A24.3.9, and A24.3.10 show successful coamplification of α_5 and β_1 sequences.

BES (2×), pH 6.95: 50 mM N,N-bis(2-hydroxyethyl)-2-aminoethanesul-
fonic acid, 280 mM NaCl, 1.5 mM Na$_2$HPO$_4$ (filter sterilized)
CaCl$_2$ (0.25 M) (filter sterilized)
Glycerol (15%, v/v) in PBS, pH 7.4 (filter sterilized)
Dialyzed fetal calf serum

Cells. dhfr$^-$ CHO cells (clone DG-44 or DUKX-B11) are grown in
α^- MEM supplemented with hypoxanthine (10 μg/ml), thymidine (10 μg/
ml), 10% (v/v) fetal calf serum (FCS), and antibiotics.

Protocol

1. Transfect the CHO-DG-44 or the CHO-DUKX-B11 simultaneously
with 10 μg of the α subunit cDNA plasmid, 10 μg of the β subunit cDNA
plasmid, and 1 μg of a plasmid encoding the dhfr enzyme, such as pSV2-
dhfr. To generate control cell lines transfect 1 μg of the plasmid encoding
dhfr. Use the Chen and Okayama calcium phosphate method and the
glycerol shock as previously described.

2. Two days after transfection split the cells in selective medium [α^-
MEM supplemented with 10% (v/v) dialyzed FCS and antibiotics, but
without nucleotides)—1 : 6 (one dish), 1 : 9 (two dishes), 1 : 12 (four dishes),
and 1 : 18 (five dishes). Culture for 10–15 days, replacing the medium
every 3 days.

3. Use cloning cylinders to pick at least 20 large, healthy, well-isolated
colonies. Avoid colonies containing many large spread-out cells because
this morphology is typical of cells synthesizing insufficient amounts of
dhfr. Subculture each colony in two separate 50-mm diameter dishes. One
dish can be used to determine the level of expression of both subunits by
FACS analysis or immunoblotting. The other can be expanded for further
analyses, such as immunoprecipitation, and for the subsequent steps of
the amplification protocol. Performing a FACS analysis immediately after
cloning allows the rapid identification of clones that express similarly high
cell surface levels of the two recombinant subunits.

4. Subject to the amplification protocol at least six distinct clones
expressing balanced levels of the α and β subunit. This is necessary
because any of the introduced genes can undergo rearrangement or mutation during amplification. Thus, a particular clone might lose expression
of one or the other integrin subunit while maintaining the expression of
dhfr or vice versa (see, for example, Fig. 4). Split a confluent dish of each
primary expressor and at least two control *dhfr*-transfected clones 1 : 15
into α^- MEM–10% (v/v) dialyzed FCS supplemented with 0.02 μM (one
plate), 0.04 μM (two plates), or 0.08 μM MTX (four plates). Replace the
medium every 3 days.

5. After about 12 days, pick healthy colonies from each dish. There
will be fewer colonies in the dishes containing the higher concentrations

of MTX, but these colonies are more likely to express high levels of the integrin. Therefore, a good general rule is to pick as many colonies as possible from the higher concentrations of MTX and proportionally less colonies from the other dishes. Select high-level integrin expressors as described in step 3. Also pick control clones growing at the same concentration of MTX. At this and at any of the subsequent steps, freeze an aliquot of each expressor line and corresponding control.

6. Repeat the procedure with concentrations of MTX 4-, 8-, and 16-fold higher than that in which the colonies are growing, and then again until overexpressor clones are isolated. Each round should take about a month, and three to four rounds should be sufficient to obtain clones with highly amplified sequences growing in 10–20 μM MTX.

Acknowledgments

Research in the authors' laboratory is supported by grants from the NIH and the American Cancer Society. F. G. Giancotti is a Fellow of the Lucille P. Markey Charitable Trust. L. Spinardi is supported by a postdoctoral fellowship of the American-Italian Foundation for Cancer Research (AIFCR). F. Mainiero is a postdoctoral fellow of the Associazione Italiana per la Ricerca sul Cancro (AIRC).

[16] Immunohistochemical Techniques to Study the Extracellular Matrix and Its Receptors

By BENJAMIN G. HOFFSTROM and ELIZABETH A. WAYNER

Classic histochemistry relies on the use of tinctorial stains to visualize components of cells and tissues. Although such stains can be used to identify anatomical structures (basement membranes and collagen fibrils) or groups of specific tissue components (amyloids, proteoglycans, and mucins), individual molecular components or cell surface receptors cannot be identified. Interest in the technique known as "immunohistochemistry" has increased primarily because utilizing antigen–antibody reactions allows for the specific visualization of previously undetectable cell or tissue components. An array of cell adhesion receptors and extracellular matrix (ECM) molecules can now be identified by monoclonal and polyclonal antibodies, which constitute a new class of histochemical stains with superior specificity and usefulness.

The goal of all immunohistochemical procedures is the identification of specific antigens *in situ* by antibodies that are visualized via the use of

Copyright © 1994 by Academic Press, Inc.
All rights of reproduction in any form reserved.

reporter molecules such as fluorochromes or enzymes. Although it might appear as though there is an endless array of protocols available for the immunohistochemical analysis of the ECM and its receptors (integrins), most techniques are variations of the antibody–secondary antibody, enzyme–anti-enzyme, and the avidin–biotin methods.[1] The present discussion is not meant to be a comprehensive review of these techniques; instead, the purpose of this chapter is to familiarize investigators with conditions that permit the precise light microscopic localization of adhesion molecules in tissue. Therefore, we describe the immunohistochemical (IHC) and immunofluorescence (IF) methods used in our laboratory to identify extracellular matrix molecules and integrin receptors in frozen (unfixed) human tissue specimens. Although we do not provide any specific examples, we also describe some common techniques used for immunohistochemical staining of formalin-fixed and paraffin-embedded tissue specimens.

Principles of Immunohistochemical Staining Techniques

Antigens and Antibodies

An antigen is a substance that, if foreign to the host, will stimulate the production of an antibody directed to each antigenic determinant or epitope. The specific recognition by an antibody of a tissue-bound antigen followed by visualization of the antibody is the basis of all IHC or IF reactions. The successful implementation of any IHC technique absolutely requires that the antigenic epitope recognized by the primary antibody be preserved through the tissue fixation, embedding, sectioning, and staining procedures. Chemical fixation with aldehydes (neutral-buffered formalin) followed by paraffin embedding dramatically reduces the antigenicity of tissue either as the result of epitope masking (cross-linking) or the destruction of protein conformation (denaturation). There are techniques available that can sometimes be used to unmask epitopes that are blocked by chemical cross-linking (see below). These techniques, however, will not restore antigenicity that is lost by denaturation. It is for this reason that many monoclonal and polyclonal antibodies known to react with adhesion molecules or integrin receptors cannot be used to stain formalin-fixed and paraffin-embedded tissue sections. The successful identification of such antigens using IHC or IF techniques is best accomplished with frozen

---

[1] J. M. Elias, "Immunohistopathology: A Practical Approach to Diagnosis." ASCP Press, Chicago, 1990.

sections derived from tissue that is preserved by quick freezing rather than by chemical fixation.

Monoclonal versus Polyclonal Antibodies. Monoclonal antibodies are produced by fusing lymphocytes derived from an immune animal with a nonsecreting immortal plasma cell line.[2] The resulting hybrid cell lines (or hybridomas) are screened and the culture secreting the desired antibody is cloned by limiting dilution. The end product is an immortal cell line that constitutively secretes a monoclonal antibody specific for an epitope on the immunizing antigen. Monoclonal antibodies are infinitely preferable to polyclonal antibodies because they are high-affinity and extremely specific reagents that are in continuous supply and, if used in purified form, exhibit background-free staining. Culture supernatant derived from an immuno-globulin-secreting hybridoma cell line can contain from 1 to 50 μg/ml whereas ascites fluid prepared from an animal inoculated with a hybridoma cell line can contain from 1 to 10 mg of specific antibody per milliliter. It is preferable to use either diluted culture supernatant or purified antibody because ascites fluid also contains irrelevant mouse immunoglobulins. Such contaminating mouse immunoglobulins will cross-react with anti-mouse reporter conjugates, causing nonspecific background staining. Although monoclonal antibodies are uniquely superior tools for a variety of tasks, including IHC and IF, they do have some negative attributes. For example, monoclonal antibodies are poor tools for direct IHC or IF reactions because conjugation with enzymes, biotin, or fluorochromes can compromise the antigen-binding site, resulting in a loss of specific activity. In addition, a major drawback to monoclonal antibodies is that chemical fixation has a negative impact on tissue antigenicity. If the single epitope recognized by the antibody is irretrievably lost, a false-negative result will be obtained. For this reason, monoclonal antibodies should be tested for utility in IHC procedures on sections derived from unfixed and quick-frozen tissue.

Unfractionated polyclonal antisera also have serious flaws such as high background, epitope cross-reactivity, lack of homogeneity between batches, and limited supply. However, affinity-purified polyclonal antibodies are superior reagents and may even be more useful than monoclonal antibodies, especially when used to detect antigens in formalin-fixed tissue. Affinity-purified polyclonal antibodies are made by isolating specific antibodies from crude antiserum on an affinity column containing antigen-coupled resin. We have had a great deal of success in using this technique to obtain affinity-purified antibodies from rabbits immunized with fibronectin, type VI collagen, and laminin. Such preparations are highly spe-

[2] G. Kohler and L. Milstein, *Nature (London)* **256,** 495 (1975).

cific, cause minimal background, and can be used in an highly dilute form (1 : 500–1 : 1000). The great advantage of a clean polyclonal antibody reagent is that multiple antigenic determinants are recognized, allowing for a greater chance of detecting epitopes preserved through chemical fixation and paraffin embedding.

Antibodies for use in IHC reactions should be stored in the cold (with preservatives) or aliquoted and frozen at $-80°$. Multiple freeze-thaw cycles should be avoided especially with purified and conjugated reagents. Purified antibody preparations should be centrifuged at 100,000 g (30 min at 4°) immediately before use to remove aggregates that can contribute background artifacts.

Antibody Structure. An antibody is composed of two heavy and two light chains. It is the heavy chain that determines the subclass of a particular immunoglobulin (Ig) molecule; IgG molecules, for example, possess a γ-type heavy chain. Many commercially available monoclonal antibodies directed to extracellular matrix molecules and integrin receptors are mouse IgGs. Therefore, in indirect staining procedures secondary antibodies specific for mouse γ chain are used. This reduces background staining due to the cross-reactivity of the reporter construct with tissue-bound immunoglobulin of host origin. The Fc portion of immunoglobulin molecules contains the constant region and is capable of interacting with Fc receptors on some cells in tissue. Fc–receptor interaction can contribute to nonspecific background staining in IHC reactions, especially in some tissues rich in Fc receptor-positive cells (tonsil, lymph node, and bone marrow). Blocking tissue-bound Fc receptors, therefore, is an important consideration in IHC and IF staining procedures. This can be achieved by the use of purified and heat-aggregated immunoglobulin (10 μg/ml) derived from the same species as the reporter conjugate. For tissues that do not contain Fc receptor-positive cells (such as skin), we simply block with a 5–10% solution of nonimmune serum (i.e., goat) derived from the same species as the secondary antibody reagent (i.e., goat anti-mouse Ig). The hypervariable regions, which contain the antigen-binding sites, are located on the Fab portions of the immunoglobulin molecule. IgG molecules are bivalent and have the ability to bind two antigen molecules, one at each Fab site. Preparation of antibodies or reporter constructs can take advantage of this and (Fab')$_2$ fragments can be used. Use of primary or secondary (Fab')$_2$ fragments greatly reduces background staining in IHC or IF procedures. (Fab')$_2$ fragments, however, may exhibit reduced antigen-binding capacity when compared to intact immunoglobulin.

Antibody Concentration. In any antibody–antigen reaction, the zone of relatively high antibody concentration within which no reaction occurs is referred to as a *prozone*. The dynamics of prozone effects in IHC reactions are not understood; however, as the antibody concentration is lowered below the prozone, the reaction can take place. Highly concentrated antibody solutions increase the risk of prozone effects in tissues with high antigen density.[3] Antigens may also be so densely packed that antigen–antibody binding is subject to steric hindrance.[4] Effects due to prozone or steric hindrance can be overcome by titrating the primary antibody.

In an IHC reaction, two criteria are evaluated to determine the optimal concentration of antibody: specific antigen staining and nonspecific background or artifactual staining. With a direct staining technique, only one antibody need be titrated to evaluate specific staining and background. However, with an indirect technique [e.g., the peroxidase–anti-peroxidase (PAP) method] there are at least two and possibly three antibodies that need to be evaluated. The optimal dilution for each antibody can be determined using a checkerboard titration analysis shown in the tabulation below.[5]

Concentration of primary antibody	Concentration of secondary antibody			
	1 : 50	1 : 100	1 : 500	1 : 1000
1 : 2				
1 : 10				
1 : 50				
1 : 100				

Staining results obtained with titrations of each reagent are tabulated relative to each other. The specific signal for each dilution is then evaluated relative to background. Many IHC staining kits are commercially available [Dako (Santa Barbara, CA), Vector Laboratories (Burlingame, CA)] that are of high quality and produce accurate and reproducible results. They vary little from lot to lot, obviating the need to titrate secondary antibodies or reporter complexes. However, we still titrate each new primary anti-

[3] E. Linder and A. Miettinen, *Scand. J. Immunol.* **5,** 514 (1976).
[4] F. Berkenbosch and F. J. H. Tilders, *Neuroscience* **23,** 823 (1987).
[5] J. A. Bourne, "Handbook of Immunoperoxidase Staining Methods." DAKO Corporation, Santa Barbara, CA, 1983.

body and our reagents for IF studies because we have found that fluorescein isothiocyanate (FITC)- or rhodamine-conjugated antibodies can vary greatly from lot to lot and from vendor to vendor.

Factors Affecting Antigen–Antibody Reactions. Several factors can affect the formation of antigen–antibody complexes in IHC or IF staining reactions. These include time, temperature, accessibility of tissue-bound antigens, pH, and the ionic strength of the buffers used. It is generally recognized that the intensity of staining can be enhanced by increasing the length of incubation with the primary antibody and by increasing the temperature to 37°. Raising the temperature or increasing the length of secondary antibody incubation usually also increases the nonspecific background staining. We routinely obtain clean, consistent results with short incubation times (30–60 min) at ambient temperature. The intensity of IHC or IF reactions can be increased by performing the whole procedure twice, from start to finish, using short incubations (30 min) at ambient temperature. This enhances IF reactions extremely well. To enhance IHC reactions, the incubation time with the reporter complex can be increased or methods can be used to intensify the chromogen reaction (see below). Permeabilization of tissue with nonionic detergents such as Triton X-100 (0.5%, v/v) can increase the accessibility of antigens by promoting the penetration of antibodies or reporter complexes [e.g., avidin–biotin complex (ABC) reagent]. This is particularly useful to enhance staining of collagens or basement membrane components or when using chemically fixed and paraffin embedded tissue. Nonspecific background binding of antibodies can be reduced by including Triton X-100 (0.5%, v/v) or 0.4 M NaCl in IHC wash buffers. It is essential, however, that if frozen tissue is used, such buffers be applied to cryostat sections after light fixation in acetone or 2% (v/v) paraformaldehyde.

The choice of a buffer is an important consideration in IHC and IF reactions. Most antigen–antibody reactions are stable at pH 7.4 in isotonic buffer (0.15 M NaCl). Highly acidic or basic buffers should be avoided when using monoclonal antibodies. Monoclonal antibody culture supernatants can be supplemented with 50 mM Tris to avoid dramatic changes in pH during primary antibody incubation on the bench at ambient temperature. Furthermore, because phosphate can inhibit alkaline phosphatase (AP) activity,[1] Tris-buffered saline should be used in AP-catalyzed IHC reactions. An important consideration when choosing a buffer for peroxidase-catalyzed IHC reactions is that it should not contain sodium azide. At high concentration (0.02%, w/v), sodium azide can inhibit peroxidase activity, and if used as a preservative in antibody solutions, the sections must be thoroughly washed before addition of peroxidase and chromogen. Thimerosal (0.05%, w/v) can be substituted for sodium azide.

Staining Methods

Enzyme-Catalyzed Reactions. A widely used immunohistochemical technique makes use of the enzyme horseradish peroxidase and the chromogen diaminobenzidine (DAB). In such IHC reactions, the enzyme (peroxidase) forms a complex with its substrate (H_2O_2), which oxidizes the chromogen (DAB) to produce the end product of the reaction, a colored molecule and water.[5] A single peroxidase molecule can catalyze many H_2O_2 molecules into colored end products. This ability to recycle peroxidase molecules contributes to the great sensitivity of IHC as opposed to IF staining reactions. There are several chromogens that can act as electron donors in peroxidase-catalyzed IHC reactions and many are commercially available. The choice of a chromogen is determined by several factors. Diaminobenzene produces a stable reddish brown reaction product which is not soluble in organic solvents and peroxidase-stained slides developed in DAB can be permanently mounted with a synthetic resin such as Permount. Some reaction products are soluble in organic solvents [e.g., 3-amino-9-ethylcarbazole (AEC; Sigma, St. Louis, MO), see below] and cannot be used with permanent mounting media. Some colored end products are more useful than others in two-color staining reactions (blue with red or red with black, as opposed to red with brown) and some chromogens (such as DAB) are known carcinogens and therefore might not be used owing to laboratory safety considerations. Other enzymes such as alkaline phosphatase (AP) and glucose oxidase can also be used as reporter molecules in IHC reactions. Alkaline phosphatase activity is demonstrated using naphthol-AS phosphate as the substrate with several chromogens that produce blue or red colors. All of the AP-catalyzed reaction products are soluble in organic solvents and require aqueous mounting medium (see below).

Immunofluorescence Reactions. Compounds known as fluorochromes emit light of a defined wavelength when excited by light of a slightly shorter wavelength. This process is referred to as *fluorescence*. Immunofluorescence is based on the direct or indirect binding of a fluorochrome-labeled antibody to an antigen in tissue. Different fluorochromes (fluorescein or rhodamine isothiocyanate) can be excited to emit light of different wavelengths. Therefore, two-color immunofluorescence studies can be carried out with a microscope equipped with a mercury vapor lamp (50–100 W) and the appropriate exciter–barrier filter combinations. Fluorescent decay of excited molecules is extremely rapid. Therefore, immunofluorescence slide preparations should be mounted in a glycerol solution containing substances such as *p*-phenylenediamine or 1,4-diazabicyclo[2.2.2]octane (DABCO; Sigma), which are used to prevent quenching of emitted light during microscopic analysis and photography. Immuno-fluorescence slides, if sealed and mounted properly and stored at 4°, can be stable for up to 6 months.

The following is a discussion of the immunohistochemical techniques used in our laboratory to detect tissue-bound adhesion molecules and integrin receptors. The most sensitive detection methods (PAP and ABC) amplify the specific signal over background by increasing the number of reporter molecules per primary antibody molecule.

Labeled Primary Method. (Please see Fig. 1A.) In this technique an enzyme or fluorochrome is conjugated to a purified primary antibody, which is then used to directly detect a tissue-bound antigen (A). This

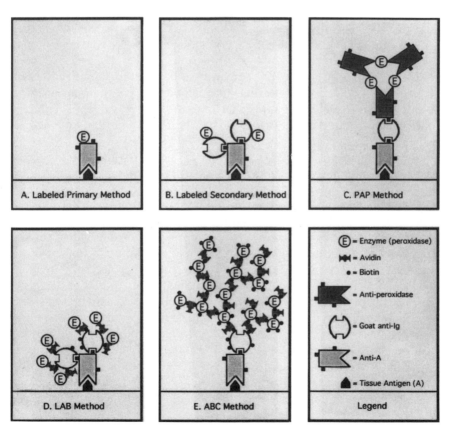

FIG. 1. Schematic illustration of IHC staining methods used to detect tissue-bound antigen, A. Ⓔ indicates conjugation with enzyme. In panels A, B, and D a fluorochrome (fluorescein or rhodamine) can be substituted for the enzyme. In panel C the enzyme peroxidase can be replaced with alkaline phosphatase or glucose oxidase.

method, although simple and fast, has several flaws. The conjugation procedure can often significantly compromise the affinity of the antibody for the tissue-bound antigen. In addition, this procedure is not as sensitive as some of the indirect methods, which can introduce several reporter molecules per primary antibody molecule. Furthermore, for each antigen of interest, a purified antibody conjugate must be prepared or purchased. This is not always feasible when multiple tissue antigens are examined simultaneously.

Labeled Secondary Method. (Please see Fig. 1B.) All other IHC or IF protocols are variations of the indirect staining method, in which the initial antibody preparation applied to the tissue section contains an unlabeled primary antibody (e.g., mouse IgG) that reacts with a tissue-bound antigen (A). The bound unlabeled primary antibody is detected with an enzyme or fluorochrome-conjugated secondary antibody raised in another species and specific for the primary host species (e.g., goat anti-mouse IgG). The indirect method has several advantages over the direct method. A single conjugated secondary reagent can be used to visualize several primary antibodies with differing specificity. Indirect staining is more sensitive than direct staining, primarily because the polyvalent secondary reagent can detect multiple sites on the Fc and Fab portions of the primary antibody.

Peroxidase–Anti-Peroxidase Method. The peroxidase–anti-peroxidase (PAP) method was originally developed by Sternberger[6,7] and utilizes three reagents: primary unlabeled antibody directed to tissue antigen A (e.g., rabbit anti-A), secondary antibody specific for the primary host species (e.g., goat anti-rabbit), and a preformed PAP complex that is composed of the enzyme peroxidase and an antibody raised against peroxidase in the primary host species (e.g, peroxidase rabbit anti-peroxidase) (Fig. 1C). The secondary antibody is referred to as a "link" reagent because it can bind both the primary antibody and the PAP complex. Absence of conjugated antibodies in this method results in increased sensitivity when compared to the indirect method using conjugated secondary reagents. This is especially useful in formalin-fixed and paraffin-embedded tissue sections in which much of the tissue antigenicity has been destroyed. Universal PAP kit systems for mouse, goat, and rabbit are available from Dako. In addition, an alkaline phosphatase–anti-alkaline phosphatase (APAAP) complex can be used for tissues high in endogenous peroxidase

[6] L. A. Sternberger, P. H. Hardy, J. J. Cuculis, and H. G. Meyer, *J. Histochem. Cytochem.* **18,** 315 (1970).
[7] L. A. Sternberger and N. H. Sternberger, *J. Histochem. Cytochem.* **34,** 599 (1986).

activity. An APAAP kit system specific for mouse antibodies has been developed by Dako.

Avidin–Biotin (LAB or ABC) Methods. These methods rely on the high affinity with which avidin binds four molecules of biotin ($K_a = 10^{15} M^{-1}$). Activated esters of biotin will react with primary amines (lysine residues) to form stable amide bonds and can therefore be used to biotinylate lysine-containing antibody molecules. Biotinylation of primary antibodies can reduce antigen-binding capacity. This problem is circumvented by biotinylating a secondary antibody specific for the primary. If biotinylated antibodies are detected with enzyme or fluorochrome-conjugated avidin, the procedure is referred to as labeled-avidin biotin (or LAB; see Fig. 1D) and was first developed for use in IHC reactions by Guedson.[8] The LAB method is not dependent on the use of a large macromolecular complex (such as ABC) and is therefore not affected by steric hindrance in sites of high antigen density or by chemical cross-linking of antigens in formalin-fixed paraffin sections.

The avidin–biotin complex or ABC method (Fig. 1E), developed by Hsu and colleagues,[9] also requires the use of three reagents: unlabeled primary, biotinylated secondary antibody specific for the primary, and a preformed complex of peroxidase-conjugated biotin and avidin (ABC). The ABC reagent is prepared fresh and allowed to form at ambient temperature before being applied to the tissue sections containing unlabeled primary and biotinylated secondary antibodies. The great sensitivity of this detection method relies on the formation of the macromolecular ABC complex, which contains multiple biotinylated enzyme molecules per avidin molecule. This preformed complex then reacts with free biotin residues on the secondary antibody, thereby introducing multiple enzyme molecules per primary antibody. Like the PAP method, the ABC method is useful for formalin-fixed and paraffin-embedded tissue specimens. ABC-peroxidase, ABC-AP, and ABC-GO kits specific for immunoglobulins derived from a variety of species are available from Vector. ABC kits can also be used with biotinylated primary antibodies.

Immunohistochemistry or Immunofluorescence. Immunohistochemistry offers results comparable to IF and enzyme-catalyzed IHC techniques have the advantage that they produce stable end products that can be viewed with an ordinary light microscope. Furthermore, owing to the development of high-quality PAP and ABC kits from vendors such as Dako and Vector, IHC is now easily accomplished and may be more

[8] J. I. Guedson, T. Ternyck, and S. Avrameas, *J. Histochem. Cytochem.* **27**, 1131 (1979).
[9] S. M. Hsu, L. Raine, and H. Fanger, *J. Histochem. Cytochem.* **29**, 577 (1981).

sensitive than IF. Some tissues may exhibit higher backgrounds with IF or IHC as the result of autofluorescence or endogenous enzyme activity. Another consideration might be presentation of IHC or IF data for publication. Although two- and even three-color IHC reactions can now be carried out with increased sensitivity and reliability, two-color immunofluorescence in our laboratory is still the method of choice when such data will be published in black and white (e.g., Fig. 5).

Procedure for Immunohistochemical Staining of Frozen Sections

Obtaining and Storing Fresh Unfixed Tissue Samples

The quality of the tissue specimens will ultimately determine the quality of the results obtained with any IHC or IF procedure carried out on cryostat-cut frozen sections. Unfixed tissue destined for IHC studies should be fresh and quick frozen in a matrix compound such as O.C.T. and embedded in a small plastic container referred to as a cryomold for storage at −80°. This adequately preserves the tissue and prepares it for sectioning in a cryostat (a cooled chamber containing a microtome). The following procedures for handling surgical specimens yield high-quality IHC data on frozen sections.

Fresh unfixed surgical specimens should not be immersed in aqueous solutions and should be embedded in a matrix compound (e.g., O.C.T.) as soon as possible for optimal preservation of antigenicity and morphology. If immediate processing is not possible (within 30 min), tissue specimens can be stored for up to 24 hr (e.g., express mail deliveries) in a sealed container at 4° on a bed of gauze moistened with *sterile* phosphate-buffered saline (PBS) or saline. Tissue should not be soaked for long periods of time (greater than a few minutes) in aqueous medium prior to mounting because retained water may, in addition to directly contributing to freezing artifacts, inhibit the infiltration and performance of the matrix compound (O.C.T.). In the event that tissue specimens are transported in aqueous solution they should be gently blotted dry on Whatman (Clifton, NJ) filter paper and then embedded in O.C.T. Alternatively, small tissue samples (4- to 6-mm punch biopsies) can be placed into a piece of aluminum foil or a Nunc (Roskilde, Denmark) freezer vial (cap loosened) and quickly frozen on dry ice or in liquid nitrogen. Such tissue specimens can be safely transported on dry ice or stored in a −80° freezer for 24 hr until used. Large surgical specimens (greater than 8 mm) should not be frozen without embedding in O.C.T. because morphological artifacts will occur owing to the formation of ice crystals as the tissue freezes. Tissue specimens frozen on dry ice or in liquid nitrogen and subsequently embedded in O.C.T. at ambient temperature may exhibit freeze-thaw artifacts, particularly at the edges of the specimen. Tissue specimens should never be allowed to freeze slowly (i.e., at $-20°$) and should never be thawed completely and refrozen. If a frozen specimen is too large to mount, it should be cut on a slab of dry ice with a cold stainless steel knife, embedded in O.C.T. within a cryomold, and then subsequently refrozen as quickly as possible. Although freezing is the preferred method for preserving fresh unfixed tissue, some investigators use an ammonium sulfate-based medium containing protease inhibitors developed by Michel[10] for storage or transportation of specimens. However, it has been our experience that transporting tissue on moist gauze at $4°$ (no more than 24 hr) or frozen on dry ice (even without embedding in O.C.T.) preserves antigenicity and morphology better than fixation in Michel's medium.

Freezing and Mounting Tissue Blocks for Cryostat Sectioning

Once the tissue samples have been obtained and have been rinsed free of blood and blotted dry they are ready for mounting. Cryomolds in various sizes are available from Tissue-Tek. A small piece of tag board labeled with the specimen identification number can be placed into a corner of the cryomold for easy identification of the frozen block. The sample should be oriented in a pool of O.C.T. Tissue-Tek #4583) in the bottom of the cryomold. With some tissues such as skin, the alignment of the specimen is critical to avoid cutting artifacts and for the interpretation of IHC data. In general, basement membranes should be oriented perpendicular to the plane of the microtome knife to minimize shear stress and separation artifacts during cryostat sectioning. The specimen should be covered with O.C.T. (no air bubbles) and frozen by floating the specimen in isopentane supercooled in liquid nitrogen or a dry ice–acetone slush. This is accomplished via the use of two stainless steel beakers. Isopentane is cooled in one beaker by placing it into a larger beaker filled half-full with liquid nitrogen. Using a large forceps the cryomold containing the specimen and O.C.T. is placed into the supercooled isopentane. The block should freeze within 1 min. The frozen tissue block embedded in O.C.T. is placed into a zip-lock plastic bag and stored at $-80°$ until use. We have successfully stored tissue specimens for several years using this technique. Use of supercooled isopentane (or a dry ice–acetone bath) freezes the O.C.T. while preserving the plasticity of the matrix; freezing O.C.T.-embedded blocks directly in liquid nitrogen should be avoided.

[10] B. Michel, Y. Milner, and K. David, *J. Invest. Dermatol.* **59**, 499 (1973).

Microscope Slide Preparation

Microscope slides should be treated to optimize adherence of frozen or paraffin tissue sections during the IHC procedure. There are several methods available for coating slides with either a charged adhesive substance [chrome alum gelatin or poly(L-lysine)] or an aminosilane group.[11] Commercially coated slides are also available. However, such slides are expensive and we have found that aminosilane coating is simple, inexpensive, and offers optimal tissue adherence with minimal background cross-reactivity (some protein coatings such as gelatin can cross-react with matrix-specific monoclonal antibodies). Aminosilane coating is accomplished by first cleaning glass microscope slides (in slide racks) in two changes of 100% ethanol (10 min each). Dry, ethanol-cleaned slides are soaked in 2% (v/v) 3-aminopropyltriethoxysilane (A-3648; Sigma) in high-performance liquid chromatography (HPLC)-grade dry acetone for 2 min at ambient temperature, followed by two passes in dry acetone and three changes of distilled H_2O. Aminosilane-coated slides are air or oven (60°) dried and stored in dust-free slide boxes. Slides prepared in this way are stable for at least 6 weeks.

Cryostat Sectioning of Frozen Tissue

A cryostat, which consists of a microtome within a cooled chamber, is used for cutting frozen, unfixed tissue. It is essential that the cryostat be adjusted properly to obtain high-quality frozen sections. The microtome blade angle is critical for obtaining uniform thin sections (2–4 μm) and should be set such that the clearance angle (or the angle at which the section is cut) is 7° from the vertical as defined by the tissue block (Fig. 2). Sections 2–4 μm in thickness are preferred because such ultrathin sections desiccate well for optimal preservation of morphology. In addition, microtome knives should be kept clean, free of rust, and sharpened after every 100–200 sections with a facet angle between 27 and 32° of center (Fig. 2). A clean microtomb knife sharpened with a facet angle of 28–30° and a properly adjusted microtomb (clearance angle of 7°) should easily yield 2- to 4-μm frozen sections.

Negative Controls

It is important that IHC experiments include the appropriate negative controls so that specific staining can be evaluated relative to background. When using polyclonal antisera the appropriate negative control would

---

[11] M. Rentrop, B. Knapp, H. Winter, and J. Schweizer, *Histochem. J.* **18**, 271 (1986).

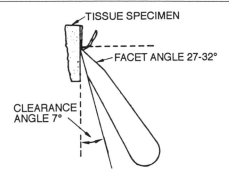

FIG. 2. Diagram showing the optimal facet and clearance angles for cryostat sectioning. Use of a protractor should aid in determining the cryostat settings for a 7° clearance angle.

be preimmune serum derived from the same species. For experiments using monoclonal antibodies it is important to use sections stained with isotype-matched negative controls. We have purchased a library of purified nonimmune mouse IgG isotypes from DAKO and routinely use these as negative controls at 1–2 µg/ml. For two-color IHC staining experiments (see below) several controls should be included: nonimmune IgG of the same isotype as each primary, first primary alone, secondary primary alone, and a slide developed with each secondary reagent without addition of a primary antibody.

Blocking Endogenous Biotin or Enzyme Activity

Endogenous peroxidase activity can be a problem in tissues rich in natural peroxidases such as the central nervous system (CNS), spleen, bone marrow, or inflamed tissues containing large numbers of infiltrating granulocytes. In such tissues, endogenous peroxidase should be blocked with dilute H_2O_2 in methanol (0.3%) or water (3.0%). However, H_2O_2, especially when applied in methanol, can have a deleterious effect on tissue antigenicity. This can be circumvented by blocking tissue sections with H_2O_2 after the polyvalent secondary antibody reagent is applied. If the choice is made not to quench endogenous enzyme activity, peroxidase-positive cells will be identifiable in the negative control sections and can be clearly evaluated relative to the positively stained sections. In addition, we use alkaline phosphatase (AP) to stain intensely peroxidase-positive tissues such as bone marrow. Endogenous AP activity should be blocked when using intestinal mucosa and can be inhibited by treatment of the sections with 1 mM levamisole during enzyme development (see below).[1] Endogenous enzyme activity is not a consideration in IF studies. Nonspecific binding of labeled avidin to endogenous biotin in tissues such as spleen, liver, lung, and kidney can be inhibited by blocking tissue sections with unlabeled streptavidin (0.1% in buffer for 30 min).

Basic ABC-Peroxidase Procedure for Staining Integrin Receptors and Extracellular Matrix Components in Frozen Sections of Human Skin

The following procedure is used in our laboratory to detect ECM components and integrin receptors in frozen human tissue and was modified from the protocol developed by Vector and provided with their Vectastain Elite ABC kit. In this protocol, mouse monoclonal antibodies are used to stain adhesion molecules in acetone-fixed frozen sections and are detected with biotinylated goat anti-mouse, ABC–peroxidase, and a DAB reaction product. The staining patterns obtained in normal human skin using commercially available antibodies (Table I) to several integrins and ECM components are shown in Figs. 3 and 4.

Materials

Cryostat sections of human tissue on aminosilane-coated glass slides
Biotin-conjugated goat anti-mouse antibody reactive with all mouse IgG
    subclasses (Vector Laboratories)
Preformed avidin–biotin–peroxidase complex provided with ABC kit
    (Vector Laboratories)
Phosphate-buffered saline (pH 7.4) or Tris-buffered saline (pH 7.6)
Diaminobenzidine (DAB) solution prepared as described below
$NiCl_2$ solution (1%, v/v) in normal saline
Meyer's hematoxylin (Rowley Biochemicals)
HPLC-grade dry acetone
Ethanol, absolute and 95% (v/v)
Xylene
Permount mounting medium

Equipment

Metal slide racks that hold 30 or 50 slides per rack
Glass staining dishes (with lids) large enough to hold the slide racks
Several Coplin jars with lids
Humidified chamber: The essential components are enough space to
    hold 30–40 slides, wet paper towels to line the bottom of the chamber,
    and a loose-fitting lid. Plastic food storage containers make useful
    humidified chambers
PAP pen (#195500, 4-mm tip; Research Products International, Mount
    Prospect, IL) or diamond tip pen for encircling the specimen. This
    is essential to prevent the antibody solutions from spreading over the

TABLE I
ANTIBODIES USED FOR IMMUNOHISTOCHEMICAL STAINING OF NORMAL HUMAN SKIN[a]

Antibody	Source[b]	Specificity	Staining pattern	Paraffin
TS2/7	T Cell Diagnostics	$\alpha 1$	M, BV, nerves	−
P1E6	Life Technologies	$\alpha 2$	SG, DC, MV, BK (not polarized)	+/−
P1B5	Life Technologies	$\alpha 3$	M, SG, BV, BK (not polarized)	+/−
P4G9	Life Technologies	$\alpha 4$	Weak staining in muscle	−
P1D6	Life Technologies	$\alpha 5$	M, DC, BV	−
GoH3	AMAC	$\alpha 6$	Nerves, MV, BK (polarized)	−
P5D2		$\beta 1$	Ubiquitous	+/−
3E1	Life Technologies	$\beta 4$	Nerves, MV, BK (polarized)	−
P1H11		FN (RGD domain)	Interstitial	+
A9		Entactin/nidogen	BM (ubiquitous)	+
F9B6		Tenascin	BM, upregulated in wounds	−
3G4	Life Technologies	Type III collagen	Interstitial	+
cIV 22	Dako	Type IV collagen	BM (ubiquitous)	+
LH7.2	Chemicon	Type VII collagen	Epidermal BM	−
P3E4		Epiligrin	Epidermal BM	−
5H2	Life Technologies	Laminin (M chain)	Epidermal BM, nerves	−
C4		Laminin (S chain)	BM (more restricted)	−
4C7	Life Technologies	Laminin (A chain)	BM (ubiquitous)	+/−
4E10	Life Technologies	Laminin (B1 chain)	BM (ubiquitous)	+/−
2E8	Life Technologies	Laminin (B2 chain)	BM (ubiquitous)	+/−

[a] As in Figs. 3 and 4. P5D2, P1H11, and P3E4 were produced in our laboratory and are not commercially available. P4C10, another anti-$\beta 1$, is identical to P5D2 and is available from Life Technologies. A9 was a generous gift from A. Fish (Department of Pediatrics, University of Minnesota Health Sciences Center), F9B6 was a generous gift from W. Carter (Fred Hutchinson Cancer Center, Seattle, WA), C4 was a generous gift from J. Sanes, and several antibodies (3E1, 4C7, 4E10, and 2E8), although available from Life Technologies, were the generous gift of E. Engvall (La Jolla Cancer Research Foundation, La Jolla, CA). BV, All blood vessels; MV, microvasculature; DC, dendritic cells; BK, basal keratinocytes; SG, sweat glands; BM, basement membrane; M, muscle. (+) or (−) indicates reactivity in formalin-fixed and paraffin-embedded tissue after protease digestion or antigen retrieval. Staining pattern in paraffin sections is never as intense or identical to frozen sections and can vary depending on the tissue used.

[b] Locations of sources are as follows: T Cell Diagnostics, Cambridge, MA; Life Technologies, Gaithersberg, MD; AMAC, Westbrook, ME; Dako, Carpenteria, CA; Chemicon, Femecula, CA.

entire surface of the slide. It is absolutely essential that the sections
     be kept moist to prevent drying and nonspecific binding of antibody
Desiccator equipped with silica gel: It is essential to avoid desiccants
     that produce dust. Desiccator must be large enough to fit slide racks
Two stainless steel beakers and a large (12-in.) pair of forceps for use
     with isopentane and liquid nitrogen. One beaker needs to fit comfort-
     ably within the other
Sharp, clean dissecting instruments (small, 4- to 6-in. forceps and scis-
     sors) dedicated to preparation of tissue for IHC

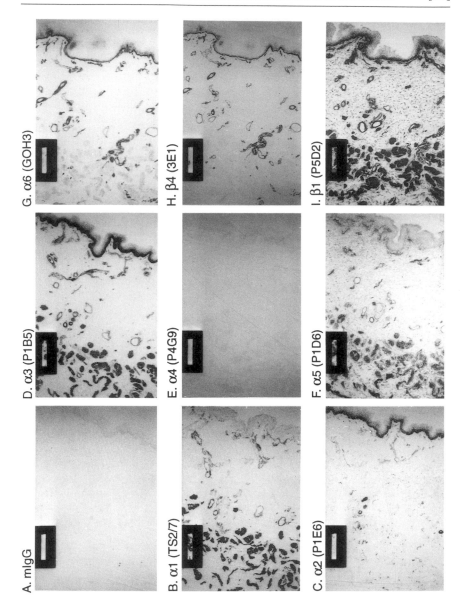

Whatman filter paper for blotting dry tissue samples

Slide boxes for storing silanized slides and for archiving finished IHC slides

Cryostat and supplies (knives, O.C.T., cryomolds)

Clean, dry, absorbant paper towels and Kimwipes

Light or fluorescence microscope

Procedure

1. Embed an 8-mm piece of fresh human foreskin in O.C.T. and freeze in liquid nitrogen-supercooled isopentane. Equilibrate the block in a cryostat for 1 hr at −20° before sectioning.

2. Cut 2- to 4-μm sections and pick up one section per silanized slide (slides should be at ambient temperature). Sections should not have folds, creases, or tears. In addition, sections should be examined microscopically before staining to be sure that the tissue is oriented properly and free from artifact.

3. Apply a PAP pen hydrophobic perimeter to the silanized slides to enclose the tissue area.

4. Desiccate the tissue sections in a sealed container with silica gel for at least 24 hr.

5. Lightly fix the sections in a bath of cold, HPLC-grade dry acetone for 10 min. Denaturing fixatives such as methanol should be avoided.

6. Drain the slides, blot them on paper towels, and redesiccate for another 24 hr. Desiccating the tissue sections aids in the preservation of morphology.

7. Rehydrate the tissue sections in a buffer bath for 5 min. Use PBS (pH 7.4) for peroxidase and Tris-buffered saline (pH 7.6) for AP. Carefully blot each slide dry with Kimwipes, being sure that each tissue section

---

FIG. 3. Immunoperoxidase localization of integrin receptors in frozen (unfixed) human foreskin. Magnification: × 100; bar: 200 μm. Acetone-fixed, 2- to 4-μ cryostat sections of normal human foreskin were reacted with the indicated monoclonal antibodies or nonimmune mouse IgG [(A) mIgG]. Specific staining was detected with a Vectastain Elite ABC kit and DAB substrate development. Sections were not counterstained for clarity and presentation of data in black and white. Basal keratinocytes are localized by specific staining with α2, α3, α6, and β4 to the right of (C), (D), (G), and (H), respectively. Note prominent staining of α6 and β4 in the basal aspect of epidermal keratinocytes, and prominent staining of muscle in the deep dermis (to the left) by α1, α3, α5, and β1. Note also the prominent staining of the microvascular endothelium located in the papillary dermis (close to the epidermis) by α1, α3, α6, and β4. Except for weak staining in muscle, α4 immunoreactivity is essentially absent from normal uninflamed skin. (A–F) and (G–I) are serial sections from the same tissue block.

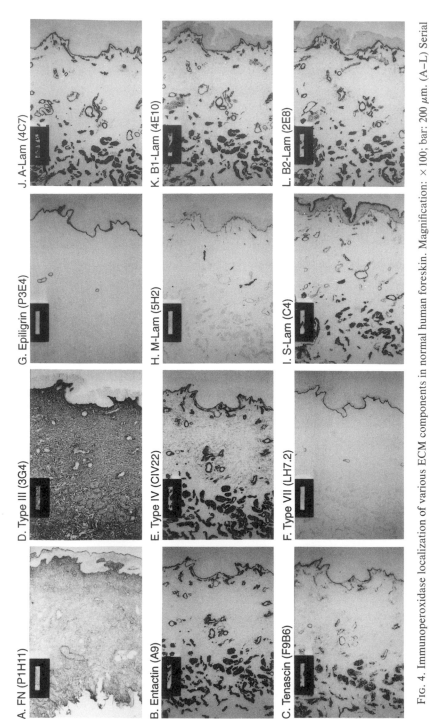

A. FN (P1H11)

B. Entactin (A9)

C. Tenascin (F9B6)

D. Type III (3G4)

E. Type IV (CIV22)

F. Type VII (LH7.2)

G. Epiligrin (P3E4)

H. M-Lam (5H2)

I. S-Lam (C4)

J. A-Lam (4C7)

K. B1-Lam (4E10)

L. B2-Lam (2E8)

FIG. 4. Immunoperoxidase localization of various ECM components in normal human foreskin. Magnification: ×100; bar: 200 μm. (A–L) Serial sections derived from the same tissue block used in Fig. 3 (same negative control). Note prominent interstitial staining of fibronectin (FN) and type III collagen throughout the dermis. Note also the staining patterns observed for entactin, tenascin, epiligrin, type VII collagen, and the laminin (Lam) isoforms. Such staining patterns may not be preserved when using formalin-fixed and paraffin-embedded tissue sections. Therefore, when evaluating such tissues for specific immunoreactivity, it is important to know the staining pattern of the antigen of interest in frozen sections.

is kept moist. At this point it is imperative that the sections be kept from drying out to prevent nonspecific binding of antibodies. Lay each slide in the humidified chamber and apply a bubble (100–200 μl) of blocking buffer containing 8% (v/v) nonimmune goat serum and 1% (w/v) bovine serum albumin (BSA) over each tissue section within the PAP hydrophobic perimeter, being sure that each section is completely covered. Block the sections for 30 min at ambient temperature. Endogenous biotin can be blocked in this step.

8. Dilute primary antibodies in blocking buffer, using the following ratios: culture supernatant (1 : 2 or 1 : 10), ascites fluid (1 : 500 or 1 : 1000), and purified antibody at 1–2 μg/ml.

9. Aspirate the blocking buffer from slides. Apply a bubble of primary antibody to each tissue section and incubate for 30–60 minutes at room temperature.

10. Blot the primary antibody from the slides by turning the slides on their sides and tapping gently on paper towels. Gently rinse off residual antibody by carefully running 2–4 ml of buffer over the surface of the slide and into a waste beaker. Blot excess buffer with Kimwipes. The first wash must be performed individually for each primary antibody used, to avoid cross-contamination. Follow with three consecutive batchwise washes in buffer carried out using slide racks and staining dishes. Avoid excessive mixing or stirring. Blot the slides dry and lay them in a humidified chamber.

11. Apply a bubble of biotinylated goat anti-mouse antibody and incubate for 30 min.

12. Blot the secondary antibody from the slides and wash batchwise with three changes of buffer.

13. If necessary, endogenous peroxidase can now be quenched by incubating the sections batchwise in a staining dish for 30 min in 0.3% (v/v) H_2O_2 in methanol or 3% (v/v) H_2O_2 in water. Wash in three changes of buffer, blot dry, and place in humidified chamber.

14. Prepare Vectastain ABC reagent according to manufacturer specifications and allow the complex to form for 30 min at ambient temperature.

15. Apply a bubble of ABC reagent to the tissue sections and incubate for 45–60 min, blot the excess, and wash batchwise in three changes of PBS.

16. Incubate with DAB (see below) for 4–8 min and wash three times in buffer.

17. Intensify DAB with $NiCl_2$ for 5 min and wash three times in buffer.

18. Counterstain with Meyer's hematoxylin and wash three times in buffer.

19. Dehydrate in graded alcohols, dip in xylene, and coverslip the slides.

To perform a PAP experiment, the method is exactly the same except that steps 14 and 15 are performed with a PAP complex instead of ABC and in step 11 an unlabeled secondary anti-mouse antibody is used. LAB procedures substitute peroxidase-conjugated avidin (LAB) or streptavidin (LSAB) in step 15 (step 14 is omitted). For alkaline phosphatase-catalyzed reactions, an ABC-AP complex (Vector) is used (steps 14 and 15) with the appropriate substrate and reaction products (steps 16–19). Steps 16–19 for peroxidase or alkaline phosphatase IHC reactions are discussed in detail below.

Peroxidase-Reactive Substrates and Reaction Products

Diaminobenzidine (Brown). Prepare a 0.1% (w/v) solution of diamino-benzidine (DAB) by dissolving 15 mg of DAB in 15 ml of PBS. Just prior to use, add 5 μl of 30% (v/v) H_2O_2 (final concentration, 0.01%). Incubate the tissue sections with 100–200 μl of the DAB substrate applied to each slide within the PAP perimeter for a period of 4–8 min (the optimal time for DAB development should be predetermined). Stop the reaction by blotting the excess DAB on paper towels and washing the slides batchwise in three changes of buffer. Increased penetration of DAB can be achieved by preincubating the sections in 0.05% (w/v) DAB without H_2O_2 for 30 min. This solution is aspirated and replaced with 0.05% (w/v) DAB in 0.01% (v/v) H_2O_2 and developed for 4–8 min. This technique is particularly useful for paraffin sections. Diaminobenzidine can be used at 0.05 to 0.1% (w/v), although it is recommended that the concentration of H_2O_2 not exceed 0.01% (v/v).[1]

Intensification of Diaminobenzidine Reaction Product. The reddish brown color of the DAB reaction product can be modified by the addition of metal ions[12] or osmium tetroxide[13] to the reaction product mixture just before adding the mixture to the tissue section. The following ions have been reported[12] to modify the color of DAB to a purplish or dark blue-black reaction product: $NiCl_2$, $CoCl_2$, and $CuSO_4$. We prefer the use of $NiCl_2$ applied after the DAB substrate solution, which results in a crisp black reaction product. This works well with the red peroxidase-catalyzed AEC or Vector Red (AP-catalyzed) reaction products in two-color studies. If metal-modified DAB is used with unmodified DAB in a two-color study, the metal ions must be used to develop the staining reaction of the first antigen. Osmium tetroxide enhancement is accomplished by a 1- to 2-min incubation of DAB-developed sections in a 0.125% (v/v) solution of osmium tetroxide in distilled water. The slides are then rinsed in tap water.

[12] S. M. Hsu and E. Soban, *J. Histochem. Cytochem.* **30**, 1079 (1982).
[13] O. Johansson and J. Bachman, *J. Neurosci. Methods* **7**, 185 (1983).

Diaminobenzidine enhancement kits are commercially available (Cat. Nos. SK-4100 and H2200; Vector). It is important to note that DAB, $NiCl_2$, and osmium tetroxide are hazardous substances and should be handled with caution. Diaminobenzidine-containing solutions can be inactivated in 3% (w/v) $KMnO_4$ and 2% (w/v) Na_2CO_3 in deionized water.

Red Reaction Product (AEC). Dissolve four AEC chromagen tablets (3-amino-9-ethylcarbazole, Cat. No. A-6926; Sigma) in 10 ml of *N,N*-diethylformamide (Cat. No. D-4254; Sigma) and add to 190 ml of acetate buffer [200 m*M* sodium acetate, 0.24% (v/v) acetic acid in distilled H_2O, pH 5.2]. Develop the slides for 4 min and wash in PBS followed by running tap water. Slides containing AEC reaction products can be permanently mounted if pretreated with Crystal Mount (Biomeda, Foster City, CA; see below).

Alkaline Phosphatase-Reactive Substrates and Reaction Products

Alkaline phosphatase activity is demonstrated with naphthol-AS phosphate as the substrate. A variety of chromogens can be oxidized in this system to provide red, blue, or black reaction products.[1] Several ABC-AP and APAAP substrate kits are available from Vector or Dako. Alkaline phosphatase-dependent substrates and reaction products can be used in two-color IHC reactions also utilizing peroxidase and DAB reaction products (see below). Endogenous AP activity is quenched by the addition of 1 m*M* levamisole directly to the substrate solution. Alkaline phosphatase-catalyzed reaction products are labile and soluble in organic solvents. For permanent mounting with synthetic resins such as Permount, slides containing AP IHC data should be preserved with Crystal Mount (see below) before coverslipping.

Counterstaining

The choice of a counterstain may be critical to the appropriate visualization of the specific IHC reaction and morphological structure. However, counterstaining may also detract from specific visualization of the IHC reaction product (blue nuclear stain and a blue or purple reaction product) and it may not be desirable to counterstain experiments destined for publication in black and white (Figs. 3–5). Furthermore, we have found that counterstaining two- or three-color IHC experiments produces something akin to fruit salad and should be avoided. A light-blue nuclear counterstain such as is produced with Meyer's hematoxylin (Rowley Biochemical) works well with brown or black peroxidase-catalyzed DAB, red peroxidase-catalyzed AEC, or alkaline phosphatase-catalyzed Vector Red reaction products. Other counterstains include nuclear fast red, methylene blue, and methyl green. However, some counterstains (methyl green and methylene blue) must be cleared in alcohols or xylene and are therefore not appropriate for soluble end products such as AEC and the alkaline phosphatase-reactive chromogens. Counterstaining for both soluble and insoluble end products can be accomplished by dipping slides in 100% Meyer's hematoxylin for 10 min. Blue in running tap water, 1% (w/v) LiCO₃, or 0.25% (v/v) ammonia water for 1–2 min. Meyer's hematoxylin is a lighter stain than Gill's No. 3 hematoxylin and works well with brown, black, or red IHC reaction products. The stain produced by Gill's No. 3 hematoxylin is dark blue and works better when used in conjunction with eosin for morphological analysis (see below).

Coverslipping Immunohistochemically Stained Slides

Dehydrate the slides in graded alcohols (two 5- to 10-sec passes in 95% ethanol followed by two 5- to 10-sec passes in absolute ethanol) and dip the slides one at a time in xylene (in a fume hood). Immediately mount with Permount (Cat. No. SP15-500; Fisher, Pittsburgh, PA) and a No. 1 coverslip (be sure the coverslips are clean; blow clean with canned air if necessary). A xylene wash minimizes the retraction of the Permount solution as it dries. For permanent mounting of soluble reaction products (e.g., peroxidase AEC, and alkaline phosphatase substrates), tissue sections must be rinsed with distilled H_2O to remove salts, and while wet completely covered with Crystal Mount (Cat. No. MO2; Biomeda) mounting medium and dried at 60° for at least 10 min. They can then be coverslipped with Permount as described above. Such slides should not be dehydrated before application of Crystal Mount.

Hematoxylin–Eosin Morphological Stain

It is sometimes desirable to perform a morphological stain on at least one slide per tissue block in each experiment, especially when counterstaining will not be used in conjunction with a specific IHC procedure. The following procedure is for frozen sections. The procedure for paraffin sections is the same except that the sections must be completely deparaffinized and rehydrated (see below) before hematoxylin–eosin staining.

Hematoxylin–Eosin Staining

1. Cut a section for hematoxylin–eosin staining from each tissue block used in the experiment and desiccate with the rest of the slides, acetone fix, and further desiccate for 24 hr.

2. Rehydrate in buffer and stain with Gill's hematoxylin #3 for 5–8 min (Cat. No. GHS-3-32; Sigma).

3. Blue for 5 min in running tap water, 1% (w/v) LiCO$_3$, or 0.25% (v/v) ammonia in water (1–2 min).

4. Wash in running tap water.

5. Stain with alcoholic eosin for 30 sec (Cat. No. HT110-2-16; Sigma).

6. Dehydrate in graded alcohols (two 5- to 10-sec passes in 95% ethanol and in absolute ethanol), clear in xylene, and coverslip with Permount as described above.

Modifications of ABC-Peroxidase Procedure for Single-Color Immunofluorescence

In single-color IF procedures, biotinylated primary or secondary antibodies are detected with fluorochrome-conjugated avidin (centrifuged at 100,000 g for 10 min immediately before use). It is essential that, before mounting, IF slides be postfixed in 2% (v/v) paraformaldehyde (in PBS) and rinsed in distilled water to remove residual salts from buffered saline solutions. Tissue sections stained by IF techniques should be mounted in glycerol containing p-phenylenediamine or DABCO (Sigma) to prevent quenching during microscopic examination and photography. The composition of the IF mounting medium used in our laboratory is 80% (v/v) glycerol, 50 mM Tris-HCl (pH 8.5), 120 mM NaCl, 0.02% (w/v) sodium azide, p-phenylenediamine (1 mg/ml) (Sigma). The glycerol should be of high quality and nonfluorescent (spectral grade, Cat. No. 1203025; Kodak, Rochester, NY). The mounting solution should be a light straw color and will be stable for about 6 months if stored in the dark at 4°. Mounted IF slides should be coverslipped and completely sealed to prevent desiccation and stored in the dark at 4°. Coverslips can be sealed with clear nail polish and allowed to air dry.

Two-Color Immunohistochemical or Immunofluorescence (Staining Protocols)

Interest in two-color techniques has increased because they allow for the dual localization of two antigens in the same tissue section or even in the same cell. The methods for IHC localization of two antigens do not differ significantly from those used to detect a single antigen. If the primary antibodies are made in the same species, then two entire immunostaining procedures are carried out sequentially. Successful implementation of two-color ABC techniques requires the complete development of a stable and insoluble primary reaction product (DAB). When properly developed, the primary DAB reaction product inhibits the antibodies and the ABC reagent from interacting with the second set of reagents. This can be achieved using a high DAB concentration (0.1%, v/v) and 0.01% (v/v)

H_2O_2 or an intensified DAB reaction product ($NiCl_2$). In peroxidase-catalyzed two-color IHC procedures (either ABC-ABC or ABC followed by PAP), the brown DAB reaction product can be paired with modified DAB reaction product (see above) or with AEC. ABC-peroxidase (DAB substrate) followed by ABC-AP (Vector Red substrate kit), or PAP (DAB substrate) followed by ABC-AP (Vector Red), also work well together in two-color experiments. One technique we have found particularly useful is to use the Vector ABC-AP kit with a Vector Red reaction product followed by Vector ABC-AP with a Vector Blue reaction product. Cells that express both antigens appear purple. Low concentrations of nonionic detergents [0.05% (v/v) Triton X-100, 0.2% (v/v) Tween] can be used in wash buffers to reduce nonspecific background staining in two-color experiments. For two-color experiments using primaries derived from different species, the reagents can be added simultaneously. In this case different enzyme kits (PAP and ABC) are used and the reaction products are developed sequentially.

For two-color immunofluorescence studies with primary antibodies derived from different species, primary antibodies (e.g., mouse and rabbit) and secondary antibodies (FITC–goat anti-mouse and rhodamine–goat anti-rabbit) are added simultaneously and the experiment is carried out as for a single-color IF study. However, for primary antibodies from the same species (e.g., two mouse monoclonal IgGs) one of the following procedures can be used.

Method 1: Both primary antibodies are directly conjugated with fluorochromes (FITC or rhodamine) and are added simultaneously

Method 2: Both primary antibodies are directly conjugated, one with FITC and one with biotin and are added simultaneously. The biotinylated primary is then detected with rhodamine-conjugated avidin (Vector)

Method 3: The first mouse IgG is unlabeled and is detected with rhodamine-conjugated goat anti-mouse (Fig. 5A). The tissue sections are then fixed with 2% (v/v) paraformaldehyde before application of the second mouse IgG. The 2% (v/v) paraformaldehyde fixation effectively cross-links the goat anti-mouse antibodies, which minimizes cross-reactivity with the second mouse IgG. The second mouse IgG, which is either directly conjugated with FITC or biotinylated (detected with FITC-conjugated avidin), is applied after cross-linking (Fig. 5B).

Method 4: This method makes use of two biotinylated mouse monoclonal antibodies. The first antibody is detected with FITC–avidin (Vector) and the second antibody is detected with rhodamine–avidin (Vector). It is essential to saturate the FITC–avidin before application of the second biotinylated antibody. This is accomplished by incubating the tissue sec-

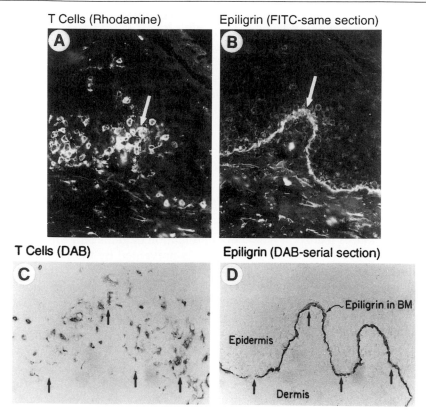

FIG. 5. Immunofluorescence (A and B) and immunoperoxidase (C and D) localization of T lymphocytes [(A) and (C) detected with anti-CD3] or epiligrin (B and D) in frozen tissue sections derived from the skin of a patient with graft-vs-host disease.[19] (A and B) Same section [(A) detected with rhodamine, (B) detected with FITC], whereas (C) and (D) are serial sections (detected with DAB). Exact comparisons regarding the colocalization of T cells with the epiligrin-rich epidermal basement membrane can be made only in (A) and (B). This illustrates the utility of two-color immunofluorescence data when compared to black and white presentation of immunoperoxidase data.[19] (Reprinted with permission from the *Journal of Cell Biology.*[19])

tions with free biotin (Sigma, 100 μg/ml in PBS) for 15–30 min at ambient temperature.

The methods we use to conjugate purified antibodies with biotin, FITC, and rhodamine are adapted from procedures described by Goding.[14] Two-

[14] J. W. Goding, "Monoclonal Antibodies: Principles and Practice." Academic Press, Orlando, FL, 1986.

color IF is particularly useful for colocalizing antigens in tissue and presenting black-and-white data for publication (Fig. 5A and B). Compare Fig. 5A and B (same section) with Fig. 5C and D (serial sections); direct comparisons can be made only with Fig. 5A and B (same section). Photomicrographs of dual-label IF studies can be published in color by double exposing single frames of film for rhodamine or fluorescein. Cells labeled with both antibodies will appear yellow.

Procedure for Single-Color Immunoperoxidase Staining of Formalin-Fixed and Paraffin-Embedded Sections

Effects of Chemical Fixation, Dehydration, and Paraffin Embedding on Immunohistochemical Reactions

Fixation. A fixative is a fluid or combination of fluids into which tissue specimens are placed so that autolysis is prevented by the processes of denaturation and cross-linking. The specimen is subsequently heated, dried, and stiffened for microtome sectioning and slide preparation by embedding in paraffin. A wide variety of agents are used to fix tissue, including aldehydes, alcohols, acetone, acetic acid, chromates, and picric acid. Although 10% (v/v) buffered neutral formalin is the most commonly used fixative solution, others include Bouin's fixative, methyl Carnoy's, and 2% (v/v) glutaraldehyde in sodium cacodylate. Aldehyde fixation with neutral-buffered formalin functions as a cross-linking agent whereas acid or alcohol (Bouin's or methyl Carnoy's) and heavy metal fixatives (B5 or Zenker's) function by precipitation. Some fixation is necessary to minimize extraction of soluble proteins and to preserve the morphology of the tissue. For this reason frozen sections are usually lightly fixed in acetone or 2% (v/v) paraformaldehyde before immunohistochemical staining. However, it has been our experience that fixation with aldehydes [10% (v/v) neutral buffered formalin], alcohols (methyl Carnoy's), or acids (Bouin's fixative) has a negative impact on the intensity of immunohistochemical and immunofluorescence staining both in terms of preserving the antigenicity of the tissue and in generating background. It is important to point out that because of the reduced antigenicity, most monoclonal antibodies that detect single epitopes and even some polyclonal antibodies cannot be used on formalin-fixed and paraffin-embedded tissue. Some tissue antigens, however, are not completely destroyed by chemical fixation and results can vary depending on the fixative used or on the anatomical location of the antigen to be studied. Some fixatives [Perfix and Omni, or 95% (v/v) ethanol or acetone] are less denaturing and can be used in place of formalin to prepare tissue for paraffin embedding. In general, fixation should be adjusted to the particular tissue antigen being studied. When a new antibody is used, it is advisable to use fresh frozen sections to determine the effects of fixation on antigenicity. Cryostat-sectioned, fresh frozen tissue (or sections) can be treated with the same fixative applied to the prefixed paraffin-embedded tissue. A small experiment can be designed in which fresh frozen, frozen but fixed, and paraffin-embedded sections can be compared. If the positive control frozen sections stain, but the chemically fixed or paraffin-embedded sections do not, then it can be concluded that either the chemical fixation or paraffin embedding have negatively impacted the antigenicity of the tissue.

Dehydration. After fixation, the tissue must be dehydrated and the tissue water replaced with organic solvents before paraffin embedding. The effects of dehydration in graded alcohols and clearing in xylene on immunoreactivity have not been systematically examined. However, it is reasonable to assume that this protocol would also result in some denaturation of tissue antigens.

Paraffin Embedding. There are conflicting views on the impact of embedding tissue in hot paraffin on antigenicity.[1] Some authors have reported that paraffin embedding decreases the sensitivity of IHC reactions and increases nonspecific background.[15] It has been suggested that infiltration time be minimized and that paraffins with low melting points be used.[1] In our experience, formalin fixation has the most detrimental effect on tissue antigenicity, although we have not systematically examined the effects of heat and paraffin. However, it is generally agreed that before an IHC procedure is initiated, the tissue must be completely deparaffinized to ensure uniform staining. This can be achieved by soaking the slides containing microtome-cut paraffin-embedded sections in xylene (see below).

Immunohistochemistry with Formalin-Fixed and Paraffin-Embedded Tissue

Formaldehyde [usually 10% (v/v) neutral buffered formalin] is the universal fixative and functions by the formation of cross-links between protein end groups. It reacts with free amino groups on lysine, arginine, and asparagine side chains.[16] The resulting molecular cross-links alter the structure of proteins and severely compromise the antigenicity of tissue. Two approaches have been used to compensate for the reduced immunoreactivity of tissue fixed in formalin: (1) IHC reactivity can be enhanced by

[15] R. C. Curran and J. Gregory, *J. Clin. Pathol.* **33,** 1047 (1980).
[16] Y. Kitamoto and H. Maeda, *J. Biochem. (Tokyo)* **87,** 1519 (1980).

increasing the sensitivity of the method used. As previously discussed, PAP and ABC procedures can often be used successfully on formalin-fixed tissue; (2) the antigenicity lost by fixation can sometimes be recovered by proteolytic digestion or antigen retrieval. The restoration of antigenicity by proteases or antigen retrieval requires that the fixative used reduce the accessibility of antibodies to their antigens by cross-linking rather than by epitope denaturation.

Protease Digestion. Treatment of aldehyde-fixed tissue sections with proteolytic enzymes may enhance the specific immunoreactivity of antigens unmasked this procedure.[17] However, protease digestion can also cleave tissue antigens and either contribute to background artifacts, or result in a negative finding because the antigen of interest was digested. Residual protease activity must be eliminated before proceeding with the IHC reaction. This can be accomplished by immersing protease-treated slides in cold buffer or by application of protease inhibitors (pepstatin, leupeptin, or soybean trypsin inhibitor). Although trypsin or protease have been widely used, several other proteolytic enzymes have also been used successfully to enhance the IHC reactivity of formalin-fixed and paraffin-embedded tissue sections.[1] The enzymes that are routinely used by the Immunohistochemistry Laboratory at the University of Minnesota are 0.6% (w/v) ficin, 0.0125% (w/v) bromelain, 0.1% (w/v) trypsin, 0.2% (w/v) pepsin, and 0.1 (w/v) pronase. In addition, in our laboratory we have successfully unmasked epitopes on fibronectin and laminin in formalin-fixed and paraffin-embedded sections with hyaluronidase (275 U/ml in 0.1 M phosphate buffer)[18] and protease type XXIV [0.025% (w/v) in 0.025% (v/v) calcium chloride].

Antigen Retrieval. Deparaffinized tissue sections can also be treated in a microwave oven to unblock chemical cross-linking. Some investigators routinely use this procedure because this circumvents the need to treat the tissue sections with a proteolytic enzyme. Antigen retrieval is accomplished by placing deparaffinized tissue sections (see below) into a Coplin jar filled with enough 10 mM citrate buffer (pH 6.0) to immerse the slides completely. The slides are microwaved for two 5-min cycles with the power setting adjusted so that the buffer boils for at least 3 min in each cycle. The number of cycles can be varied but at least two should be used. The buffer should be changed between each cycle. Cool microwaved

[17] T. Kaku, J. K. Ekem, and C. Lindayen, *Am. J. Clin. Pathol.* **80,** 806 (1983).
[18] B. Holund and I. Clemmensen, *Histochemistry* **76,** 517 (1982).
[19] E. A. Wayner, S. G. Gill, G. F. Murphy, M. S. Wilke, and W. G. Carter, *J. Cell Biol.* **121,** 1141 (1993).

tissue sections at room temperature in buffer for 20 min. Replace citrate buffer with fresh IHC buffer and begin the staining procedure.

Procedure for Staining Paraffin Sections with Anti-Integrin or Extracellular Matrix Monoclonal Antibodies

We have successfully used the following protocol to stain formalin-fixed and paraffin-embedded sections derived from human tissue. However, it should be emphasized that this technique is extremely difficult to carry out successfully and that most monoclonal and polyclonal antibodies (Table I) simply do not work in formalin-fixed and paraffin-embedded tissue regardless of whether antigen retrieval and protease digestion steps are carried out. In addition, the staining pattern obtained with paraffin sections may not be identical to frozen sections and the intensity of specific staining is usually reduced.

1. Deparaffinize the tissue sections in two 5-min passes in xylene or Histoclear (Baxter).

2. Rehydrate the tissue sections by two 5-min passes in absolute ethanol followed by two 5-min passes in 95% ethanol. Rinse in running tap water for 5 min.

3. Immerse the slides in 50 mM Tris buffer (pH 7.6).

4. Digest the sections by immersing in 0.025% (w/v) protease type XXIV (Cat. No. P-8038; Sigma) in 0.025% (w/v) $CaCl_2$ for 7 min at ambient temperature. Unblocking of cross-linked epitopes can also be attempted via antigen retrieval (see above) or the use of other proteolytic enzymes (see above). Rinse the slides in cold buffer three times, apply PAP hydrophobic perimeter, and immediately proceed with the IHC protocol.

5. Immerse the tissue sections in 50 mM Tris buffer (pH 7.6) for alkaline phosphatase reactions or PBS (pH 7.4) for peroxidase or immunofluorescence reactions. For each procedure, all washes are done with Tris or PBS.

6. Quench the slides by immersing in a bath of 3% (v/v) H_2O_2 in water or 0.3% (v/v) H_2O_2 in methanol. Wash three times in buffer bath.

7. Block the slides in 8% nonimmune goat serum in buffer for 30 min. Wipe off excess blocking solution and lay the slides flat in humidified chamber.

8. Apply primary antibody diluted in blocking buffer and incubate for 30 min, blot off excess primary, rinse once, and then wash three times in buffer bath.

9. Apply biotinylated goat anti-mouse antibody and incubate for 30 min. Blot off the excess and wash in a buffer bath three times. Prepare ABC reagent and incubate on the bench for 30 min.

10. Apply ABC reagent and incubate for 45–60 min, blot, and wash in a buffer bath three times.

11. Make up DAB substrate solution (see above) and apply to the tissue sections for 8–10 min; wash and intensify with $NiCl_2$ (see above).

12. Counterstain if desired, dehydrate in graded alcohols, dip in xylene, and coverslip with Permount.

All supplies for frozen or paraffin-embedded tissue sample preparation, and cryostat and microtome supplies, are made by Tissue-Tek and can be obtained from Baxter Scientific.

Acknowledgments

We gratefully acknowledge the expert technical assistance of Dr. Stan McCormick and Shea Merrill (Department of Pathology, United Hospital, St. Paul, MN), Betsy Spaulding (Dako Corporation, Carpentiria, CA), and Mr. Otto Anklam (The Otto Anklam Company). We also thank Dr. William Carter (FHCRC, Seattle, WA) for assistance and helpful comments during the preparation of this manuscript. Research in E. Wagner's laboratory is supported by a grant from the American Cancer Society (IM69879).

[17] Peptides in Cell Adhesion Research

By Erkki Koivunen, Bingcheng Wang, Craig D. Dickinson, and Erkki Ruoslahti

Introduction

Many of the cell adhesion receptors recognize simple sequences that can be reproduced as synthetic peptides. This circumstance has led to a widespread use of peptides as modulators of cell adhesion. Peptides capable of binding to cell adhesion receptors, such as the integrins, can be used as mediators of cell attachment to a surface by coating the surface with the peptide. Peptides bound to a solid phase can also be used to isolate adhesion receptors by affinity chromatography. Alternatively, the peptides can be used to inhibit cell attachment to the natural ligands of the adhesion receptors. In either mode, adhesion peptides have proved to be highly useful probes in cell adhesion research and they also show promise as a new class of therapeutics. This chapter reviews some of the properties and uses of adhesion peptides.

Copyright © 1994 by Academic Press, Inc.
All rights of reproduction in any form reserved.

RGD Sequence

First identified in fibronectin (Pierschbacher and Ruoslahti, 1984a), the RGD sequence is viewed as the prototype recognition motif among cell adhesion sequences recognized by integrins. The RGD sequence is the essential element of the integrin recognition site in many different adhesive proteins. The adhesive proteins with active RGD sequences include fibronectin, vitronectin, fibrinogen, laminin, entactin, denatured collagens, osteopontin, bone sialoprotein I, and, under some circumstances, tenascin and thrombospondin (Ruoslahti and Pierschbacher, 1987; Lawler *et al.*, 1988; Bourdon and Ruoslahti, 1989). In some cases, even a protein whose functions are unrelated to adhesion can function as an adhesion protein. The key to this is thought to be an appropriate presentation of the RGD sequence (Ruoslahti and Pierschbacher, 1987). Moreover, a number of integrins, at least 6 of the 20 or so integrins known at this time, recognize the RGD sequence (Table I). The RGD-directed integrins include the integrins $\alpha_5\beta_1$, $\alpha_v\beta_1$, $\alpha_{IIb}\beta_3$, $\alpha_v\beta_3$, $\alpha_v\beta_5$ (Ruoslahti, 1991), $\alpha_v\beta_6$ (Busk *et al.*, 1992), $\alpha_v\beta_8$ (Nishimura and Pytela, 1993), and possibly $\alpha_2\beta_1$ (Cardarelli *et al.*, 1992), and $\alpha_3\beta_1$ (Elices *et al.*, 1991).

RGD sequence in cell attachment is ancient; a *Drosophila* integrin also binds to it (Bunch and Brower, 1992). The multiple and overlapping specificities of the RGD-directed integrins and their protein ligands complicate the use of RGD peptides in probing it. As discussed below, progress has been made in designing peptides that bind only to one integrin among the RGD-directed integrins.

Other Integrin-Binding Sequences

A number of sequences other than RGD have been identified as integrin target sequences (Table I). One of these sequences, KQAGDV (Kloczewiak *et al.*, 1984), appears to be a mimic of the RGD sequence, because its binding to the $\alpha_{IIb}\beta_3$ integrin, for which it is essentially specific, is inhibited by RGD peptides (Lam *et al.*, 1987). However, RGD and KQAGDV are partially different in their binding to $\alpha_{IIb}\beta_3$; when used in affinity labeling of the integrin, peptides containing these sequences preferentially label different subunits of the integrin (D'Souza *et al.*, 1988, 1990). The $\alpha_4\beta_1$ integrin binds to a sequence centered around an LDV motif (Guan and Hynes, 1990; Mould *et al.*, 1990). $\alpha_2\beta_1$ is thought to recognize DGEA (Staatz *et al.*, 1991), $\alpha_1\beta_1$ a configuration of residues formed by arginine and aspartic acid residues on the surface of the type I collagen triple helix (Eble *et al.*, 1993), and $\alpha_M\beta_2$ (CD11b/CD18) a KRLDGS sequence in fibrinogen (Altieri *et al.*, 1993).

TABLE I
INTEGRIN RECOGNITION SEQUENCES

Integrins recognizing sequence				
RGD	KQAGDV	LDV	KRLDGS	DGEA
$\alpha_5\beta_1$	$(\alpha_5\beta_1)$			
$\alpha_v\beta_1$				
$\alpha_v\beta_3$				
$\alpha_v\beta_5$				
$\alpha_v\beta_6$				
$\alpha_v\beta_8$				
$\alpha_{IIb}\beta_3$	$\alpha_{IIb}\beta_3$			
$(\alpha_2\beta_1)$				$\alpha_2\beta_1$
$(\alpha_3\beta_1)$				
		$\alpha_4\beta_1$		
			$\alpha_M\beta_2$	

The LDV sequence is a particularly interesting adhesive non-RGD sequence. It is present in the alternatively spliced IIICS domain of fibronectin and binds to the $\alpha_4\beta_1$ integrin. This integrin can also recognize two other sequences in that region of fibronectin, IDA(PS) and REDV (Mould *et al.*, 1991; Mould and Humphries, 1991). Whereas the IDAPS sequence looks like a variation of LDV, REDV may be a homolog of RGD, as the position of the REDV sequence of human IIICS corresponds to an RGD in rat IIICS (Norton and Hynes, 1987). These similarities suggest a commonality in the ligand recognition specificities of $\alpha_5\beta_1$ (and the other RGD-directed integrins) and $\alpha_4\beta_1$. Indeed, peptides that bind to both $\alpha_5\beta_1$ and $\alpha_4\beta_1$ have been identified (Mould and Humphries, 1991; Koivunen *et al.*, 1993; Nowlin *et al.*, 1993). It is likely that the sites recognized by the integrins evolved from one primordial recognition site and that the specific sequences bound by the present-day integrins are still closely related.

The integrin-binding sequence motifs described above share one amino acid, the aspartic acid residue. The aspartic acid may be important because of its potential to contribute to divalent cation binding; one hypothesis regarding the binding of ligands to integrins postulates that the ligand provides a coordination site for divalent cation binding (Edwards *et al.*, 1988). The integrin α subunits contain multiple (four to five) sequences that resemble the EF-hand divalent cation-binding motif of other proteins. The integrin sequences, however, lack one of the conserved acidic amino acids that serve as coordination sites in the EF-hands of other proteins. The suggestion is that the aspartic acid of RGD and the other adhesion motifs that contain an aspartic acid residue may substitute for the missing coordination site when an integrin binds to its ligand.

In addition to the peptides that are derived from the integrin ligands, peptides active in cell adhesion systems have been fashioned after sequences contained in the divalent cation-binding region of the $\alpha_{IIb}\beta_3$ integrin. Such sequences have been garnered both from the α and β subunits (D'Souza et al., 1988, 1990; Charo et al., 1991). Like the ligand-derived RGD and KQAGDV sequences, the integrin-derived sequences inhibit the binding of fibrinogen to platelets (D'Souza et al., 1988, 1990; Gartner and Taylor, 1990; Charo et al., 1991).

Other Adhesive Motifs

Two adhesive sequences from laminin, YIGSR and IKAKV, have been described that do not contain an aspartic acid. Indeed, these sequences bind to receptors that are not integrins. The receptor for YIGSR is a 67-kDa protein (Graf et al., 1987) and IKVAV has been reported to bind to the β-amyloid protein (Nomizu et al., 1993). Other active sequences have been derived from thrombospondin (Frazier, 1993) and elastin (Hinek et al., 1988). The YIGSR peptide has been used in inhibition of metastasis and induction of differentiation of various types of cells (Iwamoto et al., 1987; Nomizu et al., 1993), but little is known about the mode of action of these peptides.

Another group of peptides active in cell attachment contains multiple basic residues. These peptides bind to heparin and may act by binding to cell surface heparan sulfate proteoglycans. Polylysine and polyornithine, which are commonly used as nonspecific mediators of cell attachment, belong to this group of compounds. Although they support cell attachment, the basic peptides generally do not support cell spreading and fail to activate the focal adhesion kinase or protein kinase C, both of which are activated by integrin-mediated cell attachment and spreading (Kornberg et al., 1991; Guan et al., 1991; Vuori and Ruoslahti, 1993).

While the cell attachment activity of basic peptides may mostly be mediated by proteoglycans, integrins can also bind to such sequences. Thus, basic peptides from a heparin-binding region in fibronectin may bind to the $\alpha_4\beta_1$ integrin (McCarthy et al., 1988; Iida et al., 1992) and a basic peptide that binds to $\alpha_3\beta_1$ has been identified in laminin (Gehlsen et al., 1988). Moreover, a peptide, RKKRRQRRR, derived from the HIV Tat protein, binds specifically to the $\alpha_v\beta_5$ integrin, and cell attachment to the Tat protein, and to the basic peptide derived from it, is inhibited by antibodies to α_v integrins (Vogel et al., 1993). The $\alpha_v\beta_5$ integrin does not bind to polylysine or polyarginine, indicating that the only nonbasic residue in the peptide, the glutamine flanked by arginine residues, may be important in the activity. In support of this possibility, a peptide derived from the vitronectin heparin-binding domain that contains both an RQK and RNR sequence also binds to $\alpha_v\beta_5$. This latter finding suggests that the heparin-binding domain of vitronectin may provide an auxiliary binding site for $\alpha_v\beta_5$, in addition to the RGD site.

Peptides Defining Fibronectin–Fibronectin-Binding Sites

Yet another group of peptides relating to cell adhesion has been identified in the first type III repeat of fibronectin (Morla and Ruoslahti, 1992). These peptides define a fibronectin–fibronectin-binding site and are capable of either inhibiting or enhancing the formation of fibronectin matrix. Because of their fibronectin-binding properties, these peptides can be used to isolate fibronectin by affinity chromatography and to prepare fibronectin-coated surfaces for cell adhesion.

Identification of Adhesion Peptides

Adhesion Peptides from Proteins

The discoveries of the RGD and LDV sequences in fibronectin are examples of protein-based identification of adhesive sequences. The process consisted of fragmentation of the protein to derive active fragments, sequencing of those fragments, and synthesis of progressively smaller peptides to define the smallest active unit. The activity was followed throughout the process with cell attachment assays. Importantly, these experiments included both the demonstration of cell attachment to the peptides and inhibition of attachment to some substrates but not others. As any number of peptides may promote cell attachment, at least to some extent, it is important to show that the part of the adhesion protein the peptide derives from exhibits that activity. Moreover, the peptide should inhibit at least some of the cell attachment activity of the intact, native protein. For example, laminin contains an RGD sequence. However, this sequence is not well exposed in the intact laminin molecule, but becomes exposed in laminin fragments (Aumailley et al., 1991). The same may be the case with tenascin and thrombospondin; fragmentation and certain treatments of the intact proteins can activate the RGD sequence of these proteins (Murphy-Ullrich and Mosher, 1987; Lotz et al., 1989; Bourdon and Ruoslahti, 1989; Aukhil et al., 1993).

Assaying for inhibition of cell attachment suffers from the drawback that impurities in peptides may make them nonspecifically active in inhibition assays, hence the need for the direct cell attachment to confirm the activity of the peptide. To guard against the nonspecific inhibition, the inhibition assays should include an attachment system that is not inhibited. For example, attachment to laminin or type I collagen is generally not sensitive to RGD peptide inhibition (although both proteins have RGD sequences that can, when the protein is denatured or fragmented, be active), and affords one with a useful control against toxicity. Other safeguards include proper purification of the peptide. Acid left from the cleavage and fractionation steps is a particular problem; the pH of the peptide solution should be checked and the neutralization performed, when necessary. Avoiding excessively high concentrations is also important. An inhibition of cell attachment seen only at peptide concentrations above 1 mM should be interpreted with great caution. In such cases, an effort can be made to increase the sensitivity of the system. The best way of accomplishing that is to reduce the amount of adhesion protein used to coat the test surface (typically microtiter wells). For example, $\alpha_5\beta_1$-mediated cell attachment to fibronectin is not easily inhibited by RGD peptides. As illustrated in Fig. 1, the fibronectin concentration must be such that clear attachment is observed but not higher than that, for the RGD inhibition to be evident. Of obvious importance also is the nature of the adhesion receptors that mediate the attachment. A cell line that uses only one receptor for the attachment event studied should be sought. Cell lines that express integrins from transfected cDNA (Giancotti and Ruoslahti, 1990; Wu et al., 1993; Bergelson et al., 1992; Zhang et al., 1993) provide well-defined test cells for this purpose. Details of the cell attachment assay have been described in this series (Ruoslahti et al., 1987).

Adhesion Peptides from Peptide Libraries

Sequences that are useful as probes of cell attachment, as well as of integrin specificities, can be isolated from peptide libraries displayed on phage. Antibody fragments displayed will also yield integrin-binding products; however, because these are proteins rather than peptides, the antibody method will not be dealt with further here. The reader is referred to articles on this topic (Zanetti et al., 1993; Barbas et al., 1993).

Phage display libraries have been used to isolate peptides that bind to the $\alpha_{IIb}\beta_3$ integrin (O'Neil et al., 1992; see [18] in this volume) or $\alpha_5\beta_1$ integrins (Koivunen et al., 1993, 1994). The principle is to express a random peptide sequence from a DNA segment inserted into one of the surface proteins of filamentous phage and to select those phage that carry a peptide with the desired activity (Smith and Scott, 1993).

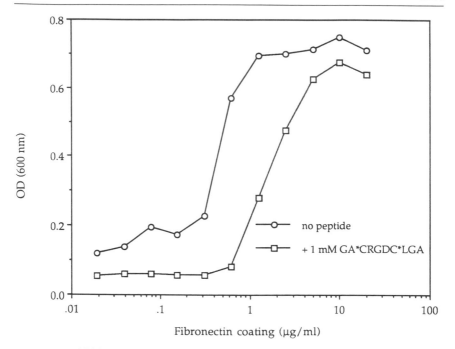

FIG. 1. Inhibition of cell attachment to fibronectin by an RGD-containing peptide. WI 38 fibroblasts (100,000 cells/well) were incubated for 1 hr at 37°, 5% CO_2 in microtiter wells coated with various concentrations of fibronectin. A competing peptide GA*CRGDC*LGA (Koivunen *et al.,* 1993) was included at a concentration of 1 mM (□). The attached cells were stained with amido black and quantitated. The results show means from duplicate wells. (○) No peptide.

The use of such libraries has confirmed the conclusion drawn from early work on synthetic peptides that cyclic RGD peptides are the most avid binders of integrin, surpassing linear peptides in potency (Piersch-bacher and Ruoslahti, 1987; see below). The peptide libraries can be designed so that they express only cyclic peptides; this is accomplished by engineering a cysteine residue on each side of the random sequence. As this system allows one to express as many as 10^{10} recombinants, it is feasible to cover all of the possible permutations of a hexapeptide in such a library. Thus, a library of the structure CX_6C should express every possible eight-amino acid ring cyclic peptide. Obviously, not all of the sequences may cyclize correctly as part of the phage protein; however, the peptides that have been isolated from such libraries appear to have had the appropriate disulfide bond, because cyclic peptides synthesized according to the phage-displayed sequences show the same enhanced integrin binding as the phage, whereas the linearized peptide is much less active (Koivunen *et al.*, 1993, 1994).

The phage display libraries have both yielded peptides that are more potent binders of a given integrin than the available RGD peptides as well as peptides that have alternative integrin-binding sequences. One interesting sequence revealed by the method is NGR (Koivunen *et al.*, 1993, 1994), because it is an almost exact reverse of the RGD sequence and because the ninth type III repeat of fibronectin has an NGR sequence at the same position as the RGD in the tenth repeat. The NGR sequence may be an auxiliary integrin-binding site in fibronectin. Cyclic peptide display libraries have yielded highly potent $\alpha_{IIb}\beta_3$-binding RGD peptides (O'Neil *et al.*, 1992; see [18] in this volume) and a non-RGD peptide that binds specifically and with high affinity to the $\alpha_5\beta_1$ integrin (Koivunen *et al.*, 1994). Table II shows sequences selected in successive pannings from a CX_7C library on $\alpha_5\beta_1$. Thus, the phage display libraries have considerable potential in cell adhesion research. Recombinant antibodies displayed on the surface of phage have also been shown to yield useful cell adhesion probes (Zanetti *et al.*, 1993; Barbas *et al.*, 1993). Another alternative is provided by libraries consisting of large numbers of synthetic peptides (Houghten *et al.*, 1991), but such libraries have not been used in cell adhesion research. The preparation of phage libraries for the display of cyclic peptides and the screening of such libraries is detailed below.

The filamentous bacteriophage display libraries are made possible by two features of the coat proteins encoded by genes III and VIII. They all have surface-exposed N-terminal domains, and they can both tolerate the insertion of foreign peptides. By exploiting these features, six different types of libraries have been developed so far (Smith and Scott, 1993). In a type 3 library, a foreign DNA fragment is inserted into the amino terminus of gene III, and the encoded peptide is displayed on each of the three to five pIII proteins on the tip of every phage (Cwirla *et al.*, 1990; Devlin *et al.*, 1990; Scott and Smith, 1990; reviewed in Dower, 1992). The construction of type 3 libraries displaying recombinant peptides that can potentially cyclize is described, followed by examples of using such libraries in screening for integrin-binding epitopes.

Preparation of Cyclic Fusion Phage Libraries

The overall strategies are modified from those described by Scott and Smith (1990). To generate cyclic libraries, two cysteine codons (TGT) are placed on the flanking sides of the degenerate oligonucleotide sequences CACTCGGCCGACGGGGCT<u>TGT</u>(NNK)₇<u>TGT</u>GGGGCCGCTGGGGC-AGAA, where N is the equimolar mixture of A, C, G, and T, and K is

TABLE II
Sequences Bound to $\alpha_5\beta_1$ Integrin from Cyclic (CX₇C) Phage Display Library[a]

Low-affinity selection: second panning			High-affinity selection: fifth panning	
L S **RGD**TP	WA **NGR**S H (3)	S T S DVGG (3)	I P **RGD**GW (3)	R R E T A WA (10)
D R**RGD**GF	F V **NGR**S F (2)	L N T NLGF (2)	L F **RGD**GW (2)	R G A P R A W
F T **RGD**AP	V L **NGR** ME	P E L F VES	Q T **RGD**GW (2)	
T S **RGD**MP	Y V **NGR**VS	FAGS LLV	V A **RGD**GW	
Q L **RGD**GW	MA **NGR**LL	R FGS HVP	L S **RGD**GW	
E G **RGD**WH	L **NGR**GLM	L GEF A FA	L F **RGD**GW	
T L **RGD**NH	AS V **NGH**T (4)	S RPS T FL	M T **RGD**GF	
H L **RGD**GW		S VANS VV	F **RGD**GFV	
M L **RGD**S F		G P CS GKS	**RGD**GFGS (2)	
M P **RGD**GF		V NVEYRN		
S **RGD**GFS (2)		AS F FAVQ		
F **RGD**HVR		L VAS MTP		
G **RGD**SVP		H V LAS AF		
S **RGD**GFR				
G **RGD**NLP				
RGDLRFN				

[a] Selection and sequencing of phage bound to the $\alpha_5\beta_1$ integrin were performed as described in text. In the second panning the coating concentration of the integrin on plastic was 50 μg/ml, and in the fifth panning it was 10 or 1 ng/ml. The number of clones encoding the same peptide is shown in parentheses. The RGD and NGR motifs are highlighted in bold.

an equimolar mixture of G and T. There are 32 possible codons at each position. Whereas there are only $32^3 = 3.3 \times 10^4$ independent clones in a CX_3C library, a fully representative CX_7C library should contain $32^7 = 3.4 \times 10^{10}$ different recombinants. Thus for construction of libraries containing six or more random amino acids, it is critical to optimize each step to achieve good representation. The following procedure was used in the construction of a CX_7C library exhibiting 5×10^9 independent clones.

Insert Preparation

DNA insert is prepared by three rounds of polymerase chain reaction (PCR) extension/amplification of synthetic oligonucleotides with a primer/oligonucleotide ratio of 10. Although biotinylated primers have been suggested for library construction (Scott and Smith, 1990), we have found that regular primers also work well. The PCR products are separated from primers and single-strand oligonucleotides on a 15% (w/v) polyacrylamide gel. Bands corresponding to the double-strand degenerate DNA are cut out. The DNA is recovered by electroelution into a Centricon 10 unit using a Centrilutor Microelectroeluter apparatus (Amicon, Danvers, MA). The eluted DNA is concentrated by centrifugation, extracted with phenol–chloroform, and precipitated with ethanol in the presence of glycogen as carrier. This procedure allows over 90% recovery of the starting material. The purified oligonucleotides are then digested with 500 units of *BgI*I restriction endonuclease (concentrated; Boehringer Mannheim, Indianapolis, IN) per microgram of DNA for 6 to 12 hr at 37°. The cleaved degenerate oligonucleotides can be easily identified on 15% (w/v) polyacrylamide gels and purified as described above.

Vector Preparation

We have been using the fUSE 5 as the vector for display library construction because of its unique features (Scott and Smith, 1990), including easy propagation and quantitation. The replicative form (double-stranded) of fUSE 5 is purified from the K802 strain of *Escherichia coli* transformed by single-stranded fUSE 5 phage DNA. Although a CsCl gradient may yield higher quality DNA, vectors purified using quick DNA isolation kits such as those from Qiagen (Chatsworth, CA) and Promega (Madison, WI) are adequate for these applications. Five units of *Sfi*I/µg DNA is used to digest fUSE 5. The "stuffer" sequence is removed by 2-propanol precipitation. Alternatively, the digested vector can be purified free from the stuffer sequence by agarose gel electrophoresis and electroeluted as detailed above.

Ligation and Electroporation

The purified vector and insert are ligated with T4 DNA ligase (Boehringer Mannheim), using buffers supplied by the manufacturer, for 4 hr at room temperature. To assure high ligation efficiency, different vector/ insert ratios are tested. The ligation products are purified with the Magic PCR kit from Promega or other equivalent systems to improve transformation efficiency. The Bethesda Research Laboratories (BRL; Gathersburg, MD) electroporation system and voltage booster are employed according to manufacturer instructions to transform electrocompetent MC1061 cells. Up to 50 ng of total ligated DNA in 1 μl of H_2O can be mixed with 25 μl of competent cells for each electroporation without affecting transformation efficiencies, which are typically in the range of $1-3 \times 10^8/\mu g$. Thus, under optimal conditions, 200 electroporations are sufficient to yield a library exhibiting over 10^9 epitopes. Bacteria from each 100 electroporations are pooled and cultured overnight in 1 liter of NZY medium plus tetracycline (20 $\mu g/ml$). Following two rounds of precipitations with 15% (w/v) polyethylene glycol 6000–3 M NaCl, phage libraries are resuspended at about 10^{13} transducing units (TU)/ml and stored at 4° (Smith and Scott, 1993).

Surveying Library for Integrin-Binding Phage

Phage recognized by integrins are selected by the affinity panning method originally exploited to identify peptide epitopes recognized by antibodies (Parmley and Smith, 1988; Scott and Smith, 1990). The integrins, which retain activity when immobilized on plastic (Staunton et al., 1989; Smith et al., 1990), are coated onto microtiter wells by incubating in Tris-buffered saline (TBS) containing 5 mM octylglucoside and 1 mM MnCl$_2$ overnight at 4°. Any remaining binding sites are blocked by bovine serum albumin, and the wells are washed two times with TBS–1 mM MnCl$_2$. MnCl$_2$ is used because it has been shown to increase the affinity of many integrins to peptide ligands (Gailit and Ruoslahti, 1988; Hautanen et al., 1989).

An aliquot of the phage display library containing 10^{12} TU is incubated for 4 hr at 25° in an integrin-coated microtiter well in TBS buffer containing 200 μl of 1% (w/v) bovine serum albumin and 1 mM MnCl$_2$. A fixed coating amount of 5 $\mu g/well$ is used in the first two pannings, but in the subsequent pannings the integrin is coated at 100- to 10,000-fold lower concentrations to select for high affinity sequences. Decreasing the binding sites available for phage on solid phase favors peptides with higher affinity (Scott and Smith, 1990; Barrett et al., 1992; Koivunen et al., 1993, 1994). Phage that remain bound after extensive washing with TBS–0.5% (v/v)

Tween 20 are eluted with 0.5 mM GRGDSP, 10 mM ethylenediaminetetra-acetic acid (EDTA), or a 0.1 M glycine buffer, pH 2.2, containing bovine serum albumin (1 mg/ml) and 0.1 mg/ml of the pH indicator phenol red. Phage are amplified by infection of F pilus-positive bacteria such as K91 or K91kan (Smith and Scott, 1993). Concentrated bacteria that adsorb phage efficiently can be prepared by growing bacteria in Terrific Broth medium until the OD at 600 nm of a 1/10 dilution reaches 0.125–0.25. The preparation of concentrated and starved bacteria that can be stored for a longer period is described elsewhere (Smith and Scott, 1993). Amplified phage are collected from cleared overnight cultures by polyethylene glycol precipitation.

To select high-affinity binding peptides, approximately 1 × 10^{10} TU of amplified phage is incubated for 1 hr at 25° in microtiter wells coated with various concentrations of the integrin, and the phage that remain bound are eluted with the pH 2.2 buffer. Phage are selected for further amplification from the well with the lowest concentration of integrin that yields phage binding over background. To sequence the peptide insert, well-separated bacterial colonies (Smith and Scott, 1993) are harvested and amplified, and the single-stranded DNA (ssDNA) of phage is sequenced by the dideoxy chain termination method using the oligonucleotide 5′ CCCTCATAGTTAGCGTAACG 3′ as primer. Sequencing reactions are carried out using the Sequenase kit according to manufacturer instructions (U.S. Biochemical, Cleveland, OH).

Determination of Affinity of Phage Insert

The sequencing of phage clones usually reveals hundreds of different promising peptides; it is valuable to know before synthesizing peptides which of those sequences have a high affinity for a target integrin. As each of the five protein III copies that occur on the surface of phage (Smith and Scott, 1993) carries the peptide insert, multivalent interaction may occur with the integrins. This problem may be even more serious with protein VIII-based vectors, as there are thousands of pVIII molecules on M13 phage. This means that the synthetic peptide may not have as high an affinity as the corresponding phage. To overcome this problem, vectors have been described that can display only one copy of the fusion protein on phage and are also able to carry longer protein domain inserts without interfering with the phage infectivity (Bass et al., 1990; Breitling et al., 1991; Kang et al., 1991). Regardless of the potential problem created by the multiple binding of the phage, we have found an excellent correlation between the avidity of phage binding and the affinity of the peptide synthesized, based on the sequence displayed by it.

The avidity of phage can be estimated from the difference of phage binding to integrin-coated wells and the background in bovine serum albumin-coated wells (Koivunen et al., 1993, 1994). Phage bound to the wells are quantitated by determining their ability to infect bacteria.

We have not found any significant differences between various phage clones in their ability to infect bacteria, at least as long as they carry a CX_7C or shorter insert. Knowledge of the titer of input phage in the assay is also not necessary. We use 10 μl of the amplified phage ($\sim 10^{10}$ TU/well) that is prepared from 1.7 ml of overnight culture by a single precipitation with polyethylene glycol (Smith and Scott, 1993). The phage pellet is dissolved in 500 ml of TBS and stored at $-20°$.

To determine the relative affinities of individual phage sequences for an integrin such as $\alpha_5\beta_1$, the phage clones are incubated for 1 hr in microtiter wells coated with the integrin or bovine serum albumin. The unbound phage are removed by washing the wells five times with TBS–0.5% (v/v) Tween 20. Twenty microliters of concentrated K91kan bacteria (Smith and Scott, 1993) diluted 1 : 10 in Terrific Broth medium is pipetted to each well and incubated for 10 min at 37°. One hundred microliters of Terrific Broth containing kanamycin (100 μg/ml) and tetracycline (0.2 μg/ml) is added, and the incubation at 37° continues for another 30 min. Finally, another 100 μl of Terrific Broth containing kanamycin (100 μg/ml) and tetracycline (80 μg/ml) is added and the microtiter plate incubated overnight (16–24 hr) at room temperature with shaking until the bacterial cultures are visibly turbid. Absorbance, which is proportional to phage number, is read at 600 nm with an enzyme-linked immunosorbent assay (ELISA) reader. This method can also be used to titer any phage solution (Koivunen et al., 1993), providing an easier way of quantitating the phage than calculating a number of bacterial colonies on a plate.

Figure 2 shows an example of the use of this method. Ten representative phage clones, five each from a linear hexapeptide and a cyclic heptapeptide library, were compared for their binding to the $\alpha_5\beta_1$ integrin. The phage displaying a linear sequence generally displayed weaker binding to $\alpha_5\beta_1$ than did the cyclic structures.

The phage bound to the target protein can also be quantitated using anti-phage antibodies (McCafferty et al., 1990; Oldenburg et al., 1992). Other possibilities include the immobilization of affinity-purified phage particles, either directly on plastic (Stephen and Lane, 1992; Scott et al., 1992) or through coating of anti-M13 antibodies (Yayon et al., 1993; Balass et al., 1993). In these systems the target molecule that binds to immobilized phage is determined. Phage can also be immobilized on a nitrocellulose filter by directly transferring bacterial colonies and washing the cells away.

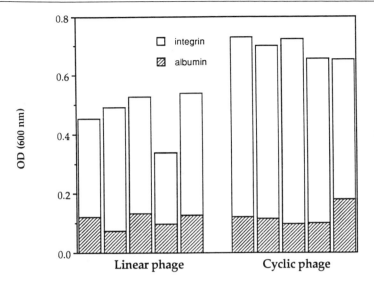

FIG. 2. Attachment of phage displaying linear and cyclic RGD-containing inserts to the $\alpha_5\beta_1$ integrin. Five phage clones displaying a linear X_6 insert and five clones displaying a CX_7C insert (10^{11} TU/well) were incubated for 1 hr in microtiter wells coated with the integrin (□) or bovine serum albumin (▨). The phage remained bound after washing were quantitated as described in text. The results show means from duplicate wells.

This method has been used to search for peptides recognized by the molecular chaperone BiP (Blond-Elguindi *et al.*, 1993).

Antibody Capture of Integrin to Microtiter Wells

The isolation of integrins is commonly accomplished by affinity purification on the ligand, such as fibronectin or RGD peptides (Pytela *et al.*, 1987). Because the specificities of the integrins are closely related, cross-contamination of one integrin with another may be a problem. For this reason, we have used antibodies to capture the desired integrin to microtiter wells to verify the binding of phage to the appropriate integrin or to screen libraries. We have found the antibodies prepared against at least the cytoplasmic domain peptides of the α_5, α_v, and β_3 subunits to be suitable for this purpose. The details of the procedure are as follows. The $\alpha_5\beta_1$ integrin is bound to wells coated with anti-α_5 antibodies (5–10 μg/ml) by incubating 300 μl of a placental extract (Pytela *et al.*, 1987) per well in TBS buffer containing 0.1 M octylglucoside, 1 mM $CaCl_2$, 1 mM $MnCl_2$, and proteinase inhibitors. Control wells are prepared by using the octylglucoside buffer instead of the placental extract. After incubation overnight at 4°, the wells are extensively washed with TBS containing 0.1% (v/v) Nonidet P-40 (NP-40). The activity of the antibody-captured integrin can be confirmed by testing its ability to bind fibronectin. ^{125}I-Labeled fibronectin (100,000 cpm/well) is incubated with or without competing peptides for 1 hr at 25° in a 100-μl volume of TBS containing 0.1% (v/v) NP-40 and 1 mM MnCl$_2$. After repeated washing, the radioactivity remaining bound is quantitated with a γ counter. Typically about 20% of the radioactivity will remain bound in the integrin wells, whereas the background is about 1%. The integrin-coated wells can be used for phage binding and screening as described above for wells coated directly with integrin. It should be noted that screening with this method will also give phage that bind to the antibody.

Chemical Peptide Design

Once candidate peptide sequences have been identified either from adhesion proteins or from phage display libraries, these peptides can be improved by means of peptide design. If the initial sequence is not cyclic, it is very likely that cyclization will improve the binding. This has been the experience with peptides binding to several integrins (Pierschbacher and Ruoslahti, 1987; Gurrath *et al.*, 1992; Koivunen *et al.*, 1993; Nowlin *et al.*, 1993).

An exhaustive screening of cyclic peptide libraries by methods that select for high-affinity peptides should ensure that the most potent binding sequence is identified. This circumstance would mean that further improvements in the affinity of such a peptide are unlikely to come from sequence variations with natural amino acids within the ring, but can be derived only from substitutions with nonnatural amino acids and other chemical groups.

The chemical design offers several possibilities that are based on previous experience with cyclic peptides designed to be potent inhibitors of the $\alpha_{IIb}\beta_3$ integrin (e.g., Tschopp *et al.*, 1993). In that case, a 20,000-fold improvement in potency relative to the fibronectin-derived linear peptide GRGDSP has been achieved. Moreover, the resulting peptides are essentially specific for $\alpha_{IIb}\beta_3$ among the RGD-directed integrins. One possibility in enhancing the potency of the peptide is by varying the residues in the exocyclic positions. The design of the $\alpha_{IIb}\beta_3$-specific peptides has shown that an exocyclic positive charge can enhance the specificity of the peptide for $\alpha_{IIb}\beta_3$ (Cheng *et al.*, 1994). To minimize the number of permutations needed, one can experiment with the addition of amino acids representing general categories of amino acids—basic, acidic, hydrophobic—in the positions next to the cysteines. Any increase or decrease of activity will then guide further efforts as in the case of the platelet peptides.

Changing the amino acids that form the cyclizing bond or changing the chemistry of the bond itself is another potentially useful change. The cysteine residues can be substituted with penicillamine either in the L or D conformation. Such a change has resulted in as large as a 10-fold change in the potencies of the $\alpha_{IIb}\beta_3$ inhibitors (Cheng et al., 1994). Additional conformational restraint can be introduced by adding side chains to the Pen residue. The Pmp (Yim and Huffman, 1983) and Pmc (Stanfield and Hruby, 1988; Yim et al., 1988) residues both contain the more lipophilic and conformationally restricting β/β cyclopentamethylene at the β-carbon atom of the side chain. These residues have been previously used to restrict the conformational freedom of other disulfide-bridged peptides (Struthers et al., 1984) and their use has resulted in two- to threefold activity changes in the $\alpha_{IIb}\beta_3$ inhibitors (Cheng et al., 1994). A peptide bond in the place of the disulfide bridge may also be a useful modification; potent binders of α_V integrins have been prepared in this manner (Gurrath et al., 1992).

A number of modifications inside the ring can also be attempted. The use of the D-amino acids is one of them. The arginine residues in particular are an appropriate target, especially because an arginine-glycine sequence may make the peptide susceptible to proteases such as thrombin. It is possible to change the arginine in linear and cyclic RGD peptides to a D-form with full retention of activity (Pierschbacher and Ruoslahti, 1987; Cheng et al., 1994). Other changes that can be attempted include replacing peptide bonds with more rigid chemical bonds (i.e., double bonds), adding conformationally restricting pendant groups to the backbone, making secondary cross-links within the cyclic peptide, and including cyclic and heterocyclic structures within the backbone ring to give the peptide more rigidity.

A second bridge may be necessary to restrict further the conformation of the peptide. This is somewhat more difficult and requires that the cysteine residues forming the second bridge be blocked at the sulfhydryl with a protecting group that is stable to HF [for example, acetamidomethyl (ACM); Stewart and Young, 1984] and can be removed after the first bridge is formed. Some ACM may be lost during cleavage. This will reduce the yield of proper peptide but should not affect the ultimate goal.

Closing the ring with the peptide bond requires more sophisticated peptide chemistry. A practical protocol for making peptides has been described by Varga et al. (1979). To make the protected linear peptide prior to cyclization, F-Moc amino acids can be used as described by Chang et al. (1980). For side-chain protection, protecting groups that are stable to trifluoroacetic acid (TFA) cleavage may be used such as F-Moc-nitroarginine, F-Moc-O-benzylserine, and F-Moc-β-benzylaspartic acid, available from Bachem (Bubendorf, Switzerland). After cyclization, the nitro- and benzyl-protecting groups can be removed by hydrogenation (Lafrenie et al., 1992). Other protocols may also be used (Kopple, 1972; Kopple et al., 1978; DeGrado and Kaiser, 1980, 1982). The peptides can be purified using high-performance liquid chromatography (HPLC) and the quality control should include sequencing and testing of reduced and unreduced peptides for biological activity.

Isolation of $\alpha_5\beta_1$ Integrin by Peptide Affinity Chromatography

Adhesive peptides are useful tools for receptor purification. For instance, the linear peptide GRGDSPK coupled to Sepharose may be used for affinity purification of $\alpha_V\beta_3$ or the platelet receptor $\alpha_{IIb}\beta_3$ (Pytela et al., 1987). Although the peptide sequence derives from fibronectin, the fibronectin receptor $\alpha_5\beta_1$ integrin does not bind this column, presumably because the affinity of this interaction is too low without additional receptor contacts that occur in the natural ligand. This was the first demonstration that differential affinity for a peptide could be used to isolate an integrin selectively.

We have designed a cyclic peptide with the sequence GA*CRRETA-WAC*GA which binds specifically to $\alpha_5\beta_1$ with high affinity (Koivunen et al., 1994). Here we describe an application for this peptide in the purification of $\alpha_5\beta_1$ from human placental tissue by a method based on the procedure for purification of the vitronectin receptor by affinity for the peptide GRGDSPK (Pytela et al., 1987). Indeed, the GRGDSPK and the GA*CRRETAWAC*GA peptide columns may be run in tandem to purify simultaneously both $\alpha_5\beta_1$ and $\alpha_V\beta_3$ from the same starting material.

Procedure

The $\alpha_5\beta_1$ affinity resin is prepared by coupling 75 mg of peptide GA*CRRETAWAC*GA to 5 ml of cyanogen bromide-activated 4B Sepharose according to manufacturer instructions (Pharmacia, Uppsala, Sweden). The peptide resin is packed into a 1-cm diameter column and equilibrated [in TBS with 1 mM CaCl$_2$, 1 mM MnCl$_2$, and 100 mM octyl-β-D-glucopyranoside (Calbiochem, La Jolla, CA)]. Two hundred grams of placental tissue stored frozen at $-20°$ is thawed and distributed into three centrifuge bottles. All subsequent steps are carried out at 4°. The tissue is washed three times by addition of 400 ml of ice-cold TBS containing proteinase inhibitors [1 mM phenylmethylsulfonyl fluoride (PMSF), leupeptin (0.5 μg/ml), and pepstatin (0.5 μg/ml)] and centrifugation for 10 min at 10,000 rpm. The washed tissue is mixed with a minimal volume (200 ml) of ice-cold extraction buffer (TBS containing 1 mM CaCl$_2$, 1 mM MnCl$_2$, 100 mM octyl-β-D-glucopyranoside, and proteinase inhibitors) and incubated for 4 hr. After centrifugation for 20 min at 10,000 g, the supernatants are pooled and passed over the peptide column that had previously been equilibrated in TBS containing 1 mM CaCl$_2$, 1 mM MnCl$_2$, and 100 mM octyl-β-D-glycopyranoside. The column is then washed with 200 ml of wash buffer (TBS containing 1 mM CaCl$_2$, 25 mM octyl-β-D-glycopyranoside, and proteinase inhibitors) and $\alpha_5\beta_1$ is eluted is with 20 ml of elution buffer (TBS containing 20 mM EDTA, 25 mM octyl-β-D-glucopyranoside and proteinase inhibitors). For most purposes, further purification is not required. After dialysis in TBS containing 1 mM MnCl$_2$ and 0.02% (w/v) NaN$_3$, the integrin may be stored at 4° for at least 1 month or aliquots may be quickly frozen with liquid nitrogen and stored at −80°. The final yield of 100 μg obtained by this method is comparable to other methods of purification (Pytela *et al.*, 1987). The peptide column may be regenerated by washing with 100 ml of 8 M urea, 50 mM Tris-HCl, pH 7.5, followed by extensive washing with storage buffer [TBS containing 0.02% (w/v) NaN$_3$].

The advantage of using peptides versus natural ligands for affinity purification of integrins is the lower cost and the potential for better purification resulting from elimination of other binding sites such as those potentially present in the type III repeat units of the fibronectin fragment previously used for the purification. Affinity purifications based on integrin antibodies have also been used (Koivunen *et al.*, 1994) but they are also expensive and generally do not select against inactivated integrins. Nowlin *et al.* (1993) described a cyclic pentapeptide, RCD(ThioP)C, that binds $\alpha_4\beta_1$ and $\alpha_5\beta_1$ with high affinity and demonstrated that it could be used to purify both these integrins. In contrast, the cyclic peptide CRRETAWAC appears to be specific for $\alpha_5\beta_1$. The usefulness of adhesive peptides for purification of integrins will likely increase as peptides with new specificities are discovered.

Promotion of Cell Attachment with Peptides

The first order of business in the use of peptides to promote cell attachment is to attach the peptide to a solid surface, usually a microtiter well. Peptides do not coat plastic particularly efficiently; coating efficiency is directly proportional to the size of the protein used in the coating (Engvall and Perlman, 1971; Salonen *et al.*, 1984). The coating efficiency falls off around a molecular weight of 10,000 and is very low for the adhesion peptides, which usually have molecular weights under 2,000.

There are several ways of making peptides coat plastic. The simplest one is to cross-link the peptide so that its effective molecular weight becomes that of a protein. The cross-linking can be accomplished with glutaraldehyde. The peptide is incubated at various concentrations ranging from 1 to 1000 μg/ml in microtiter wells in a solution containing 0.25% (v/v) glutaraldehyde in pH 7.5 phosphate-buffered saline (PBS) for 2 hr at 37°. The wells are then washed with PBS and the remaining binding sites of the plastic are blocked by incubating a 1-mg/ml solution of bovine serum albumin in the wells for another hour, after which the wells can be used for cell attachment.

A more controlled way of coating an adhesion peptide onto a surface is to coat the surface first with a protein and then couple the peptide to the protein. Pierschbacher and Ruoslahti (1984a,b) used IgG as the coating protein and N-succinimidyl 3-(2-pyridyldithio)propionate (SPDP) as the cross-linker. They added a cysteine residue to the C terminus of the peptide to facilitate the coupling, although the cross-linker used will also couple through amino groups. The details of the method are as follows.

One milligram of peptide is dissolved in 6 M urea at a concentration of 2 mg/ml at pH 8.0 and reduced by adding dithiothreitol to a final concentration of 45 mM. The peptide is then freed of the reactants by passing it through a 5-ml Sephadex G-25 column equilibrated with phosphate-buffered saline. The fractions containing the peptide are collected and pooled and used in cell attachments assays. Wells in untreated polystyrene microtiter plates are either left uncoated or coated with purified rabbit IgG by incubating a 20-μg/ml solution in the wells for 2 hr at room temperature. After washing the wells to remove unattached protein, the IgG coating is first derivatized with 3-(2-pyridyldithio)propionic acid N-hydroxysuccinimide ester, a bifunctional cross-linker, at 10 μg/ml for 30 min at room temperature, and then a solution containing the reduced peptide is added to the wells at different concentrations and allowed to react for at least 1 hr (the cross-linker reacts mainly with amino groups in the IgG and subsequently cross-links the peptide to IgG through the cysteine residue in the peptide). After repeated washing to remove unattached peptide the plate is used for cell attachment assays. Alternatively, it may be possible to use commercially available microtiter plates that have a coupling group attached to the plastic (e.g., Covalink; Nunc, Roskilde, Denmark).

While the methods described above will allow coupling of any peptide with at least one free amino group in it, a method has also been developed that depends on the design of the peptide in such a way that it binds to plastic (and to other surfaces). This is accomplished by synthesizing an RGD peptide with a hydrophobic tail that binds the peptide to the surface. In a commercial preparation, Peptite 2000 (Telios Pharmaceuticals, Inc., San Diego, CA), the tail consists of leucine residues. The details for the use of this peptide can be found in the distributor instructions (Life Technologies, Inc., Gaithersburg, MD).

Soluble Adhesion Peptides in Cell and Tumor Biology

As discussed above, RGD and other adhesion peptides can inhibit cell attachment. Such assays are useful when one wants to ascertain the involvement of certain integrins in cell attachment. Antibodies capable of inhibiting the ligand-binding function of individual integrins are also useful for this purpose.

An important question regarding the activities of the adhesion peptides is whether they act only as inhibitors or whether they might also serve as agonists of the integrins. Some results suggest that the latter can be the case: first, treatment of cells with RGD peptide can cause integrin relocation. When cells are plated on laminin, they form adhesion plaques that contain the appropriate laminin receptor integrin but not $\alpha_5\beta_1$. However, when such cells are treated with an RGD peptide that does not affect their attachment to laminin, $\alpha_5\beta_1$ moves to adhesion plaques (LaFlamme et al., 1992). The reason for this is thought to be that ligand occupancy reverses an inhibitory influence that is exerted by the α-subunit cytoplasmic domain and that prevents the β subunit from directing the integrin into adhesion plaques (Sastry and Horwitz, 1993). The translocation into adhesion plaques is important, because that appears to be where the subsequent signaling events, such as the activation of focal adhesion kinase (Guan et al., 1991; Kornberg et al., 1991; Schaller and Parsons, 1993) and possibly also that of protein kinase C (Vuori and Ruoslahti, 1993), originate. In being able to cause translocation of the $\alpha_5\beta_1$ integrin at least to preformed adhesion plaques, the soluble RGD peptides seem to be at least partially capable of functioning as integrin agonists. These effects may be important for the in vivo activities of the peptides such as inhibition of experimental metastasis.

Adhesion Peptides in Vivo

Integrin-binding peptides are useful in vitro as research tools and show promise as pharmaceuticals. It has been shown by several groups that intravenous coinjection of RGD peptides with tumor cells can reduce the colonization of tissues by the tumor cells. The transfer of hypersensitivity by lymphocytes injected intravenously has also been suppressed by administration of adhesion peptides; both RGD and LDV type peptides were active in this regard (Ferguson *et al.*, 1991). Unfortunately, it has not been possible to extend these studies to spontaneous metastasis, because the peptides have half-lives in the circulation of only a few minutes, and their affinities are low. The concentrations and quantities of peptide that would be needed under these circumstances exceed practical limits. Moreover, the RGD peptides that are currently available for modulating metastasis bind to all RGD-directed integrins, including the platelet integrin $\alpha_{IIb}\beta_3$, making the interpretation of the results difficult. For example, although there is some evidence against that possibility (Humphries *et al.*, 1988), the question remains as to whether the inhibition of experimental metastasis observed previously might be caused by inhibition of platelet aggregation around the tumor cells mediated by $\alpha_{IIb}\beta_3$ (Karpatkin *et al.*, 1988). Clearly, integrin-specific, high-affinity peptides will have to be designed for progress to be made in tackling the problem of integrins in metastasis.

All of the above-described goals in peptide development have been met with the platelet integrin, $\alpha_{IIb}\beta_3$. Several groups have designed cyclic peptides that bind only to $\alpha_{IIb}\beta_3$ among the RGD-directed integrins and that have very high affinities, 20,000-fold higher than those of the original linear RGD and KGD peptides (Samanen *et al.*, 1991; Scarborough *et al.*, 1993; Cheng *et al.*, 1994; Tschopp *et al.*, 1993). Such peptides are currently in clinical trials as a new type of antithrombotic, and efforts are underway at pharmaceutical companies to reproduce the appropriate RGD structure with a nonpeptide chemistry to derive orally active inhibitors of platelet aggregation. Similar advances with the other integrins would allow integrin modulation in metastasis and other diseases in which integrins play a role.

Acknowledgments

The author's work and the preparation of this chapter are supported by Grants CA42507, CA28896, and Cancer Center Support Grant CA30199 from the National Cancer Institute. Erkki Koivunen was supported by a grant from the Finnish Cultural Fund. We thank Dr. William Craig for help with the peptide chemistry part and Diana Lowe for help in the preparation of the manuscript.

Altieri, D. C., Plescia, J., and Plow, E. F. (1993). *J. Biol. Chem.* **268,** 1847.
Aukhil, I., Joshi, P., Yan, Y., and Erickson, H. P. (1993). *J. Biol. Chem.* **268,** 2542.
Aumailley, M., Timpl, R., and Risau, W. (1991). *Exp. Cell Res.* **196,** 177.
Balass M., Heldman, Y., Cabilly, S., Givol, D., Katchalski-Katzir, E., and Fuchs, S. (1993). *Proc. Natl. Acad. Sci. U.S.A.* **90,** 10638.
Barbas, C. F., Languino, L. R., and Smith, J. W. (1993). *Proc. Natl. Acad. Sci. U.S.A.* **90,** 10003.

Barrett, R. W., Cwirla, S. E., Ackerman, M. S., Olson, A. M., Peters, E. A., and Dower, W. J. (1992). *Anal. Biochem.* **204,** 357.

Bass, S., Greene, R., and Wells, J. A. (1990). *Proteins: Struct., Func., Genet.* **8,** 309.

Bergelson, J. M., Shepley, M. P., Chan, B. M. C., Hemler, M. E., and Finberg, R. W. (1992). *Science* **255,** 1718.

Blond-Elguindi, S., Cwirla, S. E., Dower, W. J., Lipshutz, R. J., Sprang, S. R., Sambrook, J. F., and Gething, M. J. (1993). *Cell (Cambridge, Mass.)* **75,** 717.

Bourdon, M., and Ruoslahti, E. (1989). *J. Cell Biol.* **108,** 1149.

Breitling, F., Dübel, S., Seehaus, T., Klewinghaus, I., and Little, M. (1991). *Gene* **104,** 147.

Bunch, T. A., and Brower, D. L. (1992). *Development (Cambridge, UK)* **116,** 239.

Busk, M., Pytela, R., and Sheppard, D. (1992). *J. Biol. Chem.* **267,** 5790.

Cardarelli, P. M., Yamagata, S., Taguchi, I., Gorscan, F., Chiang, S. L., and Lobl, T. (1992). *J. Biol. Chem.* **267,** 23159.

Chang, C.-D., Felix, A. M., Jimenez, M. H., and Meienhofer, J. (1980). *J. Pept. Protein Res.* **15,** 485.

Charo, I. F., Nannizzi, L., Phillips, D. R., Hsu, M. A., and Scarborough, R. M. (1991). *J. Biol. Chem.* **266,** 1415.

Cheng, S., Craig, W. S., Mullen, D., Tschopp, J. F., Dixon, D., and Pierschbacher, M. D. (1994). *J. Med. Chem.* **37**(1), 1.

Cwirla, S. E., Peters, E. A., Barrett, R. W., and Dower, W. J. (1990). *Proc. Natl. Acad. Sci. U.S.A.* **87,** 6378.

D'Souza, S. E., Ginsberg, M. H., Burke, T. A., Lam S. C.-T., and Plow, E. F. (1988). *Science* **242,** 91.

D'Souza, S. E., Ginsberg, M. H., Burke, T. A., and Plow, E. F. (1990). *J. Biol. Chem.* **265,** 3440.

DeGrado, W. F., and Kaiser, E. T. (1980). *J. Org. Chem.* **45,** 1295.

DeGrado, W. F., and Kaiser, E. T. (1982). *J. Org. Chem.* **47,** 3258.

Devlin, J. J., Panganiban, L. C., and Devlin, P. E. (1990). *Science* **249,** 404.

Dower, W. J. (1992). *Curr. Biol.* **2,** 251.

Eble, J. A., Golbik, R., Mann, K., and Kühn (1993). *EMBO J* **12,** 4795.

Edwards, J., Hameed, H., and Campbell, G. (1988). *J. Cell Sci.* **89,** 507.

Elices, M. J., Urry, L. A., and Hemler, M. E. (1991). *J. Cell Biol.* **112,** 169.

Engvall, E., and Perlman, P. (1971). *Immunochemistry* **8,** 871.

Ferguson, T. A., Mizutani, H., and Kupper, T. S. (1991). *Proc. Natl. Acad. Sci. U.S.A.* **88,** 8072.

Frazier, W. A. (1993). This series, Vol. 5, p. 212.

Gailit, J., and Ruoslahti, E. (1988). *J. Biol. Chem.* **263,** 12927.

Gartner, T. K., and Taylor, D. B. (1990). *Thromb. Res.* **60,** 291.

Gehlsen, K. R., Dillner, L., Engvall, E., and Ruoslahti, E. (1988). *Science* **241,** 1228.

Giancotti, F. G., and Ruoslahti, E. (1990). *Cell (Cambridge, Mass.)* **60,** 849.

Graf, J., Ogle, R. C., Robev, F. A., Sasaki, M., Martin, G. R., Yamada, Y., and Kleinman, H. K. (1987). *Biochem. J.* **26,** 6896.

Guan, J.-L., and Hynes, R. O. (1990). *Cell (Cambridge, Mass.)* **60,** 53.

Guan, J.-L., Trevithick, J. E., and Hynes, R. O. (1991). *Cell Regul.* **2,** 951.

Gurrath, M., Müller, G., Kessler, H., Aumailley, M., and Timpl, R. (1992). *Eur. J. Biochem.* **210,** 911.

Hautanen, A., Gailit, J., Mann, D. M., and Ruoslahti, E. (1989). *J. Biol. Chem.* **264,** 1437.

Hinek, A., Wrenn, D. S., Mecham, R. P., and Barondes, S. H. (1988). *Science* **239,** 1539.

Houghten, R. A., Pinilla, C., Blondelle, S. E., Appel, J. R., Dooley, C. T., and Cuervo, J. H. (1991). *Nature (London)* **354,** 84.

Humphries, M. J., Yamada, K. Y., and Olden, K. (1988). *J. Clin. Invest.* **81,** 782.

Iida, J., Skubitz, A. P., Furcht, L. T., Wayner, E. A., and McCarthy, J. B. (1992). *J. Cell Biol.* **118,** 431.

Iwamoto, Y., Robev, F. A., Graf, J., Sasaki, M., Kleinman, H. K., Yamada, Y., and Martin, G. R. (1987). *Science* **238,** 1132.

Kang, A. S., Barbas, C. F., Janda, K. D., Benkovic, S. J., and Lerner, R. A. (1991). *Proc. Natl. Acad. Sci. U.S.A.* **88,** 4363.

Karpatkin, S., Pearlstein, E., Ambrogio, C., and Coller, B. S. (1988). *J. Clin. Invest.* **81,** 1012.

Kloczewiak, M., Timmons, S., Lukas, T. J., and Hawiger, J. (1984). *Biochemistry* **23,** 1767.

Koivunen, E., Gay, D. A., and Ruoslahti, E. (1993). *J. Biol. Chem.* **268,** 20205.

Koivunen, E., Wang, B., and Ruoslahti, E. (1994). *J. Cell Biol.* **124,** 373.

Kopple, K. D. (1972). *J. Pharm. Sci.* **61,** 1345.

Kopple, K. D., Go, A., and Schamper, T. J. (1978). *J. Am. Chem. Soc.* **100,** 4289.

Kornberg, L. J., Earp, H. S., Turner, C. E., Prockop, C., and Juliano, R. L. (1991). *Proc. Natl. Acad. Sci. U.S.A.* **88,** 8392.

LaFlamme, S. E., Akiyama, S. K., and Yamada, K. M. (1992). *J. Cell Biol.* **117,** 437.

Lafrenie, R. M., Podor, T. J., Budanan, M. R., and Orr, F. W. (1992). *Cancer Res.* **52,** 2202.

Lam, S. C.-T., Plow, E. F., Smith, M. A., Andrieux, A., Ryckwaert, J. J., Marguerie, G., and Ginsberg, M. H. (1987). *J. Biol. Chem.* **262,** 947.

Lawler, J., Weinstein, R., and Hynes, R. O. (1988). *J. Cell Biol.* **107,** 2351.

Lotz, M. M., Burdsal, C. A., Erickson, H. P., and McClay, D. R. (1989). *J. Cell Biol.* **109,** 1795.

McCafferty, J., Griffiths, A. D., Winter, G., and Chiswell, D. J. (1990). *Nature (London)* **348,** 552.

McCarthy, J. B., Chelberg, M. K., Mickelson, D. J., and Furcht, L. T. (1988). *Biochemistry* **27,** 1380.

Morla, A., and Ruoslahti, E. A. (1992). *J. Cell Biol.* **118,** 421.

Mould, A. P., and Humphries, M. J. (1991). *EMBO J.* **10,** 4089.

Mould, A. P., Wheldon, L. A., Komoriya, A., Wayner, E. A., Yamada, K. M., and Humphries, M. J. (1990). *J. Biol. Chem.* **265,** 4020.

Mould, A. P., Komoriya, A., Yamada, K. M., and Humphries, M. J. (1991). *J. Biol. Chem.* **266,** 3579.

Murphy-Ullrich, J. E., and Mosher, D. F. (1987). *J. Cell. Biol.* **105,** 1603.

Nishimura, S. L., and Pytela, R. (1993). *Mol. Biol. Cell* **4,** Suppl., 285a.

Nomizu, M., Yamamura, K., Kleinman, H., and Yamada, Y. (1993). *Cancer Res.* **53,** 3459.

Norton, P. A., and Hynes, R. O. (1987). *Mol. Cell. Biol.* **7,** 4297.

Nowlin, D. M., Gorcsan, F., Moscinski, M., Chiang, S. L., Lobl, T. J., and Cardarelli, P. M. (1993). *J. Biol. Chem.* **268,** 20352.

Oldenburg, K. R., Loganathan, D., Goldstein, I. J., Schultz, P. G., and Gallop, M. A. (1992). *Proc. Natl. Acad. Sci. U.S.A.* **89,** 5393.

O'Neil, K. T., Hoess, R. H., Jackson, S. A., Ramachandran, N. S., Mousa, S. A., and DeGrado, W. F. (1992). *Proteins* **14,** 509.

Parmley, S. F., and Smith, G. P. (1988). *Gene* **73,** 305.

Pierschbacher, M. D., and Ruoslahti, E. (1984a). *Nature (London)* **309,** 30.

Pierschbacher, M. D., and Ruoslahti, E. (1984b). *Proc. Natl. Acad. Sci. U.S.A.* **81,** 5985.

Pierschbacher, M. D., and Ruoslahti, E. (1987). *J. Biol. Chem.* **262,** 17294.

Pytela, R., Pierschbacher, M. D., Argraves, W. S., Suzuki, S., and Ruoslahti, E. (1987). This series, Vol. 144, p. 475.

Ruoslahti, E. (1991). *J. Clin. Invest.* **87,** 1.

Ruoslahti, E., and Pierschbacher, M. D. (1987). *Science* **238,** 491.

Ruoslahti, E., Suzuki, S., Hayman, E. G., Ill, C. R., and Pierschbacher, M. D. (1987). This series, Vol. 144, p. 430.

Salonen, E.-M., Vartio, T., Miggiano, V., Stähli, C., Tacás, B., Virgallita, G., De Petro, G., Barlati, S., and Vaheri, A. (1984). *J. Immunol. Methods* **72,** 145.

Samanen, J., Ali, F., Romoff, T., Calvo, R., Sorenson, E., Vasko, J., Storer, B., Berry, D., Bennett, D., Strohsacker, M., Powers, D., Stadel, J., and Nichils, A. (1991). *J. Med. Chem.* **34,** 3114.

Sastry, S. K., and Horwitz, A. F. (1993). *Curr. Opin. Cell Biol.* **5,** 819.

Scarborough, R. M., Naughton, M. A., Teng, W., Rose, J. W., Phillips, D. R., Lannizzi, L., Arfsten, A., Campbell, A. M., and Charo, I. F. (1993). *J. Biol. Chem.* **268,** 1066.

Schaller, M. D., and Parsons, J. T. (1993). *Trends Cell Biol.* **3,** 258.

Scott, J. K., and Smith, G. P. (1990). *Science* **249,** 386.

Scott, J. K., Loganathan, D., Easley, R. B., Gong, X., and Goldstein, I. J. (1992). *Proc. Natl. Acad. Sci. U.S.A.* **89,** 5398.

Smith, G. P., and Scott, J. K. (1993). This series, Vol. 217, p. 228.

Smith, J. W., and Vestal, D. J., Irwin, S. V., Burke, T. A., and Cheresh, D. A. (1990). *J. Biol. Chem.* **265,** 11008.

Staatz, W. D., Fok, K. F., Zutter, M. M., Adams, S. P., Rodriquez, B. A. and Santoro. S. A. (1991). *J. Biol. Chem.* **266,** 7363.

Stanfield, C. F., and Hruby, V. J. (1988). *Synth. Communi.* **18,** 531.

Staunton, D. E., Dustin, M. L., and Springer, T. A. (1989). *Nature (London)* **339,** 61.

Stephen, C. W., and Lane, D. P. (1992). *J. Mol. Biol.* **225,** 577.

Stewart, J. M., and Young, J. D. (1984). "Solid Phase Peptide Synthesis," 2nd ed. Rockefeller Press, New York.

Struthers, R. S., Hagler, A. T., and Rivier, J. (1984). *ACS Symp. Ser.* 139.

Tschopp. J. F., Driscoll, E. M., Mu, D.-X., Black, S. C., Pierschbacher, M. D., and Lucchesi, B. R. (1993). *Coronary Artery Dis.* **4,** 809.

Varga, S. L., Brady, S. F., Freidinger, R. M., Hirschmann, R., Holly, F. W., and Veber, D. F. (1979). "Peptides," p. 183. Wroclaw Univ. Press, Poland.

Vogel, B. E., Lee, S.-J., Hildebrand, A., Craig, W., Pierschbacher, M., Wong-Staal, F., and Ruoslahti, E. (1993). *J. Cell Biol.* **121,** 461.

Vuori, K., and Ruoslahti, E. (1993). *J. Biol. Chem.* **268,** 21459.

Wu, C., Bauer, J. S., Juliano, R. L., and McDonald, J. A. (1993). *J. Biol. Chem.* **268,** 21883.

Yayon, A., Aviezer, D., Safran, M., Gross, J. L., Heldman, Y., Cabilly, S., Givol, D., and Katchalsii-Katzir, E. (1993). *Proc. Natl. Acad. Sci. U.S.A.* **90,** 10643.

Yim, N. C. F., and Huffman, W. F. (1983). *Int. J. Pept. Protein Res.* **21,** 568.

Yim, N. C. F., Bryan, H., Huffman, W. F., and Moore, M. L. (1988). *J. Org. Chem.* **53,** 4605.

Zanetti, M., Filaci, G., Lee, H. R., del Guercio, P., Rossi, F., Bacchetta, R., Stevenson, F., Barnaba, V., and Billetta, R. (1993). *EMBO J.* **12,** 4375.

Zhang, Z., Morla, A. O., Vuori, K., Bauer, J. S., Juliano, R. L., and Ruoslahti, E. (1993). *J. Cell Biol.* **122,** 235.

[18] Identification of Recognition Sequences of Adhesion Molecules Using Phage Display Technology

By KARYN T. O'NEIL, WILLIAM F. DEGRADO, SHAKER A. MOUSA, N. RAMACHANDRAN, and RONALD H. HOESS

Many adhesion molecules including the integrins are known to recognize relatively short peptide sequences. For instance, the integrin IIb/IIIa[1] binds to the tripeptide sequence Arg-Gly-Asp (RGD), as well as to a longer sequence found at the C terminus of fibrinogen. Classically, these recognition sequences were discovered by proteolytically cleaving the target proteins, and separating peptides that were capable of binding to an adhesion molecule of interest.[2,3] The use of peptide libraries provides an alternative method for identifying binding sequences.[4,5] Construction of these large peptide libraries has been accomplished using both chemical methods[4] and the display of peptides on the surface of filamentous bacteriophage.[5] Current methods allow for the easy creation of libraries representing every possible hexapeptide sequence and the determination of which of these bind to a receptor of interest. Biological libraries and their potential application to screening integrin receptors are the focus of this chapter.

Development of Phage-Displayed Peptide Libraries

Peptide display on the surface of bacteriophage was first demonstrated in experiments by Smith,[6] in which the coding sequence for an antigenic peptide was fused by recombinant DNA methods to the minor coat protein gIII of the filamentous bacteriophage fd. Experiments showed that the resulting phage were specifically recognized by antibodies to the antigenic peptide. Subsequently, two concepts were experimentally demonstrated that were crucial to the development of phage-displayed peptide libraries. First, using recombinant DNA methodology, a number of groups reported the construction of large libraries in which the size and complexity could theoretically encompass a complete collection of random hexapeptides

[1] E. F. Plow and M. H. Ginsberg, *Prog. Hemostasis Thromb.* **9**, 117 (1989).
[2] M. Kloczweiak, S. Timmons, and J. Haiwiger, *Thromb. Res.* **29**, 249 (1983).
[3] E. Ruoslahti and M. D. Pierschbacher, *Cell (Cambridge, Mass.)* **44**, 517 (1986).
[4] R. N. Zuckerman, *Curr. Opin. Struct. Biol.* **3**(4), 580 (1993).
[5] R. H. Hoess, *Curr. Opin. Struct. Biol.* **3**(4), 572 (1993).
[6] G. P. Smith, *Science* **228**, 1315 (1985).

Copyright © 1994 by Academic Press, Inc.
All rights of reproduction in any form reserved.

(64 million sequences).[7-9] Second, Smith and Parmley[10] devised a simple procedure, termed *biopanning,* to affinity select phage displaying a unique peptide sequence that could serve as a ligand for a given receptor molecule. Thus, the introduction of this simple technology resulted in a vast expansion of the repertoire of molecular shapes that could be screened for binding to a given target.

Types of Phage-Displayed Peptide Libraries

Phage-displayed peptide libraries are constructed by fusing an oligonucleotide encoding a peptide sequence onto the 5'end of the gene encoding the capsid protein gIII or gVIII. The main difference between gIII or gVIII fusions is the valency of the peptide displayed on the surface of the phage. For gIII fusions, there are five copies of the displayed peptide because there are five copies of gIII per phage particle. In the case of gVIII fusions the number is theoretically much higher because there are ~3000 copies of gVIII per phage particle. In practice the number appears to be considerably less, usually in the range of several hundred. Depending on the application, higher valency is not necessarily a desired property. Bass *et al.*[11] have shown that multivalency, even in the case of gIII (of which there are only five copies per phage), can obscure the discrimination of high-affinity binders (nanomolar range) from those that bind with more modest affinity (micromolar range). To identify potential lead peptides, the best strategy may be to start with peptides displayed at high valency to ensure selecting all binders, but then to reduce the valency during continued rounds of selection to find those peptides with the highest affinity. A similar problem arises with chemical peptide libraries, in which multiple peptide molecules are attached to the solid support, making it unlikely that discrimination can be made between low- and high-affinity binding peptides.

Phage-displayed peptides have been presented in a variety of conformational contexts. The first phage-displayed peptide libraries were simple fusions of linear peptides to the amino terminus of either gIII[7-9] or gVIII.[12] The resultant peptide has a free amino terminus and is tethered to the

[7] J. K. Scott and G. P. Smith, *Science* **249,** 386 (1990).
[8] S. E. Cwirla, E. A. Peters, R. W. Barrett, and W. J. Dower, *Proc. Natl. Acad. Sci. U.S.A.* **87,** 6378 (1990).
[9] J. J. Devlin, L. C. Panganiban, and P. E. Devlin, *Science* **249,** 404 (1990).
[10] S. E. Parmley and G. P. Smith, *Gene* **73,** 305 (1988).
[11] S. Bass, R. Greene, and J. A. Wells, *Proteins* **8,** 309 (1990).
[12] F. Fellici, L. Catagnoli, A. Musacchio, R. Jappelli, and G. Cesareni, *J. Mol. Biol.* **222,** 301 (1991).

phage protein via the carboxyl terminus. Such libraries have been used to discover RGD-containing sequences that bind to the integrin $\alpha_5\beta_1$.[13] In linear peptide libraries, the pendant peptide probably behaves much like a free peptide in solution in that it is free to sample many different conformational states. Although this has the advantage of increasing the number of molecular shapes that are available in the library, it has the disadvantage that the conformation required for binding may not be significantly populated. To circumvent this problem of conformational entropy, constraining the number of potential conformations is likely to increase both the affinity and specificity of the peptide for its target molecule.

Two types of conformationally constrained phage-displayed peptide libraries have been constructed and are described in this chapter. The first is a simple constrained library in which cysteine residues flank the random peptide sequence. There is ample evidence to indicate that disulfide bonds form in phage-displayed peptides and proteins.[14] This is not unexpected because phage proteins are secreted into the periplasm during phage maturation, an environment conducive to disulfide bond formation. In the cysteine-constrained library the random peptides are therefore held in a loop conformation as long as the intervening amino acids do not preclude disulfide bond formation. We have prepared cyclic, disulfide-containing hexapeptide libraries (CX_6C) and used them to select high-affinity ligands for the integrin IIb/IIIa. It is interesting to note that it is also possible to select cyclic peptides from linear peptide libraries. For instance, a stringent selection procedure was applied to a linear hexapeptide library in an attempt to find high-affinity ligands for $\alpha_5\beta_1$.[13] The peptide sequence that was most frequently obtained contained the sequence CRGDC.

A second type of constrained peptide library is one in which peptide sequences are presented in the context of a protein scaffold. Such libraries have the advantage of presenting the random peptides not only as loop structures but as part of α helices of β sheets. If the protein scaffold chosen for the libraries is one whose structure is already well characterized by either nuclear magnetic resonance (NMR) or X-ray crystallography, then presumably any binding molecules obtained by selection will be amenable to structural analysis.

We have used libraries of both disulfide-bridged random peptides as well as peptides embedded in folded proteins to select ligands for the integrin IIb/IIIA. The following section provides detailed procedures for the construction and use of these libraries.

[13] E. Koivuneu, D. A. Gay, and E. Ruoslahti, *J. Biol. Chem.* **268**, 20205 (1993).
[14] A. Luzzago, F. Felici, A. Tramontano, A. Pessi, and R. Cortese, *Gene* **128**, 51 (1993).

Methods

Construction of Cysteine-Constrained Library

As described above, phage peptide libraries can be constructed as fusions to either gIII or gVIII. For the purposes of this work only gIII fusions are described, although similar methodology can be applied for the construction of gVIII fusions. To date, two types of phage vectors have been used for phage display libraries. One is the series of fd tet vectors constructed by Parmley and Smith.[10] These phage are defective in minus-strand synthesis and in general have a low copy number of intracellular replicative form (RF) molecules. They are propagatable as plasmids and cells infected with these phage can be selected as tetracycline-resistant colonies. The second type of vector that has been used is based on M13. With these vectors, phage are recovered as plaque-forming units (PFU) rather than drug-resistant colonies. Although each vector system has its advantages and disadvantages, libraries generated with either of these vectors have been successfully used to identify ligand molecules.

Vectors for library construction have unique restriction sites introduced at or near the mature amino terminus of gIII to allow for insertion of oligonucleotides encoding random amino acids. In most cases these vectors incorporate a selection scheme such that only those phage that have obtained an oligonucleotide following ligation will grow. In the example shown in Fig. 1A an *amber* mutation has been introduced into the gIII signal sequence at position 15, resulting in a phage that can be propagated only in an su^{+3} host. The oligonucleotides that encode the random peptide sequence also contain a portion of the signal sequence including residue 15, which is synthesized with the wild-type codon for tyrosine. Following ligation of the oligonucleotides into the vector, phage are propagated on a nonsuppressing *Escherichia coli* K91 strain,[15] so that only those phage that have received the oligonucleotide are able to grow.

The M13 vectors used to prepare random peptide libraries (M13-PL9) and protein domain–gene III fusions (M13-PL10) are constructed from M13mp19[16] by the following steps. Restriction sites within the multiple cloning site of M13mp19 are removed by digestion of the vector with *Eco*RI and *Hin*dIII, followed by S1 nuclease treatment to remove the single-stranded ends. The linear DNA is then religated with T4 DNA ligase.

Unique *Kpn*I and *Bam*HI (M13-PL9) or *Bst*XI (M13-PL10) restriction

[15] L. B. Lyons and N. D. Zinder, *Virology* **49**, 45 (1972).
[16] C. Yanisch-Perron, J. Viera, and J. Messing, *Gene* **33**, 103 (1985).

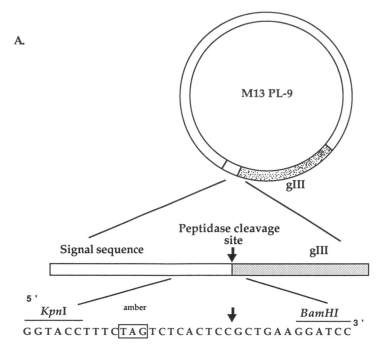

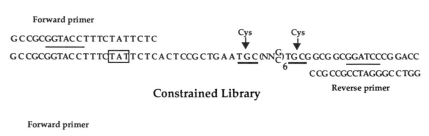

Fig. 1. (A) Schematic illustration of the M13 vector used to construct phage peptide libraries. As described in text, random sequences are inserted at the amino terminus of the mature pIII, immediately following the signal sequence in the wild-type gIII. The *amber* codon replacement in the signal sequence is boxed and the site of cleavage by the signal peptidase is indicated by an arrow. (B) Oligonucleotides used for PCR amplification of random peptide libraries. The wild-type tyrosine codon replacement for the *amber* codon in the parent vector is boxed. The cysteine codons in the CX₆C library are underlined.

sites and an *amber* mutation between these two sites are next introduced by site-directed mutagenesis, using the method of Kunkel,[17] in which the template M13 vector has been propagated in CJ236 to incorporate uracil into the DNA. Oligonucleotides for mutagenesis are purified using 10% (w/v) polyacrylamide electrophoresis containing 8 M urea. The oligonucleotide bands are excised from the gel and allowed to elute by diffusion overnight in 400 μl of 0.3 M sodium acetate, pH 5.0, followed by precipitation with 2 vol of ethanol at $-70°$. Approximately 1 μg of the purified oligonucleotide is phosphorylated in 20 μl containing 70 mM Tris-HCl (pH 7.6), 10 mM MgCl$_2$, 5 mM dithiothreitol (DTT), and 10 units of T4 polynucleotide kinase at 37° for 1 hr. For mutagenesis, the oligonucleotide is first annealed to the uracil containing template M13 DNA by mixing 2 μg of the template with 7 pmol of kinased oligonucleotide in a total volume of 10 μl containing 20 mM Tris-HCl (pH 7.5), 10 mM MgCl$_2$, 50 mM NaCl, and 1 mM dithiothreitol. The mixture is then heated to 70° for 10 min and allowed to cool slowly for approximately 1 hr. DNA extension from the mutagenic primer is begun by the addition of 1 μl of 10× buffer containing 200 mM Tris-HCl (pH 7.5), 100 mM MgCl$_2$, and 100 mM dithiothreitol, 2 μl of a 5 mM mixture of dNTPs, 1 μl of T4 DNA polymerase (3000 units/ml), 1 μl of T4 DNA ligase (400,000 units/ml), and H$_2$O to 20 μl. This mixture is first incubated on ice for 5 min, shifted to room temperature for 5 min, then incubated at 37° for 2 hr and finally at 16° for 12–16 hr. Aliquots of the reaction mixture are transformed into K91 or LE392 F' (for vectors containing an *amber* codon) as follows: 1 μl from the undiluted mix, a 1 : 10 dilution, or a 1 : 100 dilution is added to 50 μl of CaCl$_2$-competent cells and incubated on ice for 15 min. The transformation mixture is then transferred to a 37° bath for 5 min; 3 ml of top agar is added to each transformation and plated on L agar. Simultaneously, 500 ng of the single-stranded uracil-containing template DNA is transformed as a control. When equivalent amounts of double- and single-stranded DNA are transformed, the mutant double-stranded vector should transform with an efficiency several orders of magnitude higher than the single-stranded template. Successful mutagenesis to insert restriction sites is confirmed by preparing RF DNA and digesting with the appropriate enzyme. Mutants encoding the nonsense *amber* codon are identified by their inability to grow on K91 and their viability on LE392 F'. Replicative form M13 DNA from the phages is obtained by infecting 50 μl of a fresh overnight culture of K91 or LE392 F' with 100 μl of the lysate from a single plaque. The phage-infected cells are diluted into 2 ml of LB and

[17] J. Sambrook, E. F. Fritsch, and T. Maniatis, eds. "Molecular Cloning: A Laboratory Manual," 2nd ed. Cold Spring Harbor Lab., Cold Spring Harbor, NY, 1989.

are grown for 4 to 4.5 hr at 37°; cells are spun down and RF DNA is prepared from the cells by DNA minipreparation (Wizard Minipreps; Promega, Madison, WI).

The template oligonucleotide for the peptide libraries is synthesized using NNG/C codons for the random amino acids. This scheme codes for all 20 amino acids and one stop codon but does not mimic natural codon frequencies, so that some amino acids tend to be underrepresented whereas others are overrepresented. More elaborate oligonucleotide synthesis schemes have been suggested to address this bias although as a general practice the need for this measure has not been obvious. A number of different strategies have been used for assembling oligonucleotides for peptide libraries for cloning into the bacteriophage vector. Methods that have been described include (1) synthesizing the complementary strand with inosines opposite the random bases,[9] (2) synthesizing short oligonucleotides complementary to the sequences flanking the random bases so that a gapped molecule is ligated into the vector,[8] (3) designing a hairpin at the 3' terminus of the oligonucleotide containing the random bases so that the complementary strand is synthesized using DNA polymerase I Klenow fragment,[18] (4) synthesizing oligonucleotides that have short overlapping complementary sequences so that a PCR-based fill-in reaction can be done to make complementary double-stranded material,[19] and (5) using PCR primers that flank the random sequences and amplifying the template oligonucleotide.[7]

We routinely use the last method for construction of peptide libraries because we have found the cloning efficiencies to be somewhat higher than with some of the other procedures. For the synthesis of a CX_6C library, a 75-mer oligonucleotide is synthesized (Fig. 1B) containings the sequence coding for the random amino acids flanked by codons for cysteine residues and adjacent sequences that have restrictions sites necessary for cloning (including a portion of the gIII signal sequence). The template and primer oligonucleotides are gel purified as described above. For the polymerase chain reaction (PCR), 23 pmol of this template oligonucleotide is mixed with 380 pmol of the two purified PCR primers (Fig. 1B) and amplified for five cycles in a Perkin-Elmer Cetus (Norwalk, CT) GeneAmp PCR System 9600. Denaturation is performed at 95° for 2.5 min, annealing at 42° for 4 min, and extension at 72° for 20 sec using Vent DNA polymerase. Typically multiple reactions are done and the amplified

---

[18] R. B. Christian, R. N. Zuckermann, J. M. Kerr, L. Wang, and B. Malcom, *J. Mol. Biol.* **227,** 711 (1992).

[19] B. K. Kay, N. B. Adey, V. S. He, J. P. Manfredi, A. H. Mataragnon, and D. M. Fowlkes, *Gene* **128,** 59, 1993.

DNA is pooled. Following phenol extraction and ethanol precipitation, the DNA is digested with *Kpn*I and *Bam*HI restriction enzymes. A small aliquot is run on an 8% (w/v) polyacrylamide gel to test for complete digestion by the two restriction enzymes. Restriction enzymes are removed by phenol extraction and ethanol precipitation. An estimate of the concentration of the PCR-amplified insert is made by running an aliquot of the digested DNA against a known amount of ϕX174 *Hae*III digest on a 7% (w/v) polyacrylamide gel. The PCR-amplified insert is then ligated with M13-PL9 that has been digested with *Kpn*I and *Bam*HI in a reaction containing 1.5 μg of PCR material and 11.8 μg of vector in 50 mM Tris-HCl (pH 7.8), 10 mM MgCl$_2$, 10 mM dithiothreitol, 1 mM ATP, bovine serum albumin (BSA; 25 mg/ml), and 4000 units of T4 DNA ligase (New England BioLabs, Beverly, MA) in a volume of 2.0 ml. The ligation mixture is incubated overnight at 16°, phenol extracted, and ethanol precipitated. Ligated DNA is resuspended in 20 μl of distilled water and a series of five separate electroporations (4 μl of DNA and 80 μl of electrocompetent *E. coli* K91 cells) are done using a Bio-Rad (Richmond, CA) *E. coli* Pulser in 0.1-cm cuvettes at 1.80 kV, using a time constant of 5 msec. After electroporation, transformed cells are immediately diluted into 2.0 ml of SOC[17] medium and allowed to outgrow for 25 min at 37° when aliquots are removed and titered for plaque-forming units to give a measure of the complexity of the library. Phage-infected cells can now be frozen at −70° following the addition of an equal volume of sterile 40% (v/v) glycerol. To expand the library, five such frozen aliquots, representing an average complexity of ~7.0 × 10⁷ PFU/aliquot, are combined and used to inoculate 800 ml of SOC medium. Phage-infected cells are outgrown for 6 hr at 37°, and then cells are removed by centrifugation. The supernatants typically have final titers of ~10¹¹–10¹² PFU/ml. Aliquots of the expanded library can be used directly for biopanning experiments or the phage can be concentrated by polyethylene glycol (PEG) 8000 precipitation. However, further purification steps such as banding in CsCl may be required to remove excess PEG that may have a deleterious effect on the receptor being biopanned.

M13 Strp G : gpIII Fusion Vector Construction

In an effort to generate additional conformationally constrained peptide libraries a vector was constructed to allow fusion of protein domains to the amino terminus of gpIII of bacteriophage M13. Once the protein domain of choice is displayed on the phage surface it should be possible to insert segments of foreign sequence into the protein domain so that the inserted sequence is constrained to adopt a specific conformation by the scaffolding

TABLE I
IIb/IIIa-Binding Sequences Selected from
CX$_6$C Library[a]

Sequence class	Sequence
Class 1	Cys Arg Gly Asp Met Phe Gly Cys
	Cys Arg Gly Asp Phe Leu Asn Cys
	Cys Arg Gly Asp Met Leu Arg Cys
	Cys Arg Gly Asp Ala Phe Gln Cys
	Cys Arg Gly Asp Met Ala Tyr Cys
Consensus	Cys Arg Gly Asp Hb Hb Hp Cys
Class 2	Cys Asn Trp Lys Arg Gly Asp Cys
	Cys Asn Thr Leu Lys Gly Asp Cys
	Cys Phe Asn Arg Lys Gly Asp Cys

[a] From O'Neil et al.[28] Hb, Hydrophobic amino acid;
Hp, hydrophilic amino acid.

of the protein domain. The phage-displayed peptide sequences that bind to IIb/IIIa from the CX$_6$C library indicate that cyclization of the RGD sequence is required for high-affinity binding to IIb/IIIa (Table I). Thus, the protein domain chosen to introduce this sequence should have a surface-accessible turn that can be easily replaced with the RGD sequence of choice so that the structure of the remainder of the protein domain constrains the conformation of the inserted sequence. The protein domain chosen for these scaffolding experiments is the small IgG-binding domain from group G Streptococcus.[20] This domain has been well characterized by both NMR[21] and crystallography[22] and is composed of four β strands that make a single β sheet and one α helix that lies across the sheet (Fig. 2). The wealth of structural data for this domain allows for identification of specific sites where an RGD sequence is likely to be accessible for binding to IIb/IIIa and also allows easy determination of the structure of the inserted fragment within the protein scaffold.

The gene encoding Strp G is synthesized as a set of eight oligonucleotides that are each 50–60 bp in length (Fig. 3). Oligonucleotides are designed so that hybridization of complementary pairs of purified oligonucle-

[20] S. R. Fahnestock, P. Alexander, D. Filpula, and J. Nagle, "Bacterial Immunoglobulin Binding Proteins," Vol 1, p. 133. Academic Press, San Diego, 1990.
[21] A. M. Gronenborn, D. R. Filpula, N. Z. Essig, A. Achari, M. Whitlow, P. T. Wingfield, and G. M. Clore, Science 253, 657 (1991).
[22] A. Achari, S. P. Hale, A. J. Howard, G. M. Clore, A. M. Gronenborn, K. D. Hardman, and M. Whitlow, Biochemistry 31, 10449 (1992).

Fig. 2. Ribbon representation of the backbone of the Strp G IgG-binding domain. The coordinates used to generate the structure were from the Protein Data Bank (accession number IGB1).[21] In this orientation, the amino terminus is at the bottom and the carboxy terminus that is linked to gpIII is pointed up. (Generated using MOLSCRIPT [P. J. Kraulis, *J. Appl. Crystallogr.* **24,** 946 (1991)].)

otides results in 10–12 bp of single-stranded overhanging ends, which then allows adjacent pairs of oligonucleotides to anneal. The oligonucleotides are gel purified and phosphorylated with T4 polynucleotide kinase as described above. Oligonucleotides are annealed by mixing 10 pmol of each kinased oligonucleotide in a total volume of 20 μl containing 10 mM Tris, 150 mM NaCl, pH 7.5, and heating for 10 min at 70° followed by slow cooling to allow the entire gene to self-anneal. The annealed oligonucleotides are then ligated into M13-PL10 that has been digested with *Kpn*I and *Bst*XI as follows: 25 fmol of predigested vector is mixed with various amounts of the annealed oligonucleotide mixture to give an insert : vector

```
Kpn I
                                          Glu  Ile  Leu  Ala  Ala  Leu  Pro
                    CTTTCTATTCTCACTCC     GAG  ATC  TTG  GCT  GCT  CTG  CCG
                    CATGGAAAGATAAGAGTGAGG CTC  TAG  AAC  CGA  CGA  GAC  GGC


Lys  Thr  Asp  Thr  Tyr  Lys  Leu  Ile  Leu  Asn  Gly  Lys
AAA  ACC  GAC  ACC  TAC  AAA  CTG  ATC  CTG  AAC  GGT  AAA
TTT  TGG  CTG  TGG  ATG  TTT  GAC  TAG  GAC  TTG  CCA  TTT

Thr  Leu  Lys  Gly  GLu  Thr  Thr  Thr  Glu  Ala  Val  Asp
ACC  CTG  AAA  GGT  GAA  ACC  ACC  ACC  GAA  GCT  GTA  GAC
TGG  GAC  TTT  CCA  CTT  TGG  TGG  TGG  CTT  CGA  CAT  CTG
                    Pst I
Ala  Ala  Thr  Ala  Glu  Lys  Val  Phe  Lys  Gln  Tyr  Ala
GCT  GCT  ACT  GCA  GAA  AAA  GTT  TTC  AAA  CAG  TAC  GCT
CGA  CGA  ATG  CGT  CTT  TTT  CAA  AAG  TTT  GTC  ATG  CGA
                              Sal I
Asn  Asp  Asn  Gly  Val  Asp  Gly  Glu  Trp  Thr  Tyr  Asp
AAC  GAC  AAC  GGT  GTC  GAC  GGT  GAA  TGG  ACC  TAC  GAC
TTG  CTG  TTG  CCA  CAG  CTG  CCA  CTT  ACC  TGG  ATG  CTG
                                        BstE II
Asp  Ala  Thr  Lys  Thr  Phe  Thr  Val  Thr  Glu  Gly  Gly
GAC  GCT  ACC  AAA  ACC  TTC  ACG  GTT  ACC  GAA  GGT  GGC
CTG  CGA  TGG  TTT  TGG  AAG  TGC  CAA  TGG  CTT  CCA  CCG
BstX I
His  Ser
CAC  TCC  G
GTG
```

FIG. 3. Gene sequence for wild-type Strp G. The gene was constructed as described in text and cloned into M13-PL10 between the *Kpn*I and *Bst*XI restriction sites.

molar ratio between 10 : 1 and 1 : 1 in a total volume of 20 μl. The ligation mixture contains 50 mM Tris-HCl (pH 7.8), 10 mM MgCl$_2$, 10 mM dithiothreitol, 1 mM ATP, bovine serum albumin (25 μg/ml), and 400 units of T4 DNA ligase and the ligations are incubated at 16° overnight. These ligations are then transformed into CaCl$_2$-competent cells, using a nonsuppressing host.

Construction of Strp G : RGD Fusions

To facililtate construction of vectors to express peptide insertions in the Strp G scaffold, the tyrosine codon at position 54 of the protein is replaced by an *amber* codon so that the same selection strategy described above for the CX$_6$C library can be used. The mutagenic oligonucleotide 5' AGCGTCGTCCTAGGTCCAT 3' is used to introduce the *amber* codon at Tyr-54 of Strp G by site-directed mutagenesis[17] following propagation of the phage on CJ236 to incorporate uracil into the DNA as described above. The mutation can be verified by plating individual plaque isolates on K91 and LE392 F′; phage with the correct *amber* codon mutation should not grow well on the nonsuppressing host but will grow normally on the suppressor strain. The vector, M13 : SG(*Y54*Am) is then propagated on LE392 F′.

Oligonucleotides are designed to replace the existing sequence of Strp G between the *Sal*I and *Bst*EII sites with the sequence containing the RGD (or other desired sequence) insert. To introduce a constrained RGD sequence into the loop between β strands 3 and 4 of Strp G (Fig. 2) the oligonucleotides (5′ GTAACCGTGAAGGTGCCGCAGTCGCCACGTT-TCCAGTTGCAGCCGTCGTAGGTCCATTCACCG 3′) and (5′ TCGAC-GGTGAATGGACCTACGACGGCTGCAACTGGAAACGTGGCGAC-TGCGGCACCTTCACG 3′) are used. The oligonucleotides are purified, annealed, and then ligated into M13 : SG(*Y54*Am) that has been previously digested with *Sal*I and *Bst*EII as described above. The recombinant phage M13 : SG/CRGDC is verified by cleavage with *Bgl*II and *Bst*EII and comparison of the resulting DNA restriction fragment (203 bp) relative to the same fragment generated from M13 : SG (185 bp). The sequences of candidate phage are verified by sequencing, using the oligonucleotide primer described below.

Identification and Verification of Phage That Bind IIb/IIIa

To identify phage that bind to glycoprotein IIb/IIIa (GPIIb/IIIa) it is necessary to screen random peptide libraries or fusion protein phage. Screening is done using purified receptor that has been coated onto polystyrene dishes as follows. The receptor is purified from platelets as described by Fitzgerald and co-workers.[23] Briefly, the enriched platelets free of erythrocytes are recovered from outdated platelets by differential centrifugation. Platelets are lysed overnight with Triton X-100 and centrifuged to remove cytoskeletal elements. This supernatant is then loaded onto a concanavalin A–Sepharose column, washed, and IIb/IIIa receptor eluted with α-methylmannose. The presence of receptor is determined by sodium dodecyl sulfate–polyacrylamide gel electrophoresis (SDS–PAGE) and Western blot, using an antibody specific for GPIIb/IIIa (CDW41; Amec Corporation, West Brook, ME). Fractions containing receptor are pooled and dialyzed before application to a heparin–Sepharose CL-4B column, from which flowthrough fractions containing the receptor are collected. Pooled receptor fractions are concentrated and then run over a Sephacryl S-300 column for the final purification step. Receptor-containing

---

[23] L. A. Fitzgerald, B. Leung, and D. R. Phillips, *Anal. Biochem.* **151,** 169 (1985).

fractions are pooled and can be stored at −70° until use. Alternatively, the receptor can be purchased from Enzyme Research Laboratories (Indianapolis, IN).

Polystyrene dishes are coated with receptor as described.[24] Purified receptor is incubated on the plates at 4° overnight and then frozen at −70° until needed. Before incubating with phage, plates are thawed at room temperature and washed twice with buffer A [25 mM Tris, 150 mM NaCl, 0.5% (v/v) Tween 20, 0.5 mM $CaCl_2$ (pH 7.5)], blocked with 100 μl of BSA (29 mg/ml) in 0.1 M $NaHCO_3$ for 1 hr and washed three more times with buffer A.

Using a procedure similar to the biopanning procedure originally described by Parmley and Smith,[10] peptide libraries or protein domain fusions are screened to identify sequences that bind to IIb/IIIa. For peptide libraries, approximately 10^{10} phage are incubated on the receptor-coated dish overnight at 4°. Nonspecifically associated phage are washed away with 10 washes of buffer A. Specifically bound phage are generally eluted with a ligand known to be specific for the receptor (i.e., SK106760[25] or DMP 728[26]) or with glycine buffer, pH 2.2. Eluted phage are titered for plaque-forming units on K91. Multiple rounds of biopanning are done using the phage eluted during the preceding round as the input for the next round. If significant enrichment in the percentage of phage recovered after successive rounds of biopanning is observed (more than two orders of magnitude), individual plaques are isolated and used to propagate phage for preparation of single-stranded DNA. Phage are propagated by infecting 50 μl of an overnight culture of K91 with 100 μl of individual plaque lysate in 2 ml of L broth. Infected cultures are grown for 5.5–6 hr, cells are spun out, and the supernatant is reserved to prepare single-stranded DNA. DNA is prepared as described by precipitating the phage particles with polyethylene glycol, extraction of protein components with phenol, and ethanol precipitation of the DNA. Sequences of N-terminal gpIII fusions are determined using the Sanger sequencing method[27] and the oligonucleotide primer 5′ CGATCTAAAGTTTTGTCGTCT 3′.

For Strp G:RGD fusions, IIb/IIIa-specific binding can be verified

[24] I. F. Charo, L. Nannizzi, D. R. Phillips, M. A. Hsu, and R. M. Scarborough, *J. Biol. Chem.* **266**, 1415 (1991).

[25] J. Samanen, F. Ali, T. Romoff, R. Calvo, E. Sorenson, J. Vasko, B. Storer, D. Berry, D. Bennett, M. Strohsacker, D. Powers, J. Stadel, and A. Nichols, *J. Med. Chem.* **34**, 3114 (1991).

[26] S. Jackson, W. DeGrado, A. Dwivedi, A. Partha Sarathy, A. Higley, J. Krywko, A. Rockwell, J. Markwalder, G. Wells, R. Wexler, S. Mousa, and R. Harlow, *J. Amer. Chem. Soc.* **116**, 3220 (1994).

[27] F. Sanger, S. Nicklen, and A. R. Coulson, *Proc. Natl. Acad. Sci. U.S.A.* **74**, 5463 (1977).

by determining the percentage of phage recovered in a single round of biopanning. Approximately 10^9 phage are incubated on the receptor-coated dishes and treated exactly as described for the libraries. Background binding of the Strp G phase to these receptor-coated dishes is approximately 10^5, thus recoveries of $>10^7$ indicate high-affinity binding between the modified Strp G and the receptor.

To determine accurate affinities for the peptides selected from the phage peptide libraries for IIb/IIIa, peptides are synthesized by solid-phase peptide synthesis.

Determination of IIb/IIIa Affinities by ELISA

Affinities of the selected peptides for IIb/IIIa are determined by enzyme-linked immunosorbent assay (ELISA) using immobilized IIb/IIIa and various peptides in competition with biotinylated fibrinogen. The reagents required for the assay are as follows:

Purified GPIIb/IIIa (see above)
Biotinylated fibrinogen (3.0 μM)
Avidin–biotin–alkaline phosphatase conjugate (Cat. No. A7418; Sigma, St. Louis, MO)
Flat-bottom, high-binding, 96-well plates (Cat. No. 3590; Costar, Cambridge, MA)
Sigma 104 phosphatase substrate (40-mg capsules)
Bovine serum albumin (BSA) (Cat. No. A3294; Sigma)
Alkaline phosphatase buffer: 0.1 M glycine hydrochloride, 1 mM MgCl$_2$ · 6H$_2$O, 1 mM ZnCl$_2$, pH 10.4
Binding buffer: 20 mM Tris-HCl, 150 mM NaCl, 1 mM CaCl$_2$ · 2H$_2$O, 0.02% (w/v) NaN$_3$, pH 7.0
Buffer B: 50 mM Tris-HCl, 100 mM NaCl, 2 mM CaCl$_2$ · 2H$_2$O, 0.02% (w/v) NaN$_3$, pH 7.4
Blocking buffer: Buffer B plus 3.5% (w/v) BSA
Dilution buffer: Buffer B plus 0.1% (w/v) BSA
NaOH (2 N)

Fibrinogen (Sigma) is biotinylated using NHS–LC–biotin (Cat. No. 21335G; Pierce Chemical Co., Rockford, IL) without further purification. The biotinylation reaction is carried out at room temperature and excess reagent is dialyzed out; biotinylated fibrinogen is stored at $-70°$ until use. The assay is performed as follows: plates are incubated with purified GPIIb/IIIa in binding buffer (125 ng/100 μl/well) overnight at 4°, leaving the first column of wells uncoated as a control for nonspecific binding.

Plates can then be stored at −70° until use. For measuring peptide affinities, a plate is thawed for 1 hr at room temperature (or overnight at 4°). The coating solution is discarded and the plate is washed once with 200 μl of binding buffer and then blocked for 2 hr, while being shaken at room temperature, by addition of 200 μl of blocking buffer. The blocking buffer is discarded and the plates are washed once with dilution buffer (200 μl/well). Eleven microliters of the test compound (10 times the concentration to be tested) in dilution buffer is pipetted into duplicate wells, while 11 μl of dilution buffer is added to the nonspecific and total binding wells. One hundred microliters of biotinylated fibrinogen [1 : 133 (v/v) in dilution buffer; final concentration, 20 nM] is then added to each well and the plates are incubated for 3 hr at room temperature on a plate shaker. The assay solution is discarded and the plates are washed twice with 300 μl of binding buffer per well. One hundred microliters of anti-biotin–alkaline phosphatase conjugate [1 : 1500 (v/v) in dilution buffer] is added to each well and the plates are again incubated for 1 hr at room temperature on a plate shaker. Finally, the conjugate solution is discarded and the plates are washed twice with 300 μl of binding buffer per well. Color development is initiated by addition of 100 μl of phosphatase substrate (1.5 mg/ml in alkaline phosphatase buffer) to each well; plates are incubated at room temperature on the plate shaker until color develops. Color development is stopped by addition of 25 μl of 2 N NaOH per well and plates are read at 405 nm with blanking against the nonspecific binding well; total absorbance is determined from the total binding wells that lack the test inhibitor. The percent inhibition of fibrinogen binding achieved by the test compound is calculated as follows:

Percent inhibition
    = [100 − (absorbance of test compound)/(total absorbance)] × 100

Discussion

In this chapter we discuss the preparation and use of a conformationally constrained library. In addition we describe targeted loop replacements on a protein presentation scaffold to generate proteins with altered binding properties. The peptide library uses a simple cyclic disulfide to constrain the conformation of the random peptide insert, whereas the targeted loop replacement uses the structure of the protein scaffold to constrain the conformation of the peptide insert. In addition, analogous procedures can be used to provide linear peptide libraries or libraries of random sequences embedded within a protein scaffold.

As a general procedure, we advocate beginning with simple linear peptide libraries, and progressing to conformationally constrained libraries only if the linear libraries do not provide interesting results, or if some conformational information is desired. With the amount of IIb/IIIa absorbed to the microtiter plates described in this chapter, we found that phage displaying linear peptides were not bound strongly enough to allow for selection in the biopanning procedure. However, other workers have successfully identified linear RGD-containing peptides when the related integrin $\alpha_5\beta_1$ was biopanned.[13]

We next turned to the conformationally constrained CX_6C library, which allowed rapid and convenient determination of the specificity elements for IIb/IIIa. We have also used this approach to determine the affinity of other integrins of the β_3 family (K. O'Neil, unpublished results, 1993). In the case of IIb/IIIa, two major classes of RGD or KGD sequences were obtained using the CX_6C library (Table I).[28] These peptide sequences have also been prepared chemically and found to bind with high affinity to the IIb/IIIa receptor, but only if they were in the cyclic, disulfide-bonded state. The RGD or KGD sequence is not randomly distributed within the cyclic peptide, but rather is directly adjacent one of the two cysteine residues, suggesting that these orientations within the macrocycle provide a conformation favorable for interaction with the receptor. Numerous conformational studies have shown that the RGD sequence binds to IIb/IIIa in a conformationally well-defined state, and most of these studies suggest that the RGD sequence adopts a predominantly extended conformation.[29-31] Thus, the cyclic peptides might bind to the receptor in hairpin-like conformations with an extended conformation at the first, second, fifth, and sixth residues of the insert, and a turn centered at the third and fourth residues.

A second feature that appears to provide high affinity for IIb/IIIa is the presence of hydrophobic residues proximal to the RGD or KGD sequence. This is clearly seen in the class 1 sequences (Table I), in which two hydrophobic residues invariantly follow the RGD sequence.

We have also constructed multiple Strp G fusion phage that contain an RGD replacement for the loop between β strands 3 and 4. Interestingly, in this case the only structures capable of binding to IIb/IIIa contained a

[28] K. T. O'Neil, R. H. Hoess, S. A. Jackson, N. S. Ramachandran, S. A. Mousa, and W. F. DeGrado, *Proteins* **14**, 509 (1992).

[29] M. J. Bogusky, A. M. Naylor, S. M. Pitzenberger, R. F. Nutt, S. F. Brady, C. D. Cotton, J. T. Sisko, P. S. Anderson, and D. F. Veber, *Int. J. Pept. Protein Res.* **39**, 63 (1992).

[30] R. F. Nutt, S. F. Brady, C. D. Colton, J. T. Sisko, T. M. Ciccarone, M. R. Levy, M. E. Duggan, I. S. Imagire, R. J. Gould, P. S. Anderson, and D. F. Veber, *Pept.: Chem. Biol; Proc. Am. Pep. Symp. 12th, 1991;* p. 914 (1992).

[31] A. C. Bach, C. J. Eyerman, J. D. Gross, M. J. Bower, R. L. Harlow, P. C. Weber, and W. F. DeGrado, *J. Amer. Chem. Soc.* **116**, 3207 (1994).

cyclic disulfide analogous to those obtained with the CX_6C library.[32] In addition, we found that the RGD-containing loop had to be extended away from the protein surface with a peptide spacer. This requirement for extension of the RGD sequence from the protein surface has been observed for two other protein scaffolds,[33,34] suggesting that for IIb/IIIa the protein scaffold chosen to present the RGD sequence is of less importance than the requirement for optimal conformation and isolation of this peptide fragment from the bulk of the protein. Nevertheless, we expect that the Strp G framework will provide a suitable degree of conformational constraint for other applications.

[32] K. T. O'Neil, W. F. DeGrado, and R. H. Hoess, in "Techniques in Protein Chemistry V" p. 517. Academic Press, San Diego, 1994.
[33] T. Yamada, M. Matsushima, K. Inaka, T. Ohkubvo, A. Uyeda, T. Maeda, K. Titani, S. Kiyotoshi, and M. Kikuchi, *J. Biol. Chem.* **268,** 10588 (1993).
[34] G. Lee, W. Chan, M. R. Hurle, R. L. Desjarlais, F. Watson, G. M. Sathe, and R. Wetzel, *Protein Eng.* **6**(7), 745 (1993).

[19] Gene Targeting and Generation of Mutant Mice for Studies of Cell–Extracellular Matrix Interactions

By Elizabeth L. George and Richard O. Hynes

Introduction and Overview

The great advances in our understanding of the extracellular matrix (ECM) and its interactions with cells have relied on the methods used in biochemistry, cell biology, and molecular biology. These approaches have identified a large number of ECM molecules and cell surface receptors and have also provided insights into their structures, interactions, and likely effects on cell behavior. However, it is also necessary to understand the roles of these proteins in intact organisms during development and in physiological and pathological processes. To this end, previous studies have used injections of antibodies and peptides or of cells transfected with various cDNAs encoding ECM and cell surface receptor proteins to perturb the organism *in vivo*. Because of the great variety of ECM proteins and adhesion receptors, many of which may share overlapping functions, it would be desirable to have a way of selectively removing or altering individual molecules singly or in combination. The ideal way to do this is by the use of genetics. Such studies were begun in flies and worms,

Copyright © 1994 by Academic Press, Inc.
All rights of reproduction in any form reserved.

organisms in which genetic analysis was more advanced. Technological advances have made it possible to generate and study murine mutants. This approach offers great promise for the understanding of cell–matrix interactions in mammals, just as it does for many other areas of study.

Transgenic mice into which extra genes have been inserted have been available for some years, although few studies on ECM proteins or their receptors have been reported. This is perhaps partly due to the large size of many genes involved, but it may also reflect the rather complex gene regulation to which these genes are subject. Because of these features, methods for insertion of large pieces of DNA into transgenic mice are necessary and these have only recently become available (see Future Prospects, below). The opposite approach, obtaining mice with defects in genes encoding ECM proteins or cell surface receptors, has also been little used. The chance ablation of the collagen $\alpha_1(I)$ gene by retroviral insertion has provided useful insights into the roles of type I collagen[1–3] and various human mutations in ECM proteins and integrins have also been important in aiding our understanding of these molecules. These examples show how valuable it would be to obtain mutations in other molecules involved in cell–matrix interactions.

Two seminal advances achieved during the 1980s have made it possible to generate, almost at will, mice with mutations in any gene for which one has a cDNA or genomic clone. The first advance was the development of stable cultured lines of totipotential cells from mouse embryos.[4] These so-called embryonic stem (ES) cells can be grown and manipulated in tissue culture and on introduction into blastocyst embryos, can participate in the development of all tissues of the embryo, including the germ line. Therefore, ES cells that have been altered during their time in culture (e.g., by addition or alteration of genes) can give rise to lines of mice that inherit the same genetic alteration. The true potential of ES cells was made possible by the development of methods for efficient homologous recombination in mammalian cells and for selection of cells that have undergone such recombination. Particularly important was the introduction by Capecchi and colleagues of positively selectable marker genes that can be inserted into any cloned piece of genomic DNA, and transfected into ES cells; homologous recombinants can then be selected using

[1] A. Schnieke, K. Harbers, and R. Jaenisch, *Nature (London)* **304**, 315 (1983).

[2] K. Kratochwil, K. von der Mark, E. Kollar, Jaenisch, K. Mooslehner, M. Schwarz, K. Hasse, I. Gmachl, and K. Harbers, *Cell (Cambridge, Mass.)* **57**, 807 (1989).

[3] H. Wu, J. F. Bateman, A. Schnieke, A. Sharpe, D. Barker, T. Mascara, D. Eyre, R. Bruns, P. Krimpenfort, A. Berns, and R. Jaenisch, *Mol. Cell. Biol.* **10**, 1452 (1990).

[4] E. J. Robertson, ed., "Teratocarcinomas and Embryonic Stem Cells: A Practical Approach." IRL Press, Washington, DC, 1987.

the positive selection marker.[5] The method was further improved by addition of a negatively selectable marker to one end of the transfected piece of DNA, allowing selection against random insertions in parallel with the selection for the positively selectable marker.[6] With these and other technical advances it is now feasible to mutate any cloned gene by site-specific mutagenesis in ES cells and thereby generate a mouse line with the desired mutation (reviewed by Capecchi[7] and diagrammed in Fig. 1).

Several genes relevant to extracellular matrix research have already been inactivated by gene targeting, thus making null alleles (also referred to as gene knockouts). These experiments have revealed some surprising results. For example, mice with a gene knockout in tenascin develop normally, and no phenotype has been discovered as yet.[8] Mice lacking thrombospondin 1 also develop normally.[9] In contrast, the fibronectin knockout mutation results in embryonic lethality, with defects in mesoderm, neural tube, and vascular development.[10] Mice lacking collagen type IX appear normal at birth, but develop osteoarthritis by age 3 weeks.[11] The α_5 integrin receptor has also been knocked out in mice, resulting in an embryonic lethal phenotype that is similar to, but distinct from, the fibronectin-minus embryos.[12] During gestation α_4 integrin-deficient embryos also die at a later stage than either α_5 integrin- or fibronectin-deficient embryos.[12a]

These few examples illustrate both the power and the complexities of this approach. The normal development of some null mutants poses questions about functional redundancy and the embryonic lethality of other mutations, while confirming the importance of the genes involved, precludes study of their roles in adult physiology and pathology. Although null mutations represent a necessary and important first step in the analyses, it will be essential to follow these with more subtle perturbations of the functions of the genes and proteins of interest.

In this chapter we describe the now reasonably well-established methods for construction of gene targeting vectors and generation of mutations

[5] K. R. Thomas and M. R. Capecchi, *Cell* (*Cambridge, Mass.*) **51,** 503 (1987).

[6] S. L. Mansour, K. R. Thomas, and M. R. Capecchi, *Nature* (*London*) **336,** 348 (1988).

[7] M. R. Capecchi, *Science* **244,** 1288 (1989).

[8] Y. Saga, T. Yagi, Y. Ikawa, T. Sakakura, and S. Aizawa, *Genes Dev.* **6,** 1821 (1992).

[9] J. Lawler, H. Rayburn, and R. O. Hynes, unpublished.

[10] E. L. George, E. N. Georges, R. S. Patel-King, H. Rayburn, and R. O. Hynes, *Development* (*Cambridge, UK*) **119,** 1079–1091 (1993).

[11] R. Faessler, P. Schnegelsberg, J. Wausman, T. Shinya, M. McCarthy, B. Olsen, and R. Jaenisch, *Proc. Natl. Acad. Sci. USA* **91,** 5070 (1994).

[12] J. T. Yang, H. Rayburn, and R. O. Hynes, *Development* (*Cambridge, UK*) **119,** 1093–1105 (1993).

[12a] J. T. Yang, H. Rayburn, and R. O. Hynes, submitted for publication (1994).

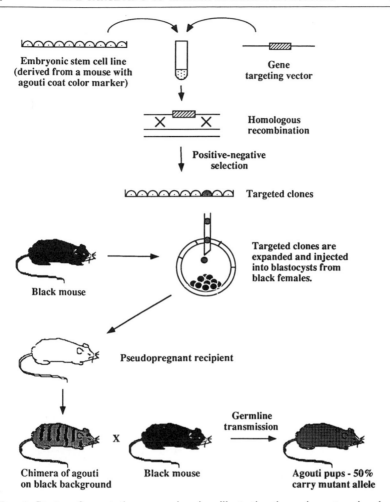

FIG. 1. Strategy for mutating a gene in mice, illustrating the various steps involved in generating mutant mice from an ES cell line. (Based on an original drawing by Joy Yang.)

in ES cells and mice, discuss some methods of analysis of the resulting mutants, and, finally, consider briefly some developments that offer the promise of further extensions of the genetic analysis of targeted mutations in mice.

Gene Targeting Vectors

The general strategy for designing targeting vectors is diagrammed in Fig. 2. In constructing the gene targeting vector, one should consider

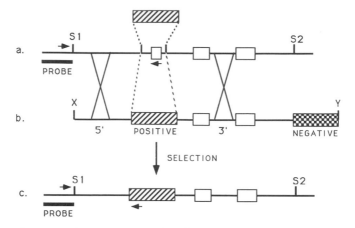

FIG. 2. Gene targeting strategy. (a) The wild-type allele. Deletion of a segment of the gene and replacement with a positively selectable marker gene (crosshatched) generates the targeting vector (b), which also has one or two copies of a negatively selectable marker gene (checkered) at one or both ends. Other significant features of the targeting vector include a unique restriction site for linearization at one or both ends (X and Y) and 5′ and 3′ flanking segments homologous with the wild-type locus between sites S1 and S2. The length of the 5′ flank depends on whether the targeted allele (c) is to be detected on Southern blots using a probe from outside the S1–S2 segment or PCR primers (arrows). If PCR is to be used, the 5′ flank should be about 1 kb; if Southern blots are to be used, it can be several kilobases long. In either case the 3′ flank should be several kilobases long.

which line of ES cells will be used, and from which mouse strain it was derived. This is because homologous recombination frequency is higher, at least at many loci, when the targeting vector is isogenic with the ES cells.[13,14] For example, the commonly used CC1.2, CCE, D3, AB1, and J1 lines of ES cells were all derived from strain 129/Sv mice, and therefore genomic clones should ideally be isolated from a library made from that strain, although many successful knockouts have been accomplished with nonisogenic DNA. Homologous recombination frequency also increases markedly with increasing length of homology.[14] With this in mind, genomic clones should be isolated that contain 3–5 kilobases (kb) of sequence homology on each side of the intended mutation. An exception to this rule is when the polymerase chain reaction (PCR) will be used to screen for homologous recombination (see below).

To make a knockout or null mutation, the most conservative approach is to make a deletion in the gene, followed by insertion of a positive

[13] H. te Riele, E. R. Maandag, and A. Berns, *Proc. Natl. Acad. Sci. U.S.A.* **89,** 5128 (1992).
[14] C. Deng and M. R. Capecchi, *Mol. Cell. Biol.* **12,** 3365 (1992).

selectable marker. An insertion alone is more likely to allow some expression of the gene by aberrant transcription and/or translation and this has been seen in some cases.[15] Generally the mutation should be made near the 5' end of the gene, including deletion of some mature protein sequence. An effective strategy is to delete the ATG, and the signal sequence, if they fall in a single exon. If the start site for transcription is known, this can often also be deleted. Such deletion of transcription, translation, and secretion control elements increases the chance of completely ablating expression of the desired protein. The positive selection marker is then inserted in place of the deleted segment.

The most common positive selectable markers confer resistance to G418 or to hygromycin B by replacing the bacterial control sequences of the neomycin resistance (*neo*) or hygromycin resistance (*hyg*) genes, respectively, with eukaryotic control sequences. Both markers are expressed at many loci in ES cells when controlled by the mouse phosphoglycerate kinase (PGK) promoter and polyadenylation signal.[16–19] The engineered pMC1 promoter has also been used successfully in ES cells for numerous gene targeting experiments[5,6,14,20] although, in our experience, this promoter is somewhat less effective in selecting homologous recombinant cells. Soriano *et al.*[20] have tested a variety of other promoters. Another potentially useful marker that has been less extensively used is the gene encoding hypoxanthine–guanine phosphoribosyltransferase (HGPRT).[21]

To enrich for targeted versus random integration of the targeting vector, a positive–negative selection procedure can be used.[6] This involves addition of a negative selectable marker at one or both ends of the targeting vector. The negative marker is lost during homologous recombination but retained during random integration of the targeting vector. The most common negative selectable marker is the herpes simplex virus thymidine kinase gene (HSV-TK), which renders cells sensitive to ganciclovir or

[15] J. E. Sligh, C. M. Ballantyne, S. S. Rich, H. K. Hawkins, C. W. Smith, A. Bradley, and A. L. Beaudet, *Proc. Natl. Acad. Sci. U.S.A.* **90**, 8529 (1993).

[16] C. N. Adra, P. H. Boer, and M. W. McBurney, *Gene* **60**, 65 (1987).

[17] M. W. McBurney, L. C. Sutherland, C. N. Adra, B. Leclair, M. A. Rudnicki, and K. Jardine *Nucleic Acids Res.* **20**, 5755 (1991).

[18] H. te Riele, E. R. Maandag, A. Clarke, M. Hooper, and A. Berns, *Nature (London)* **348**, 649, (1990).

[19] M. A. Rudnicki, T. Braun, S. Hinuma, and R. Jaenisch, *Cell (Cambridge, Mass.)* **71**, 383 (1992).

[20] P. Soriano, C. Montgonery, R. Geske, and A. Bradley *Cell (Cambridge, Mass.)* **64**, 693 (1991).

[21] M. M. Matzuk, M. J. Finegold, J.-G. J. Su, A. J. W. Hsueh, and A. Bradley, *Nature (London)* **360**, 313 (1992).

1-(2-deoxy-2-fluoro-β-p-arabinofuranosyl)5-iodouracil (FIAU). The HSV-TK gene works well under the control of either the pMC1 promoter[6] or the PGK promoter.[19] There seem to be no serious problems caused by using the same promoter for both selectable markers. In most cases, between 1 and 10% of double-resistant clones have a targeted allele. The remainder of the double-resistant cells are likely due to random breaks between the positive and negative markers. To minimize this phenomenon, the distance between the positive and negative markers should not be too large (less than 5 kb) because breaks are more likely to occur the wider the separation. The targeting vector must also have a unique restriction site, to linearize before electroporation. Several vectors are available that have eight-cutter sites (e.g., *Not*I) in their polylinkers and these should be used for the constructions.

Screening for targeted cells by genomic Southern blotting requires a probe that is outside the targeting vector. Such a probe should be tested beforehand to be sure that it hybridizes with a single restriction fragment on Southern blots of genomic DNA. That fragment must also show a detectable size difference between the wild-type and targeted alleles. Screening for targeted cells by PCR is also feasible,[10,20,22] but has some significant drawbacks. First, one side of the targeting construct is limited in length of homology, because one primer must be outside the construct. Generally, we have not been able to achieve efficient amplification over about 1 kb, using small amounts of genomic DNA as template; multiple primers may have to be tested before finding a useful pair. Second, rigorous precautions must be taken to avoid contamination of reagents with small amounts of amplification product leading to false-positive PCR signals.

One should also consider whether subsequent generation of homozygous mutant cells (double knockouts) may be desired. Homozygous cells are very likely required if the targeting experiment involves cells other than ES cells. Even if homozygous animals are to be obtained by breeding mice derived from heterozygous ES cells, homozygous cells may be useful for experiments with chimeric animals, especially in the case of embryonic lethal mutations. Generation of homozygous mutant cells can be accomplished either by consecutive gene targeting, using two different positive selectable markers,[18,23] or with a single targeting construct by growing the heterozygous cells in high concentrations of G418.[24] The high concentrations of G418 select for loss of heterozygosity at the original targeted

[22] A. L. Joyner, K. Herrup, C. A. Davis, and J. Rossant, *Science* **251,** 1239 (1991).
[23] R. M. Mortensen, M. Zubiaur, E. J. Neer, and J. G. Seidman, *Proc. Natl. Acad. Sci. U.S.A.* **88,** 7036 (1991).
[24] R. M. Mortensen, D. A. Conner, S. Chao, A. A. T. Geisterfer-Lowrance, and J. G. Seidman, *Mol. Cell. Biol.* **12,** 2391 (1992).

locus so that cells become diploid for the PGK-*neo* insertion, conferring resistance to higher concentrations (1–3 mg/ml) of G418. If the high-G418 strategy is planned, then we advise making the targeting construct from a widely used mutant form of the *neo* gene, which is somewhat reduced in phosphotransferase activity.[19,25] We and others have found that double knockouts are readily selectable when this form of the *neo* gene is used for the initial selection of heterozygous cells. The mutant PGK-*neo* gene is from plasmid pKJ1[17] and should not be confused with a corrected and slightly more active version of PGK-*neo,* known as pGEM7(KJ1)R.[19] Anecdotal results suggest that the more active form of the gene may lend itself less readily to selection for the double-knockout event when high concentrations of G418 are used to select for gene conversion. Conversely, the pMC1-*neo* gene, which is much less efficiently expressed, may be unable to overcome high G418 concentrations even when duplicated. At present there is not sufficient experience with the high-G418 selection method for double knockouts to be certain that one form of the *neo* gene is consistently more effective. Because we have used the mutant form of PGK-*neo,*[17] we can recommend that form.

Embryonic Stem Cells

The line of ES cells to be used in the procedure must be competent in order to contribute to the germ line. It is important to obtain a well-characterized ES cell line and/or to test it for germ line transmission by implanting it into blastocysts of a different coat color genotype. There are several widely used ES cell lines: CC1.2,[26] CCE,[26a] D3,[27] AB1,[28] and J1.[29] Most ES cell lines in use are from the 129/Sv strain, which is wild type (agouti) in color. Injection into C57BL/6 recipient blastocysts gives chimeric mice with a mix of black and agouti (brown) coat color. Chimeras are bred with C57BL/6. If the ES cells contribute to the germ line, one expects agouti progeny because agouti is dominant over black.

Assuming a "good" ES cell line has been obtained, it is important to maintain it. Embryonic stem cells, by their nature, tend to differentiate and this must be avoided to maintain totipotency. Embryonic stem cells should be used at low passage numbers and during use must be maintained

[25] R. Yenofsky, M. Fine, and J. W. Pellow, Proc. Natl. Acad. Sci. U.S.A. **87,** 3435 (1990).
[26] A. Bradley, M. Evans, M. H. Kaufman, and E. Robertson, *Nature (London)* **309,** 255 (1984).
[26a] E. J. Robertson, A. Bradley, M. Kuehn, and M. Evans, *Nature (London)* **323,** 445 (1986).
[27] T. C. Doetschman, H. Eistetter, M. Katz, W. Schmidt, and R. Kemler, *J. Embryol. Exp. Morphol.* **87,** 27 (1985).
[28] A. P. McMahon and A. Bradley, *Cell (Cambridge, Mass.)* **62,** 1073 (1990).
[29] E. Li, T. H. Bestor, and R. Jaenisch, *Cell (Cambridge, Mass.)* **69,** 915 (1992).

on feeder cells and/or in the presence of a growth factor known as LIF (leukemia inhibitory factor).[30] We routinely use both. Feeder cell lines have been described,[28,31] but it is generally better to use mouse embryonic fibroblast feeder cells. These obviously must be prepared (see below) and depending on the selection scheme to be used, need to contain genes for resistance to the positive selection drug (e.g., G418/neomycin, hygromycin). Thus, the feeders should be prepared from an appropriate transgenic mouse line. In the following sections we give recipes for the culture and handling of ES cells and the preparation of fibroblast feeders. Techniques for generating and handling ES cells are well covered in Hogan *et al.*[32] and Robertson.[4]

Reagents for Embryonic Stem Cell Culture

ES cell medium: One package of Dulbecco's modified Eagle's medium (DMEM) powder [with glucose (4500 mg/liter), L-glutamine, and sodium pyruvate (Cat. No. 56-499; JRH Biosciences, Lenexa, KS)], 134.8 g, is made up to 10 liters, by adding 12.0 g of $NaHCO_3$ and 62.4 g of N-2-hydroxyethylpiperazine-N'-2-ethanesulfonic acid (HEPES). Adjust the pH to 7.5 with 10 N NaOH prior to filtration. This is stable at 4° for at least 3 months. Supplement with fresh glutamine (single-use aliquots of 100× stock from GIBCO-BRL, Gaithersburg, MD) after 2 weeks. Once the following components have been added, use the medium within 2 weeks: 360 ml of medium, 65 ml of highest quality fetal bovine serum [FBS; final concentration, 15% (v/v)], 4.5 ml of nonessential amino acids (100× stock from GIBCO-BRL), 3 μl of 14.2 M 2-mercaptoethanol, 45 μl (final concentration, 1000 U/ml) of LIF (ESGRO; GIBCO-BRL)

Feeder cell medium: ES cell medium without extra components plus 10% (v/v) fetal bovine serum (quality is less important)

Phosphate-buffered saline (PBS): 0.8% (w/v) NaCl, 0.02% (w/v) KCl, 0.02% (w/v) KH_2PO_4, 0.115% (w/v) Na_2HPO_4 (pH 7.4)

Ethylenediaminetetraacetic acid (EDTA): For 1 liter use 100 ml of 10× PBS, 0.2 g of EDTA-Na, 15 mg of phenol red (sodium salt), pH 7.2

Trypsin–EDTA: 40 ml of EDTA stock [0.02% (v/v) EDTA in PBS plus phenol red] plus 1 ml of 2.5% (w/v) trypsin (GIBCO); make fresh after 1 week

[30] R. L. Williams, D. J Hilton, S. Pease, T. A. Wilson, C. L. Stewart, D. P. Gearing, E. F. Wagner, D. Metcalf, N. A. Nicola, and N. M. Gough, *Nature (London)* **336**, 684 (1988).
[31] S. Sawai, A. Shimono, K. Hanaoko, and H. Kondoh, *New Biol.* **3**, 861 (1991).
[32] B. Hogan, F. Costantini, and E. Lacy, "Manipulating the Mouse Embryo: A Laboratory Manual." Cold Spring Harbor Lab. Cold Spring Harbor, NY, 1986.

HBS: 25 mM HEPES, 134 mM NaCl, 5 mM KCl, 0.7 mM Na$_2$PO$_4$, pH 7.1

Preparation of Mouse Embryonic Fibroblast Feeder Layers

Preparation of Primary Mouse Fibroblasts. The following procedures should be performed in a dissecting hood, with sterile equipment. For G418-resistant feeders, use a transgenic mouse line containing a neomycin resistance gene (i.e., any knockout mouse line). Similarly, if hygromycin is used for selection, feeder cells with the *hyg* gene must be prepared.

1. A 14-day pregnant mouse is killed by cervical dislocation, and 70% ethanol is liberally applied to the abdomen.
2. Using a pair of scissors, cut the skin and body wall from genital area to front paws. Move the guts aside to expose the two uterine horns. The embryos will appear as bead-like bulges along the length of each uterine horn. Dissect the uterine horns by cutting below the ovaries, along the mesometrium and at the cervix. Rinse in sterile PBS.
3. Cut the uterus between the embryos to separate and rinse each embryo in fresh PBS. Dissect one embryo at a time, using two pairs of watchmaker's forceps: remove fetal membranes and placenta, pinch off the head, remove the soft tissue (liver, heart, anything dark in color), and rinse the remaining embryo in fresh PBS.
4. Place each embryo in a separate petri dish and transfer, covered, to a tissue culture hood for trypsinization. Aspirate PBS and mince tissue with two scalpels or a fine curved pair of scissors. Add 2 ml of trypsin–EDTA (fresh) to each dish and incubate at 37° for 5 min.
5. Add 8 ml of feeder cell medium to each dish and allow to settle in a conical tube for 2 min.
6. Plate out each supernatant in one 10-cm tissue culture dish. Feed with fresh medium after 24 hr. Culture until confluent (about 1–2 days) in a standard tissue culture incubator.
7. Divide each dish into ten 10-cm dishes and culture until confluent (2 days).
8. Trypsinize and resuspend each dish in 1 ml of ice-cold freezing medium [feeder cell medium, 20% (v/v) fetal bovine serum (FBS), 10% (v/v) dimethyl sulfoxide (DMSO)].
9. Freeze 1 ml per vial and store at −80° for 24 hr. For long-term storage, store at −135° or in liquid nitrogen.

Mitotic Inactivation by Exposure to γ Irradiation

1. Thaw four to eight vials from above, each to one 10-cm dish, and culture until confluent (usually overnight).

2. Divide each dish into three 10-cm dishes and culture until confluent (2 days).

3. Repeat the 1 : 3 split and culture until confluent (2 days).

4. Trypsinize all plates, resuspend in 50 ml of feeder cell medium, and count an aliquot. Expose cells to 3000-rad γ irradiation.

5. Spin cells down and resuspend in an appropriate volume to attain 2.5×10^6 cells/ml in ice-cold freezing medium.

6. Freeze feeders in convenient aliquot sizes: 2.5×10^6 cells/10-cm dish are sufficient for culture of ES cells, and frozen aliquot sizes equal to one to two dishes per vial work well.

Feeder Layers for Embryonic Stem Cells. Use mitotically inactivated embryonic fibroblasts at 2.5×10^6 cells/10-cm dish. To achieve a uniform monolayer of feeders, coat plates with 0.1% (w/v) gelatin (sterile) for at least 5 min prior to plating and aspirate off. Feeders can be either pre- or coplated with ES cells. If coplating, or if no medium change is planned before addition of ES cells, spin feeders out of DMSO, as the ES cells are sensitive to it.

Routine Culture of Mouse Embryonic Stem Cells

ES cells should always be cultured with a feeder layer. Mouse embryonic fibroblast feeders, at a density of 2.5×10^6/10-cm plate, are suitable for up to 10 days in culture. Plate ES cells at relatively high density, 3×10^6 to 1×10^7/10-cm dish. They grow rapidly, and divide every 18–24 hr. An important aspect of ES cell culture is that they be fed daily with ES cell medium. The cells need to be passaged about every third day, and will likely require a second feeding on the last day before passage (as judged by acidity of the medium). Generally, higher viability is achieved if the cells are refed 2–3 hr prior to any trypsinization (for either passage or freezing).

Passaging Embryonic Stem Cells

1. Aspirate medium and rinse each plate once with the same volume of prewarmed PBS.

2. For a 10-cm plate, add 2 ml of prewarmed trypsin–EDTA and place the plate in an incubator for 5 min until cells detach on swirling the plate.

3. Stop trypsin by adding 2 ml of ES medium.

4. Disperse cells thoroughly with a Pasteur pipette and transfer to a conical tube with an additional 6 ml of medium.

5. Spin the cells for 2 min at 500 rpm. This spinout of trypsin is important, as ES cells do not grow well in the presence of trypsin.

6. Remove the supernatant by aspiration and replace with 2 ml of fresh ES medium. Disperse the cells in this small volume first, by pipetting gently (to avoid bubbles, etc.) with a Pasteur pipette 20 times. Thorough dispersal of ES cells is important, as aggregates are more likely to differentiate.

7. The usual split is 1 : 6, but a more dense plate is often appropriate, for example, a low-density plate that has not been passaged in 3 days. The reason for the split at this time, even though the plate is not confluent, is that the colonies need to be dispersed to prevent differentiation.

Freezing and Thawing Embryonic Stem Cells. Refeed 2–3 hr ahead and trypsinize as usual until resuspension. Freeze at high density; one confluent 10-cm dish can be frozen as six 1-ml vials. As ES cells are especially sensitive to DMSO toxicity, minimize exposure by resuspending cells (with thorough dispersal) in ice-cold medium without DMSO first, and then add an equal volume of 2× freezing medium. For example, for one confluent 10-cm dish of ES cells:

1. Resuspend the pellet in 3 ml of ice-cold ES medium with 20% (v/v) FBS.

2. Add 3 ml of ice-cold ES medium with 20% (v/v) FBS, 20% (v/v) DMSO [final concentration, 20% (v/v) FBS, 10% (v/v) DMSO].

3. Keep on ice, and aliquot to six cryovials.

Freeze vials slowly in a Styrofoam container for 24 hr at $-80°$, then store at either $-135°$ or in liquid nitrogen for long-term storage. To use, thaw ES cells rapidly in a 37° water bath and quickly transfer to ice. Spin the cells out of DMSO in about 10 ml of ES medium, resuspend with thorough dispersal as for passaging, and plate on feeders.

Transfection of Embryonic Stem Cells by Electroporation

1. Linear DNA is prepared by phenol–chloroform extraction and ethanol precipitation. Pellet should be washed twice in 70% (v/v) ethanol at room temperature and vacuum dried. Resuspend DNA in sterile HBS and consider it sterile for use in tissue culture. An aliquot of the preparation should be checked for degradation by comparing restriction maps of linearized and intact plasmid DNA with multiple restriction enzymes.

2. Split cells 1 : 6 as usual, 1 or 2 days prior to electroporation. Continue to feed daily. Feed cells 2–3 hr prior to electroporation.

3. Trypsinize cells as usual but, on dispersal of cells, add about 10 ml of medium and return the plates to the incubator for 30 min. This preplating allows about 90% of feeder cells to reattach and thus the cell suspension is further enriched for ES cells. For many situations, this step is optional. Wash the cells twice in HBS. Count an aliquot of cells.

4. Resuspend the cells in ice-cold HEPES-buffered saline (HBS) at 2.5 × 10^7 cells/ml. Add linearized DNA from a concentrated stock in sterile HBS to a final concentration of 25 μg/ml. Leave on ice for 10 min. After mixing with a Pasteur pipette, transfer cells plus DNA into an electroporation cuvette. Keep on ice for 10 min. The 0.4-cm disposable cuvettes from Bio-Rad (Richmond, VA) and the Bio-Rad Gene Pulser work well. Each electroporation cuvette can hold 0.8 ml, equivalent to 2 × 10^7 cells.

5. Electroporate at 240 V, 500 μF if the Gene Pulser (Bio-Rad) with a capacitance extender is being used. Parameters could be different for other electroporators, and should be tested. Check the time constant for each cuvette, to determine the consistency of cuvettes.

6. Allow the cells to rest for 10 min at room temperature. Plate the ES cells with *neo*-resistant feeders (but no selection drugs) at about 5–7 × 10^6 ES cells/10-cm plate.

Selection of Targeted Clones

One to 1.5 days after electroporation, feed the cells daily with selection drugs until selection is complete (usually 7–8 days). Cell death is observed by 3–4 days after drug addition.

Selection Medium. Dissolve (in 0.1 *M* HEPES, pH 7.2) an appropriate weight of G418 powder (Geneticin; GIBCO-BRL) to yield 200 μg of active drug per milliliter if PGK-*neo* is in the targeting construct or to yield 150 μg of active drug per milliliter, if pMC1-*neo* is in the construct. Filter sterilize before addition to complete the ES cell medium. For double selection, add ganciclovir (Syntex Corp., Palo Alto, CA) to a 2 μM final concentration by making a fresh 1000× (2 m*M*) stock. Dissolve 5.1 mg of ganciclovir powder in 10 ml of 0.1 *M* HEPES, pH 7.2. Filter sterilize and add to complete G418 medium to give a final concentration of 2 μM. An alternative to gancyclovir is FIAU (Bristol Meyers Squibb, Wallingford, CT), which should be used at 0.2 μM. We routinely prepare the selection medium for a whole experiment and store at 4° for up to 2 weeks.

Change culture medium to selection medium after 24–36 hr. One or two plates are grown in G418 medium only, and are used to calculate the number of neomycin-resistant colonies. Feed the plates every day. Cell death is visible at about 3–4 days of selection. The morphology of the colonies should be observed carefully. Some large colonies may start flattening. Clones should be picked before they start differentiating (but flattened colonies may still contain ES cells and could be picked). We routinely pick colonies at day 7 or 8 of selection. In the plates containing G418 only, the selection takes a little longer. Plates are stained with

Giemsa, and colonies counted on day 10 of selection. This selection scheme is outlined in Fig. 3.

Other Selection Schemes. The procedures given above are suitable for the most commonly used schemes involving the neomycin resistance gene as the positive selection marker and the HSV thymidine kinase gene as the negative selection marker. If HGPRT or hygromycin or puromycin

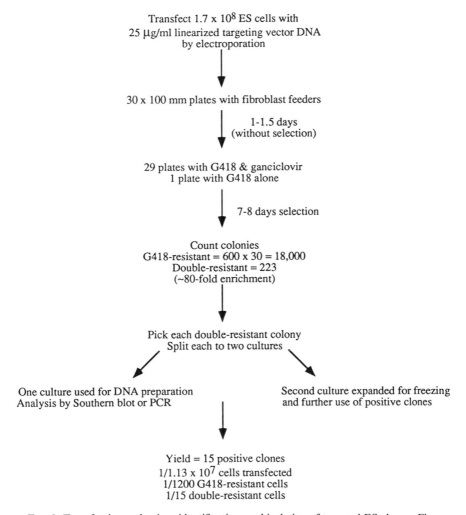

FIG. 3. Transfection, selection, identification, and isolation of targeted ES clones. Flow diagram showing the various steps involved in isolating targeted ES cell clones, using representative results from fibronectin knockout.[10] Actual numbers will vary; in particular, the enrichment observed here is unusually high.

resistance genes are used, the feeders must be modified accordingly. One can either obtain a line of mice with the appropriate transgene or use a feeder cell line just during the time of selection. A line of STO cells resistant to both neomycin (G418) and hygromycin can also be used.[31] Embryonic stem cells can, and probably should, be returned to fibroblast feeders at the time of picking the clones. Selection for HGPRT is in standard medium plus HAT (0.1 mM hypoxanthine, 0.8 μM aminopterin, and 20 μM thymidine). Hygromycin selection ranges from 100 to 150 μg/ml and should be tested to determine the effective dose for the ES cells.

Picking Drug-Resistant Embryonic Stem Cell Colonies

Equipment Required

Microscope in a dissecting hood
Sterile Pasteur pipettes that have been drawn out over flame
Mouthpiece, with tubing and a disposable disk-type filter (0.45–0.8 μm pore size)
Twelve-channel pipettor
Tissue culture plates (96 and 24 well)
Pipette tips (presterilized) in racks that fit 96-well plates
Pipette tips in racks for 24-well plates: These are assembled by placing tips in alternating columns of a 96-well rack
Clustered freezing tubes in 96-well plate format (Costar; Cambridge, MA)

Summary of Procedure. Clones are usually picked after 7 or 8 days of selection. Each clone is picked into an individual well of a 96-well plate that contains trypsin–EDTA. This is best performed in a dissecting hood, but that is not absolutely necessary. Each trypsinized colony is then split into a single well of duplicate 24-well plates, so that 1 plate can be used for freezing the colonies, and the other for DNA extraction for Southern blot or PCR analysis. The transfer of cells between 96-well plates and 24-well plates is accomplished with a 12-channel pipettor, which has pipette tips in alternating channels. This allows the transfer of clones 1, 3, 5, 7, 9, and 11 from a 96-well plate into wells 1, 2, 3, 4, 5, and 6 of a 24-well plate.

Trypsinization of Selected Clones

1. Feed selection plates 2–3 hr before picking colonies. Estimate the number of colonies to be picked, and prepare 24-well plates with mouse embryonic feeder cells (2 × 10^6 feeder cells/plate, i.e., about 10^5/well with 0.5 ml of complete medium). Each colony will be split in half, into wells of duplicate 24-well plates.

2. Wash a selection plate once with sterile PBS at 37° and replace with 10 ml of the same.

3. Place 60 μl/well of trypsin–EDTA at 37° in 1 row of a 96-well plate. Colonies will be picked in groups of 12 (one row), and the trypsin–EDTA should be maintained at 37°. Colonies should be picked fairly rapidly, so that a group of 12 takes about 5 min.

4. Pick individual clones by tearing each away from the surrounding feeder cells, using a ripping/tearing motion, and drawing each gently into the pipette. Before approaching a colony, partially fill a Pasteur pipette with PBS from the plate to prevent sticking of the colony to the glass. The volume of PBS aspirated should be small, but enough to prevent the colony from sticking.

5. Transfer each clone to a well of the 96-well plate containing trypsin–EDTA.

6. Connect six pipette tips to the multichannel pipettor (channels 1, 3, 5, 7, 9, and 11) and disperse clones 1, 3, 5, 7, 9, and 11 by pipetting up and down several times with the multipipettor. Transfer 30 μl of each cell suspension to the first row of one 24-well plate (prepared above) and the remaining 30 μl to the first row of the duplicate 24-well plate. Then disperse clones 2, 4, 6, 8, 10, and 12 in the same manner and split them into the second row of each 24-well plate.

7. Alternatively, it may be more convenient to stop the trypsin after picking 12 colonies, in the first row of the 96-well plate, by adding 60 μl of complete medium and dispersing the cells. One can then proceed to picking the next 12 colonies into the second row. This method is most helpful when the selection plates contain more than 12 colonies.

8. Feed the cells as soon as possible (the next morning is soon enough) as they do not grow as well with trypsin present. Cells should be fed daily, with 0.5 ml of complete medium.

Freezing Expanded Colonies

1. Use one of the duplicate 24-well plates for freezing. Cells should be frozen when many of the wells are subconfluent, even if some clones are still sparse. This is important to ensure that the majority of the colonies have not grown too large and begun to differentiate. Cells are frozen, at this stage only, in 10% (v/v) DMSO, fetal calf serum, and trypsin–EDTA. For this relatively short-term storage, the trypsin–EDTA does not appear to harm the cells.

2. Feed the cells 2–3 hr before freezing.

3. Place 120 μl of ice-cold 20% (v/v) DMSO–80% (v/v) FCS in each Costar cluster tube and maintain on ice [a rectangular freezing tray from Corning (Corning, NY) fits well in a tissue-culture hood].

4. Wash the cells with warm PBS and trypsinize with 60 μl of pre-warmed trypsin–EDTA for 5 min at 37°.

5. Move the plate to ice and add 180 μl of ice-cold FCS to each well. Disperse the cells by pipetting up and down with six tips on the multichannel pipettor. Make two frozen vials of each clone by transferring 120 μl of the suspension (clones 1–6) into one cluster tube (tubes 1, 3, 5, 7, 9, 11) and the remaining 120 μl into the neighboring tube in the same row (tubes 2, 4, 6, 8, 10, and 12). Thus, each row of clustered tubes corresponds to each row of the 24-well plates.

6. To freeze the cells slowly, wrap the cluster tube rack in paper towels, place at $-20°$ for about 1 hr, then transfer to $-70°$. The cluster tube racks can be maintained at $-70°$ until positive clones have been identified.

DNA Isolation for Genomic Southern Blot Analysis

1. Grow clones until most are confluent. High-quality DNA can be extracted even from overgrown clones.

2. Wash wells with PBS, add 200 μl of lysis buffer to each well, and incubate at 37° for 30 min. Plates may be stored frozen at this point.

Lysis buffer: 150 mM NaCl, 20 mM Tris (pH 7.5), 5 mM EDTA, 0.5% (w/v) sodium dodecyl sulfate (SDS), proteinase K (0.25 mg/ml) (stored frozen as single-use aliquots at 20 mg/ml; add to buffer just before use)

3. Transfer lysates to 1.5-ml tubes. Racks can be arranged to fit the format of a 24-well plate so that a 12-channel pipettor can be used to process the samples through all the steps.

4. Extract with an equal volume of phenol–chloroform. Vortex, spin, and take out the organic phase using a 12-channel pipettor.

5. Extract with chloroform–isoamyl alcohol.

6. Transfer the aqueous phase to clean tubes and precipitate with 20 μl of 3 M sodium acetate (pH 5.2) and 0.5 ml of ethanol, at $-20°$ for 30 min.

7. Resuspend the pellets in 10 mM Tris-HCl (pH 8.0), 1 mM EDTA (TE). Repeat the precipitation as described above, dry the pellets, and resuspend in TE. To dissolve DNA, leave at room temperature for several hours or at 65° for 15 min. A confluent well from a 24-well plate yields approximately 10 μg of genomic DNA. Use one-half or one-third of each sample to digest with restriction enzymes. To aid in complete digestion, use a relatively large (100-μl) volume and allow to proceed overnight. The digests can then be precipitated or reduced in volume in a Speed-Vac (Savant, Hicksville, NY) before loading on a gel.

Polymerase Chain Reaction Screen of Embryonic Stem Cell Clones. An alternative to the Southern blotting screen described above is to use the PCR, which can be done on pools of clones. The following procedures allow characterization of ES cell clones by PCR in pools of 12 (blot analysis of PCR products is usually required). Clones are maintained in culture during this quick initial screen, thus obviating the need to freeze numerous clones. Positive pools can be rescreened as individual clones by PCR, and subsequently confirmed by genomic blot analysis. The PCR screening has worked well in our hands, but two possible complications need to be discussed. First, a PCR primer must be characterized that is within the targeting vector (usually in the positive selection marker) and another in genomic DNA outside the targeting vector. These primers must be tested fairly rigorously for their lowest detection limit and extent of cross-homology. Second, we have found that, to ensure reliable amplification in small amounts of genomic DNA, the amplification distance should be only about 1 kb. This limits the homology with the locus to be targeted that can be included in the targeting vector, and thus may decrease targeting frequency.

1. Pick drug-resistant colonies by the procedures described above, except allow colonies to settle overnight in 96-well dishes with feeders. Split each colony to two 96-well dishes by trypsinization of the attached cells. One dish will be for PCR analysis and the other will be expanded. Each dish should have feeder cells and 0.2 ml of medium. Using a multichannel pipettor, routine trypsinization is performed using no more than 50 μl of trypsin–EDTA. After thorough but gentle dispersal of cells, transfer 25 μl to each 96-well dish with 0.2 ml of medium per well. The cells are still in dilute trypsin but can tolerate it overnight, which is enough time to settle. Refeed after 12 hr.

2. One to 2 days after the initial split, trypsinize cells from 1 of the duplicate 96-well dishes as usual, but do not add medium with serum.

3. Combine samples by row (12 wells/pool) by dispersing in trypsin–EDTA, followed by a rinse with PBS. Microfuge the pools for 3 min to spin down the cells.

4. Remove all but 5 μl of supernatant and resuspend the pellet in 50 μl of water. Freeze on dry ice.

5. Thaw cells at 95° for 8 min and cool. Add 10 μg of proteinase K (fresh aliquot) and incubate at 55° for 30 min. Incubate at 95° for 8 min and cool.

6. Add components for a standard 100-μl PCR reaction. Run 20 μl of each PCR reaction on an agarose gel and blot (various quick-blot procedures are adequate). Probe with subclone of partial amplification product.

7. Expand the remainder of each clone by continued culturing until the results of pooled PCR analysis are complete. Feed daily.

8. Once the PCR results are in, individually pass all clones within each positive pool to a 24-well plate with feeders. Save one-half of each sample for PCR analysis as an individual clone.

9. Feed the cells daily. After 3 days, once the PCR analysis of clones is complete, place each positive clone on a 35- or 60-mm plate with feeders to expand for freezing (see below). Feed daily.

Thawing and Expanding Targeted Clones. For each positive clone, thaw one of the cluster tubes rapidly in a 37° water bath, then cool on ice. Do not spin. Plate on a 35-mm plate with feeder cells (4×10^5 feeder cells per plate) in 2 ml of complete medium. Feed after 12 hr. At this point, cells are cultured in normal medium without G418; it has been reported that cells maintained in G418 have a lower capacity to contribute to the germ line of injected embryos.[33] Depending on the density of each clone, split up to 1 : 6. Freeze as soon as possible in small aliquots. Freeze enough aliquots for one aliquot per day of blastocyst injection. A subconfluent 60-mm plate can be frozen as eight 250-μl aliquots, in 90% (v/v) FCS and 10% (v/v) DMSO. Such an aliquot can be plated directly after thawing into one or two 35-mm plates with feeder cells, 2 days before injection.

Further Characterization of Targeted Clones

Positive clones detected by the above procedures should be retested using several different restriction enzyme digestions to ensure appropriate insertion of the selectable marker into the locus; both flanks of the insertion should be checked. To determine if targeted clones have any additional insertions of the targeting vector, test genomic DNA with a probe containing the *neo* gene. Only a single band at the expected size should be detected after long exposure.

A further simple check of the condition of targeted cells is to count chromosomes:

1. Treat a rapidly growing culture with Colcemid (0.06 μg/ml) for 4 hr.

2. Trypsinize cells, wash in PBS, and combine with washed cells from culture supernatant.

3. Resuspend the cells in 10 ml of 0.56% (w/v) KCl and incubate at room temperature for 8 min. Spin at 300 rpm for 2 min at room temperature.

[33] A. Bradley, in "Teratocarcinoma and Embryonic Stem Cells: A Practical Approach" (E. J. Robertson, ed.) p. 113. IRL Press, Oxford, 1987.

4. Remove all but 100 μl of supernatant and resuspend the cells. Add, dropwise, 10 ml of fixative [3 : 1 (v/v), methanol–glacial acetic acid]. Incubate for 10 min.

5. Spin and resuspend in 1 ml of fixative.

6. Drop the cells, from a height of 10 cm, onto glass slides (2–3 drops per slide, cleaned with ethanol and dried). Wait until the liquid has spread to the edges of each slide and blow dry.

7. After they are completely air dried, stain the slides for 15 min in 3% (w/v) Giemsa in PBS. Wash in water and air dry.

8. The easiest way to count chromosomes is to photograph the slide on print film so that counted chromosomes can be marked. Count at least 30 spreads/clone. At least 50% of the spreads should have 40 chromosomes. Watch out for spreads with too many chromosomes as well as those lacking chromosomes.

Generation and Breeding of Chimeric Mice

The animal and equipment requirements and procedures for generating chimeric mice by injection of ES cells into blastocysts have been described elsewhere.[32,33] We use strain C57BL/6 mice as embryo donors and strain CD1 as foster mothers. Contribution of embryonic stem cells, which are derived from strain 129/Sv mice and have agouti coat color, is clearly evident in contrast to the black coat color of the donor embryo.

Handling of targeted ES cells in preparation for blastocyst injection can affect ease and success of injection. The following procedure yields healthy ES cells as a single-cell suspension. Two days prior to blastocyst injection, one of the frozen aliquots [90% (v/v) serum, 10% (v/v) DMSO] is thawed rapidly in a 37° water bath and plated directly into two 35-mm dishes with feeder cells (4×10^5 feeders per plate). Replace the medium the next morning, daily, and 2 hr before trypsinization. One plate is trypsinized at the beginning of an injection session and the second about 3 hr later, in freshly prepared 0.25% (v/v) trypsin (GIBCO-BRL)–0.02% (v/v) EDTA in PBS at 37° for 5 min. Disperse cells gently after adding an equal volume of ES cell medium and pellet. Resuspend cells gently and thoroughly (by drawing them in and out of a Pasteur pipette 20 times) in 2 ml of ice-cold ES cell medium and keep on ice until transfer to the injection chamber.

The following are some general considerations and guidelines for blastocyst injection, but the references cited above should also be studied. The procedure requires an inverted microscope with interference-contrast or phase-contrast objectives. The magnification range required is $\times 40$ to $\times 200$. The injection microscope should also be equipped with a cooling stage (blastocysts are more resilient to the injection procedure when they are kept at 5–10°). A pair of micromanipulators is also required; one for the holding pipette and one for the injection pipette. The manipulators from Leitz Instruments are especially useful for embryo injections because they have joystick movement in the horizontal plane. Flow in each pipette is regulated by hand, using a micrometer to adjust a Hamilton syringe. To obtain donor embryos, a stereo dissecting microscope (total magnification, $\times 20$) is required for flushing blastocysts from the uterus and collecting them for culture. Blastocysts are cultured in a tissue culture dish, in microdrops of medium under light-weight paraffin oil, in 5% CO_2 in air at 37°. During an injection session, several small batches of blastocysts are transferred from culture to the injection chamber with trypsinized ES cells. The easiest blastocysts to inject are those that have expanded fully but have not yet hatched from the zona pellucida. Loading ES cells into the injection pipette is the most time-consuming part of the procedure, and should be done with care so as not to damage the cells. The injection pipette should be just large enough to accommodate the ES cells, but not much larger, because damage to the blastocyst would be more likely to occur. Each blastocyst should be injected with 15–20 ES cells. After injection, the blastocysts are returned to culture for at least 1 hr before surgical transfer to a pseudopregnant female. The surgery is performed on the bench with general anesthesia, using a stereo dissecting microscope. Each female should receive 10–12 blastocysts; uninjected blastocysts can be transferred along with injected ones, to help prevent complications in the pregnancy due to small litter size.

Chimeric animals are monitored by coat color. The 129/Sv-derived ES cells give rise to an agouti (brown) coat whereas the recipient C57BL/6 embryos produce black coat color. Desirable chimeras have a high proportion of brown coat. To determine if targeted ES cells have contributed to the germ line of chimeric mice, male chimeras are bred with C57BL/6 females. Agouti pups are the result of ES cell-derived sperm. If the targeted mutation is heterozygous viable, then one-half of the agouti pups will carry the targeted allele. Table I illustrates the variability observed in several aspects of germ line transmission. First, note that all four ES cell clones yielded high coat color chimerism, yet only three contributed to the germ line at reasonable frequency. Second, the extent of chimerism for a particular germ line chimera does not necessarily reflect transmission frequency to progeny. Chimera 13, for example, had only 25% agouti coat yet 49% of his progeny were ES cell derived. However, it is generally true that if none of the chimeras derived from a given ES cell clone has high coat color chimerism then germ line transmission is unlikely. We generally discard chimeras if none of the first 60 pups contains the targeted

TABLE I
VARIABILITY IN GERM LINE TRANSMISSION OF TARGETED
EMBRYONIC STEM CELL CLONES[a]

Clone	Chimera	Agouti (%)	Progeny (number of agouti/total)
3B3	13	25	51/105
	14	50	0/32
2G9	22	80	6/53
	23	60	1/25
	24	50	2/60
	26	10	37/72
	37	25	2/33
2G4	27	30	0/67
	30	90	24/81
	31	80	2/68
	32	70	4/47
	33	100	15/50
	38	90	6/22
	39	90	0/0[b]
2C10	41	40	0/47
	45	90	0/73
	46	90	0/49
	47	90	0/61
	48	70	0/0[b]
	49	70	0/79

[a] Typical results for germ line transmission from targeted ES cell clones. High coat color chimerism does not correlate well with efficient germ line transmission by individual animals. In this case (fibronectin knockout; George et al.[10]) about half the chimeras gave germ line transmission.

[b] Chimera was sterile.

allele and, because most ES clones give chimeras of which around 50% give germ line transmission, we would discard a clone if none of the first 5 or 6 chimeras gives agouti pups.

It is necessary to obtain germ line transmission from at least two independently derived ES cell clones in order to ensure that any phenotype of the resulting mice arises from the targeted mutation and not from some other mutation that occurred during handling of the ES cells.

Genotyping Mice: Tail Blots and Polymerase Chain Reaction Analysis

Genomic DNA, isolated from tail biopsies, can be analyzed by either Southern blot or PCR. Generally, mice must be at least 3 weeks old to tolerate the general anesthesia used in cutting the tail. With a fresh razor blade, cut a 1-cm length from the tip of the tail and cauterize the remaining tail with a soldering iron. Identify each mouse and corresponding DNA sample with a number, either by an ear-punch scheme[32] or by ear tags with imprinted numbers (use size 1 monel; National Band and Tag Co., Newport, KY).

DNA is isolated by a simplified procedure that is amenable to large numbers of DNA samples.[34] Transfer the tail biopsy to a 1.5-ml tube that contains 0.5 ml of tail lysis buffer: 100 mM Tris-HCl (pH 8.5), 5 mM EDTA, 0.2% (w/v) SDS, 200 mM NaCl, proteinase K (100 μg/ml) (Boehringer Mannheim, Indianapolis, IN). Continuously rotate the samples overnight at 55°. Vortex the tubes, and spin down hairs and tissue debris in a microfuge for 5–10 min. Transfer the supernatant to a fresh tube containing 0.5 ml of 2-propanol and mix thoroughly. Recover the precipitate with a pipette tip and transfer to a fresh tube containing 100 μl of 10 mM Tris-HCl, 0.1 mM EDTA, pH 7.5. Make certain that DNA is dissolved by incubation at 37° with intermittent vortexing.

To analyze tail DNA by Southern blot, a single-copy probe that hybridizes with wild-type and targeted alleles is required. Generally such a probe has already been characterized for use during screening for targeted ES cells. Generally 15 μl of DNA prepared as above is sufficient for one lane of a Southern blot. Restriction digestion should be carried out in a final volume of at least 50 μl, with at least 40 units of enzyme and in the presence of bovine serum albumin (BSA; 0.1 mg/ml) and 4 mM spermicide. Digestions usually require several hours to overnight for completion. If the DNA does not digest well, a phenol–chloroform extraction followed by reprecipitation is likely to help. A variety of standard blotting procedures suitable for genomic DNA can be used.[35] Capillary blotting to a nylon membrane, ultraviolet (UV) cross-linking, and hybridization in sodium phosphate and SDS at 65° work quite well.[36]

As an alternative, PCR amplification of unpurified tail DNA using appropriate PCR primers can be used.[37] Tail (5–10 mm) is added to 0.4 ml of PCR tail buffer [50 mM KCl, 10 mM Tris-HCl (pH 8.3), 2.5 mM MgCl$_2$, gelatin (0.1 mg/ml), 0.45% (w/v) Nonidet P-40 (NP-40), 0.45% (w/v) Tween 20]. Incubate overnight at 55° with shaking and with addition of two 25-μl aliquots of proteinase K (10 mg/ml) added at an interval of

[34] P. W. Laird, A. Zijderveld, K. Linders, M. A. Rudnicki, R. Jaenisch, and A. Berns, *Nucleic Acids Res.* **19**, 4293 (1991).
[35] J. Sambrook, E. F. Fritsch, and T. Maniatis, eds., "Molecular Cloning: A Laboratory Manual," 2nd ed. Cold Spring Harbor Lab., Cold Spring Harbor, NY, 1989.
[36] G. M. Church and W. Gilbert *Proc. Natl. Acad. Sci. U.S.A.* **81**, 1991 (1984).
[37] T. Hanley and J. P. Merlie, *BioTechniques* **10**, 56, (1991).

several hours. Heat at 95° for 10 min to denature residual proteins, cool to room temperature, and spin. Five microliters of each DNA sample should give definitive ethidium bromide signals after PCR amplification with primers for each allele.

To analyze DNA by PCR, primer pairs must be designed that indicate the presence of the wild-type allele versus the targeted allele. This can be accomplished by several strategies. First, both alleles can be assayed by a set of three primers, one of which is within the *neo* gene, as described in Polymerase Chain Reaction Screen of Embryonic Stem Cell Clones, above. Alternatively, separate primer pairs can be used to assay the alleles: one pair within the *neo* gene, and the other pair in the wild-type gene. In both these strategies, designing primers that yield amplification products of different sizes allows detection of both alleles in a single reaction tube and gel lane. Care must be taken not to contaminate reagents with amplification products. Organization of reagents into single-use aliquots is highly recommended.

Generation of Homozygotes

To determine if homozygous mutant animals are viable, heterozygous crosses are performed and all pups genotyped at weaning. The expected 1 : 2 : 1 ratio of genotypes, or lack of it, will become evident in three to four litters. However, genotype ratios should be kept for all heterozygous crosses performed. Ideally, mice derived from two or three independent targeted clones are available. Heterozygous crosses should be carried out for each targeted clone, as well as between clones. The interclonal crosses show that any phenotype is the result of the targeted mutation rather than some other mutation occurring in the ES cell clone.

For a homozygous viable mutation, the next job is to show that the targeted mutation is truly a null allele. Procedures will depend on the gene/protein of interest. This has been performed quite thoroughly for the tenascin-deficient mice[8] and for P-selectin-deficient mice[38] at both the mRNA and protein level.

Analysis of Embryonic Lethal Phenotype

Determining the approximate time of death is accomplished by genotyping embryos at various stages during gestation. If death is actually perinatal, pups may not be observed because the mother eats them. However, homozygous mutant fetuses should still be observed by embryonic

[38] T. N. Mayadas, R. C. Johnson, H. Rayburn, R. O. Hynes, and D. D. Wagner, *Cell* (*Cambridge, Mass.*) **74,** (1993).

day 17 or 18 (E17 or E18). The yolk sac is an excellent source of genomic DNA, and can easily be dissected without contaminating maternal tissue. Genomic DNA can be isolated for analysis either by Southern blot or by PCR. This leaves the embryo for a variety of analyses, including expression of mRNA and protein, histology, or morphology. Genotyping of yolk sac DNA can be performed on embryos as young as E8.5, during gastrulation. Even at this early stage, the yield from yolk sac alone is enough for two lanes on a Southern blot.

Yolk sacs are lysed in 0.1 to 0.5 ml of 100 mM Tris-HCl (pH 8.5), 5 mM EDTA, 0.2% (w/v) SDS, 200 mM NaCl, proteinase K (0.2 mg/ml) at 55° for 2 hr (E8.5) or overnight (later than E10.5) with vortexing or rotation. Lysate is extracted once with phenol–chloroform and precipitated in ethanol.

If periimplantation lethality is suspected, then preimplantation blastocysts can be genotyped by PCR. A single blastocyst can be digested and assayed in two separate PCR reactions and amplification products visualized via Southern blot.

1. Flush blastocysts from the uterus on E3.5.

2. Culture the blastocysts for 24 hr in M2 embryo culture medium[33] at 37° in 5% CO_2 and air to assay for expansion and hatching from the zona pellucida.

3. Transfer individual blastocysts (using a drawn-out capillary pipette) to PCR tubes with 50 μl of 10% (v/v) PBS.

4. Freeze on dry ice, and incubate at 95° for 8 min.

5. Digest with 10 μg of proteinase K at 55° for 30 min, followed by incubation at 95° for 8 min. Chill on ice. Split sample for separate reactions and add PCR components.

Histological Techniques for Postimplantation Embryos

Analysis of sectioned embryonic tissue is necessary initially for the morphological description of the mutant phenotype as well as for antibody staining. As characterization of the mutant phenotype progresses, sectioned material will be necessary for a variety of other analyses, including, for example, assay of other matrix proteins and cell proliferation rates. One important point is that sectioned tissue can be genotyped by PCR (see below). This allows correlation of a phenotype with genotype, even in embryos as young as E7.5. As for most other tissues, cell morphology of postimplantation mouse embryos is preserved better when embedded in paraffin. Cryostat sections may be necessary, however, to preserve antigen recognition.

The procedures required for handling early embryos vary depending on stage and tissues of interest. Useful references regarding handling of embryos, along with many descriptions, diagrams, and photographs, are given in Hogan *et al.*[32] and in Cockroft.[39] *The Atlas of Mouse Development* prepared by Kaufman[40] is an invaluable resource. Individual embryos are within maternal tissue known as the decidual swelling, which can be dissected easily from the uterus. Individual decidua are larger and easier to handle than dissected embryos. Sectioning embryos while still in the decidua is especially helpful prior to E8.5, when mouse embryos begin the turning process characteristic of many rodent embryos. Before E8.5, the orientation of the embryo is fixed. Thus, to make transverse sections from a presomite embryo at E7.5, the pear-shaped deciduum can be embedded simply by standing it on its mesometrial (wide) end. Transverse orientation is also quite informative for E8.0 embryos, especially for analysis of early heart development.

Mouse embryos initiate the turning sequence at about E8.5, with 8–10 pairs of somites.[40] During turning, which is complete by about E9.0 with 15–20 pairs of somites, the orientation of the embryo within the decidual swelling is not known prior to sectioning. To obtain a particular orientation, E8.5 embryos must be dissected from the decidual swelling and usually from the yolk sac. Again, transverse orientation is generally the most informative. To orientate small dissected embryos, a dissecting microscope is required at the embedding station. Alternatively, a dissecting microscope equipped with a warming stage can be used. Sagittal orientation is easier to achieve, as the embryo can be embedded while lying on its side. This too can be complicated, however, if the embryo happens to have been dissected while in the midst of turning.

While sectioning in the deciduum may obscure embryonic orientation, it also has advantages. Such sections leave the extraembryonic membranes and umbilical cord intact. At E8.5 the yolk sac is the site of primitive erythrocyte proliferation and supplies the embryo with blood on fusion of the embryonic and extraembryonic vascular systems.[40] Decidual sections revealed the defects in yolk sac vasculature in embryos lacking fibronectin[10] and α_5 integrin.[12]

Embryo Dissection

1. Kill a pregnant mouse by cervical dislocation (noon of the day of vaginal plug is defined as E0.5).

---

[39] D. L. Cockroft, *in* Postimplantation Mammalian Embryos: A Practical Approach'' (A. J. Copp and D. L. Cockroft, eds.), p. 15. Oxford Univ. Press, Oxford, 1990.

[40] M. H. Kaufman, ''The Atlas of Mouse Development,'' p. 26. Academic Press, San Diego, 1992.

2. Expose the uterine horns by abdominal incision through skin and body wall.

3. Remove the uterus by cutting below the ovaries, at the cervix, and along the mesometrium. The decidual swellings appear as beadlike swellings along the length of each uterine horn. Wash in PBS.

4. In PBS, dissect the decidua by ripping uterine muscle with two pairs of watchmaker's forceps (Dumont No. 5). If decidual sections are planned, transfer the decidua to fixative.

5. Embryo dissection methods are described and illustrated in Hogan *et al.*[32] As early as E8.0, the yolk sac can be dissected from the embryo and used for genotyping (see below).

Fixation. Different fixatives should be used, depending on the type of information desired. Bouin's fluid [75 ml of saturated picric acid, 25 ml of 37% (w/v) formaldehyde, 5 ml of glacial acetic acid], combined with embedding in paraffin is superior for maintenance of cell morphology. However, Bouin's fluid frequently destroys antigen recognition and also interferes with PCR genotyping of sectioned material (see below). Carnoy's fixative (60% ethanol, 30% chloroform, 10% glacial acetic acid, v/v/v) is also excellent for maintenance of morphology. The PCR genotyping works well with Carnoy's-fixed tissue, as do some polyclonal antibodies, for example, those that cross-react with mouse fibronectin.[10] Paraformaldehyde [PFA; 4% (w/v) in PBS, made fresh] is generally used only for cryostat sections; it is also compatible with PCR analysis. Carnoy's and PFA fixations were done at 4° and Bouin's at room temperature.

Fixation times for whole decidua are shown in the tabulation below.

Stage	Fixation times (hr)		
	Bouin's	Carnoy's	PFA
E7.5	6	2	6
E8.0	6	2	6
E8.5	6	4–6	6
E9.5	6–24	6	6–24
E10.5	6–24	6	6–24

Fixation times for dissected embryos are less variable. In both Bouin's and Carnoy's, E7.5 and E8.0 embryos need only 1 hr and E8.5, E9.5, and E10.5 embryos need 2 hr. In PFA, E7.5, E8.0 and E8.5 embryos need two hours and older embryos need from four hours to overnight.

Paraffin Sections

1. After fixation, tissue should be stored in 70% ethanol at 4°.

2. Dehydrate twice in 95% ethanol on ice for 10 min, then twice in 100% ethanol on ice for 10 min.

3. Clear in 50% ethanol, 50% xylene for 15 min.

4. Clear in 100% xylene at room temperature for 30 min.

5. Infiltrate in 50% xylene–50% wax (Paraplast extra) at 55–60° for 1 hr.

6. From this point on, a Reichert–Jung histostat embedding center is helpful. In an embedding cassette, infiltrate embryos overnight in 100% paraffin at 56° under vacuum.

7. Transfer embryos to an embedding mold with molten paraffin on the warming surface.

8. Orientate embryos in paraffin by moving them between the warm and cold surfaces rapidly. This hardens the wax slightly, thus supporting the embryo while still allowing adjustment of its orientation. Avoid allowing the wax to cool too much, as the necessary melting damages the embryo. Orientation of small dissected embryos requires a dissecting microscope, situated either at the warm plate of the embedding station or separately and equipped with a warming unit.

9. Allow the wax to cool completely on the cold surface and store the blocks at room temperature.

10. Trim the blocks to reduce the section area and cut sections 6–8 μm thick.

11. Transfer the sections to a water bath at 37° to reduce curling.

12. Pick up sections from the water bath with a polylysine-coated slide and immediately transfer the slide to a slide warmer at 45°. Allow to dry overnight.

13. Store sections at 4°.

14. Dewax and rehydrate sections for staining, by soaking them for 5 min in each of the following: xylene, 100% ethanol, 95% ethanol, 70% ethanol, 30% ethanol, water, or PBS.

15. Histological staining: hematoxylin, 5 min; running water, 1 min; eosin, 10 min; running water, 1 min.

16. Dehydrate sections by the same treatments as in step 14, but in reverse order. Mount in Permount and coverslip. Store at room temperature.

Polymerase Chain Reaction Genotyping of Paraffin-Embedded Sections. Tissue can be scraped from wax sections for PCR analysis. Use slides from step 13 described above.

1. Use every third or fourth section for a particular embryo, so as not to destroy serial section information.

2. Under a dissecting microscope, with a fresh disposable scalpel, scrape embryonic tissue and transfer to 60 μl of water in a PCR tube. For E7.5 and E8.0 embryos, the allantois is the most likely source of pure embryonic tissue and three or four sections are needed per embryo; for E8.5 embryos, two sections should be sufficient. If the scalpel slips into maternal tissue, discard and begin again with a new scalpel.

3. Scrape maternal decidual tissue as a positive control for both PCR primer sets.

4. Add 20 μg of proteinase K (fresh aliquot) and digest at 55° for 30 min, followed by further digestion at 95° for 10 min. Cool on ice, and split the sample for separate PCR reactions. Add PCR components.

5. For E7.5, E8.0, E8.5, and E9.0 embryos, a Southern blot must be performed to visualize amplification products. By E9.5 the embryos are big enough that several sections can be sacrificed and PCR products can be detected by ethidium bromide staining.

Cryostat Sections

1. After fixation and two washes with cold PBS, infiltrate the embryos in 0.6 M sucrose in PBS at 4° overnight (or until they sink in the sucrose).

2. Transfer the embryos to an embedding mold containing OCT (Polysciences, Warrington, PA).

3. Orientate the embryos, using a dissecting scope if necessary. Freeze the embryos in OCT slowly, so as not to trap gas, by placing the mold on top of ground dry ice.

4. Store blocks at $-80°$.

5. The optimal cutting temperature is generally $-25°$. Equilibrate the blocks to cryostat chamber temperature, then cut sections 6–8 μm thick.

6. Melt the sections to polylysine-coated slides, and dry overnight at room temperature.

7. Store the sections at $-80°$.

8. To stain the sections, dissolve the OCT by soaking in PBS for 15 min. Do not allow the sections to dry during staining.

9. For histological staining, use Multiple Stain (Polysciences) for about 1 min, followed by running water, and mount in Permount.

Immunofluorescence Staining of Sectioned Embryos

1. Dewax the paraffin sections or dissolve the OCT from the cryostat sections and soak in PBS.

2. Around the embryos to be stained, create wells with a PAP Pen (Polysciences). Do not allow the sections to dry (return them to PBS while working with other slides).

3. Block for 30 min at 37° in blocking buffer [10% (v/v) serum of the

same species as the second antibody, 0.05% (v/v) Tween 20, 0.02% (w/v) azide, in PBS]

4. React with the first antibody (diluted in blocking buffer) overnight at 37° in a humid chamber. Wash three times (5 min each) with PBS.

5. Reblock for 30 min at 37°.

6. React with second antibody (usually diluted at least 1 : 200 in blocking buffer) for 1 to 2 hr at 37°.

7. Wash three times (5 min each) with PBS. Mount in Gelvatol (Monsanto, St. Louis, MO) or Vinol (Air Products and Chemicals, Allentown, PA) at 20% w/v with antifade (1,4-diazabicyclo(2.2.2)octane Aldrich Chemical, Milwaukee, WI) at 15 mg/ml. Store in the dark.

Culture and Analysis of Embryonic Cells

An embryonic lethal phenotype does suggest that a true null allele has been generated, but may actually be misleading. To show that the protein in question is indeed absent, metabolic or surface labeling of cultured embryonic cells followed by immunoprecipitation may be necessary. The following procedures work well for E7.5 to E8.5 embryos.

1. Dissect decidua from the uterus under sterile conditions in a dissecting hood. Transfer them to sterile PBS.

2. Set up a sterile plate (6-cm plastic for tissue culture works well) with two separate drops of PBS (about 50 μl) and two separate drops of ES cell medium. A fresh plate will be used for each embryo.

3. Dissect each embryo individually and clean and sterilize watchmaker's forceps between embryos.

4. Wash each embryo, with extraembryonic membranes and ectoplacental cone, in the first drop of PBS.

5. Transfer to the second drop of PBS. Remove and discard the Reichert's membrane and ectoplacental cone. Save the yolk sac for genotyping as described above.

6. Transfer each embryo with a sterile, cutoff pipette tip to the first drop of medium to wash.

7. Transfer each embryo to the second drop of medium and mechanically dissociate into pieces that are as small as possible, using the watchmaker's forceps.

8. Transfer the pieces to 1 well of a 24-well tissue culture plate that has been pretreated with 0.1% (w/v) gelatin and contains 0.5 ml of fresh ES cell medium plus antibiotics and 0.5 ml of conditioned ES cell medium plus antibiotics. Conditioned medium: Culture a 10-cm plate of mouse embryonic feeder cells (irradiated) with 10 ml of ES cell medium for 2–3 days. Filter the medium, replenish the glutamine, and use on the same day.

9. Culture for 3–4 days. The cultures will have many clumps of cells, with individual cells progressing out from the clumps. Replace the medium and culture for an additional 1–2 days.

At this point there are a variety of options. The primary cultures can be processed for immunofluorescence if the original plating was on coverslips. Cells can be labeled with [^{35}S]methionine or surface labeled by lactoperoxidase-catalyzed iodination. Labeled culture medium or cell lysates can be harvested for analysis of protein content. Alternatively, the cell layer can be extracted for analysis of DNA or RNA, or both.

10. [^{35}S]methionine labeling medium

ES cell medium minus methionine (90%, v/v)
ES cell medium including methionine (10%, v/v)
[^{35}S]methionine (100 μCi/ml)

Label for 24 hr with 200 μl of labeling medium per well.

After labeling, collect media on ice and bring them to 10 mM EDTA and 2 mM phenylmethylsulfonyl fluoride (PMSF). Spin them in a microfuge at 15,000 g for 5 min and transfer them to a fresh tube. Freeze on dry ice. These samples are ready for immunoprecipitation.

11. Surface labeling:
   a. Wash cells and incubate in 100 μl of PBS per well, with 1 mM Ca^{2+}, 1 mM Mg^{2+}, 10 mM glucose, at room temperature for 10 min.
   b. Replace PBS solution from step a with 100 μl of the same plus 100 μCi of ^{125}I.
   c. Add 5 μl of LP/GO labeling mix [lactoperoxidase (1 mg/ml) glucose oxidase (20 units/ml), both from Calbiochem, La Jolla, CA, in PBS with Ca^{2+}/Mg^{2+}]. Label for 10 min at room temperature. Keep the plate covered.
   d. Remove the labeling mix from the cells. Wash the cells five times with cold 50 mM NaI in PBS plus Ca^{2+}/Mg^{2+} (made fresh). Wash the cells once with cold lysis buffer without NP-40.
   e. Add 400 μl of cold lysis buffer to each well and incubate on ice for 15 min. [Lysis buffer suitable for membrane proteins: 50 mM Tris-HCl (pH 8.0), 150 mM NaCl, 0.5 mM CaCl$_2$, 0.5% (v/v) NP-40, 2 mM PMSF, aprotinin (20 μg/ml), leupeptin (12.5 μg/ml).] Some matrix proteins will require stronger detergents for solubilization.
   f. Disperse cells from the plate and transfer the cell lysate to a microfuge tube and spin for 10 min. These samples are ready for immunoprecipitation.

12. Harvest for PCR, protein, DNA, or RNA extraction: Wash the cells in PBS and aspirate off. Scrape 3–5 μl of cell into a pipette tip and freeze on dry ice (for PCR genotyping of culture). If the cell layer is to be analyzed for protein, the rest can be lysed in the buffer chosen, depending on subsequent analyses to be performed.

13. Extract DNA from the cultures if desired: Add 100 μl of lysis buffer to the remaining cells and incubate at 37° for 30 min. Phenol–chloroform extract once and precipitate in sodium acetate and ethanol. Lysis buffer: 150 mM NaCl, 20 mM Tris (pH 7.5), 5 mM EDTA, 0.5% (w/v) SDS, proteinase K (0.25 mg/ml).

14. Alternatively, extract RNA from the cultures:
   a. Wash once with PBS.
   b. Lyse the cells in 200 μl of 4 M guanidinium isothiocyanate.
   c. Extract once with phenol–chloroform–isoamyl alcohol.
   d. Precipitate in 2.5 vol of 100% ethanol.
   e. Resuspend in 100 μl of water and treat with 10 units of RNase-free DNase (Promega, Madison, WI) at 37° for 30 min.
   f. Extract with phenol–chloroform–isoamyl alcohol.
   g. Precipitate with sodium acetate and ethanol.

15. As a final alternative, the cultures can be subcultured by trypsinization and replated. If the trypsinized cells are plated in uncoated tissue culture dishes for 1 hr at 37°, unattached cells removed, and the attached ones refed, one obtains fibroblastic cells. If plating is extended for several hours, additional, more rounded and clumped cells attach as well. The subcultures can be made on coverslips for immunofluorescence or cells can be analyzed in other ways; for example, for migration.[12]

Future Prospects

Using the methods reviewed and described above, it is now relatively routine to generate mice mutated in a gene of choice. In a laboratory where the methods are established, it is feasible to go from the DNA clone to a homozygous mutant mouse in less than 1 year. Analyzing and interpreting the mutant phenotype follows. Of course, the most readily interpretable results come from mutations that produce homozygous viable and fertile mice with an obvious phenotype. Viable fertile homozygotes with no obvious phenotype such as the tenascin-deficient mice[8] raise questions about functional redundancy, although it seems likely that further analyses will reveal subtle defects. It seems improbable that genes are well conserved during evolution if they have absolutely no function. It is more reasonable to suppose that the functions are not revealed by the initially rather crude analyses of the mutant mice.

An obverse problem is presented by mutations that produce embryonic lethality when homozygous, such as fibronectin,[10] and α_5 and α_4 integrins.[12,12a] Although these mutations produce valuable insights into functions of the genes in development, they provide no information about roles in adult physiology and pathology. Establishment of the null phenotype is an important and essential first step in the genetic analysis of a given gene. However, further analyses require generation of more subtle defects. Several approaches are available and others are being developed.

Transgenes

One obvious possibility is to generate lines of mice expressing the relevant gene as a transgene and intercross these transgenic strains with the knockout strain. By use of transgenes encoding variant forms of a given protein, one can therefore hope to test the functions of these variants in rescuing the mutant phenotype of the null. The methods for generating transgenic mice are not reviewed here as they have been covered elsewhere.[32,41,42] A significant problem with transgenes can arise in obtaining adequate levels and appropriate distributions of expression (both tissue and temporal) of the transgenes. In most cases the regulatory elements (promoter elements, enhancers, etc.) of the gene in question have not been well studied and can be widely dispersed. Therefore insertion of a genomic DNA fragment containing the body of the gene or of a cDNA minigene coupled to a 5′ flanking presumptive promoter segment often fails to produce adequate expression. This problem can frequently be overcome by use of large DNA segments but, given the large size of many ECM proteins and their genes, the requisite DNA segments can be very large; too large to be encompassed in a single λ phage genomic clone ($\leq$20 kb) or cosmid ($\leq$40 kb). Advances in the use of yeast artificial chromosomes (YACs) in generating transgenic mice[43-45] offer promise in this regard.

More Subtle Mutations

Another approach is to use homologous recombination methods in ES cells to generate more subtle mutations—alterations rather than knock-

[41] N. D. Allen, S. C. Barton, M. A. H. Surani, and W. Reik, in "Mammalian Development: A Practical Approach" (M. Monk, ed.), p. 217. IRL Press, Oxford and Washington, DC, 1987.
[42] D. Hanahan, *Science* **246,** 1265 (1989).
[43] T. K. Choi, P. W. Hollenbach, B. E. Pearson, R. M. Ueda, G. N. Weddell, C. G. Kurahara, C. S. Woodhouse, R. M. Kay, and J. F. Loring, *Nat. Cenet.* **4,** 117 (1993).
[44] A. Schedl, L. Montoliu, G. Kelsey, and G. Schütz, *Nature (London)* **362,** 258 (1993).
[45] W. M. Strauss, J. Dausman, C. Beard, C. Johnson, J. B. Lawrence, and R. Jaenisch, *Science* **259,** 1904 (1993).

outs. Examples exist in which an alternatively spliced exon has been inactivated by insertion of a neomycin resistance cassette and mice derived by the methods reviewed above. These mice still express other splice isoforms.[46] However, similar attempts have failed with other genes, including fibronectin.[47] In the latter case, insertion of a neomycin resistance cassette within an alternatively spliced region of the fibronectin gene led to loss of all fibronectin expression, not simply that of the desired splice isoform. Clearly the efficacy of this method varies and the failures presumably result from the fact that these mutations leave the *neo* gene in place in the altered gene. Several papers have described alternative methods for introducing changes within a gene without leaving selectable markers in place at the end of the manipulation. Hasty *et al.*[48] and Valancius and Smithies[49] describe methods that involve insertion by homologous recombination of a partial duplication containing the desired mutation followed by a second round of selection leading to excision of the wild-type copy of the duplicated segment, leaving the mutated version in place in the ES cell genome. These "hit and run" or "in-out" targeting procedures offer the promise of introducing subtle mutations into genes. Such procedures require both a positive and a negative selection step and this is rendered easier by the development of a *neo*-TK fusion gene that possesses both neomycin phosphotransferase and thymidine kinase activity and can therefore be selected either for or against.[50] This fusion gene has been successfully used in a two-step procedure in which the targeted gene is first knocked out by homologous recombination using the *neo*-TK gene casette as a positive selection marker. The desired mutation is then introduced by a second homologous recombination event selecting against the *neo*-TK fusion gene.[51] This so-called tag and exchange strategy opens the way to generation of multiple mutations in a given locus using different mutated gene segments in the second-round exchange with a single tagged cell line. It seems clear that a variety of such ingenious strategies for generation of mutations in ES cells will soon be available, allowing one to produce subtle mutations in genes. These should allow analysis of specific features of given genes/proteins and their roles *in vivo*.

Another likely future development is the establishment of methods to knock out genes selectively, in particular cell types, at specific times or

[46] M. W. Appleby, J. A. Gross, M. P. Cooke, S. D. Levin, X. Qian, and R. M. Perlmutter, *Cell* **70,** 751 (1992).

[47] E. N. Georges-Labouesse, H. Rayburn, and R. O. Hynes, (in press) (1994).

[48] P. Hasty, R. Ramirez-Solis, R. Krumlauf, and A. Bradley, *Nature (London)* **350,** 243 (1991).

[49] V. Valancius and O. Smithies, *Mol. Cell. Biol.* **11,** 1402 (1991).

[50] F. Schwartz, N. Maeda, O. Smithies, R. Hickey, W. Edelmann, A. Skoultchi, and R. Kucherlapati, *Proc. Natl. Acad. Sci. U.S.A.* **88,** 10416 (1991).

[51] G. R. Askew, T. Doetschman, and J. B. Lingrel, *Mol. Cell. Biol.* **13,** 4115 (1993).

under specific stimuli. Such localized or conditional mutations will be extremely valuable in analyzing the functions of genes whose widespread roles lead to embryonic lethality. This class includes several, probably many, genes encoding ECM proteins or their receptors.

In conclusion, it is clear that most genes of interest to scientists working on cell–ECM interactions will be knocked out within the next few years; it is merely a matter of time and effort using already available methods. These null mutations will themselves provide significant information about functions of the genes in question and will provide the basis for future, more elaborate genetic manipulations producing subtle, local, or conditional mutations. As more of the mutant mouse strains become available, further details of their phenotypes will be uncovered and, by intercrossing the different strains, yet further information will be obtained. These analyses will undoubtedly add new insights into the roles of ECM proteins and their receptors in mammals with clear implications for the understanding of human physiology and disease.

Acknowledgments

The work in the authors' laboratories was supported by a grant from the NHLBI Program of Excellence in Molecular Biology and by the Howard Hughes Medical Institute. E.L.G. is a Special Fellow of the Leukemia Society of America and R.O.H. is an Investigator of the Howard Hughes Medical Institute.

We gratefully acknowledge many helpful suggestions and ideas from our colleagues Elisabeth Georges-Labouesse, Tanya Mayadas, Glenn Radice, Helen Rayburn, and Joy Yang, and from Reinhard Faessler, Rudolf Jaenisch, Peter Laird, En Li, and Mike Rudnicki.

[20] Polymerase Chain Reaction Cloning with Degenerate Primers: Homology-Based Identification of Adhesion Molecules

By Robert Pytela, Shintaro Suzuki, Johannes Breuss, David J. Erle, and Dean Sheppard

Sequence information on matrix proteins, secreted adhesion proteins, and adhesion receptors is rapidly accumulating. More than 100 different proteins in this category have been characterized by cloning and sequencing of the corresponding cDNAs. It is likely that many more remain to be identified. Increasingly, novel protein-coding sequences are identified on the DNA level, before the corresponding protein is characterized.

Copyright © 1994 by Academic Press, Inc.
All rights of reproduction in any form reserved.

Improved recombinant DNA techniques have greatly accelerated the pace of progress in this area. In this chapter we discuss applications of the polymerase chain reaction (PCR) to the identification and cloning of adhesion receptors. We summarize cloning experiments and sequence data on the integrin α and β subunit families, the cadherin family, and the selectin family.

Homologous Domain Families in Matrix Proteins and Adhesion Receptors

Advances in the cloning and sequencing of matrix proteins and adhesion receptors have revealed that many of these proteins are organized in families of related gene products. Many matrix proteins are composed of modular arrangements of domains that can occur in various combinations (see Table I). For example, fibronectin is composed of three types of homologous repeat domains,[1] one of which (FN type II domain) is also found in the tenascins,[2] type VI[3] and XIV collagens[4]/undulin,[5] and in the $\beta4$ integrin.[6] Some adhesion receptors contain combinations of domains that are part of superfamilies, such as the epidermal growth factor (EGF)-like domains or complement receptor domains found in selectins.[7] Some types of homologous domains occur both in matrix proteins and in receptors; for example, von Willebrand factor A domains are also found in type VI and XIV collagens,[3,4] and in some integrins (I domains).[8] Adhesion receptors are typically organized in families of homologous proteins, such as the integrin α subunits (15 known members), integrin β subunits (8 known members), or cadherins (more than 20 known members). On the basis of these findings, it is now possible to identify DNA molecules encoding novel matrix proteins or receptors by taking advantage of their

[1] T. E. Petersen, H. C. Thogersen, K. Skorstengaard, K. Vibe-Pedersen, P. Sahl, L. Sottrup-Jensen, and S. Magnusson, *Proc. Natl. Acad. Sci. U.S.A.* **80**, 137 (1983).

[2] J. Bristow, M. K. Tee, S. E. Gitelman, S. H. Mellon, and W. L. Miller, *J. Cell Biol.* **122**, 265 (1993).

[3] M. L. Chu, R. Z. Zhang, T. C. Pan, D. Stokes, D. Conway, H. J. Kuo, R. Glanville, U. Mayer, K. Mann, and R. Deutzmann, *EMBO J.* **9**, 385 (1990).

[4] D. R. Gerecke, J. W. Foley, P. Castagnola, M. Gennari, B. Dublet, R. Cancedda, T. F. Linsenmayer, M. van der Rest, B. R. Olsen, and M. K. Gordon, *J. Biol. Chem.* **268**, 12177 (1993).

[5] M. Just, H. Herbst, M. Hummel, H. Durkop, D. Tripier, H. Stein, and D. Schuppan, *J. Biol. Chem.* **266**, 17326 (1991).

[6] S. Suzuki, W. S. Argraves, H. Arai, L. R. Languino, M. D. Pierschbacher, and E. Ruoslahti, *J. Biol. Chem.* **262**, 14080 (1987).

[7] L. A. Lasky, *J. Cell. Biochem.* **45**, 139 (1991).

[8] R. Pytela, *EMBO J.* **7**, 1371 (1988).

TABLE I

HOMOLOGOUS DOMAINS IN MATRIX PROTEINS AND ADHESION RECEPTORS[a]

Domain type	Occurs in:	Consensus sequences
EGF-like[9]	Laminin, tenascins, fibulin,[10] selectins, etc.	$CX_4CX_3GXCX_8CXCX_2GX_5C$
FN type III	FN, tenascins, collagen XIV, integrin $\beta 4$	$PX_4LtvtX_5sltvsWtvp$
Collagen	Collagens, ficolin[11]	GXY, PGPPG
vWF-A	vWF, collagens VI and XIV, CMP,[12] integrins	DivfliDGSXS, KilvviTDG
Fibrinogen-like	Fibrinogen, tenascins, ficolin[11]	$WXXYXXGFGX_4EFWLGNdXihXlTXqG$
TSP type III	Thrombospondins[13]	$NXXQXDXDXDXXGDXCX_4DXDXD$
CR-like	Selectins, CR1, proteoglycans[14]	$CX_5PX_5CX_{13}cX_3CX_2GX_4gX_5CX_3gXWX_4pXC$
C-type lectin	Selectins, proteoglycans[14]	$WIGLX_{8-10}WVX_{7-11}NWX_4PX_7EDCVX_{10}GkWND$
GAG attachment	Proteoglycans[14]	(E/D)GSG(E/D), SGXG
Link/CD44	Proteoglycans,[14] link protein,[15] CD44[16]	GvVFhy, CdaGwl, YDvyCf
Integrin β ligand-binding	Integrins	$PvDLYYLMDLSXSMXDDl-X_{19}-iGFGSFVdK$
Integrin α		
Metal-binding	Integrins	$FGX_7DXdXdgXXDllvGAp$
Cytoplasmic	Integrins, steroid receptors[17]	wkcGFFKR
Cadherin EC repeats	Cadherin superfamily	KXXDyE, DXdEXPXF, DXNDNAP
Cadherin cyto	Cadherins	DPTAPPYDS, WGPRF

[a] FN, Fibronectin; vWF, von Willebrand factor; CMP, cartilage matrix protein; TSP, thrombospondin; CR, complement receptor; GAG, glycosaminoglycan; EC, extracellular; cyto, cytoplasmic. Consensus sequences: X, any amino acid; upper case letters, predominant residues (>90% identical); lower case letters, preferred residues (>60% identical). Single-letter amino acid code: A, alanine; C, cysteine; D, aspartic acid; E, glutamic acid; F, phenylalanine; G, glycine; H, histidine; I, isoleucine; K, lysine; L, leucine; M, methionine; N, asparagine; P, proline; Q, glutamine; R, arginine; S, serine; T, threonine; V, valine; W, tryptophan; Y, tyrosine.

relatedness to established protein families. Some of these domain families (e.g., immunoglobulin, EGF, and FN type III domains) are defined by conserved patterns of cysteine, glycine, and tryptophan residues, but are otherwise very divergent in sequence (see Table I).[9-17] In contrast, other

[9] J. Engel, *FEBS Lett.* **251,** 1 (1989).

[10] W. S. Argraves, H. Tran, W. H. Burgess, and K. Dickerson, *J. Cell Biol.* **111,** 3155 (1990).

[11] H. Ichijo, U. Hellman, C. Wernstedt, L. J. Gonez, L. Claesson-Welsh, C. H. Heldin, and K. Miyazono, *J. Biol. Chem.* **268,** 14505 (1993).

types of homologous domains contain continuous stretches of very high sequence conservation, alternating with regions of low or insignificant similarity. These highly conserved sequences probably correspond to amino acid residues that play essential roles in overall folding of the polypeptide chain, or to residues with conserved functional importance (e.g., in binding divalent cations). Table I lists some of the most typical consensus sequences found in the major types of homologous domains.

Identification of Novel Members of Gene Families by Homology Probing

Several different approaches have been used to identify novel members of gene families by cross-hybridization. Conventionally, cDNA probes covering large portions of the coding sequence have been employed to screen cDNA libraries under low-stringency conditions. Although this technique has been successful in many cases, it is generally not possible to detect related cDNAs with less than 50–60% overall identity. A more powerful version of this technique takes advantage of the presence of continuous consensus sequences in some domain families by using a shorter probe (often a synthetic oligonucleotide 50–70 nucleotides in length) that corresponds to a highly conserved domain (e.g., Moyle *et al.*[18]). Synthetic DNA probes can be designed to be "degenerate," by including either inosine (which does not contribute to base pairing) or equimolar mixtures of several nucleotides at nonconserved positions. As a result, the probe becomes less biased toward known members of the gene family, reducing the problem of preferential detection of already known sequences. However, use of highly degenerate probes for library screening may be problematic owing to a high frequency of false-positives. Nevertheless, this approach has been successfully applied in some instances.[19] The approach discussed in this chapter is also based on using

---

[12] W. S. Argraves, F. Deak, K. J. Sparks, I. Kiss, and P. F. Goetinck, *Proc. Natl. Acad. Sci. U.S.A.* **84**, 464 (1987).
[13] J. Lawler, M. Duquette, C. A. Whittaker, J. C. Adams, K. McHenry, and D. W. DeSimone, *J. Cell Biol.* **120**, 1059 (1993).
[14] D. R. Zimmermann and E. Ruoslahti, *EMBO J.* **8**, 2975 (1989).
[15] F. Deak, I. Kiss, J. Sparks, W. S. Argraves, G. Hampikian, and P. F. Goetinck, *Proc. Natl. Acad. Sci. U.S.A.* **83**, 3766 (1986).
[16] L. A. Goldstein, D. F. Zhou, L. J. Picker, C. N. Minty, R. F. Bargatze, J. F. Ding, and E. C. Butcher, *Cell (Cambridge, Mass.)* **56**, 1063 (1989).
[17] M. V. Rojiani, B. Finlay, V. Gray, and S. Dedhar, *Biochemistry* **30**, 9859 (1991).
[18] M. Moyle, M. A. Napier, and J. W. McLean, *J. Biol. Chem.* **266**, 19650 (1991).
[19] G. Singh, S. Kaur, J. L. Stock, N. A. Jenkins, D. J. Gilbert, N. G. Copeland, and S. S. Potter, *Proc. Natl. Acad. Sci. U.S.A.* **88**, 10706 (1991).

degenerate oligonucleotides; however, the oligonucleotides are used as primers in the polymerase chain reaction (PCR).[20] With this technique, termed homology PCR, it is possible to target PCR primers toward relatively short (5–10 amino acid residues) regions of high sequence conservation. The selectivity of detection is enhanced because an amplification product of predicted size is unlikely to be generated unless both of the PCR primers specifically hybridize to the target DNA.

The homology PCR approach is especially useful for the analysis of domain families containing several short, highly conserved consensus sequences (see Table I), but sharing a low (<50%) overall degree of sequence identity. As a further advantage, this method does not depend on the availability of cDNA libraries; instead, relatively small amounts of RNA purified from the tissue or cells of interest can be used as starting material. An important limitation of the technique is that the target sequence needs to contain at least two conserved regions that are spaced closely enough to allow efficient amplification. In most cases, this means that the distance between the two primers should not exceed 800–1000 nucleotides. Another potential problem is that PCR amplification may favor detection of abundant cDNAs, while novel cDNAs present at low levels may be overlooked. The PCR product obtained usually contains a mixture of DNA fragments corresponding to different members of the gene family expressed in the cell type or tissue used to isolate the template RNA. The composition of this mixture is then determined by subcloning and sequencing of individual subclones. Because there is a practical limit to the number of subclones that can be analyzed in this way, DNA species that are present at low levels may not be detected. The chances for detecting minor products can be significantly enhanced by using a method that we have termed restriction/reamplification (see Detailed Methods, below).

Use of Homology Polymerase Chain Reaction for Cloning Cross-Species Homologs

Sequences that are strongly conserved among different members of a gene family are likely to be conserved over a wide range of species. Therefore, homology PCR provides a convenient method to identify rapidly and clone most or all members of a gene family expressed in a particular cell type or tissue obtained from a species for which only limited information on that family exists. For example, a wide range of integrin α and β subunits were identified and cloned in *Xenopus* embryos, using

[20] R. K. Saiki, D. H. Gelfand, S. Stoffel, S. J. Scharf, R. Higuchi, G. T. Horn, K. B. Mullis, and H. A. Erlich, *Science* **239**, 487 (1988).

only two pairs of degenerate PCR primers.[21,22] The cDNA clones that are obtained in this manner can then be used as probes for Northern blotting, RNase protection analysis, or *in situ* hybridization. The spectrum of sequences amplified by PCR also provides semiquantitative information on the expression of different members of the gene family in the tissue of origin. However, this information must be interpreted with caution because, owing to the extreme sensitivity and potential bias inherent in the PCR cloning method, it is possible that molecules present at low levels are detected preferentially. For example, we have consistently been able to amplify selectin-specific sequences from epithelial cells, which do not express selectins as analyzed by any other method. Another caveat is that trace amounts of previously amplified or cloned DNA may contaminate the PCR reaction, resulting in artifactual, and sometimes preferential, amplification of the contaminant. It is important to perform stringent control experiments and to follow the general recommendations for avoiding PCR contamination.

Primer Design

Homology PCR has been used to identify novel cDNAs encoding members of a variety of protein families. Table II lists some examples and shows the corresponding target sequences and primer designs. Degenerate primers containing either inosine or mixed bases have been employed.[23–27] Theoretically, the use of inosines reduces the degeneracy of the primer, because inosine does not contribute to base pairing. Therefore, the probability of coincidental cross-hybridization with nontarget DNA should be reduced by including inosines in poorly conserved positions. On the other hand, inosine-containing probes hybridize more weakly than mixed probes, because even a highly degenerate mixed probe is likely to contain a small percentage of molecules that perfectly match the target sequence. Even though this issue has not been systematically studied, it is likely that the advantages of inosine relative to mixed primers are dependent

[21] D. G. Ransom, M. D. Hens, and D. W. DeSimone, *Dev. Biol.* **160**, 265 (1993).
[22] C. A. Whittaker and D. W. DeSimone, *Development (Cambridge, UK)* **117**, 1239 (1993).
[23] Γ. Libert, M. Parmentier, A. Lefort, C. Dinsart, C. Van Sande, C. Maenhaut, M. J. Simons, J. E. Dumont, and G. Vassart, *Science* **244**, 569 (1989).
[24] A. Kamb, M. Weir, B. Rudy, H. Varmus, and C. Kenyon, *Proc. Natl. Acad. Sci. U.S.A.* **86**, 4372 (1989).
[25] G. R. Pellegrino and J. M. Berg, *Proc. Natl. Acad. Sci. U.S.A.* **88**, 671 (1991).
[26] M. T. Murtha, J. F. Leckman, and F. H. Ruddle, *Proc. Natl. Acad. Sci. U.S.A.* **88**, 10711 (1991).
[27] T. H. Chang, J. Arenas, and J. Abelson, *Proc. Natl. Acad. Sci. U.S.A.* **87**, 1571 (1990).

TABLE II

PROTEIN FAMILIES ANALYZED BY HOMOLOGY POLYMERASE CHAIN REACTION[a]

Family	Consensus sequences	Primers used
G protein-coupled receptors[23]	LcalavDRY	CTG TGY GYS ATY GCI ITK GAY MGS TAC
	FilCWLPffi	TTC RYS ITC TGC TGG CTG CCC TWC TWC MT
Tyrosine kinases[24]	IKWTAPE	RTH AAR TGG ACC GCN CCN GA
	DVWsFGv	GAY ATN TGG RYG TTY GGN GT
Potassium channels[24]	AVYFAEA	GCA GTN TAY TTY GCN GAR GC
	PDAFWWA	CCN GAY GCN TTY TGG TGG GC
Zinc finger domains[25]	CPECGKS	TGY CCN GAR TGY GGN AAR TCN
	HTGEKP	CAY ACN GGN GAR AAR CCN TT
Homeobox domains[26]	yQtIELEkEFL	TAC CAG ACS YTG GAR CTG GAG AAR GAR TTY C
	IWFQNRRmKwK	ATC TGG TTY CAG AAY CGN MGS ATG AAR TGG AAR
RNA helicases[27]	VLDEAD	GTN CTN GAY GAR GCN GA
	YIHRIG	TAY ATH CAY AGR ATH GG

[a] Degenerate DNA code: N, A or C or G or T; H, A or C or T; R, A or C or G or T; Y, C or T; R, A or G; S, G or C; W, A or T; M, A or C; K, G or T. I, Inosine. For each pair of primers, the second one corresponds to the reverse complement of the sequence shown.

on the primer sequence and the target DNA used. In practical terms, it may be advisable to try both approaches. We have generally adopted the strategy used by Libert *et al.*[23] (see Table II), using mixed bases in positions that are twofold degenerate, but inosines in positions that are three- or fourfold degenerate. The choice of a suitable target sequence depends not only on the conservation of the sequence, but also on the amino acid composition; sequences that are rich in R and L (six codons) should be avoided, whereas sequences rich in W, M, D, E, N, Q, K, H, C, F, and Y (one or two codons) can be targeted with much less degenerate primers. In the case of serine (six codons), the nucleotide sequences of all members of the gene family should be inspected, because serines are often encoded invariably by either AG(C/T) or TCN [Note that two point mutations are required for converting an AG(C/T) codon into a TCN codon, or vice versa.] It should also be taken into account that the 3' end of the primer is extended in the PCR, and therefore it is important that the 3' region of the primer provides a perfect match with the target. Mismatches in the 5' half of the primer often do not interfere with specific amplification. However, in some cases targets have been amplified with primers containing mismatches near the 3' end (see Fig. 2). If the target sequence is highly degenerate, longer primers (>30 nucleotides) may be required to accomplish specific amplification. However, short, highly degenerate primers (e.g., 17-mers with 2^8-fold degeneracy) have also been successfully used (see Table II and Cadherins, below). The template DNA is usually obtained by reverse transcription of mRNA or total RNA. Genomic DNA has also been used,[25,26] but in many cases this approach is problematic because large introns may be present in the region targeted by PCR primers, preventing efficient amplification owing to the large size of the amplified product. Even if the intron locations in some members of a gene family are known, it is possible that novel members of the same family contain additional introns. For example, the genomic organization of different integrin genes can vary substantially (see the next section).

Identification and Cloning of Integrin β Subunits

The first integrin β subunits (β1–3) were identified at the protein level, and the cDNAs cloned by screening expression libraries with antibody probes and/or with synthetic oligonucleotides based on partial protein sequences.[28–33] Comparison of the sequences revealed 56 conserved cyste-

[28] J. W. Tamkun, D. W. DeSimone, D. Fonda, R. S. Patel, C. Buck, A. F. Horwitz, and R. O. Hynes, *Cell (Cambridge, Mass.)* **46,** 271 (1986).
[29] W. S. Argraves, S. Suzuki, H. Arai, K. Thompson, M. D. Pierschbacher, and E. Ruoslahti, *J. Cell Biol.* **105,** 1183 (1987).

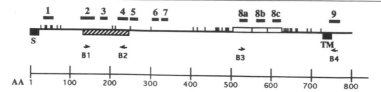

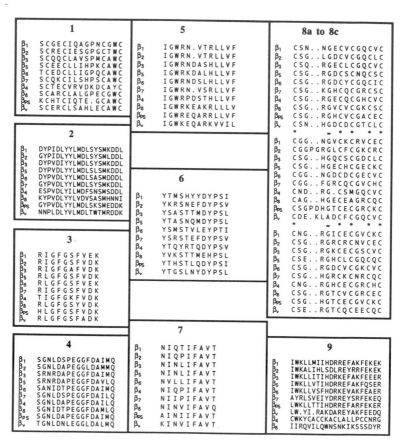

FIG. 1. Alignment of amino acid sequences conserved among integrin β subunit sequences. In the diagram, conserved domains are represented by horizontal bars numbered 1 through 9. Short vertical lines represent the 56 conserved cysteine residues. Cysteines in the three homologous cysteine-rich repeats (open boxes) are not shown individually. Each repeat contains eight cysteine residues. Filled boxes represent the signal peptide (S) and the transmembrane domain (TM). The crosshatched box represents the conserved putative ligand-binding domain. Locations of PCR primers are shown as arrows numbered B1 through B4. The diagram is drawn to scale, based on the $\beta 1$ sequence. The scale shown represents numbers of amino acid residues counted from the initiator methionine. Sequence alignments of regions 1–9 are shown below the diagram. Regions 8a–c represent the conserved portion

ine residues, 24 of which are organized in 3 homologous repeats in the C-terminal part of the extracellular domain (Fig. 1).[34-44] The overall degree of identity is 44–50%, and the alignment is characterized by regions of high sequence conservation alternating with highly divergent sequences. Nine highly conserved regions are shown in Fig. 1. All of these could potentially be used to design primers for homology PCR; however, so far only regions 2, 3, 4, 7, 8a, and 9 have been tested. Long continuous stretches of nearly invariant sequence, which are believed to be involved in ligand binding,[45,46] are located in the N-terminal half of the molecule (regions 2, 3, and 4). Primers based on regions 2 and 4 have been used extensively in the cloning of β4, β5, β6, and β7.[34-42] Most recently, β8

[30] L. A. Fitzgerald, B. Steiner, S. C. J. Rall, S. S. Lo, and D. R. Phillips, *J. Biol. Chem.* **262,** 3936 (1987).

[31] S. K. A. Law, J. Gagnon, J. E. K. Hildreth, C. E. Wells, A. C. Willis, and A. J. Wong, *EMBO J.* **6,** 915 (1987).

[32] T. K. Kishimoto, K. OConnor, A. Lee, T. M. Roberts, and T. A. Springer, *Cell (Cambridge, Mass.)* **48,** 681 (1987).

[33] J. P. Rosa, P. F. Bray, O. Gayet, G. I. Johnson, R. G. Cook, K. W. Jackson, M. A. Shuman, and R. P. McEver, *Blood* **72,** 593 (1988).

[34] F. Hogervorst, I. Kuikman, A. E. von dem Borne, and A. Sonnenberg, *EMBO J.* **9,** 765 (1990).

[35] S. Suzuki and Y. Naitoh, *EMBO J.* **9,** 757 (1990).

[36] R. N. Tamura, C. Rozzo, L. Starr, J. Chambers, L. F. Reichardt, H. M. Cooper, and V. Quaranta, *J. Cell Biol.* **111,** 1593 (1990).

[37] J. W. McLean, D. J. Vestal, D. A. Cheresh, and S. C. Bodary, *J. Biol. Chem.* **265,** 17126 (1990).

[38] H. Ramaswamy and M. E. Hemler, *EMBO J.* **9,** 1561 (1990).

[39] S. Suzuki, Z. S. Huang, and H. Tanihara, *Proc. Natl. Acad. Sci. U.S.A.* **87,** 5354 (1990).

[40] D. Sheppard, C. Rozzo, L. Starr, V. Quaranta, D. J. Erle, and R. Pytela, *J. Biol. Chem.* **265,** 11502 (1990).

[41] Q. A. Yuan, W. M. Jiang, G. W. Krissansen, and J. D. Watson, *Int. Immunol.* **2,** 1097 (1990).

[42] D. J. Erle, C. Rüegg, D. Sheppard, and R. Pytela, *J. Biol. Chem.* **266,** 11009 (1991).

[43] A. J. MacKrell, B. Blumberg, S. R. Haynes, and J. H. Fessler, *Proc. Natl. Acad. Sci. U.S.A.* **85,** 2633 (1988).

[44] G. H. Yee and R. O. Hynes, *Development (Cambridge, UK)* **118,** 845 (1993).

[45] J. W. Smith and D. A. Cheresh, *J. Biol. Chem.* **265,** 2168 (1990).

[46] S. E. DSouza, M. H. Ginsberg, T. A. Burke, S. C. Lam, and E. F. Plow, *Science* **242,** 91 (1988).

of the cysteine-rich homologous repeat sequence. The three repeats are aligned with each other (8a to 8c), based on the positions of cysteines (*). All sequences shown are human, except βPS and βv, which are from *Drosophila*. References to sequences are as follows: β1,[29] β2,[31,32] β3,[30] β4,[34-36] β5,[37-39] β6,[40] β7,[41,42] β8,[18] βPS,[43] βv.[44]

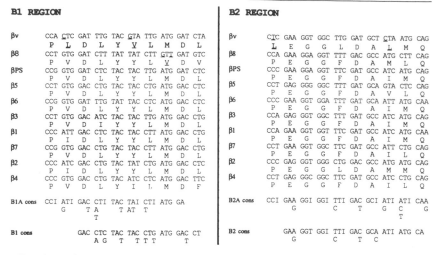

FIG. 2. Design of β integrin primers. At the bottom of the alignment, the consensus sequences corresponding to the B1/B2 and B1A/B2A primers are shown. The B1A and B2A primers are more degenerate and can be used to amplify β4 and βv. B1 or B1A both are unlikely to amplify β8, owing to two mismatches in the five residues closest to the 3' end of the primer (mismatches are underlined). However, it is interesting to note that the βv sequence was amplified[44] using a primer essentially identical to B2A, in spite of a mismatch with the second nucleotide from the 3' end of the primer. Note that the B2 and B2A primers are synthesized according to the reverse complement of the consensus sequence shown.

was cloned due to its cross-hybridization with a long oligonucleotide probe based on the β3 sequence.[18]

The PCR approach that we have used to identify novel integrin β subunits is outlined in Figs. 1 and 2 (these experiments were performed at a time when only β1, β2, and β3 had been cloned and sequenced). The primers B1, B2, B3, and B4 were based on alignments of highly conserved domains in β1, β2, and β3. After the sequences of β4–β7 were determined, we designed an optimized, more degenerate pair of primers, termed B1A and B2A (Fig. 2). As templates, we used cDNA prepared from guinea pig tracheal epithelial cells, and from the cell lines HeLa, MeWO (human melanoma), and WR2.3 (mouse lymphoma). The PCR amplification products were subcloned in a plasmid vector and characterized by dideoxy sequencing. In addition to the known sequences of β1, β2, and β3, we found three novel sequences (Fig. 3), corresponding to the subunits β5,[47] β6,[40] and β7.[42] The β4 and β5 sequences were identified[35,39] using slightly

[47] D. J. Erle, D. Sheppard, J. Breuss, C. Ruegg, and R. Pytela, Am. J. Respir. Cell Mol. Biol. **5,** 170 (1991).

different primers, also hybridizing to regions 2 and 3 (see Fig. 1). The β5 and β7 cDNAs were also identified by others,[34,36–38,41] using similar approaches. Additional PCR experiments using the B1A and B2A primers showed that these primers also work well for amplifying the known integrin cDNAs, but did not result in cloning of any novel integrins.

For cloning the full-length coding sequence of the novel integrins, we have used two alternative approaches: PCR cloning or conventional cDNA library screening, using the original PCR fragment as a probe. Approximately 75% of the guinea pig β6 sequence was cloned by PCR, using combinations of the primers B1, B2, B3, and B4, along with two specific primers based on the initial β6 sequence obtained. However, PCR cloning of the 3' and 5' portions proved problematic, even when various modifications of the RACE protocol (rapid amplification of cDNA ends) was applied. This may be explained in part by the fact that the β6 mRNA, like most integrin mRNAs, contains a large 3' untranslated portion. Therefore, using oligo(dT) in combination with an integrin-specific forward primer to amplify the 3' end of the cDNA is not likely to succeed. Consequently, we have relied on cDNA library screening to obtain the full-length sequences of β6 and β7.[40,42] However, this approach depends on the availability of appropriate cDNA libraries, and can be time consuming because 5' regions of large mRNAs are often poorly represented in cDNA libraries.

In addition to identifying novel members of gene families, homology PCR is a powerful method for cloning cross-species homologs. Sequence regions that are conserved among gene families are almost certain to be conserved across species, at least among vertebrates. Using the B1/B2 primer pair, we have amplified and cloned β subunit sequences from pig, sheep, rabbit, guinea pig, and mouse (Fig. 3). Also shown are the sequences of *Xenopus* β2, β3, and β6, which have been amplified by Ransom *et al.*,[21] using primers similar to B1 and B2. For comparison, Fig. 3 includes all other published β subunit sequences from different species. In each group of cross-species comparisons, amino acid residues that differ from the human sequence are double underlined. It is clear that cross-species variation among different vertebrate integrins (70–99% identity) is much less than the variation between different β subunits in one species (30–56%

[48] D. W. DeSimone and R. O. Hynes, *J. Biol. Chem.* **263**, 5333 (1988).

[49] D. L. Zeger, N. Osman, M. Hennings, I. F. McKenzie, D. W. Sears, and P. M. Hogarth, *Immunogenetics* **31**, 191 (1990).

[50] T. A. G. Bisland and T. A. Springer, Genbank accession number X71786 (1993).

[51] S. J. Kennel, V. Godfrey, L. Y. Chang, T. K. Lankford, L. J. Foote, and A. Makkinje, *J. Cell Sci.* **101**, 145 (1992).

[52] Q. Yuan, W. M. Jiang, E. Leung, D. Hollander, J. D. Watson, and G. W. Krissansen, *J. Biol. Chem.* **267**, 7352 (1992).

```
                  •  .•  ••••                      •• *           • • ***  •• *        • • *         *   *     • *    **            • • *    *      • • •  *
β1 human          PIDLYYLMDLSYSMKDDLENVKSLGTDLMNEMRRITSDFRIGFGSFVEKTVMPYISTTPAK-LRNPC-TSEQ--NCTTP-FSYKNVLSLTNKGEVFNELVGKQRISGNLDSPEGGFDAIMQ
β1 pig#               SYSMKDDLENVKSLGTDLMNEMRRITSDFRIGFGSFVEKTVMPYISTTPAK-LRNPC-TSEQ--NCTSP-FSYKNVLSLTDKGEVFNELVGKQRISGNLDS
β1 sheep#                               FVEKTVMPYISTTPAK-LRNPC-TNEQ--NCTSP-FSYKNVLSLTDKGEVFNELVGKQRISGNLDS
β1 mouse#              SYSMKDDLENVKSLGTDLMNEMRRITSDFRIGFGSFVEKTVMPYISTTPAK-LRNPC-TSEQ--NCTSP-FSYKNVLSLTDKGEVFNELVGQQRISGNLDS
β1 guinea pig#        SYSMKDDLENVKSLGTELMYEMKRITSDFRIGFGSFVEKTVMPYISTTPAK-LRNPC-TSEQ--NCTSP-FSYKNVLSLTDKGEVFNELVGQQRISGNLDS
β1 chicken        PIDLYYLMDLSYSMKDDLENVKSLGTALMREMEKITSDFRIGFGSFVEKTVMPYISTTPAK-LRNPC-TGDQ--NCTSP-FSYKNVLSLTSEGNKFNELVGKQHISGNLDSPEGGFDAIMQ
β1 xenopus        PIDLYYLMDLSFSMKDDLENVKSLGTALMTEMEKITSDFRIGFGSFVEKTVMPYISTTPAK-LINPC-TSDQ--NCTSP-FSYKNVLNLTKDGKLFNDLVGKQQISGNLDSPEGGFDAIMQ

β2 human          PIDLYYLMDLSYSMLDDLRNVKKLGGDLLRALNEITESGRIGFGSFVDKTVLPFVNTHPDK-LRNPCPNKEK--ECQPP-FAFRHVLKLTNNSNQFQTEVGKQLISGNLDAPEGGLDAMMQ
β2 mouse          PIDLYYLMDLSYSMLDDLNNVKKLGGDLIQALNEITESGRIGFGSFVDKTVLPFVNTHPEK-LRNPCPNKEK--ACQPP-FAFRHVLKLTDNSNQFQTEVGKQLISGNLDAPEGGLDAIMQ
β2 sheep#                SYSMVDDLANVKLGGDLLRALNDITESGRIGFGSFVDKTVLPFVNTHPEK-LRNPCPNKEK--ECQPP-FAFRHVLKLTDNSKQFETEVGKQLISGNLDAPEGGLDAILQ
β2 chicken        PIDLYYLMDLSYSMLDDLENVKKLGGQLLRALESTTPSRRIGFGSFVDKTVLPFVNTHPEK-LKNPCPNKDS--NCQPP-FAFKHLSLTDNAEKFESEVGKQEISGNLDAPEGGLDAMMQ
β2 xenopus**                             ALNGITKSAQIGFNSFVDKTVLPFVNTHPEK-LKNPCPEKNE--NCQPP-FSFKHILNLTANGKEFQDQVGKQLISGNLDR

β3 human          PVDIYYLMDLSYSMKDDLWSIQNLGTKLATQMRKLITSNLRIGFGAFVDKPVSPYMYISPPEALENPCYDMKT--TCL-PMFGYKHVLTLTDQVTRFNEEVKKQSVSRNRDAPEGGFDAIMQ
β3 sheep#                                               VSPYMYIFPPEAPENPCYDMKT--TCL-PMFGYKRVVTLTDQVTRFNEEVKKQSVSRNRDA
β3 xenopus**      PVDIYYLMDLSYSMNDDLIKIQTLGTSLSERMPRLITSNLRIGFGAFVDKPMSPYMFMSPPEVIKNPCYEFNT--ECM-PTFGYKHVLTLTEEVLRFNEEVQKQVSRNRDSPEGGFDAVLQ

β4 human          PVDIYIMDFSNSMSDDLDNLKKMGQNLARVLSQLTSDYTIGFGKFVDKVSVPQTDMRPEK-LKEPWPNSD------PPFSFKNVISLTEDVDEFRNKLQGERISGNLDAPEGGFDAILQ
β4 mouse          PVDIYIMDFSNSMSDDLDNLKQMGQNLAKILRQLTSDYTIGFGKFVDKVSVPQTDMRPEK-LKEPWPNSD------PPFSFKNVISLTENVEEFWNKLQGERISGNLDAPEGGFDAILQ

β5 human          PVDLYYLMDLSLSMKDDLDNIRSLGTKLAEEMRKLITSNFRLGFGSFVDKDISPFSYTAPRY-QTNPCIGYKLFPNCV-PSFGFRHLLPLTDRVDSFNEEVRKQRVSRNRDAPEGGFDAVLQ

β6 human          PVDLYYLMDLSASMDDDLNTIKELGSGLSKEMSKITSNFRLGFGSFVEKPVSPFVKTTPEE-IANPCSSIPYF--CL-PTFGFKHILPLTNDAERFNEIVKNQKISANIDTPEGGFDAIMQ
β6 guinea pig#        SASMDDDLNTIKELGSLLSKEMSKITSNFRLGFGSFVEKPVSPFMKTTPEE-IANPCSSIPYI--CL-PTFGFKHILPLTNDAERFNEIVKNQKISANIDT
β6 xenopus**            MDDDLKTIKELGSSLSREMSKLITNNFQLGFGSFVEKPVSPYIKTVPKD-IENPCHSIPYY--CL-PTFGYKHVLSLTPNAQNFNEIVTKQRIVSGNIDTPEGGFDAIMQ

β7 human          PVDLYYLMDLSYSMKDDLERVRQLGHALLVRLQEVTHSVRIGFGSFVDKTVLPFVSTVPSK-LRHPCPTRLER--CQSP-FSFHHVLSLTGDAQAFEREVGRQSVSGNLDSPEGGFDAILQ
β7 mouse          PVDLYYLMDLSYSMKDDLERVRQLGHALLVRLQEVTHSVRIGFGSFVDKTVLPFVSTVPSK-LHHPCPSRLER--CQPP-FSFHHVLSLTGDAQAFEREVGRQNVSGNLDSPEGGFDAILQ
β7 rabbit#           SYSMKDDLERVRSLGHHLLVQLQNVTQSVRIGFGSFVDKTVLPFVSTVPAK-LRHPCPSRLER--CQPP-FSFHHVLSLTGDAQAFEREVGRQSVSGNLDS

β8 human          PVDLYYLVDVSASMHNNIEKLNSVGNDLSRKMAFFSRDFRLGFGSYVDKTVSPYISIHPER-IHNQCSDYNLD--CMPP-HGYIHVLSLTENITEFKAVHRQKISGNIDTPEGGFDAMLQ
β8 rabbit         PVDLYYLVDVSASMHNNIEKLNSVGNDLSRKMAFFSRDFRLGFGSYVDKTVSPYISIHPER-IHNQCSDYNLD--CMPP-HGYIHVLSLTENITEFKAVHRQKISGNIDTPEGGFDAMLQ

βPS drosophila    PVDLYYLMDLSKSMEDDKAKLSTLGDKLSETMKRITNNFHLGFGSFVDKVLMPYVSTIPKK-LEHPCEN------CKAP-YGYGNHMPLMNNTESFSNEVKNATVSGNLDAPEGGFDAIMQ
βv drosophila     PLDLYVLMDLTWTMRDDKKTLEELGAQLSQTLKNLTGNYRLGFGSFADKPTLPMIL--PQH-RENPCAAERAT--CE-PTYGYRHQLSLTDDIPAFTSAVANSKITGNLDNLEGGLDALMQ
```

identity).[44] In contrast, the two known β subunits from *Drosophila* cannot be unambiguously classified as homologs of individual mammalian subunits. The βPS subunit is more closely related to β1 (46%) and β3 (41%) than to other β subunits (34–39%), and therefore may be the homolog of β1, or of a common evolutionary ancestor of β1 and β3. Another β subunit from *Drosophila,* termed βv,[44] is strongly divergent from the vertebrate integrins, and cannot be classified as the clear homolog of any known individual mammalian subunit. The overall degree of homology between βv and human β subunits is 29–33%. The evolutionary relationship between the *Drosophila* β subunits and the mammalian β subunits remains to be determined, but it appears possible that βv is the homolog of a currently unknown vertebrate β subunit.

Genomic DNA Templates

Our initial attempts to amplify integrin sequences from human genomic DNA using the B1/B2 primer pair were unsuccessful, probably due to the presence of large introns in the area targeted by these primers. In agreement with this notion, published genomic structures of several β subunit genes[53–56] show that in all cases, at least one intron is present in the region targeted by B1/B2. We have also used B1 in combination with a more closely spaced reverse primer, B5, which is based on conserved region 3 shown in Fig. 1 (consensus sequence GFGSFVDK). Polymerase chain reaction using B1/B5 and genomic DNA as a template resulted in amplifi-

[53] J. B. Weitzman, C. E. Wells, A. H. Wright, P. A. Clark, and S. K. Law, *FEBS Lett.* **294,** 97 (1991).

[54] A. B. Zimrin, S. Gidwitz, S. Lord, E. Schwartz, J. S. Bennett, G. C. White, 2nd, and M. Poncz, *J. Biol. Chem.* **265,** 8590 (1990).

[55] F. Lanza, N. Kieffer, D. R. Phillips, and L. A. Fitzgerald, *J. Biol. Chem.* **265,** 18098 (1990).

[56] W. M. Jiang, D. Jenkins, Q. Yuan, E. Leung, K. H. Choo, J. D. Watson, and G. W. Krissansen, *Int. Immunol.* **4,** 1031 (1992).

FIG. 3. Alignment of integrin β subunit amino acid sequences. The region amplified by the B1/B2 primer pair is shown (see Fig. 1). Asterisks denote residues that are identical in all the sequences, dots denote residues identical in all sequences except β4, β8, or βv (these are the most divergent β subunits). Dashes represent gaps in the alignment. Within each group of species homologs, amino acid residues that differ from the human sequence are double underlined. #, Sequences identified by homology PCR in mouse lymphoma cells, pig aortic endothelial cells, sheep lung tissue, or guinea pig tracheal epithelial cells; **, sequences identified by Ransom *et al.*[21] in *Xenopus* embryos, using homology PCR with primers similar to B1 and B2. The same primers were used to identify the novel *Drosophila* βv subunit.[44] References to sequences: human sequences, see Fig. 1; β1 chicken,[28] β1 *Xenopus,*[48] β2 mouse,[49] β2 chicken,[50] β4 mouse,[51] β7 mouse,[52] β8 rabbit.[18]

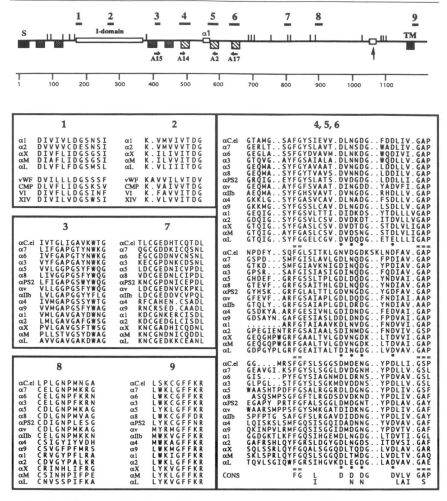

FIG. 4. Alignment of amino acid sequences conserved among integrin α subunit sequences. In the diagram, conserved domains are represented by horizontal bars numbered 1 through 9. Regions 4–6 represent the metal-binding repeat domains. Sequence alignments of regions 1–9 are shown below the diagram. Short vertical lines represent conserved cysteine residues. Filled boxes represent the signal peptide (S) and the transmembrane domain (TM). Open bars represent domains present only in a subset of α subunits; I domain, additional domain in α1, α2, αL, αM, α1, 20-amino-acid insert in α1. Vertical arrow denotes the site of proteolytic processing in α3, α5, α6, α7, α8, αv, and αIIb. The PCR primers are shown as horizontal arrows. The diagram is drawn to scale, based on the α1 sequence. The scale shown represents numbers of amino acid residues counted from the initiator methionine. All sequences shown are human, except α7 (rat), α8 (chicken), αPS2 (*Drosophila*), and αC.el (*C. elegans*). For regions 1 and 2, which are located in the vWF-like I domains, the corresponding homologous sequences of von Willebrand factor,[64] cartilage matrix protein

cation of sequences corresponding to $\beta3$ and $\beta6$.[57] Others have obtained similar results,[56] except that $\beta5$ was also amplified. Indeed, the published gene structure of $\beta3$[54,55] shows that the B1 and B5 regions are located within one exon of the $\beta3$ gene. However, the genes encoding $\beta1$,[55] $\beta2$,[53] and $\beta7$[56] do contain an additional intron closely preceding the B5 consensus region. It is possible that the loss of this intron is a typical feature of the $\beta3/\beta5/\beta6$ subfamily. Therefore, it may be impossible to amplify genomic sequences corresponding to β subunits other than $\beta3/\beta5/\beta6$ using the B1/B5 primers. When more complete information on integrin gene organization becomes available, it may be feasible to design new primers that are invariably located within one exon. These primers could be used to obtain more definitive information on the total number of integrin genes in the genome.

Identification and Cloning of Integrin α Subunits

The first α subunit sequences that were determined ($\alpha5$,[29] αv,[6] αIIb,[58] αX,[59] αM,[8,60] αL,[61] $\alpha2$,[62] and $\alpha4$[63]) revealed the general structural features of this protein family (see Fig. 4).[64–71] The N-terminal half is composed of

[57] A. Postigo, D. Sheppard, D. J. Erle, and R. Pytela, unpublished (1992).
[58] M. Poncz, R. Eisman, R. Heidenreich, S. M. Silver, G. Vilaire, S. Surrey, E. Schwartz, and J. S. Bennett, J. Biol. Chem. 262, 8476 (1987).
[59] A. L. Corbi, L. J. Miller, K. OConnor, R. S. Larson, and T. A. Springer, EMBO J. 6, 4023 (1987).
[60] A. L. Corbi, T. K. Kishimoto, L. J. Miller, and T. A. Springer, J. Biol. Chem. 263, 12403 (1988).
[61] R. S. Larson, A. L. Corbi, L. Berman, and T. Springer, J. Cell Biol. 108, 703 (1989).
[62] Y. Takada and M. E. Hemler, J. Cell Biol. 109, 397 (1989).
[63] Y. Takada, M. J. Elices, C. Crouse, and M. E. Hemler, EMBO J. 8, 1361 (1989).
[64] K. Titani, S. Kumar, K. Takio, L. H. Ericsson, R. D. Wade, K. Ashida, K. A. Walsh, M. W. Chopek, J. E. Sadler, and K. Fujikawa, Biochemistry 25, 3171 (1986).
[65] R. Briesewitz, M. R. Epstein, and E. E. Marcantonio, J. Biol. Chem. 268, 2989 (1993).
[66] Y. Takada, E. Murphy, P. Pil, C. Chen, M. H. Ginsberg, and M. E. Hemler, J. Cell Biol. 115, 257 (1991).
[67] W. K. Song, W. Wang, R. F. Foster, D. A. Bielser, and S. J. Kaufman, J. Cell Biol. 117, 643 (1992).
[68] B. Bossy, E. Bossy-Wetzel, and L. F. Reichardt, EMBO J. 10, 2375 (1991).
[69] E. L. Palmer, C. Rüegg, R. Ferrando, R. Pytela, and D. Sheppard, J. Cell Biol. 123, 1289 (1993).
[70] T. Bogaert, N. Brown, and M. Wilcox, Cell (Cambridge, Mass.) 51, 929 (1987).
[71] J. Sulston, Genbank accession number Z19155, Locus CEF45 (1993).

(CMP),[12] and collagens type VI[3] and XIV[4] are shown. For each one of these proteins, the repeat domain sequence most closely related to the integrin I domains is shown. References to integrin sequences are as follows: $\alpha1$,[65] $\alpha2$,[62] $\alpha3$,[66] $\alpha4$,[63] $\alpha5$,[29] $\alpha6$,[36] $\alpha7$,[67] $\alpha8$,[68] $\alpha9$,[69] αv,[6] αIIb,[58] αL,[61] αM,[60] αX,[59] $\alpha PS2$,[70] $\alpha C.el.$[71]

a seven-repeat structure, containing the conserved motives syFG, livGAp, and G(a/q)vyvy. C-terminal repeats invariably contain EF-hand-like binding sites for divalent cations, sharing the consensus sequence DX(D/N)X(D/N)GXXD.

These conserved motives alternate with regions that are highly divergent both in length and in sequence, and may correspond to surface loops owing to their hydrophilic nature. Some α subunits (α1, α2, αL, αM, and αX) contain an additional domain (I domain) that is inserted between the second and third homologous repeats in the N-terminal portion of the extracellular domain. The I domains belong to the von Willebrand factor (vWF)-A type of homologous domains (see Table I), and contain the consensus sequences DivfliDGSXS and KilvviTDG (regions 1 and 2 in Fig. 4). The C-terminal half of the extracellular domain is characterized by a conserved pattern of cysteine residues, but a low degree of overall sequence conservation. Cytoplasmic domains are highly divergent, with the exception of the segment joining the transmembrane and cytoplasmic domains, which contains the nearly invariant sequence (L/M)(W/Y)K(L/C)GFFKR (region 9 in Fig. 4).

Generally, homology-based identification of novel α subunits is more difficult than is the case with β subunits, owing to the lack of extensive, universally conserved areas of sequence. Nevertheless, the chicken α8 cDNA was identified and cloned owing to its cross-hybridization with a human αv cDNA probe.[68] For homology PCR, we have designed a series of degenerate oligonucleotides based on consensus sequences that are conserved among most or all known α subunits. A primer based on the GFFKR sequence (region 9 in Fig. 4) would be expected to amplify selectively all integrin α subunits, but use of this primer has been hampered by the fact that no other strongly conserved sequences are present within 1000 nucleotides. In our experience, it is difficult to amplify very large DNA fragments using highly degenerate primers. In contrast, the primers A14 and A2A (Figs. 4 and 5) are designed to hybridize to conserved sequences that are separated by approximately 300 nucleotides, spanning the highly conserved fifth and sixth homologous repeats, and including the YFG and DLLVGAP consensus sequences (regions 4 and 5 in Fig. 4). Using this primer pair, we have observed efficient amplification of the previously known α1, α2, α3, α4, α5, and αv cDNA sequences from a variety of mammalian species. However, owing to several mismatches with the A14 sequence, this primer cannot be used to amplify α6 and α7 sequences. It is also unlikely that the A2A primer will amplify αL, αM, and αX sequences (see Fig. 5), although this has not been formally tested. It should be noted that the α1 sequence contains an approximately 54-nucleotide insert in the fragment amplified by A14/A2A, and therefore gives rise to a significantly larger PCR product (350 vs 300 nucleotides).

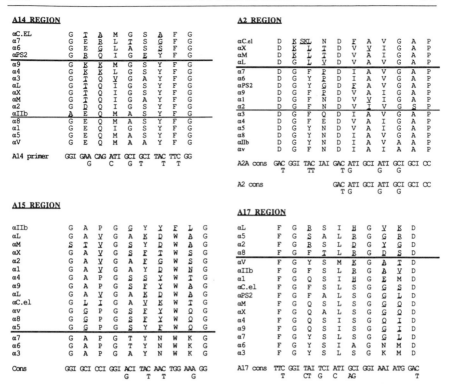

FIG. 5. Design of α integrin primers. Consensus nucleotide sequences used to synthesize primers are shown underneath aligned amino acid sequences. Restriction sites (*Eco*RI or *Sal*I) were added to the 5' ends of all primers to facilitate cloning of PCR products (not shown). Residues that produce mismatches with the primer sequence are underlined. Sequences that conform most closely with the consensus sequence are shown at the bottom, sequences with one or two mismatches with the 5' half of the primer are shown in the middle. Sequences with mismatches near the 3' end of the primer are shown on top, above the bold line. The latter group of sequences is unlikely to be amplified by the respective primer pair. The A14/A2A primer pair has been shown to amplify α1, α2, α3, α4, α5, α8, α9, and αv, and should also amplify αIIb. The A15 primer was designed specifically to amplify the α3, α6, and α7 subfamily. It has been used in combination with A2A and A17. The A17 primer was designed to amplify α3, α6, and α7 preferentially, but combined with A14 it should also amplify α1, α4, α5, α9, αv, αIIb, αM, αX, αPS2, and αC.el. However, this has not yet been tested.

We have used the A14/A2A primers to amplify cDNA sequences from guinea pig tracheal epithelial cells, sheep lung, pig aortic endothelial cells, and mouse lymphoma cells (see also Erle *et al.*[47]). It should be noted that the sequences of α7, α8, and α9 were unknown when these experiments were performed. The sequences we obtained are marked with the symbol # in Fig. 6, and are aligned with all published α sub-unit sequences.

```
               *   *   ***                  * *                    * *                             **      *
α6 human##     FGYDVAVMDLNKDGWQDIVI-GAPQYFD----RDG-EVGGAVYVYMNQQGRWNNVKPIR--LNGTKD----------------SMFGIAVKNIGDINQ
α6 sheep##                 NKDGWQDIVI-GAPQYFD----RSG-EVGGAAYVYINQQGRWNNVKPIR--LNGAKD----------------SMFGIAVKNIGDINQ
α6 chicken     FGYDVAVVDLNSDGWQDIVV-GAPQYFD----RSG-DIGGAVYIYINQRGKWEGIKPIR--LNGTAD----------------SMFGLAVENVGDINQ
α6 xenopus**   YDVAVVDLNRDRWNDIVV-GAPQYFDKNN-R---DIGGAIYVYINKNGKWNEVAPVR--IVGTKD----------------SMFGISVKNIGDINQ

α7 human##     FGYSLAVADLNSDGWPDLIV-GAPYFFE----RQ-EELGGAVYVYLNQGGHWAGISPLR--LCGSPD----------------SMFGISLAVLG
α7 rat         FGYSLAVTDLNSDGWADLIV-GAPYFFE----RQ-EELGGAVYVYMNQGGHWADISPLR--LCGSPD----------------SMFGISLAVLGDLNQ
α7 mouse       FGYSLAVTDLNNDGWADLIV-GAPYFFE----RQ-EELGGAVYVYMNQGGHWRDISPLR--ICGSPD----------------SMFGISLAVLGD

α3 human       FGSAIALADLNNDGWQDLLV-GAPYYFE----RK-EEVGGAIYVFMNQAGTSFPAHP-SLLLHGPSG----------------SAFGLSVASIGDINQ
α3 sheep#      SAIALADLNNDGWQDLLV-GAPYYFE----RK-EEVGGAIYVFMNQAGTSFPTHP-SLLLHGPSG----------------SAFGLSVASIGDINQ
α3 pig#        SAIALADLNNDEWQDLLV-GAPYYFE----RK-EEVGGAVYVFMNQAGTSFPDDP-SLLLHGPSR----------------SAFGFSVASIGDINQ
α3 hamster     FGSAIALADLNNDGWQDLLV-GAPYYFE----RK-EEVGGAVYVFMNQAGTSFPDQP-SLLLHGPSR----------------SAFGISIASIGDINQ
α3 guinea pig# SAIALADLNNDGWQDLLV-GAPYYFD----RK-EEVGGAVYVFMNQAGTSFPAEP-SLLLHGPSR----------------SGFGFSMASIGDINQ
α3 mouse#      SAIALADLNNDGWQDLLV-GAPYYFE----RK-EEVGGAVYVFMNQAGASFPDQP-SLLLHGPSR----------------SAFGISIASIGDVNQ
α3 xenopus**   YDVAVVDLNRDRWQDIVV-GAPYYFD----RK-EEIGGAVYVYNNVAGF-FIDKA-AMVLHGTSF----------------SGFGFALANIGDINQ

α C.elegans    FGYSIEVVDLNGDGFDDLIV-GAPFEHRSGIDG-N--FGGIVVYVYFSQGVQRKQHESHLVFHPPKILKNPDFY----------SQFGLSITKLGNVDG

α5 human       FGYAVAATDVNGDGLDDLLV-GAPLLMDRTPDGRPQEV-GRVYVYL-QHPAGIEPTPTLT-LTGHDEF---------------GRFGSSLTPLGDLDQ
α5 mouse#      YAVAATDTNGDGLDDLLV-GAPLLMERTADGRPQEV-GRVYVYL-QRPAAIDPTPTLT
α5 guinea pig# YAVAATDINGDGLDDLLV-GAPLLMERTADGRPQEV-GRVYIYM-QHPAGIDPTPTLS-LTGYDEF---------------GRFGSSLTPLGDLDQ
α5 pig#        YAVAATDINRDGLDDLLV-GAPLLMERTADGRPQEV-GRVYIYL-QSPAGMEPAPART-LTGHDEF---------------GRFGSSLTPLGDLDQ
α5 xenopus**   YSVSATDLNSDGLDDLLI-GAPLFMDRTHDGRVQEV-GRVYVYL-QGDH-MESTPHLI-LTGMEEY---------------GRFGSSIASLGDLDQ

α8 human       FGYTVVRSDVNSDGLDDVLV-GAPLFMEREFESNPREV-GQIYLYL-QVSSLLFRDPQI--LTGTDEF--------------GRFGSAMAHLGDLNQ
α8 chicken     FGYTVAVSDVNNDGLDDILV-GAPLFMEREFESKPKEV-GQVYLYL-QESAFLFRDPQI--LTGTEVF--------------GRFGSAITHLGDLNQ
α8 sheep#      YTVAVSDVNNDGMDDILI-GAPLFMEREFESSPREV-GQVYLYL-QESALVFRDPQV--LAGTEVF--------------GRFGSAVAHLGDLNQ
α8 guinea pig# YTVVVSDVNNDG?DDLLV-GAPLFMEREFESSPREV-GRVYLYL-QVNALLFQTPRS--LTGTEIF--------------GRFGSAVVHLGDLNQ

αv human       FGFSVAATDINGDDYADVFI-GAPLFMDRGSDGKLQEV-GQVSVSL-QRASGDFQTT---KLNGFEVF-------------ARFGSAIAPLGDLDQ
αv pig#        FSVAATDINGDDYADVFI-GAPLFMDRGSDGKLQEV-GQVSVSL-QKASGNFQTT---KLNGFEVF-------------ARFGSAIAPLGDLDQ
αv mouse#      FSVAATDINGDDYSDVFI-GAPLFMDRGSDGKLQEV-GQVSVSL-QRAVGDFQRT---KLNGFEVF-------------ARFGSAIAPLGDLDQ
αv guinea pig# FSVAATDINGDDLADVFI-GAPLFMDRGSDGKLQEV-GQVSLSL-QAASGGFQTS---KLTGFEVF-------------ARFGSAIAPLGDLDQ
αv chicken     FGYSVATTDINGDDYTDLFI-GAPLFMDRGSDGKLQEV-GQVSICL-QRASGGFQIA---KLNGFEIF-------------ARFGSAIAPLGDLDQ
αv? xenopus**  HSVAVTDVNNDGKDDVLV-GAPLFMERRTGGKLQEV-GRVYVYL-QRTYSRFSNDHP-ILRGSRVY-------------GQFGSSIAPIGDIDQ

αIIb human     FGHSVAVTDVNGDGRHDLLV-GAPLYMESRADRKLAEV-GRVYLFL-QPRGPHALGAPSLLLTGTQLY-------------GRFGSAIAPLGDLDR
αIIb rat       FGHSVAVTDVNGDGRHDLLV-GAPLYMESRVDRKLAEV-GRVYLFL-QPKGLQALSSPTLVLTGTQVY-------------GRFGSAIAPLGDLNR

αPS2 drosoph.  FGYSLATSDVDGDGLDDLLI-GAPMYTDPDNVEGKYDV-GRVYILL-QGGPTEEKRWTT--EHIRDGYHSK-----------GRFGLALTTLGDVNG

α4 human       FGASVCAVDLNADGFSDLLV-GAPM--QSTI--R-EE--GRVFVYINSGSGAVMNAMETN-LVGSDKYA------------ARFGESIVNLGDIDN
α4 sheep#      ASICVVDLNADGFSDLLV-GAPM--QSTI--R-EE--GRVFVYINSGSGAVMNEMETE-LIGSDKY
α4 mouse#      ASVCAVDLNADGFSDLLV-GAPM--QSTI--R-EE--GRVFVYINSGMGAVMVEMERV-LVGSDKYA------------ARFGESIANLGDIND
α4 xenopus**   AAVCAADLNGDGLSDLLV-GAPI--QSTI--R-EE--GRVFVYMNTGSGA-MEELKFE-LSGSDLYA------------ARFGETIANLGDIDN

α9 human       FGSSLCAVDLNGDGLSDLLV-GAPM--FSEI--R-DE--GQVTVYIN-RGNGALEEQLA--LTGDSAYN------------GAFGESIASLDDLDN
α9 guinea pig# SSLCAVDLNRDGLSDLLV-GAPM--FSEV--R-DE--GQVTVYIN-RGNGVLEEQLA--LTGDRAYN------------AHFGESIAGLGDLDD

α1 human       FGSILTTTDIDKDSNTDILLVGAPMYMGT---EK-EEQ-GKVYVYALNQTRFEYQMSLEPIKQTCCSSRQHNSCTTENKNEPCGARFGTAIAAVKDLNL
α1 sheep#      SELTTIDIDKDSNTDILLVGAPMFMGS---EK-EEQ-GKVYVYALNQTRFEYQMSLEPIKQTCCSSPKHNSCTKENKNEPCGARFGTAIAAVKDLNL
α1 pig#        SVLTTVDIDKDSNTDILLVGAPMFMGT---EK-EEQ-GKVYVYAVNQTRFEYQMSLEPIKRTCCSSLKHNSCTKENKNEPCGARFGTAIAAVKDLNL
α1 mouse#      SVLTTIDIDKDCYTDLLLVGAPMYMGT---EK-EEQ-GKVYVYALNPTRFEYQMSLEPIKQTCCSSLKDNSCTKENKNEPCGARFGTAVAAVKDLNV
α1 guinea pig# SVLTTIDIDNDSNTDILLIGAPMYMGT---EK-EEQ-GKVYVYTLNQTKFEYQMSLEPIKQTCCASLKHNSCTKENKNEPCGARFGTAIAAVKDLNL
α1 rat         FGSVLTTIDIDKDSYTDLLLVGAPMYMGT---EK-EEQ-GKVYVYAVNQTRFEYQMSLEPIRQTCCSSLKDNSCTKENKNEPCGARFGTAIAAVKDLNV

α2 human       FGSVLCSVDVDKDTITDVLLVGAPMYMSD-L-KK-EE--GRVYLFTIKKGILGQHQF----LEGPEGIEN-------------TRFGSAIAALSDINM
α2 guinea pig# SVLCSVDVNKDTITDVLLVGAPTYMND-L-KK-EE--GRVYLFTIAKGILNQHQF----LEGPEGIEN-------------ARFGSAIAALSDINM
α2 sheep#      DVNKDTITDVLLVGAPMYMND-L-KK-EE--GRVYLFTITKGILNWHQF----LEGPKGLEN-------------ARFGSAIAALSDINM
α2 mouse#      SVLCSVDVDKDTITDVLLVGAPTYMND-L-KK-EE
α2 xenopus**   SVLCSVDVNRDSITDVLLVGAPTFMNE-Y--KK-EE--GQVYMFSIRDGILVQREQ----LEGPKSLEN-------------TRFGSAIVELSDIDL

αM human       FGASLCSVDVDSNGSTDLVLIGAPHYYEQTR-------GGQVSVCPL-PRGORARWQCDAVLYGEQGQPW-------------GRFGAALTVLGDVNG
αM mouse       FGASLCSVDMDADGNTNLILIGAPHYYEKTR-------GGQVSVCPL-PRG-RARWQCEALLHGDQGHPW-------------GRFGAALTVLGDVNG

αX human       FGASLCSVDVDTDGSTDLVLIGAPHYYEQTR-------GGQVSVCPL-PRGWR-RWWCDAVLYGEQGHPW-------------GRFGAALTVLGDVNG

αL human       FGGELCGVDVDQDGETELLLIGAPLFYGEQR-------GGRVFIY--QRRQLGFEEVSE--LQGDPGYPL-------------GRFGEAITALTDING
```

FIG. 6. Alignment of integrin α subunit amino acid sequences. The region amplified by the A14/A2A primer pair is shown (see Fig. 4). Asterisks denote residues that are identical in all the sequences, dots denote residues identical in 90% of the sequences. Dashes represent gaps in the alignment. Within each group of species homologs, amino acid residues that differ from the human sequence are double underlined. #, Sequences amplified using the

Results of a typical experiment were as follows. mRNA from guinea pig tracheal epithelial cell cultures was used as a template in random-primed cDNA synthesis followed by 30 cycles of PCR, using the A14/A2A primers (for details, see below). The PCR products were subcloned in a plasmid vector and a total of 40 clones was analyzed by dideoxy sequencing and/or restriction analysis. Thirty-five of the clones corresponded to guinea pig αv, 1 to $\alpha 1$, and 1 to $\alpha 5$. The 3 other clones all contained the same sequence that was 24–60% identical to known integrin α subunit sequences, and thus was presumably a novel integrin α subunit. We provisionally called this subunit αA. The initial PCR product was then digested with two rare-cutting restriction endonucleases (*PvuII* and *PstI*) that recognize sequences in the amplified fragments of guinea pig αv and $\alpha 5$, the digested product was reamplified with the same primers under the same conditions, and the amplified product was subcloned into pBluescript (Stratagene, La Jolla, CA). Eleven clones containing inserts were analyzed by dideoxy sequencing. Five of these clones corresponded to αv, one to $\alpha 2$, one to a nonintegrin protein, and two to another sequence that appeared to be a novel integrin. We provisionally designated this sequence αB. αB was also amplified from human endothelial cells and sheep lung (Fig. 6), and is closely related to the $\alpha 8$ subunit identified in the chicken.[68] We have now cloned the complete human cDNAs encoding αA and αB, and have raised specific antibodies that define the corresponding proteins as distinct human integrin subunits widely distributed in mammalian tissues.[69,78] Both can associate with $\beta 1$, and therefore, according to conven-

[72] I. de Curtis, V. Quaranta, R. N. Tamura, and L. F. Reichardt, *J. Cell Biol.* **113**, 405 (1991).

[73] B. L. Ziober, M. P. Vu, N. Waleh, J. Crawford, C. S. Lin, and R. H. Kramer, *J. Biol. Chem.* **268**, 26773 (1993).

[74] T. Tsuji, F. Yamamoto, Y. Miura, K. Takio, K. Titani, S. Pawar, T. Osawa, and S. Hakomori, *J. Biol. Chem.* **265**, 7016 (1990).

[75] B. Bossy and L. F. Reichardt, *Biochemistry* **29**, 10191 (1990).

[76] H. Neuhaus, M. C. Hu, M. E. Hemler, Y. Takada, B. Holzmann, and I. L. Weissman, *J. Cell Biol.* **115**, 1149 (1991).

[77] M. J. Ignatius, T. H. Large, M. Houde, J. W. Tawil, A. Barton, F. Esch, S. Carbonetto, and L. F. Reichardt, *J. Cell Biol.* **111**, 709 (1990).

[78] L. M. Schnapp, J. Breuss, D. M. Ramos, D. Sheppard, and R. Pytela, *J. Cell Sci.* (in press) (1994).

primers A14/A2A and template cDNA from mouse lymphoma cells, pig aortic endothelial cells, sheep lung tissue, or guinea pig tracheal epithelial cells; ##, amplified using the primer pair A15/A2A and cDNA from sheep lung or the human melanoma cell line, MeWo; **, sequences identified by Whittaker and DeSimone[22] in *Xenopus* embryos, using primers similar to the A14/A2A primers. References to sequences: Human sequences, see Fig. 1; $\alpha 6$ chicken,[72] $\alpha 7$ mouse,[73] $\alpha 3$ hamster,[74] αv chicken,[75] αIIb rat,[58] $\alpha 4$ mouse,[76] $\alpha 1$ rat,[77] αM mouse.[8]

tion, are now designated with numbers (αA = α9, αB = α8). Figure 6 also includes the sequence of mouse α7, which was cloned by PCR amplification using the A2A primer in combination with a primer based on the N-terminal protein sequence of α7.[73] Also shown is the human α7 sequence, which was amplified using the A15/A2A primer pair.[79] Primers similar to A14/A2A were used by Whittaker and DeSimone[22] to clone the *Xenopus* homologs of α2–α6 and αv (marked with double asterisks in Fig. 6). It is likely that homology PCR will also be applicable to the cloning of invertebrate α subunits, because a *Caenorhabditis elegans* α subunit sequence determined in the course of the *C. elegans* genome sequencing project shares significant homology with mammalian integrins.[71]

Comparison of all the known α subunit sequences reveals that they can be divided into subfamilies. It is clear that the fibronectin/vitronectin/fibrinogen-binding integrins α5, αv, and αIIB, and also the novel α8 subunit, share a much higher degree of homology with each other (45–55% identity) than with other integrins (20–30% identity). Similarly, the laminin/epiligrin(kalinin)-binding integrins, α3, α6, and α7, are 45–50% identical with each other, but only 20–25% with other α subunits. Two additional subfamilies are defined by the I domain-containing integrins α1, α2, αL, αM, and αX, and by the α4/α9 subunits. It is likely that the integrins within these subfamilies are similar not only in sequence, but also in function. The emergence of these subfamilies suggests a new approach to homology cloning, because several consensus sequences are found within subfamilies, but not shared with other α subunits. For example, the N-terminal domains of α3, α6, and α7 contain the invariant sequence GTYNWKG, which is not found in other known α subunits (see Fig. 4, region 3). We have used the primer A15 (Figs. 4 and 5) based on this sequence for homology PCR, and have shown that, in combination with either A2A or A17, it specifically amplifies α3, α6, and α7 sequences.

Cadherins

Cadherins are a family of membrane proteins that mediate calcium-dependent homophilic cell–cell adhesion. Until recently, only three cadherins (E-, N-, and P-cadherin) had been known. The cDNA sequences of these three cadherins revealed a closely homologous structure (for review, see Takeichi[80]). The extracellular domain consists of five homologous repeats of a cadherin-specific motif, and the C-terminal cytoplasmic

---

[79] M. Chabanon, N. Waleh, R. H. Kramer, and R. Pytela, unpublished (1992).
[80] M. Takeichi, *Science* **251**, 1451 (1991).

domain is highly conserved. Homology PCR using primers based on conserved sequences in the cytoplasmic domains has led to the identification of 10 additional cDNAs encoding novel cadherins.[81-83] Several novel cadherins were also identified on the protein level.[82,84-87] It has also become clear that the cadherins are part of a larger superfamily, characterized by cadherin-like extracellular domains, but distinctive cytoplasmic domains. This superfamily includes the desmosomal proteins, desmoglein and desmocollins,[88-91] T-cadherin,[92] and the protocadherins,[93] which also include the *fat* protein found in *Drosophila*.[94] Protocadherins contain more than five repeats of the extracellular cadherin motif, and have variable cytoplasmic domains. Homology PCR using primers based on the extracellular cadherin motif has revealed that a large number of previously unidentified protocadherins are expressed in mammalian tissues.[93]

The highly conserved cytoplasmic domain of the typical cadherins provides several good candidate sequences for the synthesis of degenerate primers.[81] Figure 7[95-97] shows aligned sequences from the C-terminal region of cadherin cytoplasmic domains. Polymerase chain reaction using

[81] S. Suzuki, K. Sano, and H. Tanihara, *Cell Regul.* **2**, 261 (1991).

[82] M. Donalies, M. Cramer, M. Ringwald, and A. Starzinski-Powitz, *Proc. Natl. Acad. Sci. U.S.A.* **88**, 8024 (1991).

[83] H. Tanihara, K. Sano, R. Heimark, T. St John, and S. Suzuki, *Cell Adhes. Commun.* **2**, 15 (1993).

[84] D. Ginsberg, D. DeSimone, and B. Geiger, *Development (Cambridge, UK)* **111**, 315 (1991).

[85] E. W. Napolitano, K. Venstrom, E. F. Wheeler, and L. F. Reichardt, *J. Cell Biol.* **113**, 893 (1991).

[86] H. Inuzuka, S. Miyatani, and M. Takeichi, *Neuron* **7**, 69 (1991).

[87] M. G. Lampugnani, M. Resnati, M. Raiteri, R. Pigott, A. Pisacane, G. Houen, L. P. Ruco, and E. Dejana, *J. Cell Biol.* **118**, 1511 (1992).

[88] L. Goodwin, J. E. Hill, K. Raynor, L. Raszi, M. Manabe, and P. Cowin, *Biochem. Biophys. Res. Commun.* **173**, 1224 (1990).

[89] J. E. Collins, P. K. Legan, T. P. Kenny, J. McGarvie, J. L. Holton, and D. R. Garrod, *J. Cell Biol.* **110**, 1575 (1991).

[90] P. J. Koch, M. J. Walsh, M. Schmelz, M. D. Goldschmidt, R. Zimbelmann, and W. W. Franke, *Eur. J. Cell Biol.* **53**, 1 (1990).

[91] M. Amagai, V. Klaus-Kovtun, and J. R. Stanley, *Cell (Cambridge, Mass.)* **67**, 869 (1991).

[92] B. Ranscht and M. T. Dours-Zimmermann, *Neuron* **7**, 391 (1991).

[93] K. Sano, H. Tanihara, R. L. Heimark, S. Obata, M. Davidson, T. St John, S. Taketani, and S. Suzuki, *EMBO J.* **12**, 2249 (1993).

[94] P. A. Mahoney, U. Weber, P. Onofrechuk, H. Biessmann, P. J. Bryant, and C. S. Goodman, *Cell (Cambridge, Mass.)* **67**, 853 (1991).

[95] A. Nagafuchi, Y. Shirayoshi, K. Okazaki, K. Yasuda, and M. Takeichi, *Nature (London)* **329**, 341 (1987).

[96] W. J. Gallin, B. C. Sorkin, G. M. Edelman, and B. A. Cunningham, *Proc. Natl. Acad. Sci. U.S.A.* **84**, 2808 (1987).

[97] K. Hatta, A. Nose, A. Nagafuchi, and M. Takeichi, *J. Cell Biol.* **106**, 873 (1988).

FIG. 7. Homology PCR of cadherin cytoplasmic domains. Alignment of amino acid sequences near the C terminus of various cadherin cytoplasmic domains is shown. Consensus sequences used to synthesize PCR primers are shown below the alignment. Primer 1 is identical to the consensus sequence, primer 2 is the reverse complement of the consensus sequence. The sequence GAATTC (encoding the *Eco*RI recognition site) was added to the 5' end of both primers (not shown). B, B-Cadherin[85]; E, E-cadherin[95]; EP, EP-cadherin[84]; L, L-cadherin[96]; M, M-cadherin[82]; N, N-cadherin[97]; 4–12, cadherins -4 through -12.[81,83] Degenerate nucleotide code used in consensus sequences: N, A/C/G/T; Y, C/T; R, A/G; S, G/C; W, A/T; M, A/C; K, G/T.

primer 1 and primer 2 amplifies a product of approximately 160 nucleotides, which usually does not contain significant amounts of cadherin-unrelated by-product (Fig. 8, lane b). These primers work well for most of the known typical cadherins, with the exception of cadherin-12 and M-cadherin. On the basis of the alignment shown in Fig. 7, replacement of primer 2 with a primer based on the sequence WGPRF may result in amplification of most of the known cadherins, including M-cadherin and cadherin-12. However, we have not tested this hypothesis.

To amplify sequences corresponding to desmosomal cadherins and protocadherins, primers based on the extracellular domains are required. In designing these primers, two problems are encountered. The extracellular domains do not contain long stretches of highly conserved sequences, and the extracellular repeats are similar to one another. Nevertheless, the primers described below (see also Sano *et al.*[93] and Tanihara *et al.*[83]) work reasonably well in most cases, and at least some of them show limited subdomain specificity. Primers 3 and 4 (Fig. 9) are based on conserved sequences near the middle and the end, respectively, of the extracellular repeat EC-3 of typical cadherins and desmosomal cadherins. A PCR using primer 3 and primer 4 amplifies a product of about 160 nucleotides corresponding to the C-terminal portion of EC-3 of typical cadherins and some desmosomal cadherins (Fig. 8, lane c). The corresponding region of most EC domains other than EC-3 contains similar, but distinct, sequences (middle part of Fig. 9). These sequences were used to design primers 5

a   b   c   d   e

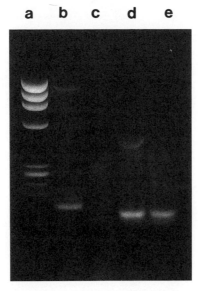

FIG. 8. Agarose gel electrophoresis of PCR products. PCR was carried out as described in text, using 100 μg of rat brain cDNA as a template and various primer sets. The reaction products were electrophoretically separated on a 4% agarose gel. Lane a, size markers (*Hae*III-digested φX174); lane b, primers 1 and 2; lane c, primers 3 and 5; lane d, primers 3 and 6; lane e, primers 5 and 6.

and 6. A PCR amplification of typical cadherins and protocadherins using primers 3 and 6 yields a 450-nucleotide fragment extending from the middle of EC-3 to the end of EC-4. However, primers 3 and 6 also yield a 140-nucleotide fragment (Fig. 8, lane d) amplified from a new class of cadherins, termed protocadherins.[93] Protocadherins are found in a variety of species ranging from *C. elegans* to human. They contain more than five EC domains, most of which contain the C-terminal consensus sequence NDNAP, and either LDRE or LD(F/Y)E in the middle of the domain. The *Drosophila fat* protein[94] contains 34 cadherin EC domains of the protocadherin type. A consensus sequence of the 34 *fat* domains is shown in Fig. 9. Primers 3 and 6 amplify a subset of protocadherin sequences of the LD(F/Y)E type. In contrast, PCR using primer 5 (based on the common LDRE sequence) combined with primer 6 (based on the common NDNAP sequence) amplifies a broad range of cadherin sequences from different EC domains of all known classes of cadherins. The product is about 140 nucleotides in size (Fig. 8, lane e). Because this set of primers does not show subdomain specificity, detailed sequence comparison is necessary to identify the obtained sequences. In our experience, all of

```
                                                  Primer 4   3'-TRCTYSGNGGNNNNAAR-5'
   Primer 3:  AARSSNNTNGAYTWYGA-3'                 CONSENSUS     AYGARSCNCCNNNNTTY

B-cadherin    K  G  L  D  Y  E  AKRQFVLHVAVVNEAPFAIKL......PTATATVMVSVEDV  N  E  A  P  V  F
E-cadherin    D  G  L  D  F  E  AKQQYILHRVENEEPFEGSL.......VPSTATVTVDVVDV  N  E  A  P  I  F
EP-cadherin   K  G  L  D  F  E  LRKQYVLQITVENAEPFSVPL......PTSTATVTVTVEDV  N  E  A  P  F  F
L-cadherin    Q  G  L  D  Y  E  TKSRYDLVVTVENKVPLSVPI......TLSTASVLVTVLDV  N  E  P  P  V  F
P-cadherin    D  G  L  D  F  E  AQDQHTLYVEVTNEAPFAVKL......PTATATVVVHVKDV  N  E  A  P  V  F
N-cadherin    K  P  I  D  F  E  TNRMFVLTVAAENQVPLAKGIQHP...PQSTATVSVTVIDV  N  E  N  P  Y  F
DC            K  P  L  N  Y  E  VNRQVVLQIGVLNEAQFAKAVNSKTTTTMCTTCCTVKVKDH D  E  G  P  C  F
DG            K  P  L  D  F  E  AMNNLQLSLGVRNKAEFHQSIMSQY..KLTATAISVTVLNY  Y  E  G  S  V  E
Cadherin 4    K  A  V  D  Y  E  LNRAFMLTVMVSNQAPLASGIQMS...FQSTAGVTISIMDI  N  E  A  P  Y  F
Cadherin 5    K  P  L  D  Y  E  YIQQYSFIVEATDPTIDLRYMSPP..AGNRAQVIINITDV  D  E  P  P  I  F
Cadherin 6    K  G  L  D  F  E  KKKVYTLKVEASNPYVEPRFLYLGP..FKDSATVRIVVEDV  D  E  P  P  A  F
Cadherin 8    K  P  L  D  F  E  TKKSYTLKDEAANVHIDPRFSGRGP..FKETATVKIHVEDA  D  E  P  P  V  F
Cadherin 9    K  G  C  D  Y  E  AKTSYTLRIEAANRDADPRFLSLGP..FSDTTTVKIIVEDV  D  E  P  P  Y  S
Cadherin 10   K  P  L  D  Y  E  NRRLYTLKVEAENTHVDPRFYYLGP..FKDTTIVKISIEDV  D  E  P  P  P  F
Cadherin 11   K  P  V  D  F  E  TERAYSLKVEAANVHIDPKFISNGP..FKDTVTVKISVEDA  D  E  P  P  M  F
Cadherin 12   K  P  L  D  F  E  TKKAYTFKVEASNLHLDHRFHSAGP..FKDTATVKISVLDV  D  E  P  P  V  F
Cadherin 13   K  P  L  D  Y  E  ISAFHTLLIKVENEDPLVPDVSYGP...SSTATVHITVLDV  N  E  G  P  V  F

N-EC1         K  P  L  D  R  E  QIASFHLRAHAVDVNGNQ.........VENPIDIVINVIDM  N  D  N  R  P
E-EC2         S  G  L  D  R  E  SYPTYTLVVQAADLQGEG.........LSTTAKAVITVKDI  N  D  N  A  P
P-EC4         G  I  L  D  R  E  DVQRVKNNVYEVMVLATHSGNPP....TTGTGTLLLTLTDI  N  D  H  G  P
Fat consensus r  p  L  D  R  E  ----Y-L-V-A-D--G-p.........------v-v-v-D-  N  D  N  a  P

   Primer 5:  AARSSNNNNGAYMGNGA                     CONSENSUS     AAYGAYAACGCICCI
                                                    Primer 6   3'-TTRCTRTTGCGNGGN-5'

PC-consensus   -  L  D  -  E  ----y-L-V-A-D.-G-p..........------V-V-VVD-  N  D  N  A  P
Rat-123                       EQPELSLILTALD.GGTPS........RSGTALVQVEVIDA
Rat-212                       ESSSYQIYVQATD.RGPVP........MAGHCKVLVDIIDV
Rat-214                       TLQTFEFSVGATD.HGSPS........LRSQALVRVVVLDH
Rat-216                       ALQSFEFYVGATD.GGSPA........LSSQTLVRMVVLDD
Mouse-321                     DTKLHEIYIQAKD.KGANP........EGAHCKVLVEVVDV
Mouse-322                     DQREFQLTAHIND.GGTPV........LATNISVNVFVTDR
Mouse-324                     ESNNYEIHVDATD.KGYPP........MVAHCTVLVGILDE
Mouse-326                     KVKDYTIEIVAVD.SGNPP........LSSTNSLKVQVVDV
Human-11                      DTKLHEIYIQAKD.KGANP........EGAHCKVLVEVVDV
Human-13                      VSPRLRLVLQAES.RGA..........FAFT.VLTLTLQDA
Human-42       T  P  L  D  Y  E  KVKDYTIEIVAVD.SGNPP........LSSTNSLKVQVVDV  N  D  N  A  P
Human-43       V  P  L  D  Y  E  DRREFELTAHISD.GGTPV........LATNISVNIFVTDR  N  D  N  A  P
Xenopus-21                    AIREYSLRIKAQD.GGRPPL.......SNTTGMVTVQVVDV
Xenopus-23                    KASEYEIYVQAAD.KGAVP........MAGHCKVLLEIVDV
Drosophila-12                 SVRSYRLVIRAQD.GGSPS........RSNTTQLLVNVIDV
Drosophila-13                 LTHLYEIWIEAAD.GDTPS........LRSVTLITLNVTDA
C. elegans-41                 ATRNYKLRVKATD.LGIPP........RSSNMTLFIHVLDV
C. elegans-42                 TQRVHSLSIKCVD.NQGREPH......HEVFASVTVTVIDV
```

FIG. 9. Homology PCR of cadherin extracellular (EC) domains. *Top:* Alignments of the amino acid sequences of the C-terminal regions of EC-3 of various cadherins. Primer sequences are shown above the alignment. *Middle:* Alignment of EC-1 (N-cadherin), EC-2 (E-cadherin), and EC-4 (P-cadherin), with a consensus sequence derived from the 34 cadherin repeats of the *Drosophila fat* protein.[94] Primers 3, 5, and 6 are based on this alignment. *Bottom:* Sequences of various protocadherin clones obtained from different organisms, using primers 3 and 6. Symbols are the same as in Fig. 7. DC, Desmocollin[89]; DG, desmoglein[90]; 13, cadherin-13.[83]

these primers work well and do not produce significant amounts of by-products.

Selectins

Selectins are a class of adhesion receptors expressed on leukocytes, endothelial cells, and platelets.[98] They mediate dynamic interactions between leukocytes and vascular walls, resulting in a "rolling" movement of circulating blood cells. The three members of this family are L-selectin[99]

[98] L. A. Lasky, *Science* **258,** 964 (1992).

[99] L. A. Lasky, M. S. Singer, T. A. Yednock, D. Dowbenko, C. Fennie, H. Rodriguez, T. Nguyen, S. Stachel, and S. D. Rosen, *Cell (Cambridge, Mass.)* **56,** 1045 (1989).

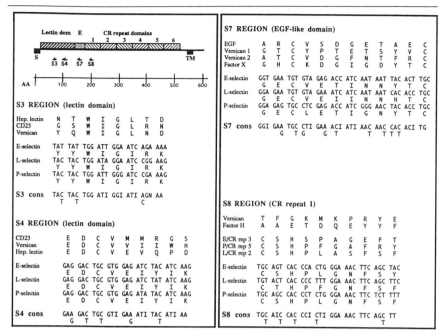

FIG. 10. Design of selectin primers. The diagram shows the domain structure of selectins, consisting of a lectin domain, an EGF-like domain (E), and several repeats of a complement receptor-like domain (CR). The diagram shown is based on E-selectin, which contains six CR repeats. L-selectin and P-selectin contain two or nine CR repeats, respectively. S, Signal peptide; TM, transmembrane domain. Arrows denote location of primers used in homology PCR. The alignments show partial nucleotide and amino acid sequences of the three human selectins, and the consensus sequences (cons) that were used to design primers. I, Inosine; N, any nucleotide. Amino acid sequences of other proteins sharing partial homology with selectins are shown for comparison. Note that these other sequences have significant mismatches with the primer sequences. For the S8 region, the corresponding amino acid sequences of L-selectin CR repeat 2 (L/CR rep 2), P-selectin repeat 5, and E-selectin repeat 3 are shown. These sequences may hybridize with the S8 primer, giving rise to larger than expected PCR products. All sequences shown are human. References: E-selectin,[100] L-selectin,[104] P-selectin,[101] versican,[14] CD23 (IgE receptor),[102] hepatic lectin,[103] EGF,[105] factor X,[106] complement factor H.[107]

(Mel-14, LECCAM-1), E-selectin (ELAM-1),[100] and P-selectin (gmp-140).[101] These receptors share a high degree of sequence homology and a conserved domain structure, consisting of a C-type lectin domain, an EGF-like domain, several repeats of a complement receptor-like domain, a transmembrane domain, and a short cytoplasmic domain (see Fig. 10).

[100] M. P. Bevilacqua, S. Stengelin, M. J. Gimbrone, and B. Seed, *Science* **243**, 1160 (1989).
[101] G. I. Johnston, R. G. Cook, and R. P. McEver, *Cell* (*Cambridge, Mass.*) **56**, 1033 (1989).

We have designed PCR primers based on conserved sequences in the lectin domains (S3 and S4), EGF-domains (S7), and CR domains (S8). Even though lectin domains occur in a variety of other proteins, such as proteoglycans (versican),[14] the IgE receptor (CD23),[102] and hepatic lectin,[103] it is unlikely that the S3/S4 primers will cross-react with the cDNAs encoding these proteins (see Fig. 10).[104-107] In our experiments using cDNA templates from leukocytes, epithelial cells, or total lung tissue, the S3/S4 primers exclusively amplified sequences corresponding to the three known selectins (Fig. 11).[108-114] The primer pairs S3/S7 and S3/S8 are also specific for selectins. The size of the amplified products is 160 nucleotides (S3/S4), 285 nucleotides (S3/S7), and 405 nucleotides (S3/S8), respectively. All of these primers work well in a variety of mammalian species. Figure 11 shows all published selectin sequences, along with sequences that we identified in alveolar lavage cells from rabbit (**) or guinea pig (*), and sheep lung (***). It is likely that these primers will be useful for identifying selectins in nonmammalian species; however, this has not been tested.

In the experiments described above, we have not identified any sequences that would indicate the existence of novel members of the selectin family. If such molecules exist, their identification in cDNA prepared from cells or tissues may be hampered by the presence of much larger amounts of the known selectins. Therefore, we have amplified selectin sequences from nuclear DNA, using the S3/S4 primers, as well as additional, more degenerate primers based on the lectin domain sequences.

---

[102] H. Kikutani, S. Inui, R. Sato, E. L. Barsumian, H. Owaki, K. Yamasaki, T. Kaisho, N. Uchibayashi, R. R. Hardy, and T. Hirano, *Cell (Cambridge, Mass.)* **47,** 657 (1986).

[103] M. Spiess and H. F. Lodish, *Proc. Natl. Acad. Sci. U.S.A.* **82,** 6465 (1985).

[104] M. H. Siegelman and I. L. Weissman, *Proc. Natl. Acad. Sci. U.S.A.* **86,** 5562 (1989).

[105] A. Gray, T. J. Dull, and A. Ullrich, *Nature (London)* **303,** 722 (1983).

[106] M. R. Fung, R. M. Campbell, and R. T. MacGillivray, *Nucleic Acids Res.* **12,** 4481 (1984).

[107] T. Kristensen and B. F. Tack, *Proc. Natl. Acad. Sci. U.S.A.* **83,** 3963 (1986).

[108] A. Weller, S. Isenmann, and D. Vestweber, *J. Biol. Chem.* **267,** 15176 (1992).

[109] A. M. Manning, C. L. Lane, J. A. Auchampach, G. L. Kukielka, C. L. Rosenbloom, and D. C. Anderson, unpublished. GenBank Accession No. L23087 (1993).

[110] J. D. Larigan, T. C. Tsang, J. M. Rumberger, and D. K. Burns, *DNA Cell Biol.* **11,** 149 (1992).

[111] T. Watanabe, Y. Song, Y. Hirayama, T. Tamatani, K. Kuida, and M. Miyasaka, *Biochim. Biophys. Acta* **1131,** 321 (1992).

[112] B. Walcheck, M. White, S. Kurk, T. K. Kishimoto, and M. A. Jutila, *Eur. J. Immunol.* **22,** 469 (1992).

[113] A. M. Manning, J. A. Auchampach, M. G. Oliver, C. L. Rosenbloom, and D. C. Anderson, unpublished. GenBank Accession No. L23088 (1993).

[114] N. A. Strubel, M. Nguyen, G. S. Kansas, T. F. Tedder, and J. Bischoff, *Biochem. Biophys. Res. Commun.* **192,** 338 (1993).

LECTIN DOMAINS (partial)

```
              <- S3 ->                          <- S4  ->
E (human)     YYWIGIRKVNNVWVWVGTQKPLTEEAKNWAPGEPNNRQKDEDCVEIYIKREKDVGMWNDERCSKKKLALCYTA
E (mouse)     YYWIGIRKVNNVWIWVGTGKPLTEEAQNWAPGEPNNKQRNEDCVEIYIQRTKDSGMWNDERCNKKKLALCYTA
E (dog)       YYWIGIRKVNKKWTWIGTQKLLTEEAKNWAPGEPNNKQNDEDCVEIYIKRDKDSGKWNDERCDKKKLALCYTA
E (rabbit)    YYWIGIRKVNNVWIWVGTHKPLTEGAKNWAPGEPNNKQNNEDCVEIYIKRPKDTGMWNDERCSKKKLALCYTA

L (human)     YYWIGIRKIGGIWTWVGTNKSLTEEAENWGDGEPNNKKNKEDCVEIYIKRNKDAGKWNDDACHKRLAALCYTA
L (mouse)     YYWIGIRKIGKMWTWVGTNKTLTKEAENWGAGEPNNKKSKEDCVEIYIKRERDSGKWNDDACHKRKAALCYTA
L (rat)       YYWIGIRKIGKTWTWVGTNKTLTKEAENWGTGEPNNKKSKEDCVEIYIKRERDSGKWNDDACHKRKAALCYTA
L (bovine)    YYWIGIRKVEGVWTWVGTNKSLTEEAKNWGAGEPNNRKSKEDCVEIYIKRNKDSGKWNDDACHKAKTALCYTA
L (guinea pig)* ----------IWTWVGTNKSLTKEAENWGAGEPNNKSKEDCVEIYIKRERDSGNWNDDACHKRKAALCYTA
L (rabbit)*     --------IGNIWTWVGTNKSL?AEAENWGEGEPNNKKTK-----------------------------

P (human)     YYWIGIRKNNKTWTWVGTKKALTNEAENWADNEPNNKRNNEDCVEIYIKSPSAPGKWNDEHCLKKKHALCYTA
P (mouse)     YYWIGIRKINNKWTWVGTNKTLTEEAENWADNEPNNKKNNQDCVEIYIKSNSAPGKWNDEPCFKRKRALCYTA
P (rat)       YYWIGIRKINNKWTWVGTNKTLTAEAENWADNEPNNKRNNQDCVEIYIKSNSAPGKWNDEPCFKRKRALCYTA
P (bovine)    YYWIGIRKINNKWTWVGTKKTLTEEAENWADNEPNNKRNNQDCVEIYIKSLSAPGKWNDEPCWKRKRALCYRA
P (rabbit)**    ----------KWTWVGTKKPLT?EAENWADNEPNNKKNNEDCVEMYIKRSTSPGKWNDDRCRYKKRALCYTA
P (sheep)***    --------IDNKWTWVGTKKTLTEEDENWADNEPNNKKNNQDCVEIYIKSPSAPGKWNDEHCLKKKHALCYTA
```

EGF-LIKE DOMAINS   CR DOMAIN #1 (partial)

```
              <-   S7   ->                      <-   S8   ->
E (human)     ACTNTSCSGHGECVETINNYTCKCDPGFSGLKCEQ     IVNCTALESPEHGSLVCSHPLGNFSYNSSCSISC
E (mouse)     SCTNASCSGHGECIETINSYTCKCHPGFLGPNCEQ     AVTCKPQEHPDYGSLNCSHPFGPFSYNSSCSFGC
E (dog)       ACTPTSCSGHGECVETVNNYTCKCHPGFRGLRCEQ     VVTCQAQEAPEHGSLVCTHPLGTFSYNSSCFVSC
E (rabbit)    ACTEASCSGHGECIETINNYSCKCYPGFSGLKCEQ     VVTCEAQVQPQHGSLNCTHPLGNFSYNSSCSVSC

L (human)     SCQPWSCSGHGECVEIINNHTCNCDVGYYGPQCQL     VIQCEPLEAPELGTMDCTHPFGNFSFSSQCAFSC
L (mouse)     SCQPGSCNGRGECVETINNHTCICDAGYYGPQCQY     VVQCEPLEAPELGTMDCIHPLGNFSFQSKCAFNC
L (rat)       SCQPESCNRHGECVETINNNTCICDPGYYGPQCQY     VIQCEPLKAPELGTMNCIHPLGDFSFQSQCAFNC
L (bovine)    SCKPWSCSGHGQCVINNYTCNCDLGYYGPECQF       VTQCVPLEAPKLGTMACTHPLGNFSFMSQCAFNC
L (guinea pig)* SCQPGSCNGRGECVETINNHTCICDEGYYGPQCQY
```

```
P (human)     SCQDMSCSKQGECLETIGNYTCSCYPGFYGPECEY     VRECGELELPQHVLMNCSHPLGNFSFNSQCSFHC
P (mouse)     SCQDMSCSNQGECIETIGSYTCSCYPGFYGPECEY     VKECGKVNIPQHVLMNCSHPLGEFSFNSQCTFSC
P (rat)       SCQDMSCNSQGERIETIGSYTCSCYPGFYGPECEY     VQECGKFDIPQHVLMNCSHPLGDFSFSSQCTFSC
P (bovine)    SCQDMSCSKQGECIETIGNYTCSCYPGFYGPECEY     VRECGEFDLPQHVHMNCSHPLGNFSFNSHCSFHC
P (rabbit)**  SCQDTSCSKQGECIETIGNYT--------------
P (sheep)***  SCQDMSCSKQ---------------------
```

FIG. 11. Alignment of partial selectin amino acid sequences. The region between the S3 and S8 primer sites (see Fig. 10) is shown. Dashes represent undetermined portions of a sequence. *, Sequences amplified from guinea pig alveolar lavage cells; **, sequences amplified from rabbit alveolar lavage cells; ***, sequence amplified from sheep lung. References to sequences: E (human),[100] E (mouse),[108] E (dog),[109] E (rabbit),[110] L (human),[104] L (mouse),[99] L (rat),[111] L (bovine),[112] P (human),[101] P (mouse),[108] P (rat),[113] P (bovine).[114]

In the three known selectin genes, the lectin domain is not interrupted by introns,[115-117] suggesting that amplification of novel selectin gene sequences may be possible. We found that sequences corresponding to the known selectins were easily amplified from genomic DNA, but no novel sequences were observed, even after extensive restriction/reamplification.[118] This makes it unlikely that the selectin family contains additional

[115] G. I. Johnston, G. A. Bliss, P. J. Newman, and R. P. McEver, *J. Biol. Chem.* **265,** 21381 (1990).

[116] T. Collins, A. Williams, G. I. Johnston, J. Kim, R. Eddy, T. Shows, M. J. Gimbrone, and M. P. Bevilacqua, *J. Biol. Chem.* **266,** 2466 (1991).

[117] M. L. Watson, S. F. Kingsmore, G. I. Johnston, M. H. Siegelman, B. M. Le, R. S. Lemons, N. S. Bora, T. A. Howard, I. L. Weissman, and R. P. McEver, *J. Exp. Med.* **172,** 263 (1990).

[118] S. Hemmerich, S. Rosen, R. Pytela, and D. J. Erle, unpublished (1992).

members that share a similar degree of homology as the known members share with each other. However, our data do not rule out the existence of more divergent members of the family, or novel selectin genes containing large introns interrupting the lectin domain.

Detailed Methods

Oligonucleotide Synthesis

An Applied Biosystems (Foster City, CA) DNA synthesizer is used. Oligonucleotides are cleaved from the resin and deprotected using concentrated ammonia. Following evaporation of the ammonia, oligonucleotides are used as PCR or sequencing primers without further purification.

RNA Purification

Total RNA is prepared from cultured cells or tissue using the LiCl/urea method. Cells or tissues are lysed in 3 M LiCl–6 M urea, homogenized, stored overnight at 4°, and then centrifuged (11,000 g for 60 min at 4°). The pelleted material is dissolved in 10 mM N-2-hydroxyethylpiperazine-N'-2-ethanesulfonic acid (HEPES)–1 mM ethylenediaminetetraacetic acid (EDTA)–1% sodium dodecyl sulfate (SDS), and then extracted first with phenol and then with chloroform. RNA is recovered by precipitation in sodium acetate–ethanol and redissolved in RNase-free water. For purification of poly(A)$^+$ RNA directly from cell lysates, we use the Fast-Track kit (Invitrogen, San Diego, CA).

Homology-Based RT-PCR

Single-stranded cDNA is synthesized from either total RNA or poly(A)$^+$ RNA using modified Moloney murine leukemia virus reverse transcriptase (MoMLV-RT) (Superscript; GIBCO-Bethesda Research Laboratories, Gaithersburg, MD). Twenty to 50 μg of total RNA is used as template for a 20-μl reverse transcription reaction. The reaction mixture includes dATP, dGTP, dCTP, and dTTP (0.5 mM each) and 9 $\mu$$M$ random hexamers, and the reaction is performed for 45 min at 44°. One to 5 μl of the resultant cDNA is used as template for the PCR. The PCR is carried out in a reaction volume of 25–200 μl. In addition to the template cDNA, each PCR reaction contains 1.5 mM $MgCl_2$, 0.01% (w/v) gelatin, 0.1% (v/v) Triton X-100, dATP, dGTP, dCTP and dTTP (0.2 mM each), Taq DNA polymerase (0.05 units/μl), and the buffer provided by the supplier of the enzyme [obtained from either United States Biochemical Corporation (Cleveland, OH) or from Promega (Madison, WI)]. For each reaction, two oligonucleotide primers are also added to obtain a final concentration of 1 μM. Each reaction mixture is overlaid with mineral oil, heated to 95° for 4 min in a thermal cycler (Ericomp, San Diego, CA), and then subjected to 30 cycles of PCR. For β subunit amplifications, each cycle consists of 45 sec at 95°, 45 sec at 53°, and 1 min at 72°. For α subunit amplifications using A14/A2A, the conditions are the same except that the temperature during the annealing step is 48° rather than 53°. For selectin amplifications, the annealing temperature is 53°. Immediately after the last cycle, the sample is maintained at 72° for 10 min.

The results of each PCR reaction are analyzed by gel electrophoresis in 1.5% (w/v) agarose. Reactions that produce the expected size fragments (350 bp for β subunit amplifications and 300–350 bp for α subunit amplifications) are electrophoresed in 1.5% (w/v) low gel temperature agarose (Bio-Rad, Richmond, CA). The appropriate-sized band is then excised and melted at 68° and the DNA is purified by sequential extraction with phenol and then chloroform, and precipitated in ethanol and ammonium acetate.

Cloning of Fragments Obtained by Polymerase Chain Reaction

Purified PCR products are cloned into pBluescript (Stratagene) for further analysis. We use two different strategies for cloning PCR products. Initially, the PCR primers we used did not include recognition sites for restriction endonucleases, therefore PCR products are blunt-ended and phosphorylated for cloning into blunt-ended, dephosphorylated vector. Purified fragments are resuspended in distilled water and treated with 2.5 U of DNA polymerase I, large (Klenow) fragment (Promega) to fill in any 3' recessed ends left after the last cycle of PCR. Next, the 5' ends are phosphorylated with 5 U of T4 polynucleotide kinase (New England BioLabs, Beverly, MA). An aliquot of the above reaction mixture, containing approximately 100–200 ng of DNA, is then ligated into pBluescript that has been cut with *Eco*RV (Promega) and dephosphorylated with calf intestinal alkaline phosphatase (Boehringer Mannheim, Indianapolis, IN). Ligations are performed at 22° for 1 hr with T4 DNA ligase (Bethesda Research Laboratories, Gaithersburg, MD). However, because of the low efficiency of these blunt-end ligation reactions, we subsequently incorporated recognition sites for restriction endonucleases into the 5' ends of our PCR primers. For this purpose we have obtained good results with *Eco*RI, *Bam*HI, *Sal*I, and *Xho*I sites, generally protected by inclusion of five additional nucleotides at the 5' end. For cloning of these fragments, the PCR reaction mixture is first extracted with phenol and then chloroform to remove *Taq* polymerase, and then precipitated with ethanol and ammonium acetate. The digested product is electrophoresed in low melting temperature agarose and the appropriately sized bands are excised and purified as described above. Purified fragments are then ligated into digested, dephosphorylated pBluescript at 14° for 2 hr.

The ligation mixture thus obtained is used to transform competent JM109 *Escherichia coli* (Clontech, Palo Alto, CA). Plasmids containing inserts are purified using a minipreparation lysis kit (Pharmacia-LKB, Pleasant Hill, CA), denatured in 0.3 M NaOH, further purified over spun columns containing Sephacryl S-400 (Pharmacia-LKB), and then sequenced using the Sequenase version 2.0 sequencing kit (United States Biochemical Corporation) and [^{35}S]dATP (Amersham, Arlington Heights, IL). Alternatively, plasmids are purified over S-400 columns without denaturation and analyzed by restriction digestion with enzymes chosen to detect integrin subunits of known sequence.

Restriction and Reamplification

Once the predominant cDNA sequences in a given PCR product are determined, additional sequences are sought using a combination of digestion with rare-cutting restriction endonucleases followed by reamplification of the digested product using the same PCR conditions as in the initial amplification. For this procedure, restriction endonuclease recognition sites are identified in the predominant sequences obtained from the initial PCR reaction. The PCR reaction mixture is then digested with one or more endonuclease, by addition of the enzyme(s) and the appropriate buffer concentrate. A portion of the digested mixture is then analyzed by electrophoresis in low melting temperature agarose, and the regions of the gel corresponding to undigested integrin subunits are excised and diluted at least fivefold in water. One microliter of melted gel slice and 1 μl of digested PCR reaction mixture are then used as templates for separate PCR reactions, using the same conditions and primers as in the initial amplification. The products of these reamplification reactions are analyzed, digested, subcloned into pBluescript, and sequenced, as described above.

Specific Methods Used for Homology Polymerase Chain Reaction of Cadherins

Because the cadherin primers are relatively short (17 nucleotides), a low annealing temperature should be used.[81,93] Typical reaction conditions are as follows: denaturation, 94° for 1.5 min; annealing, 45° for 2 min; polymerization, 72° for 3 min; reaction cycles, 35; template DNA, 100 ng; primers, 20μM. As a template for RT-PCR, we have used first-strand cDNA obtained by reverse transcription of poly(A)$^+$ RNA. For subcloning of the PCR product, we have used various commercially available kits, and also the following simple method. The reaction products are separated by agarose gel electrophoresis, the band of expected size is excised, and the DNA eluted using the freeze-thaw method as follows. The agarose gel is crushed with a spatula, mixed with an equal volume of phenol, subjected to several cycles of freezing-thawing, centrifuged, and the DNA in the supernatant collected by ethanol precipitation. The DNA is phosphorylated at the 5' end using T4 polynucleotide kinase, and inserted by blunt-end ligation into the *Sma*I site of M13 or pBluescript (Stratagene). Typically, more than 50% of the resulting clones have inserts that correspond to cadherin sequences.

Acknowledgments

This work was supported in part by National Institutes of Health Grants CA53259 and HL191551 (R.P.), HL/A133259, HL47412, and HL25816 (D.S.), and EY08106 (S.S.), and by University of California Tobacco-Related Disease Research Program Grants RT338 (R.P.) and 1KT71 (D.J.E).

[21] Solution Structures of Modular Proteins by Nuclear Magnetic Resonance

By MICHAEL J. WILLIAMS and IAIN D. CAMPBELL

Introduction

Extracellular matrix (ECM) components are mostly mosaic proteins, assembled from a variety of repeating protein modules. The modules are typically 40–100 residues in size and are defined by a consensus sequence of conserved residues, encoded on discrete exons and bordered by introns of identical phase.[1] In general, they form autonomously folded homologous structures that seem to be used either as "spacers" for producing molecules of the correct length or to mediate a variety of protein–protein interactions.

There is considerable interest in the tertiary structures of ECM proteins owing to their various fundamental biological roles. Unfortunately, they have generally proved difficult to crystallize, which has restricted struc-

---

[1] L. Patthy, *Curr. Opin. Struct. Biol.* **1,** 351 (1991).

Copyright © 1994 by Academic Press, Inc.
All rights of reproduction in any form reserved.

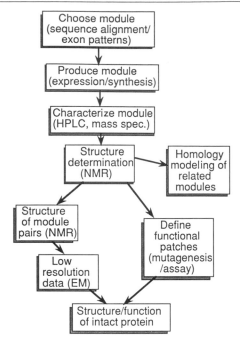

FIG. 1. Strategy for the study of modular proteins.

tural studies by X-ray crystallography. This is probably because the intact proteins are elongated, segmentally flexible, and often glycosylated.

A viable approach to study their structure utilizes a "dissect and build" strategy (see Fig. 1) developed in our laboratory and elsewhere. The structural component of the strategy can be considered in three stages: (1) the conserved "consensus" structure of one or more examples from each module family is determined; (2) structural studies on module pairs provide insight into the way consecutive modules come together; and (3) this atomic resolution information is combined with low-resolution structural data to yield a model of the intact protein.

We concentrate here on fibronectin, a classic mosaic protein found as a soluble component in plasma and laid down as part of the ECM.[2-4] The solution structures of single isolated members of all three fibronectin

[2] E. A. Ruoslahti, *Annu. Rev. Biochem.* **57,** 375 (1988).
[3] D. F. Mosher, "Fibronectin." Academic Press, San Diego, 1989.
[4] R. O. Hynes, "Fibronectins." Springer-Verlag, New York, 1990.

module families[5-7] have previously been reported and we have determined the structure of a pair of fibronectin type 1 modules.[8,9] We describe the methods that we typically employ in protein structural studies, with particular reference to the tenth type 3 module ([10]F3) and the fourth and fifth type 1 module pair ([4]F1 · [5]F1).

Nuclear Magnetic Resonance Methodology

Since the first introduction of two-dimensional nuclear magnetic resonance (NMR) spectroscopy methods in the early 1980s, the technique has grown into a powerful partner to X-ray crystallography for the determination of three-dimensional structures of biomolecules up to 25 kDa in size. Key to this rapid maturation has been the development of more sensitive spectrometers, high-field magnets, access to high-powered computers, and the use of recombinant protein expression for large-scale protein production and isotopic (^{15}N/^{13}C) labeling. The basic principles that govern the practical use of multidimensional NMR for structure determination have been discussed in numerous texts[10] and reviews.[11,12] For this discussion the experimental procedures can be considered in six basic stages: sample preparation, optimization of NMR conditions, data acquisition, sequence-specific resonance assignment, collation of NMR spatial restraints [nuclear Overhauser effect (NOE), coupling constant, and slow exchanging amide data], and structure calculation.

Sample Preparation

NMR structural studies require tens of milligrams of folded protein, soluble at concentrations of over 1 mM. Moreover, for proteins larger than about 10 kDa, resonance assignment often requires uniform isotopic labeling with ^{15}N and/or ^{13}C to overcome the problems of resonance overlap and line broadening in the NMR spectrum. Many of the protein mod-

[5] M. Baron, D. G. Norman, A. Willis, and I. D. Campbell, *Nature (London)* **345,** 642 (1991).

[6] K. L. Constantine, M. Madrid, L. Banyai, M. Trexler, L. Patthy, and M. Llinas, *Biochemistry* **30,** 1663 (1992).

[7] A. L. Main, T. S. Harvey, M. Baron, J. Boyd, and I. D. Campbell, *Cell (Cambridge, Mass.)* **71,** 671 (1992).

[8] M. J. Williams, I. Phan, M. Baron, P. C. Driscoll, and I. D. Campbell, *Biochemistry* **32,** 7388 (1993).

[9] M. J. Williams, I. Phan, T. S. Harvey, A. Rostagno, L. I. Gold, and I. D. Campbell, *J. Mol. Biol.* in press.

[10] K. Wüthrich, "NMR of Proteins and Nucleic Acids." Wiley, New York, 1986.

[11] G. M. Clore and A. M. Gronenborn, *Science* **252,** 1390 (1991).

[12] G. M. Clore and A. M. Gronenborn, *Annu. Rev. Biophys. Chem.* **20,** 29 (1991).

ules that we have studied to date were produced by recombinant expression in either a yeast system based on the α-factor secretion pathway or in an *Escherichia coli* glutathione transferase (GST) fusion system. These efficient expression systems allow the isolation of large quantities of purified (>95%) protein samples and make uniform isotopic labeling feasible.

Production of protein for NMR samples using the GST fusion system is relatively standard.[13] The yeast secretion system is less widely used but has proved particularly successful for the expression of correctly folded protein modules, including those containing disulfide bonds. The system makes use of the yeast mating pheromone α-factor (MFα) secretion pathway.[14,15] In yeast, the MFα gene encodes for the 165 α-factor precursor, consisting of an 83-amino acid leader and four α-factor repeats, each separated by short spacers of between 6 and 8 amino acids. At the aminoterminal end of the leader is a hydrophobic signal sequence, followed by approximately 60 residues of prosequence. The short spacers consist of the dibasic pair Lys-Arg, followed by two or three Glu-Ala or Asp-Ala repeats. These create the proteolytic sites required to separate the α-factor peptides. In the adapted expression system, the protein of interest is expressed as a fusion with the α-factor leader sequence, linked via a single Lys-Arg (see Fig. 2a). Thus, during secretion, the leader sequence is efficiently cleaved off from the recombinant protein at the C terminus of the dibasic pair, by KEX2 cathepsin B-like protease.

We have produced several fragments of fibronectin using this yeast α-factor system, including ^{1}F1, ^{1}F1 · ^{2}F1, ^{4}F1 · ^{5}F1, ^{7}F1, ^{10}F3, and ^{10}F1. In addition, we have expressed a number of other modules in yeast, such as the single fibronectin type 1 module from t-PA,[16] a number of CCP modules from factor H,[17,18] and several growth factor-like domains.[19,20]

[13] G. W. Booker, I. Gout, A. K. Downing, P. C. Driscoll, J. Boyd, M. D. Waterfield, and I. D. Campbell, *Cell* (*Cambridge, Mass.*) **73**, 813 (1993).

[14] J. Kuryan and I. Herskiwitz, *Cell* (*Cambridge, Mass.*) **30**, 933 (1982).

[15] A. J. Blake, J. P. Merryweather, D. G. Coit, U. A. Heberlein, F. R. Masiarz, G. T. Mullenbach, M. S. Urdea, P. Valenzuela, and P. J. Barr, *Proc. Natl. Acad. Sci. U.S.A.* **81**, 4642 (1984).

[16] A. K. Downing, P. C. Driscoll, T. S. Harvey, T. J. Dudgeon, B. O. Smith, M. Baron, and I. D. Campbell, *J. Mol. Biol.* **225**, 821 (1992).

[17] P. N. Barlow, M. Baron, D. G. Norman, A. J. Day, A. C. Willis, R. B. Sim, and I. D. Campbell, *Biochemistry* **30**, 997 (1991).

[18] P. N. Barlow, D. G. Norman, A. Steinkasserer, T. J. Horne, J. Pearce, P. C. Driscoll, R. B. Sim, and I. D. Campbell, *Biochemistry* **31**, 3626 (1992).

[19] P. A. Handford, M. Baron, M. Mayhew, A. Willis, T. Beesley, G. G. Brownlee, and I. D. Campbell, *EMBO J.* **9**, 475 (1990).

[20] T. J. Dudgeon, R. M. Cooke, M. Baron, I. D. Campbell, R. M. Edwards, and A. Fallon, *FEBS Lett.* **261**, 392 (1990).

In the standard protocol, a DNA fragment encoding the target module is amplified by the polymerase chain reaction (PCR) from a cloned fragment of cDNA. The PCR oligonucleotides are designed such that the 5' and 3' ends generally correspond to the position of intron–exon boundaries, at the N and C termini of the amino acid consensus sequence. The sense-strand oligonucleotide primer provides a blunt end to the DNA fragment, while the antisense-strand oligonucleotide primer contains a stop codon, followed by a *Bam*HI site and a short four-nucleotide tail to aid efficient restriction digestion.[8] The DNA fragment is ligated between unique *Stu*I/ *Bam*HI restriction sites in the pMB50 vector (Fig. 2b), downstream and in phase with the α-factor leader sequence. The α-factor–insert combination is then excised as a *Bgl*II/*Bam*HI fragment and is subsequently ligated into the single *Bgl*II site of the pMA91 yeast expression plasmid (Fig. 2c).[21] Competent yeast cells [*Saccharomyces cerevisiae* MD50 (a/α/leu2/ leu2pep4-3/+his3/+)] are then transformed with the pMA91 construct and selected by their ability to grow on leucine-minus medium. Cultures (0.5 liter) are grown in 2-liter baffled flasks for 48–60 hr at 30° in a rotary shaking incubator, using 0.17% (w/v) yeast nitrogen base medium without amino acids (Difco, Detroit, MI) supplemented with 0.5% (w/v) ammonium sulfate and 2% (w/v) glucose. The processed protein is secreted into the supernatant, where it is isolated by binding to C_{18} reversed-phase silica beads, present in the medium during fermentation. The beads are filtered from the yeast culture and the crude protein is eluted off in 60% (v/v) acetonitrile/0.1% (v/v) trifluoroacetic acid (TFA) and lyophilized. The crude material is reconstituted in a few milliliters of water and purified by high-performance liquid chromatography (HPLC). In the case of ^{4}F1 · ^{5}F1, this involved two reversed-phase separations on a C_8 semiprepar-ation column, using a gradient of acetonitrile in 0.1% (v/v) TFA. ^{10}F3 was purified by one C_8 reversed-phase separation, followed by cation exchange and a final reversed-phase separation.[22] Typical yields of purified protein for ^{4}F1 · ^{5}F1 and ^{10}F3 were on the order of 1 mg/liter of yeast culture.

Uniformly ^{15}N-labeled samples of ^{4}F1 · ^{5}F1 and ^{10}F3 were also produced using this system. Here the growth medium was supplemented with only 0.05% (w/v) [^{15}N] ammonium chloride to limit the expense of the isotope and the cultures were grown for no more than 48 hr. These conditions caused little reduction in the level of protein expression and suitable quantities of purified proteins were expressed from 24 liters of yeast culture in each case.

[21] J. Mellor, M. J. Dobson, N. A. Roberts, M. F. Tuite, J. S. Emtage, S. White, P. A. Lowe, T. Patel, A. J. Kingsman, and S. M. Kingsman, *Gene* **24,** 1 (1983).
[22] M. Baron, A. L. Main, P. C. Driscoll, H. J. Mardon, J. Boyd, and I. D. Campbell, *Biochemistry* **31,** 2068 (1992).

(a)

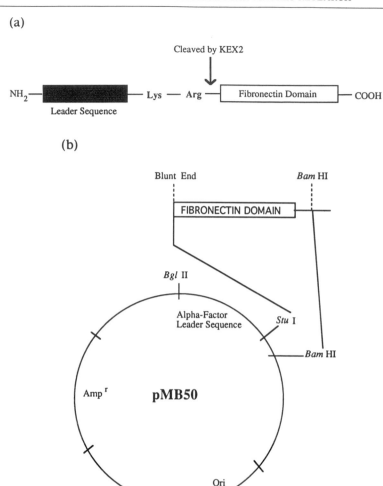

(b)

FIG. 2. (a) Cleavage of fusion protein: The recombinant fusion protein is cleaved at the C-terminal end of the dibasic Lys-Arg by the yeast KEX2 cathepsin B-like protease during secretion. (b) The pMB50 plasmid: The DNA fragment is ligated between unique StuI/BamHI sites, in phase with the α-factor prepro-leader sequence. Amp^r, Ampicillin resistance gene; ori, E. coli origin of replication. (c) The pMA91 yeast expression plasmid[21]: The leader sequence–recombinant gene construct is ligated into the unique BglII site. Expression is directed by the phosphoglycerate kinase promoter (pPGK) and is terminated by the PGK terminator sequence (tPGK). The leu2 gene allows selection of positive yeast transformants on leucine-minus medium. 2 micron (2 μ), Yeast origin of replication.

(c)

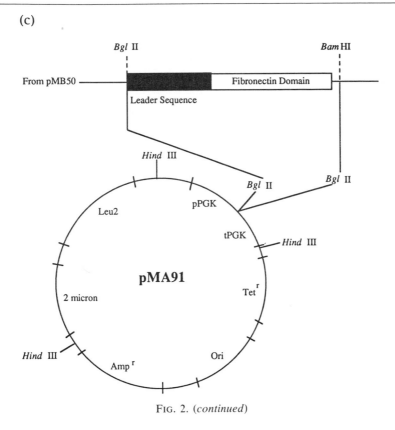

FIG. 2. (*continued*)

Optimization of Nuclear Magnetic Resonance Conditions

The fact that reversed-phase HPLC is used as the final stage of purification means that the samples are effectively desalted before preparation of each NMR sample. Although NMR experiments can be recorded on proteins with small quantities of salt, improved results are obtained in the absence of salt and, in our experience, salt buffers are often not required to maintain sample integrity and pH. Lyophilized samples are reconstituted in 0.5-ml solutions of 99.96% D_2O and 90% $H_2O/10\%$ D_2O, before a series of one-dimensional (1D) NMR spectra are recorded to identify suitable ranges of temperature and pH (uncorrected meter readings) at which the protein is folded and free of aggregation. At this stage it may prove necessary to alter the sample concentration or add salt to overcome aggregation effects, although we have encountered few such problems with protein modules. One important consideration is that the rate of amide proton exchange with the bulk solvent is at a minimum near pH 4.0. This assists the assignment of these protons and is one reason why most NMR studies are performed below physiological pH. Line broadening is reduced in spectra recorded at higher temperatures and increased sample concentration improves the signal-to-noise ratio. However, the values of pH and temperature chosen must be consistent with the protein sample being stable and free of aggregation for periods of time between a few days and about 2 weeks, if complete data sets of multidimensional spectra are to be recorded. Samples are normally stored at 4° between experiments.

A 1D spectrum of [10]F3 in D_2O solution, recorded on a Bruker AM series 600-MHz spectrometer, is shown in Fig. 3. Certain characteristic features of the folded protein are recognizable from this simple spectrum. In particular, there are a number of upfield-shifted methyl proton resonances near 0 ppm and numerous $C_\alpha H$ resonances are found shifted downfield of the water resonance. These are both indications of a folded struc-

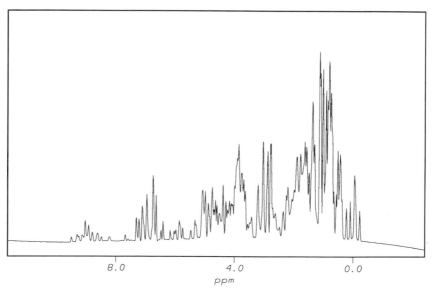

FIG. 3. A 1D spectrum of [10]F3 fibronectin module. Spectrum of a 2–3 mM sample of [10]F3 at pH 3.8 and 312 K, recorded in D_2O solution in under 1 min. The water resonance (at approximately 4.7 ppm) has been suppressed by presaturation. The NMR spectrum provides a precise fingerprint of the folded protein. Particularly important features are the four upfield methyl resonances around 0 ppm, CαH resonances shifted downfield of the water resonance, and the presence of amide proton resonances that undergo slow exchange with deuterons from the bulk solvent.

ture. The methyl protons experience an upfield shift due to proximal aromatic residues in the core of the protein, whereas the $C_\alpha H$ protons are shifted downfield as a consequence of the β-strand backbone conformation.

A useful experiment in NMR studies of proteins is to change the solvent from ^1H_2O to ^2H_2O. This allows amide protons (observed between 7.5 and 9.5 ppm in this spectrum) that exchange slowly with solvent to be identified. Slow exchange is characteristic of protons that are involved in hydrogen-bonding interactions within the core of the protein, where they are sheltered from the bulk solvent. Hence these provide further evidence of protein folding.

Data Acquisition

NMR studies in this laboratory are performed on 500- and 600-MHz spectrometers. Data processing is performed using one of the off-line software packages such as FELIX (Biosym, San Diego, CA) or TRIAD (Tripos, St. Louis, MO) on SUN Microsystems SPARC workstations.

We usually record a combination of two-dimensional homonuclear ^1H nuclear Overhauser effect (NOE) spectroscopy, homonuclear Hartmann–Hann (HOHAHA), and correlated spectroscopy (COSY) experiments, under a variety of temperature and pH conditions. The HOHAHA and COSY spectra identify "through-bond" proton cross-peak connectivities. In a COSY spectrum, cross-peaks connect protons that are separated by three or fewer bonds, whereas in a HOHAHA spectrum it is possible to trace connectivities along entire side chains. On the other hand, NOESY spectra are used to identify "through-space" NOE connectivities between protons separated by less than 5 Å. These experiments generally take between 4 and 20 hr to record and spectra are acquired using samples dissolved in D_2O and in 90% H_2O/10% D_2O.

The majority of the spectra are recorded with the 90% H_2O solvent because many labile amide protons exchange rapidly with solvent and cannot be observed in pure D_2O (the small quantity of D_2O is required in the water sample to provide the deuterium "lock" signal that stabilizes the magnetic field). It is also essential to record spectra in 90% H_2O to allow their assignment. In this case, the large water resonance is suppressed, where possible, by the application of "jump-return" read pulses,[23] for example, in NOESY and HOHAHA experiments. Such a procedure avoids transfer of saturation from water to labile protons. Water suppression for COSY experiments relies solely on presaturation as the

[23] P. Plateau and M. Guéron, *J. Am. Chem. Soc.* **104,** 7310 (1982).

jump-return pulse sequence is not applicable. Use of pure D_2O solvent reduces the water proton resonance and allows the observation of proton resonances that have chemical shifts similar to the water resonance. A priority with this sample is to prepare it with the minimium water content by prior lyophilization of the protein from a D_2O solution and then reconstituting it into ultrapure (99.96%) D_2O.

For ^{15}N-labeled samples such as the [10]F3 module, several 4- to 8-hr 2D heteronuclear ([1]H–[15]N) single quantum correlation (HSQC) experiments are recorded over a range of temperature and pH conditions. Heteronuclear 3D ([1]H–[15]N) NOESY-multiple quantum correlation (HMQC) spectra and ([1]H–[15]N) HOHAHA-HMQC spectra are then acquired, with spin-lock purge pulses applied during the heteronuclear pulse sequence for water suppression.[24] These 3D experiments are generally recorded over 2–3 days and therefore sample stability is an important consideration, especially for an expensive labeled sample.

Sequence-Specific Resonance Assignment

The aim of sequence-specific resonance assignment is to assign "spin systems" of cross-peak connectivities between proton resonances, which are characteristic of each type of amino acid side chain. Afterward these spin systems can be assigned to specific residues within the protein sequence. In practice there is no single best method for this resonance assignment process and generally spin system and sequential assignments are performed in tandem. The 2D ^{1}H NMR spectrum of a protein is made up of three main regions: Amino acid side-chain $C_\alpha H$, $C_\beta H$, $C_\gamma H$, and $C_\varepsilon H$ protons are found upfield of about 5.0 or 6.0 ppm for regions of α-helix or β-strand secondary structure, respectively. Aromatic ring protons are located between approximately 5.5 and 8.5 ppm, whereas amide protons are shifted downfield of about 6.5 ppm, so that the important $NH–C_\alpha H$ fingerprint region is found downfield of 6.5 ppm in the F2 ^{1}H dimension and generally between 6.0 and 3.5 ppm in the F1 ^{1}H dimension. We use spectra recorded in D_2O solution to identify side-chain spin systems, including the aromatic ring spin systems. However, as discussed earlier, it is necessary to use water spectra to assign amide proton connectivities.

To illustrate the assignment process we have included three figures from NMR spectra recorded with 2–3 mM samples of [4]F1 · [5]F1. Figure 4 shows a two-dimensional ^{1}H HOHAHA spectrum recorded in D_2O at 37° and pH 5.5. The excellent dispersion of cross-peaks in the NMR spectrum of this protein allowed almost complete resonance assignment of all 93

[24] B. A. Messerle, G. Wider, G. Otting, C. Weber, and K. Wüthrich, *J. Magn. Reson.* **85,** 608 (1989).

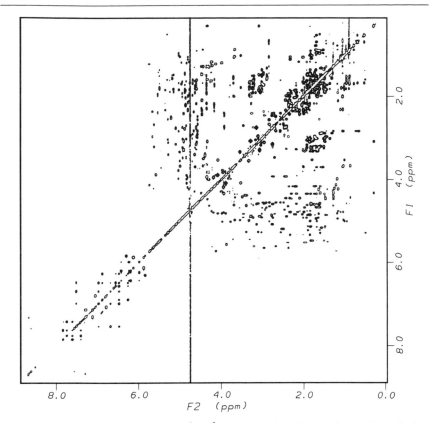

FIG. 4. A 2D HOHAHA spectrum of ^{4}F1 · ^{5}F1 recorded in D$_2$O. The diagonal is equivalent to a 1D spectrum of the protein. In this experiment, cross-peaks off the diagonal correspond to through-bond connectivities between proton resonances that lie along the diagonal. Individual amino acid spin systems can be mapped from characteristic patterns of cross-peaks. Amide proton connectivities are not observed in this spectrum, owing to deuterium exchange. The residual water resonance is observed as a line running parallel to the F1 axis, at approximately 4.6 ppm.

amino acid spin systems. A section of this spectrum has been expanded in Fig. 5 so that the through-bond connectivities for the tryptophan, tyrosine, and phenylalanine ring proton spin systems can be highlighted. Finally, the sequential assignment of the NH-C$_\alpha$H fingerprint region of a water NOESY spectrum, for residues 1–21 from ^{4}F1 · ^{5}F1, is shown in Fig. 6. It is necessary to use COSY, HOHAHA, and NOESY spectra to distinguish between the cross-peaks, which correspond to through-space connectivities and through-bond connectivities of either three bonds or less or more than three bonds. At this stage many of the assignment

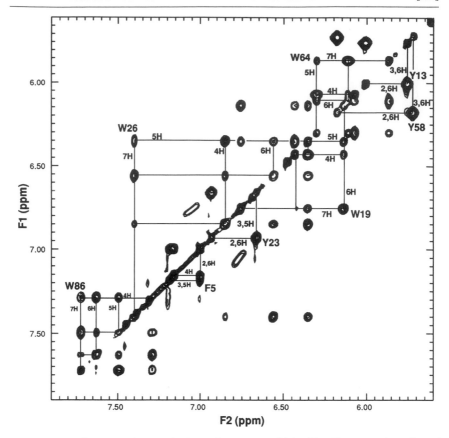

FIG. 5. Assignment of aromatic ring spin systems of ^{4}F1 · ^{5}F1. The aromatic region of the HOHAHA spectrum from Fig. 5 is shown, illustrating the ring spin system assignments for the four tryptophans, three tyrosines, and single phenylalanine. The protein is numbered from 1 to 93, corresponding to residues 151–244 from the amino terminus of human fibronectin.

ambiguities, caused by resonance overlap, can be overcome by comparing equivalent spectra recorded at different temperatures and pH. However, for ^{10}F3 it was still necessary to acquire 3D ^{15}N heteronuclear NMR spectra to complete the sequence-specific resonance assignment. Thus, the amide connectivities could be spread out over three dimensions rather than two, with the two proton dimensions (F1 and F2) being supplemented by a third ^{15}N dimension (F3).

Nuclear Magnetic Resonance Data Collation

NOE-derived distance restraints form the most important type of experimental data for the determination of NMR structures. As described

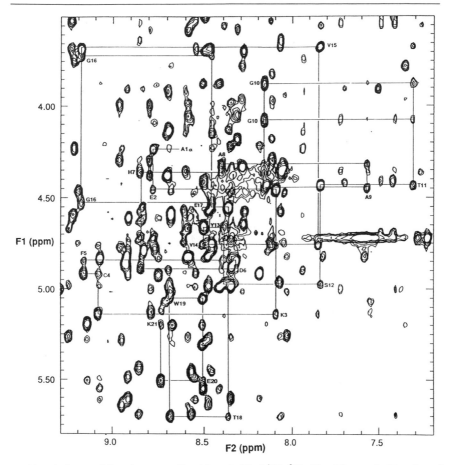

FIG. 6. Sequential assignment of residues 1–21 of ^{4}F1 · ^{5}F1. The "fingerprint" region of an NOESY spectrum recorded in water at pH 4.5 and 310 K is shown. Amide proton resonances lie parallel to the F2 axis and C_α-proton resonances lie parallel to the F1 axis. The spiral of sequential connectivities link "through-bond" cross-peaks between HN_i (where i corresponds to residue number) and $C_{\alpha,i}H$ resonances (also observed in the equivalent HOHAHA and COSY spectra) and "through-space" NOE cross-peaks between HN_i and $C_{\alpha,i+1}H$ resonances (only observed in NOESY spectra). The spiral commences from the NOE between $C_\alpha H$ of A1 and NH of E2 and terminates at K21, which precedes a proline. The backbone amide of A1 is not observed, owing to rapid proton exchange.

above, NOE cross-peaks are observed between protons that are separated by less than 5 Å. We usually classify these as strong, medium, or weak, corresponding to distance restraints of 0–2.7, 0–3.5, or 0–5.0 Å, respectively. These classifications are based on relative NOE peak intensities either estimated by eye or through peak volume integration.

Appropriate pseudoatom corrections are made for equivalent protons in the absence of stereospecific assignments or when these are degenerate.[10]

Dihedral angles derived from vicinal coupling constants between amide and C_α protons ($^3J_{NH-C_\alpha H}$) and between C_α and C_β protons ($^3J_{C_\alpha H-C_\beta H_2}$) provide a further source of structural restraints. The $^3J_{NH-C_\alpha H}$ coupling constants can be measured from F2 cross-sections of high-resolution COSY or DQF-COSY spectra recorded in water. Two cross-sections are taken through each well-resolved, nonglycine NH–C_αH cross-peak and these are best fitted to simulated line shapes that incorporate the effects of the data processing functions used.[25] For ^{15}N-labeled samples, the F1 doublet lineshapes for backbone amide protons from an HMQC-J spectrum[26] provides a second source of $^3J_{NH-C_\alpha H}$ coupling constants.[27] Approximate values for ϕ torsion angle restraints can then be derived by comparing the $^3J_{NH-C_\alpha H}$ coupling constants to the Karplus curve.[28] The $^3J_{C_\alpha H-C_\beta H_2}$ coupling constants are measured directly from the fine structure of each well-resolved C_αH–C_βH cross-peak in the F2 dimension of a primitive exclusive-COSY (PE-COSY) spectrum.[29] χ^1 torsion angles are then determined by comparison of the two $^3J_{C_\alpha H-C_\beta H_2}$ coupling constants and intraresidue NH–C_βH NOE intensities for each AMX spin system.[30] These dihedral angle estimates are used as restraints in the structure calculations within reasonably broad minimum ranges. Thus, for ^{4}F1 · ^{5}F1 the ranges allowed were ±30° and ±60° for ϕ and χ^1 angles, respectively. We use the STEREOSEARCH program developed by Michael Nilges[31] for the stereospecific assignment of β-methylene protons. These assignments are compared to the tertiary structure before inclusion as restraints during the final stages of structural refinement.

Amide NH protons, which exchange slowly with the solvent, are identified by reconstituting a protein sample, lyophilized from H_2O solution, in D_2O and recording a short (ca. 7 hr) 2D HOHAHA spectrum. As mentioned earlier, slow solvent exchange is characteristic of backbone amide protons, which are involved in hydrogen bonding interactions. Thus once the secondary structure of the protein has been defined, it is possible to assign tentatively hydrogen bond donor–acceptor pairs. Each hydrogen

[25] C. Redfield and C. M. Dobson, *Biochemistry* **29,** 7201 (1990).

[26] L. E. Kay and A. Bax, *J. Magn. Reson.* **86,** 110 (1990).

[27] T. J. Norwood, A. Crawford, M. E. Steventon, P. C. Driscoll, and I. D. Campbell, *Biochemistry* **31,** 6285 (1992).

[28] A. Pardi, M. Billeter, and K. Wüthrich, *J. Mol. Biol.* **180,** 741 (1984).

[29] L. Müller, *J. Magn. Reson.* **72,** 191 (1987).

[30] P. C. Driscoll, A. M. Gronenborn, and G. M. Clore, *FEBS Lett.* **243,** 223 (1989).

[31] M. Nilges, G. M. Clore, and A. M. Gronenborn, *Biopolymers* **29,** 813 (1990).

bond can then be incorporated into the structure calculation process as a pair of distance restraints ($d_{NH-O} < 2.3$ Å, $d_{N-O} < 3.3$ Å).

An ^{15}N-labeled sample allows for further studies of backbone dynamics in solution through the measurement of ^{15}N T_1, T_2, and heteronuclear ^{1}H–^{15}N NOE and time-resolved amide NH–solvent exchange kinetics.[32] One such study is discussed in the next section with respect to the ^{10}F3 module.

Three-Dimensional Structure Calculation

NMR solution structures are calculated and selected on the basis of agreement with the experimentally observed NOE-derived distance restraints and spin–spin coupling constant-derived dihedral angle restraints. We employ a dynamic, simulated annealing approach to structure calculation using the X-PLOR program.[33] The protocol involves using high-temperature molecular dynamics to fold the protein from a conformation with randomized backbone torsion angles and extended side chains. The calculation is performed under the influence of forces that maintain correct covalent geometry and chirality of the chemical structure, whereas soft square-well NOE and dihedral angle potentials are introduced to drive the protein chain toward the correctly folded structure. A repulsive term is used to represent the van der Waals, nonbonded component of the target function. This force constant is initially kept low to allow atoms to approach each other closely during several picoseconds of high-temperature dynamics at 1000 K. This allows the structure to rearrange under the influence of the experimental restraints without being caught in local energy minima. The van der Waals force constant is then slowly increased to the target value and the system is cooled in stages to 300 K, followed by restrained energy minimization.

Initially we perform structure calculations using a small set of relatively unambiguous experimental restraints. These initial structures can then be used to identify consistently violated distance and torsion angle restraints that can be reassigned by returning to the NMR spectra. At this stage it is often possible to assign other previously ambiguous NOEs and to introduce these as distance restraints, along with stereospecific assignments of β-methylene protons. The experimental data set is thus continuously refined until the convergence rate for a large number of structures calculated from different randomized protein configurations reaches a satisfactory level. Final structures are selected that exhibit no large violations for one or more NOE distances (>0.5 Å) or dihedral angle ($>3°$) restraints.

[32] L. E. Kay, D. A. Torchia, and A. Bax, *Biochemistry* **28**, 8972 (1989).
[33] A. T. Brunger, "X-PLOR Manual." (Yale University, New Haven, CT, 1988.

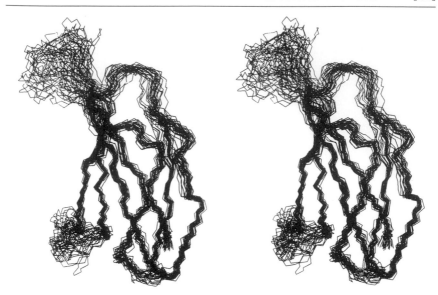

FIG. 7. The solution structure of [10]F3 from fibronectin. The 36 final calculated structures are shown, superimposed on the backbone atoms (N, Cα, and C) of β-strand residues from one of the structures. The disordered RGD-containing loop is found in the top left-hand side of the illustration. (Reproduced with alterations from Main *et al.*[7])

These structures should exhibit good covalent geometry and have low values for the quadratic van der Waals term.

We usually represent the ensemble of final structures by superimposing backbone (N, C_α, and C) atoms of regions of well-defined secondary structure on the equivalent atoms of the lowest energy structure. This type of representation is highlighted for the fibronectin [10]F3 module in Fig. 7. Regions that are well defined show a relatively small root mean square difference (RMSD), while regions that appear more disordered in Fig. 7 are characterized by increased RMSDs. The ϕ and ψ dihedral angle parameters[34] provide a further indication of the spread of each torsion angle over the ensemble of final structures, with a small value indicating a more disordered region of the backbone and a value of 1.0 corresponding to an angle which is identical for all the final structures.

In Fig. 7, it is apparent that conformations of the two turns at the top left and bottom left of the overlaid structures are more divergent. The turn at top left contains the important integrin-binding RGD ligand, and so it was of interest to determine whether this disorder reflected a simple

[34] S. G. Hyberts, M. S. Goldberg, T. F. Havel, and G. Wagner, *Protein Sci.* **1**, 736 (1992).

lack of localized experimental restraints or represented the real dynamic nature of the backbone. Order parameters defining the dynamic properties of the amide group can be derived from analysis of the ^{15}N relaxation parameters $(T_1, T_2,$ and NOE).[32] A complete analysis using these experiments can be time consuming and heteronuclear ^{1}H–^{15}N NOE measurements are a relatively convenient way of probing protein dynamics in solution. The ^{1}H–^{15}N NOE is sensitive to the effective correlation time, with relatively mobile residues within the molecule having a smaller value of NOE.[27] The ^{15}N NOE measurements for ^{10}F3 clearly indicated that the RGD turn was more flexible than the core of the module.[7] They also confirmed that residues of the second disordered loop were more conformationally dynamic than the β strands, although to a lesser degree.

One can calculate a single restrained minimized average structure from the collection of superimposed final structures, by averaging the backbone coordinates of the superimposed structures and then performing restrained energy minimization on the resulting structure. The averaged structure of the ^{4}F1 · ^{5}F1 module pair with overlaid side chains is illustrated in Fig. 8. The individual modules have been separated, so that their conserved structures can be compared directly in similar orientations. The core side chains are particularly well defined for the final structures. However, solvent-exposed side chains generally appear more disordered. This is an important feature of NMR solution structures, and although the conformations of surface side chains are typically restrained by less experimental data in the structure calculations it is probable that these are in fact more flexible in solution than the tightly packed side chains of the protein core.

Discussion

Protein modules have proved highly amenable to NMR structural studies and consensus solution structures of isolated members of some seven or eight different families have been reported to date. A detailed understanding of how different consecutive modules link up is vital to any attempt to rebuild accurate structural models of large mosaic proteins. To this end, we have extended our strategy to the study of pairs of modules. The structures of the fifteenth and sixteenth CCP module pair from factor H[35] and the ^{4}F1 · ^{5}F1 pair from fibronectin[8,9] have been reported. Furthermore, NMR studies of the F1–G module pair from t-PA and a second F1 pair from fibronectin are now well advanced. We anticipate that this research will provide evidence for any preferred types of

[35] P. N. Barlow, D. G. Steinkasserer, D. G. Norman, B. Kieffer, A. P. Wiles, R. B. Sim, and I. D. Campbell, *J. Mol. Biol.* **232,** 268 (1993).

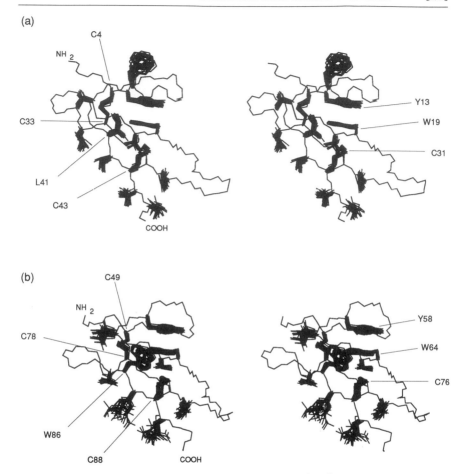

FIG. 8. The conserved type 1 module structures from ⁴F1·⁵F1. Stereo views of the averaged minimized structures of both (a) ⁴F1 and (b) ⁵F1 are shown, including the overlaid side chains of selected residues. The averaged minimized structure was calculated by coordinate averaging from 32 superimposed final structures, followed by restrained minimization. The side chains were obtained by overlaying backbone atoms (N, Cα, and C) of well-defined β-strand regions of the 32 final structures onto the lowest energy structure. Seven conserved core residues have been labeled from each module. Note that the local positions and relative orientations of these residues are particularly well conserved. (Reproduced with minor alterations from Williams et al.[9])

intermodule interfaces for each module family. This may allow models of intact modular proteins to be rebuilt from the structures of relatively few module pairs, together with data obtained for larger fragments and entire molecules using low-resolution techniques such as electron microscopy.

Acknowledgments

The strategy described here has been developed through the contributions of numerous members of the Campbell laboratory, both past and present. We would particularly like to acknowledge Alison Main, who provided Fig. 7 and the NMR data for Fig. 3. This is a contribution from the Oxford Centre for Molecular Sciences, which is supported by the UK SERC and MRC. We also acknowledge support from the Wellcome Trust.

[22] Electron Microscopy of Extracellular Matrix Components

By Jürgen Engel

Introduction

In 1987 Heinz Furthmayr and I reviewed[1] electron microscopic techniques that had been used for the elucidation of proteins and proteoglycans of the extracellular matrix (ECM). We focused on successful applications of the newly introduced glycerol spraying/rotary shadowing technique[2] for the determination of the structure of large multidomain proteins such as laminin, collagen IV, collagen VI, and some proteoglycans.

The present chapter[3] deals with several new applications of the rotary shadowing and negative staining techniques. Electron microscopy has become increasingly important for the elucidation of the domain organization of large multidomain proteins because of the information now available on the structure of individual domains as derived by nuclear magnetic resonance (NMR) analysis or X-ray crystallography.[4,5]

Electron microscopy has also been used to monitor interactions such as the unfolding and refolding of collagen, the assembly of laminin from its three chains,[6] and the binding of proteoglycan-binding regions to hyaluronate.[7] A second area in which high-resolution electron microscopy has been successful is in the investigation of supramolecular organizations

---

[1] J. Engel and H. Furthmayr, this series, vol 145, p. 3.

[2] D. M. Shotton, B. E. Burke, and D. Branton, *J. Mol. Biol.* **131,** 303 (1979).

[3] This work was supported by the Swiss National Science Foundation. Critical reading of the manuscript by M. Mörgelin (Lund) is gratefully acknowledged.

[4] M. Baron, D. G. Norman, and I. D. Campbell, *Trends Biochem. Sci.* **16,** 13 (1991).

[5] R. Doelz, J. K. Engel, and K. Kühn, *Eur. J. Biochem.* **178,** 357 (1988).

[6] I. Hunter, T. Schulthess, and J. Engel, *Biochem J.* **267,** 6006 (1992).

[7] M. Mörgelin, M. Paulsson, T. E. Hardingham, D. Heinegard, and J. Engel, *Biochem. J.* **253,** 175 (1988).

Copyright © 1994 by Academic Press, Inc.
All rights of reproduction in any form reserved.

such as fibrillin microfilaments,[8] the association of collagen IX to collagen II,[9] and the assembly of many components in the vitreous body[10] and in basement membranes.[11] In many cases it was possible to identify individual domains in a protein or assembly product by immunolabeling.[12] Furthermore, new electron microscopic techniques have gained importance, such as in the determination of mass distributions in collagen fibers by scanning transmission electron microscopy[13] (STEM), and in structure determinations without staining,[14] in the frozen hydrated state or of specimens prepared by conventional methods. Last but not least, initial results have been obtained with scanning probe microscopy such as tunneling microscopy or atomic force microscopy[15,16] Not included in the present review are histological techniques or immunoelectron microscopy, which are extensively used for localization of ECM components in tissues.

Improved Resolution of Structure of Laminin

Laminin, with its typical cruciform shape and its many different domains,[17–20] is a good example for electron microscopy of isolated ECM proteins. Its basic electron microscopic structure was reviewed previously.[1,18,19] By careful application of the negative staining technique (see Procedures, below) it was possible to visualize the protein at high resolution (Fig. 1). Using this technique, but also after rotary shadowing with platinum/carbon (see Procedures, below), it was possible to detect a third globular domain in one of its short arms. By comparison with the amino acid sequence of murine Engelbreth-Holm–Swarm (EHS) tumor laminin it was possible to identify this arm as the N-terminal region of its

---

[8] D. R. Keene, K. Maddox, H.-J. Kuo, L. Y. Sakai, and R. W. Glanville, *J. Histochem. Cytochem.* **39,** 441 (1991).

[9] L. Vaughan, M. Mendler, S. Huber, P. Bruckner, K. Winterhalter, M. I. Irwin, and R. Mayne, *J. Cell Biol.* **106,** 991 (1988).

[10] R. G. Brewton and R. Mayne, *Exp. Cell Res.* **198,** 237 (1992).

[11] P. D. Yurchenco and J. C. Schittny, *FASEB J.* **4,** 1577 (1990).

[12] T. M. Schmid and T. F. Linsenmayr, *Dev. Biol.* **138,** 53 (1990).

[13] D. A. Holmes, P. A. Mould, and J. A. Chapman, *J. Mol. Biol.* **220,** 111 (1991).

[14] K. Kobayashi, J. Niwa, T. Hishino, and T. Nagatani, *J. Electron Microsc.* **41,** 235 (1992).

[15] G. F. Cotterill, J. A. F, Fergusson, J. S. Gani, and G. F. Burns, *Biochem. Biophys. Res. Commun.* **194,** 973 (1993).

[16] D. Anselmetti, M. Dreier, R. Lüthi, T. Richmond, E. Meyer, J. Frommer, and H.-J. Güntherodt, *J. Vac. Sci. Technol., B***12(3),** 1500 (1994).

[17] R. Timpl, *Eur. J. Biochem.* **180,** 487 (1989).

[18] K. Beck, I. Hunter, and J. Engel, *FASEB J.* **4,** 148 (1990).

[19] J. Engel, *Biochemistry* **31,** 10643 (1992).

[20] M. Bruch, R. Landwehr, and J. Engel, *Eur. J. Biochem.* **185,** 271 (1989).

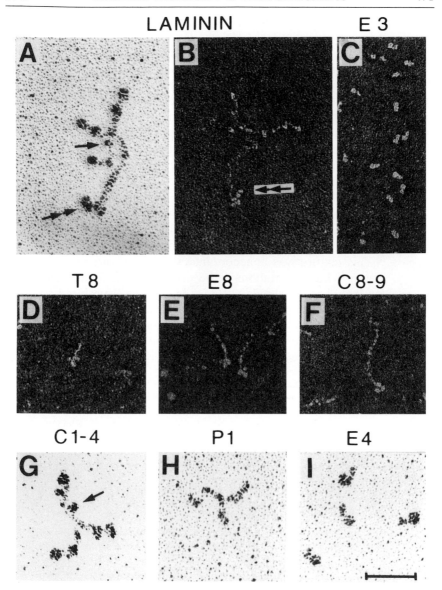

LAMININ   E 3

T 8   E 8   C 8-9

C 1-4   P 1   E 4

FIG. 1. Electron micrograph of EHS laminin and of laminin fragments after rotary shadowing (A, G–I) and negative staining (B–F). Fragments E3, T8, and C8–9 originate from the long arm of laminin, fragment C1–4 comprises the entire, short-arm structure, and fragments P1 and E4 are the inner and terminal regions of the short arms, respectively.[18,19] Arrows indicate the third globular domain in the α chain and double arrows indicate region E3 in intact laminin. Bar: 50 nm. (Reproduced with permission from Beck et al.[18])

α1 chain.[21] The short arm formed by the α1 chain is longer than those of the β1 and γ1 chains. Several EHS-like laminins of different organs and species were found to have similar shapes. This essentially also holds true for laminin isoforms with α2 and β2 chains of merosin[23] and S-laminin.[24] It was found, however, for such laminins isolated from bovine heart and human placenta that one of its short arms contains a spoonlike structure instead of two well-separated terminal globular units.[25] Y-Shaped laminins with only two short arms were also detected by electron microscopy.[26,27] Their chain composition is not clear yet, but from the presence of a globular domain at the terminus of the long arm it may be concluded that they contain a truncated α chain without the N-terminal short arm region. This conclusion is based on the finding that in all laminins studied up to now the terminal globular region at the long arm was contributed by the α chain. Kalinins, which are now well-characterized members of the laminin family,[28,29] exhibit unusually short arms with only a single globular domain.[28] This feature was in accordance with the shorter sequences of their γ2 chains.[30] Large variations in the structure of laminins from different species including *Drosophila, Hydro medusa,* sea urchin, and leech were also demonstrated by electron microscopy.[19,31]

Electron microscopy has also been used for the characterization of recombinantly prepared human B1 chains.[32] In the isolated recB1 chains the short-arm regions were found to be properly structured as far as it

[21] The new nomenclature for laminin chains according to Burgeson *et al.*[22] has been adopted. Old designations are A for α, B1 for β, and B2 for γ. Arabic numbers indicate isoforms: α1, β1, and γ1 for EHS laminin, α2 for merosin, β2 for S-laminin, and so on.

[22] R. E. Burgeson, M. Chiquet, R. Deutzmann, P. Ekblöm, J. Engel, H. Kleinman, G. R. Martin, J.-P. Ortonne, M. Paulsson, J. Sanes, R. Timpl, K. Tryggvason, Y. Yamada, and P. D. Yurchenco, *Matrix Biol.* (in press) (1994).

[23] E. Engvall, D. Earwicker, T. Haaparanta, E. Ruoslahti, and J. R. Sanes, *Cell Regul.* **1,** 731 (1990).

[24] D. D. Hunter, V. Shah, J. P. Merlie, and J. R. Sanes, *Nature (London)* **338,** 229 (1989).

[25] A. Lindblom, T. Marsh, C. Fauser, J. Engel, and M. Paulsson, *Eur. J. Biochem.* **219,** 383 (1994).

[26] D. Edgar, R. Timpl, and H. Thoenen, *J. Cell Biol.* **106,** 1299 (1988).

[27] G. E. Davis, M. Mathorpe, E. Engvall, and S. Varone, *J. Neurosci.* **5,** 2662 (1985).

[28] P. Rouselle, G. P. Lunstrum, D. R. Keene, and R. E. Burgeson, *J. Cell Biol.* **114,** 567 (1991).

[29] M. P. Marinkovich, P. Verrando, D. R. Keene, G. Meneguzzi, G. P. Lunstrum, J. P. Ortonne, and R. E. Burgeson, *Lab. Invest.* **69,** 295 (1993).

[30] P. Kallunki, K. Sainio, R. Eddy, M. Byers, T. Kallunki, H. Sariola, K. Beck, H. Hirvonen, T. B. Shows, and K. Tryggvason, *J. Cell Biol.* **119,** 679 (1992).

[31] K. Beck, R. A. McCarthy, M. Chiquet, L. Masuda-Nakagawa, and W. K. Schlage, *in* "Cytoskeletal and Extracellular Proteins" (U. Aebi and J. Engel, eds.), p. 102. Springer-Verlag, Heidelberg, 1989.

[32] T. Pikkarainen, T. Schulthess, J. Engel, and K. Tryggvason, *Eur. J. Biochem.* **209,** 571 (1992).

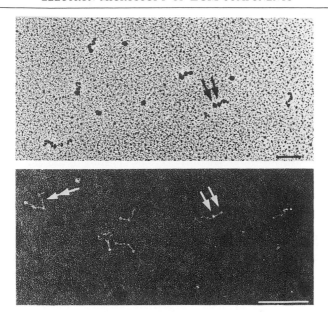

FIG. 2. Electron microscopy of recombinantly prepared β1 chains after rotary shadowing. The long arm is not visible because it is unfolded in the absence of the other partner chains. The short-arm fragments appear to be properly folded. Bar: 100 nm. (Reproduced with permission from Pikkarainen et al.[32])

can be derived by electron microscopy, but the long-arm regions remained unfolded in the absence of a partner chain (Fig. 2).[32] This was manifested by their essential invisibility in the electron micrographs. The long arms of all laminin chains contain heptad repeats that are designed to interact with related repeats of the partner chains to form double- or triple-stranded α-helical coiled coil regions. This process of chain combination is highly specific and the formation of homoassociates is not possible.[6] Double- and triple-stranded coiled coil domains are clearly visualized by electron microscopy as rodlike structures of somewhat different diameter. Therefore electron microscopy was used extensively as an analytical tool to monitor chain assembly.[6,33] It was even possible to monitor the kinetics of the association of an α chain to a preformed double-stranded B1–B2 dimer by this technique.[33] By fast spraying of the specimen solution[6] the kinetics of the assembly was monitored with a dead time of 10 sec. A half-time of 10 min was determined for the assembly of the α chain, in

[33] I. Hunter, T. Schulthess, M. Bruch, K. Beck, and J. Engel, Eur. J. Biochem. 188, 205 (1990).

agreement with the half-time determined spectroscopically by circular dichroism measurements in solution.[6]

Procedures

Rotary shadowing after spraying from buffer–glycerol mixtures was performed as described earlier.[1] In contrast to earlier reports, it is not always necessary to use volatile buffers such as bicarbonate, acetate buffer, or acetic acid. Dilute buffers of less than 10 mM salt concentrations could also be used without visibility of salt crystals in the electron micrographs. Urea concentrations of up to 2 M were tolerable with a sacrifice of resolution. Kinetic measurements were performed by fast mixing with glycerol and quick spraying[5,6] with the nebulizer for small volumes as described.[1] Both steps could be performed within 10 sec. From comparisons with spectroscopic measurements, it was concluded that the process of laminin chain assembly[6] (or collagen triple-helix refolding[5]; see below) was stopped by the absorption to the mica surface. The time of handling of the mica chips before drying in vacuum did not have an effect on the results of the kinetic analysis.

Two modifications of the glycerol spraying/rotary shadowing technique were proposed. In the mica sandwich squeezing technique[34] a drop of the solution in glycerol–buffer was squeezed between two freshly cleaved mica plates instead of being sprayed by a vaporizer. This method was developed to minimize shear forces. We were unable to observe shear fragmentation by our vaporizer[1] even for the very long cuticle collagen of worms,[35] but we observed different spreading of glycosaminoglycan chains and a better preservation of proteoglycan aggregates by the squeezing technique.[36] Mica centrifugation[37] was also less destructive than was spraying in the case of protoglycan aggregates.[36]

Negative staining with uranyl acetate or uranyl formate is a classic technique used, with several modifications, by many workers. In our laboratory we use the following procedure: A fresh uranyl formate solution is prepared by dissolving 37.5 g of the stain in 5 ml of boiling water. The solution is stirred in the dark for 20 min, after which 5 to 10 μl of 5 M NaOH solution is added. A small color change from yellow to dark yellow is observed. A solution with greenish color should be discarded. The solution is filtered through a 0.22-μm pore size Millipore (Bedford, MA)

[34] P. Mould, D. Holmes, K. Kadler, and J. Chapman, *J. Ultrastruct. Res.* **91**, 66 (1985).
[35] F. Gaill, H. Wiedemann, K. Mann, K. Kühn, R. Timpl, and J. Engel, *J. Mol. Biol.* **221**, 209 (1991).
[36] M. Mörgelin, J. Engel, D. Heinegard, and M. Paulsson, *J. Biol. Chem.* **267**, 14275 (1992).
[37] R. Nave, D. O. Fürst, and K. J. Weber, *Cell Biol.* **109**, 2177 (1989).

filter and used for 1 day only. Glow-discharged ultrathin carbon-coated grids are held with a pair of tongs. First, 5 to 10 μl of the sample solution is applied, followed by two washes (30 sec) with a drop of water each, a short wash with a drop of the staining solution, and staining for 10 sec with a drop each of the uranyl formate solution. Between steps and at the end surplus solution is removed by soaking with filter paper.

Negative staining is less suited for tracing elongated thin structures, for example, collagen triple helices, over their entire length probably because of uneven stain deposition. The technique, however, provides a higher resolution than the conventional rotary shadowing technique. Diameters are represented approximately correctly. They appear enlarged by about 2.5 nm owing to metal crystallite deposition after rotary shadowing (compare Fig. 1A and B).

Domain Organizations of Collagens

Electron microscopy was of great help for the elucidation of the domain organization of a number of newly discovered collagens.[38] As an example, a negatively stained image of the large form of type XII collagen[39] is shown (Fig. 3) together with a diagram of its structure, mainly derived by comparing the sequence[40] with electron microscopic information.[39,41]

Type XII collagen together with types IX and XIV collagens belong to the subgroup of proteins called fibril-associated collagens with interrupted triple helices (FACIT).[43] Their collagenous domain contributes only a small weight fraction and other globular domains such as fibronectin type III domains and von Willebrand factor type A domains form the larger part.

Note that by negative staining (Fig. 3) the collagen triple helix appears with a realistic diameter of only 1.5 nm. Also, the noncollagenous domain appears with a diameter of about 2.5 nm, which is close to that determined by NMR.[4] In contrast, in images produced after rotary shadowing[38,41] all elongated structures appear with diameters about 2.5 nm thicker, owing to metal crystallite decoration. The von Willebrand type A domains are detectable by negative staining (Fig. 3), but not by rotary shadowing.[41]

[38] M. Van der Rest and R. Garrone, *FASEB J.* **5**, 2814 (1991).
[39] M. Koch, C. Bernasconi, and M. Chiquet, *Eur. J. Biochem.* **207**, 847 (1992).
[40] M. Yamagata, K. M. Yamada, S. S. Yamada, T. Shinomera, H. Tanaka, Y. Nishida, M. Obara, and K. Kimata, *Cell Biol.* **115**, 209 (1991).
[41] G. P. Lundstrum, N. P. Morns, A. M. McDonough, D. R. Keene, and R. E. Burgeson, *Cell Biol.* **113**, 963 (1991).
[42] L. W. Murray, J. Engel, and M. L. Tanzer, *Ann. N.Y. Acad. Sci.* **460**, 478 (1985).
[43] L. M. Shaw and B. R. Olson, *Trends Biochem. Sci.* **16**, 191 (1991).

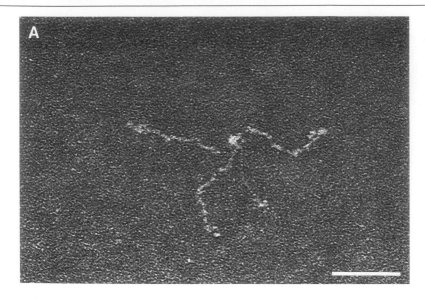

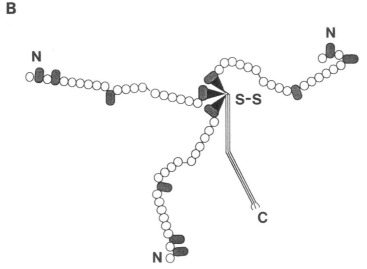

FIG. 3. (A) Electron micrograph of a negatively stained molecule of the long form of collagen XII. Bar: 50 nm. (B) Domain organization according to the sequence data of Yamagata *et al.*[40] Open circles designate fibronectin type III domains, closed ovals are von Willebrand factor type A domains, triangles indicate regions homologous to the NC3 domain of collagen IX, and the three parallel lines stand for the collagen triple helix with a region homologous to NC1 of collagen IX at its C terminus. (Original work was by Koch *et al.*[39] and the figure was provided by these authors.)

Electron micrographs were also taken from the cuticle collagens of several deep sea worms[35] (Fig. 4). They revealed triple-helical domains of unusually long length (2.4 μm), which were earlier found for the collagen of a sandworm.[42] Interestingly, the collagens after nondenaturing isolation revealed terminal globular domains, and dimerization and multimerization of these domains was frequently observed. These processes

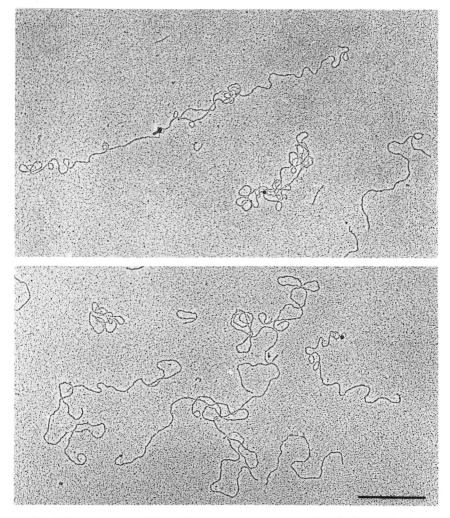

FIG. 4. Cuticle collagen of *Alvinella pompejana*, a worm that lives in deep sea hydrothermal vents. Electron micrographs were taken after rotary shadowing. Bar: 500 nm. (Original work was by Gaill *et al.*[35])

may be important for the formation of the complex cuticle organization.

Interactions between Collagen Molecules

The noncollagenous domains of FACIT molecules may mediate the interaction of collagen fibers with other ECM components. The first collagen region (COL1) near the C terminus apparently assembles to a quarter-staggered collagen II fibril and the other (COL2) domain may serve as a rigid arm that projects out of the fibril. These functional features were suggested by investigating the best studied member of the FACIT class, namely collagen IX,[9,38,42] in a complex with fibrillar type II collagen by electron microscopy (Fig. 5). The Col1 domain of collagen IX is homologous to that of collagen XII. It is clearly seen in Fig. 5 that this domain attaches to the type II collagen fibril. The molecules are aligned along the fibril surface, being separated by a regular spacing of 65 nm, corresponding to the D period of quarter-staggered fibers[38] (Fig. 5). The electron micrograph of the complex of type IX collagen with type II collagen is an interesting example of a specimen that was obtained after mechanical separation (in this example by vigorous vortexing of cartilage) instead of biochemical separation. In the present case the complex was stable because of disulfide linkage between collagens IX and II. It even persisted treatment with 8 M urea, which was used to free the cartilage fibrils of single collagen molecules.

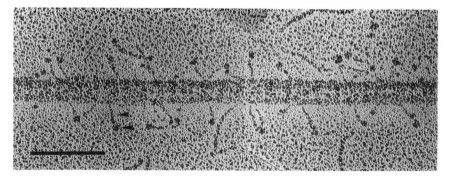

FIG. 5. Rotary shadowed cartilage fibril with collagen IX attached. Arrowheads indicate kinks in the otherwise relatively rigid collagenous domain. The terminal globe represents the N-terminal globular domain of collagen IX. Bar: 100 nm. (Original work was by Vaughan et al.[9] and the figure was kindly provided by L. Vaughan.)

A new helical model of nucleation and propagation was proposed for the formation of fibrils of collagen I.[44] This model is based on the experimental observation of symmetrical pointed tips of different mass-per-length ratios. The accurate determination of these mass-per-length ratios by STEM[45] was essential for establishing the details of this model.[46] It was found that the α tip was nearly paraboloidal, as indicated by a decrease in mass toward the tip that was an average of 17 collagen molecules per D period (640 nm) and that was linear over 100 or more D periods. The slope of the β tips varied from 55 to 200 molecules per D period. Mass determinations of unstained particles by STEM were also essential for the determination of sheetlike assemblies of pN-collagen, pC-collagen, and procollagen.[13]

In vitro studies on the interactions of collagen IV molecules in the formation of the chickenwire-like network in basement membranes have been reviewed.[1] The visualization of this network in the almost intact organization of the amnion at high resolution[47] is perhaps the most advanced study of a complex extracellular matrix by electron microscopy. The applied method was similar to the quick-freeze, deep-etch electron microscopy of Heuser.[48] With this method excellent results were obtained on the organization of elastic fibers in tissues.[49]

Monitoring Folding of Collagen Triple Helices by Electron Microscopy

Native and unfolded collagens can be conveniently distinguished by the rotary shadowing technique. Unfolded, randomly coiled polypeptide chains with a molar mass-per-length ratio of only 300 Da nm^{-1} are not (or only partially) detectable by this technique, whereas the triple helix (1000 DA nm^{-1}) is clearly seen. It was therefore possible[5] to distinguish intact and unfolded type IV collagen in partially folded intermediates during the time course of the refolding process (Fig. 6) The C-terminal noncollagenous hexameric domain NC1, by which two type IV collagens are connected,[1] is visible even before refolding of the triple helix has started (Fig. 6, $t = 0$). The electron micrographs taken at different times after the start of refolding revealed that triple-helix formation only started

[44] D. Silver, J. Miller, R. Harrison, and D. J. Prockop. *Proc. Natl. Acad. Sci. U.S.A.* **89,** 9860 (1992).

[45] A. Engel and C. Colliex, *Curr. Opin. Biotechnol.* **4,** 403 (1993).

[46] D. F. Holmes, J. A. Chapman, D. J. Prockop, and K. Kadler, *Proc. Natl. Acad. Sci. U.S.A.* **89,** 9855 (1992).

[47] P. D. Yurchenco and G. C. Ruben, *J. Cell Biol.* **105,** 2559 (1987).

[48] J. E. Heuser, *Methods Cell Biol.* **22,** 97 (1981).

[49] R. P. Mecham and J. E. Heuser, *Connect. Tissue Res.* **24,** 83 (1990).

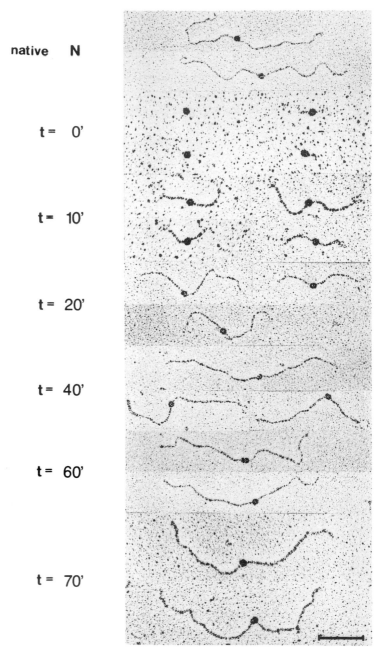

native   N

t =   0'

t =   10'

t =   20'

t =   40'

t =   60'

t =   70'

FIG. 6. Refolding of collagen IV as monitored by electron microscopy. Selected molecules are shown that were observed at the indicated times of refolding. They are compared with the native starting material (N). Samples were heated to 50° for 40 min for unfolding and refolding was performed at 25°. Bar: 100 nm. (Reproduced with permission from Doelz *et al.*[5])

at the NC1 domain, which apparently serves as a nucleus. The kinetics of refolding measured by electron microscopy agreed with that measured by the time course of circular dichroism, demonstrating the possibility of using electron microscopy for monitoring relatively slow kinetic processes (see Procedure, below). A similar approach was successfully applied to monitor the unfolding of the muscle protein myosin.[50]

Visualization of Collagen Molecules by Scanning Probe Microscopy

In scanning probe microscopy a sensor designed as a pointed needle traces over the probe. A constant distance from the probe surface is maintained by small up-and-down movements of the needle, which are controlled by keeping a distance-sensitive signal constant. This signal is either the tunnel current in tunneling microscopy or short-range interaction forces in atomic force microscopy. A review about these and other scanning probe methods and their biological application is available.[51] These techniques are not electron microscopic but results obtained match and expand electron microscopy results. Therefore it appears to be justified to deal with the few results obtained for ECM proteins in the present chapter. Some applications to the extracellular matrix systems are presented in Cotterill et al.,[15] Guckenberger et al.,[52] and Chernoff and Chernoff.[53] In these studies the resolution of single proteins (e.g., collagen IV[52] or collagen I[53]) was lower than in comparable electron micrographs. Much improved pictures of a mutant of procollagen I were, however, obtained by dynamic force microscopy, which is a modification of atomic force microscopy (see Procedure, below). The small globular domain at the N terminus and the larger one at the C terminus are clearly represented (Fig. 7).[54] The diameter of the 300-nm long triple helix is 13 nm, which is larger by a factor of 8 than the true diameter of 1.5 nm. Overly large diameters of thin structures in horizontal directions are frequently observed in scanning probe microscopy. The effect may be attributed to the relatively large size of the needle tip and to resulting convolution of the image[16,51] The method also allows determination of the vertical height of the samples. A value of 0.3 nm was obtained for the collagen triple helix in Fig. 7. The reduced height compared to the true thickness was explained by nesting of the amino acid residues to the mica surface.

[50] M. Katsuhide, J. Struct. Biol. 103, 249 (1990).
[51] A. Engel, Annu. Rev. Biophys. Chem. 20, 79 (1991).
[52] A. Guckenberger, W. Wiegräbe, and W. Baumeister, J. Microsc. (Oxford) 152, 795 (1988).
[53] E. A. Chernoff and D. A. Chernoff, J. Vac. Sci. Technol., A [2] 10, 596 (1992).
[54] A. M. Romanic, L. Spotila, E. Adachi, J. Engel, Y. Hojima, and D. J. Prockop, J. Biol. Chem. 269, 11614 (1994).

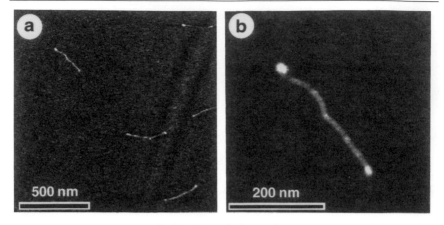

FIG. 7. Procollagen I visualized by dynamic force microscopy. The human procollagen was isolated from cultured skin fibroblasts of proband homozygous for a mutant that substituted serine for glycine at position 661 in the a2(I) chain.[54] (a) Field of molecules in which the height of the molecules is gray scale encoded. (b) One of the procollagen molecules at higher magnification. (Original work was by Anselmetti et al.[16] and the figures were kindly provided by D. Anselmetti.)

Scanning probe microscopic methods have not yet contributed key information for ECM proteins that had not already been obtained by electron microscopy. Their potential is, however, high because of the possibility to scan directly and *in situ* hydrated samples at normal pressure, and even to work under water. It can also be hoped that it will be possible to develop the methods further and to obtain as high resolutions for biological molecules as for metals and semiconductors.

Procedure

For visualization of procollagen I by dynamic force microscopy a 10-μg/ml solution of the protein in a 1 : 1 v/v mixture of 0.2 M ammonium bicarbonate and glycerol or in dilute (50 mg/ml) acetic acid is sprayed onto freshly cleaved mica. The samples are dried in high vacuum (10^{-6} torr) overnight. The mica chips are mounted with double-sided adhesive tape to the sample stage of a home-built noncontact dynamic force microscope. Comparable instruments are manufactured by Burleigh Instruments, Fisher (Pittsburgh, PA), Digital Instruments (Santa Barbara, CA) Omicron Vacuumphysik (Taunusstein, Germany), and by other companies. The sensor consists of a cantilever with an integrated tip (Nanoprobe, Wetzlar-Blankenfeld, Germany) and is driven close to its mechanical resonance frequency (150 to 600 kHz) by means of a piezoelectric transducer

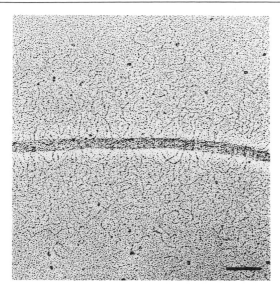

FIG. 8. Proteoglycan coating of collagen fibrils. Rotary shadowing of a collagen fibril isolated from the chick vitreous humor of an adult chicken demonstrated an extensive coat of chondroitin sulfate chains that extended for about 200 nm from the surface of the fibril. (Reproduced with permission from Wright and Mayne.[55])

with an amplitude of 0.5 to 2 nm. For further instrumental details, see Anselmetti *et al.*[16] Samples are investigated at normal pressure and at room temperature. Because of the low forces in dynamic force microscopy no staining or stabilization of the samples is necessary.

Interactions of Collagens with Proteoglycans

It is often difficult to define the interaction partners of ECM components under physiological conditions. Electron microscopy may be helpful if applied to mixtures of components that were not or were only partially purified. An example for this approach is the study of the complex between collagens II and IX (Fig. 5). By similar techniques complexes between collagen and chondroitin sulfate proteoglycans were demonstrated and thus defined in the vitreous body.[10,55] The complexes (Fig. 8) were seen in mixtures obtained after crude sonication and extraction of the vitreous. The authors proposed that the ~200-nm-long chondroitin sulfate chains originate from highly ordered arrays of collagen IX, which contains a

[55] D. W. Wright and R. Mayne, *J. Ultrastruct. Mol. Struct. Res.* **100**, 224 (1988).

glycosaminoglycan chain and is thus a proteoglycan.[10,55] It should be mentioned that complexes between tenascin and collagen were demonstrated[55] in a similar way. In the case of the interaction of decorin with collagen I the authors were able to define the interaction site in the collagen gap region by specific staining of collagen fibers in tissues. The labels used were either a dermatan sulfate-specific dye[56] or a decorin-specific antibody.[57] The relatively high resolution obtained for the supramolecular assemblies typical for the selected examples. Therefore not only was complex formation demonstrated but valuable information on the mode of assembly was obtained.

Interaction of Aggrecan with Hyaluronate

Electron microscopy of proteoglycans has revealed a wealth of structural information on this class of ECM components. A most intensively studied example is aggrecan, which is the major aggregating proteoglycan of cartilage, and this work has been reviewed.[58] It led to an elucidation of the domain structure of aggrecan with two globular domains, G1 and G2, at the N terminus, an extended domain E1 connecting them, and a second extended domain, E2, which connects the C-terminal globular domain G3 with G2. G1 and G2 are both homologous to link protein but only G1 was found to bind the hyaluronate. No interactions with other matrix molecules have been observed so far for G2 and the lectin-like domain G3.

In this chapter a rather unusual application of electron microscopy for the titration of hyaluronate by the hyaluronate binding fragment G1 is demonstrated. In Fig. 9 images of the complexes are shown that were obtained by rotary shadowing at different molar ratios of hyaluronate to binding fragment.[7] Using large excesses of hyaluronate, isolated binding regions decorate the hyaluronate strands. Distances between them were shown to be statistical. Increasing concentrations of binding regions increase the density of bound fragments and free binding regions also became apparent. With a large excess of binding region the hyaluronate strands were completely decorated and many free binding regions were visible in their surrounding. The center-to-center distances between binding regions at the hyaluronate corresponded to 12 disaccharide units. Therefore hyaluronate concentrations are expressed in Fig. 9 as concentrations of seg-

[56] J. E. Scott and C. R. Orford, *Biochem. J.* **197,** 213 (1981).

[57] R. Fleischmajer, L. W. Fischer, E. D. MacDonald, J. R. L. Jacobs, J. S. Perlish, and J. D. Termine, *J. Struct. Biol.* **106,** 82 (1991).

[58] M. Mörgelin, D. Heinegard, J. Engel, and M. Paulsson, *Biophys. Chem.* **50,** 113 (1994).

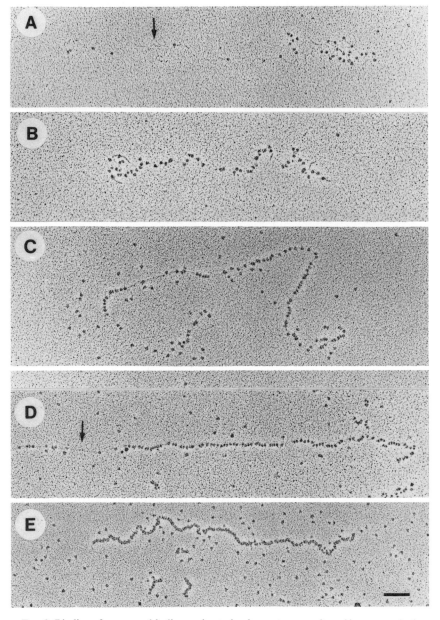

FIG. 9. Binding of aggrecan-binding region to hyaluronate as monitored by rotary shadowing electron microscopy. The molar ratios of binding region fragments to hyaluronate segments (12 disaccharide units) were 0.07 : 1 (A), 0.17 : 1 (B), 0.34 : 1 (C), 1 : 1 (D) and 4.8 : 1 (E). Arrows represent free hyaluronate strands. Bar: 100 nm. (Reproduced with permission from Mörgelin *et al.*[7])

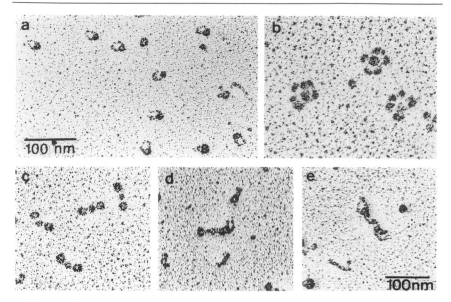

FIG. 10. Structure of integrin $\alpha IIb\beta 3$ and its complexes with fibrinogen. The electron micrographs were taken after rotary shadowing. (a) Monomeric integrin in the presence of Triton X-100 (10 mg/ml). (b) Multimeric integrin in the absence of detergent. (c) Monomeric fibrinogen. (d) Integrin with fibrinogen bound at one side. (e) Integrin with fibrinogen bound at both sides. (Reproduced from the Ph.D. thesis of Beate Mueller, University of Basel, 1993.)

ments with 12 units. It was possible to estimate a binding constant from the experiment shown in Fig. 9 that was in reasonable agreement with values determined by more conventional methods. The reasonable agreement may point to the fact that the equilibrium state between reaction partners was frozen by absorption to the mica without major disturbances of the equilibrium. Furthermore, by the spraying technique all species are absorbed and their ratios can therefore be quantitated in the pictures. For the evaluation of the binding constants only ratios were taken from the electron micrographs and concentrations were those in solution before spraying. The same study[7] also found that in the presence of link protein binding regions bind cooperatively to hyaluronate and the structure of the protein filament is changed.

Structure of Integrin in Solubilized and Membrane-Bound Form

The most detailed electron microscopic investigation of an integrin (fibronectin receptor $\alpha 5\beta 1$) in solubilized form was performed by Nermut

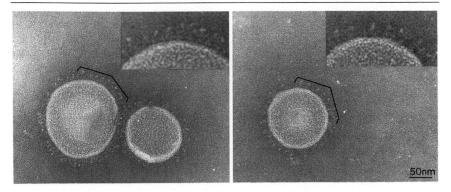

FIG. 11. Negatively stained electron micrographs of integrin αIIbβ3 reconstituted into lipid vesicles. The integrin receptor was reconstituted into DMPC/DMPG vesicles (1 : 1, mol/mol) and the preparation was negatively stained at neutral pH with sodium phosphotungstate. Insets are twofold enlargements of the indicated border region of the vesicles. (Original work was by Müller *et al.*[62] and the figures were kindly provided by the authors.)

et al.[59] The receptor consists of an 8- to 12-nm-thick globular head and two ~0.2-nm-thick tails 18 to 20 nm in length. The globular regions of the two subunits α and β are apparently intimately connected but their taillike regions are separated. These are terminated by the two membrane-spanning regions of the receptor. Frequently these appeared to be connected in solubilized integrins, demonstrating their hydrophobic character. In particular, when the detergent was removed the membrane-spanning regions of different molecules join, giving rise to the formation of integrin aggregates that were called rosettes.[59]

In Fig. 10a and b, electron micrographs of the platelet fibrinogen receptor αIIbβ3 are shown, with features resembling those found for the fibronectin receptor. It was possible to visualize the interaction of integrins with their ligand proteins, as demonstrated for αIIbβ3–fibrinogen in Fig. 10c and d. A detailed study of the same system was performed by Weisel *et al.*[60] Only a small fraction of complexes was observed in the presence of detergent at the low concentrations (0.01 to 0.1 μM) used in electron microscopy. This indicates a low binding constant for the detergent-solubilized monomers. The extent of binding was much increased when rosettes were formed after removal of detergent. Another interesting feature is the observation of the binding of (at maximum) two fibrinogen molecules to an integrin (see Fig. 10 and Weisel *et al.*[60]). According to biochemical

[59] M. V. Nermut, N. M. Green, P. Eason, S. S. Yamada, and K. M. Yamada, *EMBO J.* **7**, 4093 (1988).

[60] J. W. Weisel, C. Nagaswami, G. Vilaire, and J. S. Bennet, *J. Biol. Chem.* **267**, 16637 (1992).

data fibrinogen contains four to six potential binding sites, of which two contain an RGD sequence.[61] These are apparently not all used, probably because of differences in binding affinity.

Integrin $\alpha IIb\beta 3$ was reconstituted into lipid bilayers and could as such be visualized in membrane-bound form[62] (Fig. 11). For optimal results the reconstitution procedure had to be optimized and negative staining had to be performed at neutral pH (see Procedure, below). It can be seen from Fig. 11 that the integrin molecules point out from the membrane with their rather rigid tails thus keeping the globular domains at a distance of about 15 nm from the membrane surface. This geometry is probably of advantage for an easy interaction with the large protein ligands. Also, in a natural membrane steric protection of the ligand-binding site in the globular integrin domain by other membrane proteins or proteoglycans may thus be avoided. There are no strong mutual interactions between the integrins that would lead to cluster formation. Only infrequently are integrin-coated vesicles attached to each other. This indicates a lack of potential for homoaggregation.

Procedure

Equal amounts of vacuum-dried 1,2-dimyristoylglycero-3-phospho-choline (DMPC) and 1,2-dimyristoylphosphatidylglycerol (DMPG) are dissolved in an 10-mg/ml solution of Triton X-100 in 20 mM Tris-HCl (pH 7.4), 50 mM NaCl, 0.5 mM CaCl$_2$. Integrin dissolved in the same solvent is added. Triton X-100 is removed by two additions of Bio-Beads SM-2 (Bio-Rad, Richmond, CA), which are washed before use with methanol and water as described.[63] Vesicles are separated from free protein by loading the sample on top of a stepwise sucrose gradient [2, 1, 0.6 and 0.4 M in 20 mM Tris-HCl (pH 7.4), 50 mM NaCl, 0.5 mM CaCl$_2$] and centrifuging at 275,000 g at 4° in an ultracentrifuge for 24 hr. The vesicle band at about 1 M sucrose is detected by visual inspection, collected, and dialyzed against the same buffer. For negative staining the vesicle suspension is diluted to about 10 μ/ml. About 10 μl of the diluted solution is put on a glow-discharged collodium and carbon grid. After about 4 min, 5 μl of sodium phosphotungstate solution (0.2 g/ml, pH 7) is added for 0.5 min. After removal of the first stain, incubation is repeated with 10 μl of the same solution for about 5 min.

[61] I. F. Charo, L. Nannizzi, D. R. Phillips, M. A. Hsu, and R. M. Scarborough, *J. Biol. Chem.* **266,** 1415 (1991).
[62] B. Müller, H.-G. Zerwes, K. Tangemann, J. Peter, and J. Engel, *J. Biol. Chem.* **268,** 6800 (1993).
[63] P. W. Holloway, *Anal. Biochem.* **53,** 304 (1973).

[23] Basement Membrane Assembly

By Peter D. Yurchenco and Julian J. O'Rear

Introduction

Mass action-driven self-assembly, in which macromolecular protomers interact to form higher ordered structures, is a paradigm for the formation of the architectural frameworks of animal tissues. This paradigm is illustrated by the basement membranes, specialized cell-associated extracellular matrices that form perm-selective barriers between cell compartments and that provide cell-interactive surfaces. By electron microscopy these structures appear as fine mesh networks. Compositional analysis has revealed that basement membranes contain one or several isoforms of the laminin glycoprotein family, type IV collagens, entactin/nidogen, heparan sulfate proteoglycan (perlecan), and other components. A challenge therefore has been to explain how these protomers join together to form mesh-like molecular sieves in glomerular and other basement membranes capable of excluding proteins the size of albumin or smaller proteins with net negative charges.

From studies of basement membrane architecture *in situ*, the reconstitution of polymers *in vitro*, and the molecular morphology and binding repertoire of basement membrane fragments, hypotheses for the assembly and supramolecular organization of these matrices have evolved.[1,2] One hypothesis, which we consider a working model, holds that both strong and weak molecular interactions contribute to final architecture and that both laminin and type IV collagen form network-like polymers[2,3] bridged by nidogen.[4] In this model, the laminin network is formed by reversible interactions in which the short arms join with each other in the presence of divalent cation.[5] Furthermore, the collagenous network, which is the principal basis for structural stability, is one in which an irregular branching array of filaments is formed not only by covalently stabilized N-terminal 7S bonds (joining of four triple-helical ends) and C-terminal

[1] R. Timpl, *Eur. J. Biochem.* **180,** 487 (1989).
[2] P. D. Yurchenco and J. C. Schittny, *FASEB J.* **4,** 1577 (1990).
[3] P. D. Yurchenco, Y.-S. Cheng, and H. Colognato, *J. Cell Biol.* **117,** 1119 (1992).
[4] J. W. Fox, U. Mayer, R. Nischt, M. Aumailley, D. Reinhardt, H. Wiedemann, K. Mann, R. Timpl, T. Krieg, J. Engel, and M.-L. Chu, *EMBO J.* **10,** 3137 (1991).
[5] P. D. Yurchenco and Y.-S. Cheng, *J. Biol. Chem.* **268,** 17286 (1993).

Copyright © 1994 by Academic Press, Inc.
All rights of reproduction in any form reserved.

head-to-head dimeric bonds,[6] each separated by about 800 nm of contour length, but by lateral associations as well, which provide for a much tighter network.[7,8] Whereas the N- and C-terminal bonds become covalently cross-linked, lateral joinings appear to be formed only through noncovalent and thermally reversible interactions. Many important biochemical and architectural features, such as the interacting sequences for laminin polymerization and collagen lateral associations, the bonds that immobilize perlecan firmly in the network, the mechanism of collagenous binding to nidogen/entactin, and the structural variations caused by protomeric isoforms and other components, remain to be worked out.

In this chapter we review some of the experimental methods that have been used to study basement membrane components, their interactions, and their supramolecular structure. Many of these methods have been tailored to study large, conformationally active multidomain protomers with extensive posttranslational modification whose functions can be lost following denaturation and whose binding interactions span a wide range of affinities. They include transmission electron microscopy (TEM), gel filtration, sedimentation, turbidity, solid phase binding, and antibody binding and inhibition assays. Although the fine network structure of basement membranes is partially visualized by conventional TEM, TEM used in conjunction with platinum metal replication has been invaluable to study both protein monomeric and polymer structure. This method, however, has been most powerful when complemented by biochemical studies of the interactions of individual components. Evaluation of binding interactions has generally been easier when these interactions have been of high affinity. The study of higher ordered structure resulting from low-affinity interactions, however, has required special approaches to circumvent the problems of dissociation of species during analysis. Equilibrium approaches permitting quantitation of these interactions have included isotopic modifications of the original equilibrium gel filtration developed by Hummel and Dryer,[9] affinity retardation chromatography,[10] affinity coelectrophoresis,[11] and classic equilibrium centrifugation. Localization of binding activities has presented its own special challenges. A common strategy has been to identify the interactions with the complete protein and then proceed to follow the interaction through smaller and smaller

[6] R. Timpl, H. Wiedemann, V. van Delden, H. Furthmayr, and K. Kühn, *Eur. J. Biochem.* **120**, 203 (1981).

[7] P. D. Yurchenco and H. Furthmayr, *Biochemistry* **23**, 1839 (1984).

[8] P. D. Yurchenco and G. C. Ruben, *J. Cell Biol.* **105**, 2559 (1987).

[9] J. P. Hummel and W. J. Dryer, *Biochim. Biophys. Acta* **63**, 530 (1962).

[10] J. C. Schittny and P. D. Yurchenco, *J. Cell Biol.* **110**, 825 (1990).

[11] M. K. Lee and A. D. Lander, *Proc. Natl. Acad. Sci. U.S.A.* **88**, 2768 (1991).

fragments, with final localization using small synthetic peptides. Activity can be lost during fragmentation and small natural or synthetic peptides, although providing site-specific reagents, often lack proper conformation and frequently cannot be used to identify active sites in proteins. Genetic engineering of recombinant proteins is proving to be a powerful tool with which to investigate structure and self-assembly.[4,12,13] If a correctly folded glycoprotein can be generated, then molecular genetics can be employed to map and characterize structural and functional binding sites.

Our focus is on components found in the Engelbreth-Holm–Swarm (EHS) tumor,[14] because it is an abundant source[15] that has been subjected to intensive study. However, the reader should realize that the EHS tumor possesses only one each of the genetic variants of laminin and type IV collagen. We discuss laminin, nidogen/entactin, perlecan, and type IV collagen.

Basement Membrane Components

Purification of Basement Membrane Components from Engelbreth-Holm–Swarm Tumor

The EHS tumor is grown and passaged subcutaneously in mice. Although C57BL/6 mice have generally been used, we find the tumor yields are higher (10–20 g/mouse) when maintained in Swiss-Webster mice. For isolation of type IV collagen the animals are made lathyritic with 0.3% (w/v) β-aminopropionitrile in their drinking water.

Laminin, Nidogen/Entactin, and Fragments. Laminin is a major and ubiquitous component of basement membranes. The "classic" variant of laminin, first characterized in the mouse EHS tumor, consists of three polypeptide chains (A, B1, and B2) that are linked together through noncovalent and disulfide bonds to form an asymmetrical four-armed molecule with terminal and included globular domains in the three "short" arms and a larger oblong globule at the end of the long arm (Figs. 1 and 2). Variant forms of the three chains, which are expressed in different tissues at different times of development, include B1s in the glomerulus and

[12] R. Nischt, J. Pottgiesser, T. Krieg, U. Mayer, M. Aumailley, and R. Timpl, *Eur. J. Biochem.* **200**, 529 (1991).

[13] P. D. Yurchenco, U. Sung, M. Ward, Y. Yamada, and J. J. O'Rear, *J. Biol. Chem.* **268**, 8356 (1993).

[14] R. W. Orkin, P. Gehron, E. B. McGoodwin, G. R. Martin, T. Valentine, and R. Swarm, *J. Exp. Med.* **145**, 204 (1977).

[15] H. K. Kleinman, M. L. McGarvey, L. A. Liotta, P. G. Robey, K. Tryggvason, and G. R. Martin, *Biochemistry* **21**, 6188 (1982).

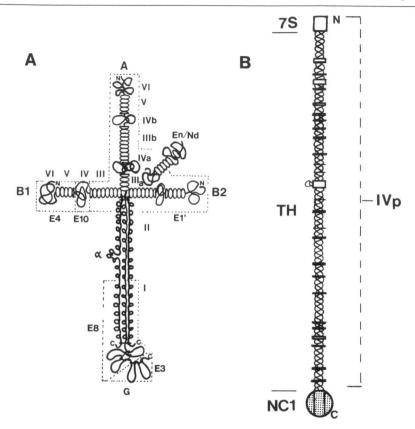

Fig. 1. Laminin and type IV collagen. (A) Diagram of laminin and its fragments. EHS laminin contains B1, A, and B2 chains joined together to form a four-armed structure. Entactin/nidogen (En/Nd) is commonly found bound to the B2 short arm. Elastase, cathepsin G, and pepsin have each been used to generate a series of fragments corresponding to different regions of laminin. The boundaries of the major elastase (E) fragments are shown. Domains indicated by roman numerals and G. Amino and carboxyl termini indicated with N and C, respectively. (B) Diagram of type IV collagen. 7S-forming domain (7S), triple-helical region (TH) containing interruptions (horizontal lines and open bars), and noncollagenous globular domain (NC1) indicated. A large pepsin fragment (IVp) that lacks the globular domain and forms tetramers is shown.

neuromuscular synapse, Am in muscle and placenta, B2t near hemidesmosomes in epithelia,[16] and B1-2 discovered in eye.[17]

Purification steps are carried out at 0–5° from lathyritic EHS tumor, using a procedure modified from that of Paulsson[18] and scaled up to permit

[16] E. Engvall, *Kidney Int.* **43,** 2 (1993).
[17] J. J. O'Rear, *J. Biol. Chem.* **267,** 20555 (1992).
[18] M. Paulsson, *J. Biol. Chem.* **263,** 5425 (1988).

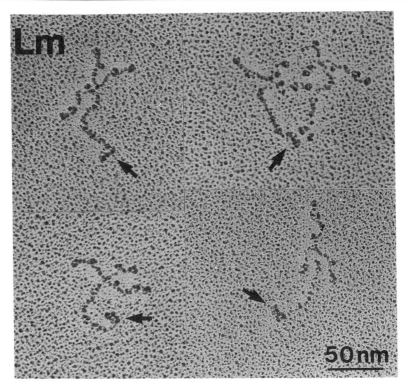

FIG. 2. Electron micrograph of laminin. Laminin (Lm) purified from the EHS tumor sprayed onto mica in glycerol and rotary shadowed at a low angle with Pt/C. Arrows indicate location of G domain at end of long arm.[13]

sufficient protein for the generation of milligram amounts of fragments (Fig. 3). From 150 to 200 g of frozen tumor is homogenized in 0.1 M NaCl, 50 mM Tris-HCl, pH 7.4, containing 0.5 mM diisopropylfluorophosphate (DFP), 0.5 mM phenylmethylsulfonyl fluoride (PMSF), and p-hydroxymercuribenzoic acid (HMB; 10 μg/ml). After washing the pellet by centrifugation [12,000 rpm for 30 min in a refrigerated Sorvall (Newtown, CT) GSA rotor], the tumor matrix is twice extracted (several hours) with 150 ml of the above buffer containing 10 mM ethylenediaminetetraacetic acid (EDTA) and centrifuged (20,000 rpm for 20 min in a Sorvall SS-34 rotor) to remove insoluble residue. The combined extracts are chromatographed (Fig. 3A) on a Sephacryl HR-400 (Pharmacia, Uppsala, Sweden) column (11.3 × 90 cm) in 0.1 M NaCl, 50 mM Tris-HCl (pH 7.4), 2 mM EDTA, 0.5 mM PMSF, and HMB (10 μg/ml). The first and major peak is then pooled and loaded onto a DEAE-Sephacel (Pharmacia) column (11.3 × 5 cm) equilibrated in 50 mM Tris-HCl (pH 7.4), 2.5 mM EDTA and 0.5

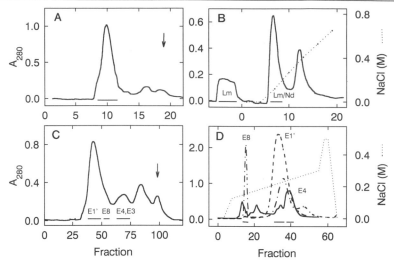

FIG. 3. Purification of laminin and its elastase fragments. (A) Gel filtration (Sephacryl HR-400) of an EDTA extract of the EHS tumor. The large peak near the void volume is pooled (bar). Arrow indicates total column volume determined with phenol red. (B) DEAE-Sephacel purification of laminin. Unbound fraction is laminin without intact nidogen/entactin. First bound peak is laminin/nidogen complex. Second bound peak contains laminin, proteoglycan, and nucleic acid. The unbound and first bound peaks are pooled together for generation of fragments. (C) Gel filtration (Sepharose CL6B) of elastase digest of laminin. Fractions pooled (bars) from left to right are enriched in E1', E8 (shoulder), and E3/E4, respectively. Arrow indicates total column volume. (D) HPLC (DEAE-5PW column) purification of the E1'-enriched (–––), E8-enriched (–·–), and E3/E4-enriched (—) fractions pooled from the gel-filtration column shown in (C). Fragment E3 (peak not shown) is recovered in the unbound peaks, which elute prior to the start of the elution program. Salt gradient shown is that applied to the column.

mM PMSF (Fig. 3B). After collection of protein in an unbound peak [which contains laminin without intact nidogen/entactin species as determined by sodium dodecyl sulfate–polyacrylamide gel electrophoresis (SDS–PAGE)], the column is subjected to a linear 0 to 0.8 M NaCl gradient. The first bound peak (eluting at 0.1 to 0.22 M NaCl), consisting of intact laminin/nidogen complex, and the unbound protein are separately dialyzed into 50 mM Tris-HCl (pH 7.4), 90 mM NaCl (TBS) containing 0.1 mM EDTA and stored at or below $-70°$. The unbound and bound fractions can be combined and used for the generation of defined proteolytic fragments, as described below. Yields are about 800 mg/200 g of tumor. Nidogen/entactin can be isolated from laminin under dissociative buffer conditions. To accomplish this, the DEAE-bound fraction is adjusted to 2 M guanidine hydochloride on ice, and subjected to Sepharose CL-4B gel filtration in 2 M guanidine hydrochloride. The first peak is laminin and the second peak is nidogen/entactin. Although good yields of nidogen/entactin can be achieved, the chaotropic conditions appear to cause a partial loss of function such that the laminin polymerizes independent of divalent cation and with an altered critical concentration and the nidogen/entactin binds less well to laminin or (especially) collagen.[4] To evaluate some of the functions of nidogen/entactin properly, it has been necessary to express recombinant glycoprotein.

Laminin variants containing the Am chain have been purified (with lower yields per unit mass of tissue) from both human placenta[19] and bovine heart,[20] using similar EDTA extraction methods. Other variants, until suitable productive cell lines are found or the chains are expressed as recombinant molecules, are not available for conventional analysis.

Elastase, pepsin, cathepsin G, and other enzymes have been used to generate various characterized and defined fragments of classic laminin that can be used to localize function.[5,21–26] We describe some of the commonly used fragments (see Figs. 3 and 4). Elastase fragment E1′ is a short-arm complex of all three chains and a portion of nidogen/entactin that contains the inner EGF region of the B1 chain and the A and B2 short arms. Although there are a few internal cleavages (e.g., between domains A-VI and A-V) the complex behaves as a single species under nonreducing and physiological salt conditions. This fragment inhibits laminin polymerization and supports the adhesion of some cells. Pepsin fragment P1′ is similar but somewhat smaller, lacking short-arm globules. Unlike E1′, it has little ability to inhibit laminin polymerization. Fragment E4 consists of N-terminal B1 chain short-arm domains VI and V. It also inhibits laminin polymerization and binds to E1′. Fragment E10 (purification not described here; see Mann et al.[23]) includes most of B1 domain IV and lies between E4 and E1′. Its function is unknown. Fragment E3 consists of the distal moiety (fourth and fifth subdomains) of long-arm G domain, binds to heparin, and can support cell adhesion. Fragment E8 consists of the distal moiety of the rod of the long arm and adjacent proximal subdomains of G. It supports cell adhesion and/or migration by a large variety

[19] K. Ehrig, I. Leivo, W. S. Argraves, E. Ruoslahti, and E. Engvall, *Proc. Natl. Acad. Sci. U.S.A.* **87**, 3264 (1990).

[20] M. Paulsson and K. Saladin, *J. Biol. Chem.* **264**, 18726 (1989).

[21] U. Ott, E. Odermatt, J. Engel, H. Furthmayr, and R. Timpl, *Eur. J. Biochem.* **123**, 63 (1982).

[22] M. Paulsson, *J. Biol. Chem.* **263**, 5424 (1988).

[23] K. Mann, R. Deutzmann, and R. Timpl, *Eur. J. Biochem.* **178**, 71 (1988).

[24] R. Deutzmann, J. Huber, K. A. Schmetz, I. Oberbäumer, and L. Hartl, *Eur. J. Biochem.* **177**, 35 (1988).

[25] M. Bruch, R. Landwehr, and J. Engel, *Eur. J. Biochem.* **185**, 271 (1989).

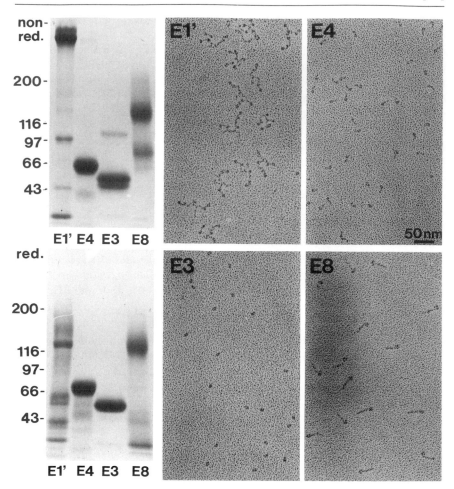

FIG. 4. Characterization of elastase fragments of laminin. *Left:* Coomassie blue-stained SDS–polyacrylamide gels (3.5–12% linear gradient) of nonreduced (upper left) and reduced (lower left) laminin elastase fragments E1' (short arm complex with globular domains), E4 (B1 short arm domains VI and V), E3 (long arm globular subdomains G4–5), and E8 (distal long arm moiety lacking globular subdomains G4–5). *Right:* Electron micrographs of Pt/C replicas of glycerol spreads (rotary shadowed at a 10° angle) of the same fragments.

of cells. Cathepsin G fragment C1–4 is a large complex consisting of all three short arms with portions of entactin/nidogen bound to the B2 short arm.[25] It polymerizes and supports some cell adhesion. Cathepsin G fragment C8–9 (purification not described here; see Bruch *et al.*[25]) consists of the entire long-arm rod and proximal portion of G domain.

Elastase fragments E1′, E4, E8, and E3 of laminin are prepared from laminin/nidogen by digestion in neutral digestion buffer [50 mM Tris-HCl (pH 7.4) at 25°), 100 mM NaCl, 2 mM EDTA] at 25° for 23 hr with elastase at a 1 : 200–1 : 300 enzyme : substrate ratio (the exact ratio is determined for each lot of enzymes in a trial digest) followed by gel filtration (Sepharose CL-6B; 2.6 × 90 cm column). The first main peak (Fig. 3C), close to the void volume, is primarily E1′. Because any residual intact laminin will coelute, it is important that digestion with enzyme has consumed all intact protein without overdegrading E1′. The second smaller "peak," seen as a shoulder on the first, is enriched in E8. The third peak contains E3, E4, E10, and other laminin and entactin/nidogen fragments. The fourth peak, nearest to the included volume, contains small uncharacterized fragments. The first three peaks are separately pooled and subjected to DEAE-5PW (Toso-Haas, Philadelphia, PA) high-performance liquid chromatography (HPLC) ion-exchange chromatography in 50 mM Tris-HCl, pH 7.4, eluting with a 0–0.5 M NaCl gradient (Figs. 3D and 4). For preparation of fragment P1′, laminin/nidogen complex is dialyzed into 10% (v/v) acetic acid and digested with pepsin [1 : 15 (w/w) enzyme : substrate ratio] for 24 hr at 15°, mixed with a twofold mass excess of pepstatin (relative to enzyme), dialyzed against 1 M calcium chloride in 50 mM Tris-HCl, pH 7.4, and purified by gel filtration through an agarose Bio-Gel A −1.5m (Bio Rad, Hercules, CA) column (2.6 × 90 cm) in the same buffer. Fragment C1–4 is generated by incubation of laminin with cathepsin G in TBS containing 1 mM calcium at an enzyme : substrate mass ratio of 1 : 300 (w/w) for 5 hr at 37°. The C1–4 precipitate is then pelleted by centrifugation, resuspended in TBS containing 10 mM EDTA, and dissolved by incubation overnight on ice.

Low-Density Heparan Sulfate Proteoglycan (Perlecan). Perlecan is a ubiquitous component of many basement membranes. It typically possesses three heparan sulfate chains of variable length that extend from the N-terminal domain of the core protein. The mass of the core is approximately 400 kDa. The complete amino acid sequence is known for both mouse and human perlecan.[26–28] This proteoglycan is thought to be involved in the maintenance of a negative charge barrier and in the interactions of cells with basement membranes.

Perlecan is extremely sensitive to proteolytic degradation. Higher yields of intact proteoglycan are favored by buffers containing a mixture

[26] D. M. Noonan, A. Fulle, P. Valente, S. Cai, E. Horigan, M. Sasaki, Y. Yamada, and J. R. Hassel, *J. Biol. Chem.* **266,** 22939 (1991).

[27] P. Kallunki and K. Tryggvason, *J. Cell Biol.* **116,** 559 (1992).

[28] A. D. Murdoch, G. R. Dodge, I. Cohen, R. S. Tuan, and R. V. Iozzo, *J. Biol. Chem.* **267,** 8544 (1992).

of inhibitors of proteolysis, with freshly harvested or liquid nitrogen-frozen tissue, and by the avoidance of excessive sample warming during purification steps. One procedure (Paulsson et al.[29]; method B) is to homogenize tumor in 3.4 M NaCl in 50 mM Tris-HCl, pH 7.4, containing 1 mM DFP, 1 mM PMSF, 1 mM EDTA, and 8 mM N-ethylmaleimide (NEM),[30] washing five times by suspension/centrifugation with a total of 5 liters of buffer. The tumor is then extracted twice with 300 ml of 1.7 M NaCl in the same buffer. If laminin is to be purified from the same batch of tumor, one can alternatively start with the tumor residue remaining after the EDTA treatment step that removes most laminin. With either approach, this residue can be extracted with 5 M urea in 50 mM Tris-HCl, pH 7.4, containing 1 mM EDTA, 8 mM NEM, 50 mM ε-amino caproic acid, 1 mM DFP, and 0.5 mM PMSF. The extract, after centrifugation, is then dialyzed against 5 M urea in 125 mM Tris-HCl, pH 7.4, containing 0.1% (w/v) 3-[(3-cholamidopropyl)-dimethyl-ammonio]-1-propanesulfonate (CHAPS) and inhibitors. After centrifugation to remove aggregates, the supernatant is loaded onto a DEAE-Sephacel (11.3 cm i.d. × 4 cm) column equilibrated in the same buffer, and eluted with a 0 to 0.8 M NaCl linear gradient (total, 1.5 liter). The second carbazole-positive peak to elute (at 0.32 M NaCl) is enriched in perlecan. After pooling and concentration, this fraction is adjusted to 0.55 g/ml with CsCl and centrifuged at 10°, 40,000 rpm for 68–72 hr in a type 65 fixed angle rotor (Beckman, Fullerton, CA). The proteoglycan is found below the protein plug at a density of 1.48 g/ml. This fraction is collected, dialyzed into 4 M guanidine hydrochloride in 50 mM Tris-HCl, pH 7.4, containing inhibitors and 0.1% (w/v) CHAPS, and chromatographed on a Sephacryl S400 column (2.6 × 90 cm). Purified perlecan elutes in the void volume. Yields are 15–20 mg of perlecan per 250 g of tumor. Using this or similarly purified perlecan, self-interactions,[31] binding to nidogen/entactin,[32] and cellular interactions[33–35] have been described. A general concern with functional studies carried out with perlecan, however, is that the use of chaotropic agents and cesium for purification might lead to loss or alteration of native function.

[29] M. Paulsson, P. D. Yurchenco, G. C. Ruben, J. Engel, and R. Timpl, *J. Mol. Biol.* **197,** 297 (1987).
[30] If the tumor is used for the preparation of laminin as well, we prefer to use hydroxymercuribenzoic acid as an alkylating reagent.
[31] P. D. Yurchenco, Y.-S. Cheng, and G. C. Ruben, *J. Biol. Chem.* **262,** 17668 (1987).
[32] C. Battaglia, U. Mayer, M. Aumailley, and R. Timpl, *Eur. J. Biochem.* **208,** 359 (1992).
[33] B. Clement, B. Segui-Real, J. R. Hassell, G. R. Martin, and Y. Yamada, *J. Biol. Chem.* **264,** 12467 (1989).
[34] K. Hayashi, J. A. Madri, and P. D. Yurchenco, *J. Cell Biol.* **119,** 945 (1992).
[35] C. Battaglia, M. Aumailley, K. Mann, U. Mayer, and R. Timpl, *Eur. J. Cell Biol.* **61,** 92 (1993).

Several fragments of perlecan can be prepared. Digestion with V8 protease releases numerous large peptides: 44-kDa (P44), 46-kDa (P46), and 200-kDa (P200) fragments can be purified from such a mixture as described.[36] The two smaller fragments are contained in the larger one and all are derived from the interior of the core. Another fragment, representing the N-terminal moiety, can be prepared by digestion with trypsin on ice followed by gel filtration and DEAE ion-exchange chromatography.[31]

Type IV Collagen and Fragments. Type IV collagen forms the major structural framework of many basement membranes. The EHS tumor basement membrane collagen has an $\alpha1(IV)_2\alpha2(IV)$ chain composition with respective chain molecular masses of about 185 and 170 kDa. The threadlike molecule consists of an N-terminal domain (7S-forming region), a collagenous region containing multiple interruptions, and a C-terminal globular domain. This collagen monomer forms dimers joined in a head-to-head fashion through disulfide bonds at the C-terminal globules. The dimers unite through low-affinity lateral associations involving the collagenous segment[7] and high-affinity N-terminal interactions[6] that become covalently stabilized. The last interaction requires four end-overlapping N termini.

The following purification procedure is based on the method of Kleinman *et al.*[37] as described in Yurchenco and Furthmayr.[7] Type IV collagen, mostly as dimers, can be purified from the previous urea or guanidine-treated residue by extracting with 500 ml of 2 M guanidine hydrochloride in 50 mM Tris-HCl, pH 7.4, containing 2 mM dithiothreitol (DTT), 1 mM EDTA, 1 mM PMSF by stirring overnight in the cold. After centrifugation to remove remaining insoluble material, the extract is dialyzed overnight into 1.7 M NaCl, 50 mM Tris-HCl, pH 7.4, containing 1 mM DTT, 1 mM EDTA, and 1 mM PMSF. The precipitate that forms is redissolved in the guanidine extraction buffer and precipitated again in the same manner. After again dissolving in guanidine extraction buffer, the crude collagen preparation is dialyzed into 5 M urea in 50 mM Tris-HCl, pH 7.4, with DTT and inhibitors, and passed through a DEAE-Sephacel or DEAE-cellulose column equilibrated in the same buffer. The unbound fraction is then dialyzed into guanidine extraction buffer and stored on ice. Purely monomeric intact type IV collagen can be purified from PF-HR9 cultured cell medium as described.[7]

Limited pepsin digestion of type IV collagen can be used to generate monomers that lack the C-terminal globular domain but that are capable

[36] S. R. Ledbetter, L. W. Fisher, and J. R. Hassell, *Biochemistry* **26**, 988 (1987).

[37] H. K. Kleinman, M. L. McGarvey, L. A. Liotta, P. Gehron Robey, K. Tryggvason, and G. R. Martin, *Biochemistry* **21**, 6188 (1982).

of forming "7S" tetramers *in vitro* in the absence of network formation.[7] Type IV collagen purified from EHS tumor (0.6–1 mg/ml), obtained after the ion-exchange step but before gel filtration, and mixed with trace radio-iodinated collagen as a marker, is dialyzed against 0.1 M acetic acid in the cold. The protein is then mixed with pepsin at an enzyme : substrate ratio of 1 : 50 (v/v) and incubated at 28° for 70 min [reaction stopped by adding pepstatin A (Sigma, St. Louis, MO)]. The digested protein is dialyzed into 2 M urea, 50 mM Tris-HCl, pH 7.4, containing 2 mM DTT, 5 mM glycine, 1 mM EDTA, and 0.5 mM PMSF and chromatographed on a Sephacryl S-1000 column (2.5 × 95 cm) in the same buffer (K_{av} 0.36). The monomeric peak is pooled, concentrated, dialyzed into the 2 M guanidine hydrochloride buffer used for intact collagen, and stored on ice or frozen. The NC1 noncollagenous globular domain of type IV collagen, which has been found both to bind to the triple-helical region of type IV collagen and to inhibit lateral associations,[38] can be prepared from EHS collagen on the basis of the method of Timpl *et al.*[6] Briefly, the collagen is extensively dialyzed against 0.2 M NaCl in 50 mM Tris-HCl, pH 7.6, containing 2 mM CaCl$_2$, and digested with purified bacterial collagenase (CLSPA grade; Worthington, Freehold, NJ) at an enzyme : substrate ratio of 1 : 100 (w/w) at 37° for 48 hr.[39] The protein is then centrifuged at 20,000 rpm for 30 min to remove aggregates, dialyzed against 0.2 M ammonium bicarbonate (pH 8.5) in the cold, and chromatographed on a Sephacryl S-300 column in the same buffer. The dimeric NC1 peak elutes with a K_{av} of 0.43. Isolated cross-linked 7S domain ("short form"), a 28-nm long rod formed by the binding of four N-terminal segments, can be purified from placenta by a combination of pepsin and collagenase digestion as described.[40] Other fragments have been described. *Pseudomonas* elastase, for example, can be used to release fragments with intact NC1 and 7S domains from lens capsule and glomerular basement membrane as described,[41] allowing the analysis of cross-linked complexes not obtainable with pepsin or collagenase.

Generation of Basement Membrane Glycoproteins with Recombinant Technology

The application of recombinant DNA technology to the study of basement membrane structure and function offers great promise. For many

[38] E. C. Tsilibary and A. S. Charonis, *J. Cell Biol.* **103**, 2467 (1986).
[39] A. S. Charonis, E. C. Tsilibary, P. D. Yurchenco, and H. Furthmayr, *J. Cell Biol.* **100**, 1848 (1985).
[40] R. Qian and R. W. Glanville, *Biochem. J.* **222**, 447 (1984).
[41] S. Gunwar, N. E. Noelken, and B. G. Hudson, *J. Biol. Chem.* **266**, 14088 (1991).

basement membrane components, obtaining purified forms for biochemical and ultrastructural studies is difficult because of the limited availability of basement membrane material and because basement membranes contain complex mixtures of large proteins. This is particularly true with respect to the genetic variants of collagen IV and laminin. In those cases in which purified components are available, mapping of functions to individual domains, and the molecular characterization of these functions, has been hampered by the inability to isolate many domains using limited proteolysis. Recombinant DNA expression vectors not only offer the promise of producing variant proteins and previously unobtainable domains, but should also permit the application of molecular genetics to aid in the study of basement membrane structure and function.

The generation of recombinant basement membrane components requires special consideration. These components are secreted glycoproteins, making the use of *Escherichia coli* expression vectors in most cases unattractive, because carbohydrate may be required for proper folding and cell–matrix interactions, and because signal sequence cleavage may be important for proper function of N-terminal domains. Other modifications found on basement membrane components may also be important for proper structure and function. Collagen IV contains hydroxylproline residues, perlecan has large glycosaminoglycan chains, and nidogen/entactin may be O-sulfated on tyrosine. For these reasons, we have limited our use of *E. coli* recombinant proteins to the production of domain-specific antibodies. With the bacteriophage T7 RNA polymerase-based pET vector system developed by Studier and co-workers,[42] we have generated large quantities of recombinant domains, frequently easy to purify because of their formation of insoluble complexes.

Basic procedures for use of baculovirus vectors have been described[43] and therefore we only briefly mention advantages and disadvantages of the system. Baculovirus expression vectors are capable of producing large quantities of protein (milligrams per liter) that can be secreted and glycosylated by the host cells, although carbohydrate is mostly of the high-mannose type rather than the complex carbohydrate found on secreted mammalian proteins. Other posttranslational modifications are also performed including signal sequence cleavage, proteolytic processing, and myristylation. Importantly, the transfer vectors are capable of expressing large proteins such as basement membrane components, and the host cells lack reactivity to antibodies directed against mammalian basement membrane com-

[42] F. W. Studier, A. H. Rosenberg, J. J. Dunn, and J. W. Dubendorff, this series, Vol. 185, p. 60.
[43] M. D. Summers and G. E. Smith, *Tex. Agric. Exp. Stn.* [Bull.] **1555** (1987).

ponents. Molecular genetic manipulation of expressed proteins is simplified because the complete nucleotide sequence of the transfer vectors is known. The primary disadvantage of the early vector systems, the requirement of cloning out the recombinant virus from a 100- to 1000-fold excess of wild-type virus, has largely been eliminated. The use of linearized viral DNA in transfections greatly reduces the background of wild-type virus (4- to 10-fold excess; Kitts *et al.*[44]), and a report describes recombination between a transfer vector and infectious viral DNA in *E. coli*[45] so that cloning of recombinants from wild-type viral DNA can be accomplished in days rather than weeks.

To direct internal and C-terminal domains to the secretory pathway, we modified the baculovirus transfer vectors VL1392 and VL1393 by inserting a recombinant cDNA encoding the rat fibronectin signal sequence (kindly provided by J. Schwarzbauer, Princeton University, Princeton, NJ) between the polyhedron promoter and the first restriction endonuclease cleavage sites in the polylinker sequences (Fig. 5). The resulting 9.8-kb plasmids, VL1392SS and VL1393SS, have multiple unique restriction endonuclease cleavage sites and permit fusions in all three reading frames. Recombinant fusion proteins expressed from these transfer vectors will therefore contain the signal sequence and the 19 adjacent amino-terminal amino acids of rat fibronectin, and linker residues varying with individual constructs. VL1392SS and VL1393SSA are pUC-based plasmids carrying an ampicillin resistance gene and an origin of replication for manipulations in *E. coli*. Insertion of the cDNA of interest can be carried out with standard recombinant DNA techniques.[46] Plasmid DNA for recombination with wild-type viral DNA is purified by CsCl–ethidium bromide equilibrium density gradient centrifugation.

Recombinant virus is generated using $CaPO_4$ or lipofectamine (GIBCO-BRL, Gaithersburg, MD) transfection to introduce wild-type viral and plasmid DNAs into adherent *Spodoptera frugiperda* (Sf9) cells (derived from fall army worm ovary). Five to 6 days after transfection recombinant virus is separated from wild-type virus by end-point dilution cloning in 96-well microtiter plates containing 10^4 cells/well, assaying for recombinant protein expression 6–9 days after infection. A virus titer of 10^8 plaque-forming units (PFU)/ml of transfected stock is assumed, diluting to an expected 0.5 and 5 PFU/well. Positive wells are identified by filtration of an aliquot of medium onto a nitrocellulose membrane through a 96-well

[44] P. A. Kitts, M. D. Ayres, and R. D. Possee, *Nucleic Acids Res.* **18,** 5667 (1990).
[45] V. A. Luckow, S. C. Lee, G. F. Barry, and P. O. Olins, *J. Virol.* **67,** 4566 (1993).
[46] J. Sambrook, E. F. Fritsch, and T. Maniatis, eds., "Molecular Cloning: A Laboratory Manual," 2nd ed. Cold Spring Harbor Lab., Cold Spring Harbor, NY, 1989.

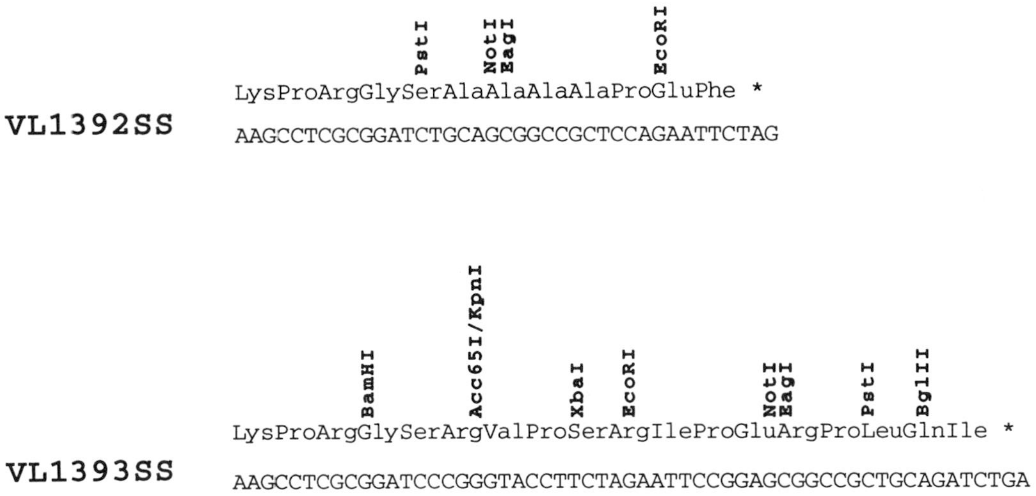

VL1392SS

```
                    Pst I
                    Not I
                    Bgl I
                                                    EcoR I

LysProArgGlySerAlaAlaAlaAlaAlaProGluPhe *
AAGCCTCGCGGATCTGCAGCGGCCGCTCCAGAATTCTAG
```

VL1393SS

```
              BamH I
                    Acc65I/Kpn I
                            Xba I
                            EcoR I
                                    Not I
                                    Bgl I
                                        Pst I
                                            Bgl II

LysProArgGlySerArgValProSerArgIleProGluArgProLeuGlnIle *
AAGCCTCGCGGATCCCGGGTACCTTCTAGAATTCCGGAGCGGCCGCTGCAGATCTGA
```

Fig. 5. Sequences of the polylinkers in the baculovirus transfer vectors VL1392SS and VL1393SS. The positions of unique restriction endonuclease cleavage sites are indicated above the first base in their recognition sequence. The first two amino acid residues shown are from rat fibronectin (corresponding to amino acid residues 50 and 51). *, Translation termination codon. The *SmaI* cleavage sites in the polylinkers are no longer unique, and the *BglII* cleavage site in VL1392 was lost on insertion of the rat fibronectin signal sequence.

manifold, followed by specific antibody and ^{125}I-labeled protein A. Alternatively, recombinant virus can be cloned using a DNA hybridization procedure. Cloned virus stocks are expanded in tissue culture flasks, followed by suspension culture.

Production of recombinant protein is begun by sedimenting 2×10^9 Sf9 cells (>98% viability as determined by trypan blue dye exclusion) grown in suspension culture at 28° in Sf-900 11 SFM serum-free insect medium (GIBCO-BRL, Grand Island, NY) supplemented with antibiotics [gentamicin (50 μg/ml) and amphotericin B (Fungizone, 2.5 μg/ml; GIBCO-BRL]. Cells are resuspended in medium containing approximately 5×10^9 PFU of recombinant virus. After incubating at room temperature for 1 hr, infected cells are diluted to a final volume of 1 liter in the same medium and transferred to a 2-liter spinner flask (Bellco, Vineland, NJ). Cultures are incubated at 28° for 60–72 hr at 100 rpm, and the medium harvested for protein purification by low-speed (1000 rpm, 15 min, Sorvall GSA rotor) and high-speed (20,000 rpm, 30 min, Sorvall SS-34 rotor) centrifugation. It is important that the magnetic stir plate used with the spinner cultures not generate an excessive amount of heat.

Several independent approaches have been used to assess whether the secreted recombinant glycoprotein corresponding to the G domain of EHS laminin is properly folded. In low-angle platinum replicas, purified recombinant G domain has a size and shape similar to the G domain of native laminin. Elastase cleaves the recombinant protein into two fragments of the expected size (70 and 50 kDa) and antibody reactivity, the smaller fragment having the same N-terminal amino acid sequence found for elastase fragment E3 of laminin. Circular dichroism spectra of the recombinant elastase fragment and fragment E3 are essentially indistinguishable. Finally, the larger elastase fragment and recombinant G domain are sufficient for C2C12 myoblast adhesion and spreading. These results suggest that this recombinant protein is representative of the G domain in intact laminin. The G domains of the variant laminin A chain merosin and *Drosophila* laminin A have also been produced with baculovirus expression vectors and are secreted.

Some basement membrane proteins are not secreted by baculovirus-infected cells. Entactin/nidogen is expressed at high levels but forms an insoluble complex requiring a combination of reducing and denaturing agents to solubilize the protein.[47] The complete human laminin B2 chain also fails to be secreted but is soluble and properly folded in its amino-

[47] T. Tsao, J.-C. Hsieh, M. E. Durkin, C. Wu, S. Chakravarti, L.-J. Dong, M. Sewis, and A. E. Chung, *J. Biol. Chem.* **265,** 5188 (1990).

terminal half.[48] Likewise, the complete murine laminin B1 and B2 chains and amino-terminal constructs derived therefrom also fail to be secreted from baculovirus-infected cells.[49] The failure of these proteins to be secreted is probably not a function of the signal sequence because replacement of the natural signal sequence of murine laminin B1 chain with the rat fibronectin signal sequence does not result in the secretion of recombinant domain VI. In the case of entactin/nidogen, secretion has been achieved[50] with a mammalian expression vector.

Approaches to Evaluate Assembly Interactions

Metal Replication and Electron Microscopy

Electron microscopy can be used to visualize the molecular surface of extracellular matrix components, evaluate their purity and conformation, study binding interactions, and evaluate supramolecular organization. We discuss the approaches of low-angle replication of glycerol spreads of macromolecules and high-angle replication of proteins, polymers, and tissue following freeze-etching or freeze-drying.

Glycerol-Based Techniques. Low-angle platinum/carbon replication of macromolecules has been widely used to study extracellular matrix components. The method, developed by Shotton *et al.,*[51] is one in which macromolecules stabilized with 50–70% (v/v) glycerol are spread onto mica, evacuated, and rotary coated at a low angle with vaporized platinum or platinum/carbon. We normally dilute or dialyze the macromolecule into either 0.15 M ammonium acetate or bicarbonate (or a mixture of the two at pH 7.4), dilute the protein concentration to 40 to 100 nM (for molecular masses between about 1 million and 50,000 daltons, respectively), and add glycerol to a final concentration of 55–60% (v/v). The sample is then nebulized onto freshly cleaved mica disks (0.25-in. diameter or less) such that multiple fine droplets can be seen on the surface. Alternatively, the sample can be spread on mica by centrifugation[52] after application with a pipette. The protein-coated mica is then attached onto a replication table (e.g., with double-stick tape) and placed into the vacuum cham-

[48] T. Pikkarainen, T. Schulthess, J. Engel, and K. Tryggvason, *Eur. J. Biochem.* **209**, 571 (1992).

[49] O'Rear and Yurchenco, unpublished data (1993).

[50] J. W. Fox, U. Mayer, R. Nischt, M. Aumailley, D. Reinhardt, H. Wiedemann, K. Mann, R. Timpl, T. Krieg, J. Engel, and M.-L. Chu, *EMBO J.* **10**, 3137 (1991).

[51] D. M. Shotton, B. Burke, and D. Branton, *J. Mol. Biol.* **131**, 303 (1979).

[52] R. Nave, D. O. Fürst, and K. Weber, *J. Cell Biol.* **109**, 2177 (1989).

ber of a metal shadowing device. Common instruments include the Balzers models 300 and 400 freeze-etch/shadow systems (now serviced by Bal-Tec, Middlebury, CT) pumped either with oil diffusion or turbomolecular pumps. The mica is evacuated until the vacuum is stable (2×10^{-7} matm or better). For visualization of glycosaminoglycan chains, longer evacuation times seem to improve imaging. The stage, set at an angle of 8 to 10° relative to the evaporation source, is rotated at a rate of 80–100 rpm, and the sample shadowed with about 1 nm of Pt/C (95 : 5) using an electron beam gun with deposition measured with a quartz crystal thickness monitor. The sample is then stabilized with an 8- to 10-nm coat of carbon applied at right angles. We find a modest improvement in fineness of metal grain size and evenness by replicating under ultrahigh vacuums (1 to 3 × 10^{-9} matm) at a stage temperature of 11–13 K. The mica is then removed from the vacuum chamber, the replica floated onto water, and lifted onto 400-mesh copper grids. If the replica does not easily separate from the mica, it can first be placed in a moist chamber overnight or the mica can be dissolved with hydrofluoric acid. The replicas can be examined with an electron microscope (we use a Philips model 420, Eindhoven, Nether-lands) at 80 kV with a 30-μm objective aperture at magnification of ×20,000 to ×80,000. Shadowing at low angles increases contrast and somewhat exaggerates molecular contours (e.g., laminin globular domains are more obvious relative to adjacent rod domains), all at the expense of resolution. Low-angle shadowing produces self-decoration effects as well as true shadowing. Whereas the former is cause of decreased resolution, decoration can sometimes aid visualization, as in the case of the thin glycosaminoglycan chains. In examining replicas, certain artifacts need to be avoided, for example, molecular aggregation induced at the edge of glycerol droplets, replicas of salt crystals, and regions in which the background grains are large and/or irregular in pattern. It is also important, when evaluating interactions, to dilute the sample sufficiently to minimize spurious contacts due to molecular crowding. Examples of interactions identified for laminin, perlecan, and type IV collagen are shown in Figs. 6–8.[52a]

Freeze-Etch, High-Angle Replication Techniques. Several variations of high-angle metal replication of quick-frozen, deep-etched macromole-cules, biological polymers and tissues provide a useful analytical approach to evaluate basement membrane structure. Most of the studies we have carried out have used a stage-to-evaporation source angle of 45 to 60°. Further improvements in resolution can be achieved at even higher

[52a] P. D. Yurchenco and J. J. O'Rear, in "Molecular and Cellular Aspects of Basement Membranes" (D. Rohrbach and R. Timpl, eds.). pp. 19–47. Academic Press, New York, 1993.

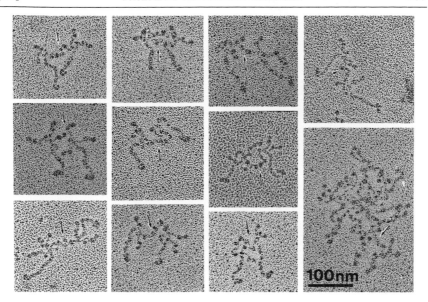

Fig. 6. Laminin short arm self-interactions. Electron micrographs of low-angle rotary shadowed replicas of laminin dimers and one oligomeric aggregate are shown.[5] Small arrows indicate sites of interaction.

angles.[53] (1) Individual collagen molecules can be sprayed onto mica in 0.1 M acetic acid, plunge-frozen in a liquid propane slush cooled with liquid nitrogen, and evacuated on a stage cooled to 133 K or less. The stage is then gradually warmed to 190 K and freeze-dried for about 0.5 hr. After cooling the stage to the temperature of liquid nitrogen, the sample is shadowed with 1 nm of Pt/C at a 45° angle without rotation, and then backed with about 12 nm of carbon split between two depositions, one at about 100 K, and the other after warming the sample to 190 K or higher. The replica is lifted in water as described above. There are limitations in the application of this simple technique. Most proteins are not soluble in acetic acid and most buffers do not sublime. Even such volatile neutral salts as ammonium acetate and ammonium bicarbonate, commonly employed in shadowing, can leave small deposits on mica that detract from the images. Furthermore, many large proteins are not as stable as the collagens and freeze-drying may lead to some degree of structural collapse; (2) an alternative method is the mica slurry technique of Heuser.[54] The technique permits one to evaluate macromolecules and their complexes

[53] G. C. Ruben, *J. Electron Microsc. Tech.* **13,** 335 (1989).
[54] J. E. Heuser, *J. Mol. Biol.* **169,** 155 (1983).

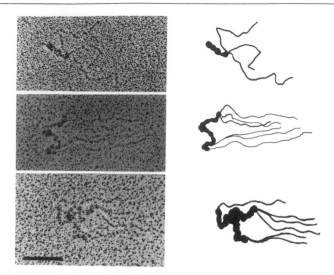

FIG. 7. Perlecan self-interactions. Electron micrographs of Pt/C replicas (low-angle rotary shadow) of low-density heparan sulfate proteoglycan (perlecan) isolated from EHS tumor matrix. *Top:* A monomer. Note thicker elongated core protein and three slender heparan sulfate chains extending from one end. *Middle and bottom:* A dimer and trimer, respectively, each joining at what appears to be the C-terminal region of the core.[31,52a] Bar, 100 nm.

in native buffers with minimal freeze-drying and can overcome artifacts of salt and ice for a portion of the replica. However, proteins that do not bind well to mica (e.g., laminin) pose a special problem. Typically the macromolecules of interest are adhered onto small mica flakes, washed by centrifugation, quick frozen with a slam-freezer or other suitable device, etched, and replicated at a high angle; (3) reconstituted and tissue basement membrane polymer slurries can be sandwiched between copper planchets, quick-frozen (88 K) to a vitreous ice depth of up to 10 μm with a propane jet freezer (Bal-Tec), inserted into a liquid nitrogen-cooled double-replica table, and fractured in an ultrahigh vacuum (1–3 $\times$ 10^{-9} matm) at a stage temperature of 133 K, deep-etched at 170 K for 30 min, and then replicated with 1 nm of Pt/C at 11–13 K with or without rotation at a 45 or 60° angle.[3] The specimen is then backed with carbon as described above. Laminin polymers are difficult to study because low ionic strength, formaldehyde fixation, and some procedures of glutaraldehyde fixation, can disrupt the network. We have found that a brief glutaraldehyde fixation carried out at high concentration [2–3% (v/v) for a few minutes] can preserve network structure. Such fixation permits the subsequent removal

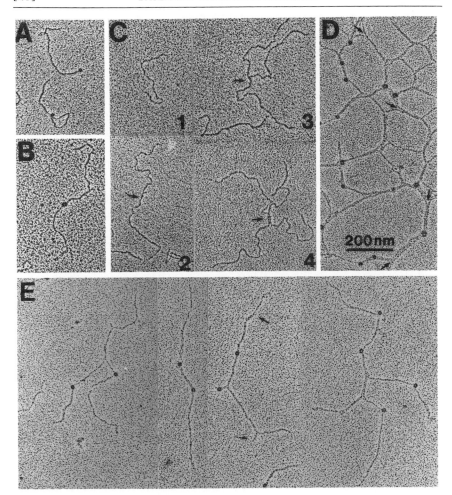

FIG. 8. Type IV collagen self-interactions. Electron micrographs of rotary shadowed Pt/C replicas of type IV collagen and its complexes. A monomer is shown in (A). Note the globular domain located at the end of the threadlike collagenous domain. A dimer joined at the C-terminal globule is shown in (B). (C) Pepsin fragments of type IV collagen (consisting of nearly all of the collagenous domain, but lacking the globule). A monomer (1), dimer (2), trimer (3), and tetramer (4), each joined at the N-terminal "7S" domain, arc shown. A laterally and end-associated network of type IV collagen is shown in (D). Laterally associated oligomers of collagen are shown in (E).[2,52a]

of salt by washing with water prior to fixation and can therefore reduce salt accretions; (4) if the network under study is sufficiently stable to withstand chemical fixation, washing, and freeze-drying, it becomes easier to obtain clean, high-resolution replicas. The type IV collagen network can be visualized in thin tissue specimens (e.g., placental amniotic basement membrane,[8] EHS tumor,[55] and lens capsule[56] after depleting the basement membrane of other components by extraction with chaotropic agents (2–6 M guanidine hydrochloride or selective proteolytic digestion (plasmin) followed by fixation [4% (v/v) paraformaldehyde dissolved in phosphate-buffered saline with tissue incubated overnight at 0–4°]. The tissue is then washed with water followed by 25–30% (v/v) ethanol for increased cryoprotection, placed on filter paper, and then quick frozen by plunging it into a propane slurry as above. Because it is the surface of the specimen that is being studied, and fracturing is not required, the freezing method is adequate to prevent ice crystal damage. The specimen is then loaded onto a cold table and stage, evacuated, gradually freeze-dried at 190 K, and then coated with 1 nm of Pt/C as above. The appearance of freeze-etched EHS basement membrane following high-angle (60°) rotary deposition is shown in Fig. 9.

Biochemical/Biophysical Evaluation of Binding Interactions

Moderate to high-affinity interactions (less than about 0.1 μM) can be evaluated by a variety of techniques including standard gel filtration, zonal velocity sedimentation, immunoprecipitation assays, native gel electrophoresis, affinity chromatography, and solid-phase binding assays. Radioligand-binding assays of mixtures of two basement membrane components, in which an antibody specific for the unlabeled species is used to precipitate any labeled species, has been successfully used to characterize interactions of the basement membrane, for example, between nidogen/entactin and laminin, and nidogen/entactin and type IV collagen.[4] The antibody reagents used for immunoprecipitation should not, of course, interfere with the reaction under consideration. Low-affinity interactions (approximate range, 0.1–10 μM), a focus of this discussion, are thought to play a number of important roles in determining the molecular architecture of basement membranes.[2] These types of interactions appear to act both in concert with higher affinity bonds and as part of cooperative nucleation–propagation polymerizations. To measure such readily reversible interactions between reacting pairs, methods that prevent separation

[55] P. D. Yurchenco and G. C. Ruben, *Am. J. Pathol.* **132,** 278 (1988).
[56] G. C. Ruben and P. D. Yurchenco, *Microsc. Res. Tech.* **28,** 13–28 (1994).

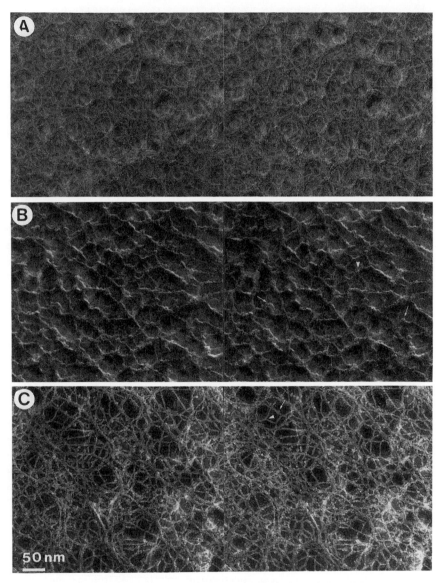

FIG. 9. Electron micrographs of the EHS tumor basement membrane. After freezing the matrix samples with a propane jet freezer, they were fractured, deep-etched, and rotary replicated with Pt/C at a 60° angle either untreated (A), following removal of type IV collagen with bacterial collagenase to reveal the laminin polymer (B), or following removal of laminin by extraction with guanidine hydrochloride to expose the type IV collagen network (C). Electron micrographs are shown as stereo pairs and are contrast reversed. Arrowhead in (B) indicates joining of struts or lateral fusion of filaments, whereas small arrows indicate globules in network. The tighter mesh of laminin/collagen network (a) is likely formed by intermeshing of laminin (B) and collagen polymers (C).[3]

of bound from free species are required. Solution conditions in which the thermodynamics are understood are preferred. Furthermore, it is often important that the reacting species be studied under physiologically relevant conditions of pH, ionic strength, and temperature, and with macromolecules that possess conformation that reflects native structure. Laminin and laminin fragments, for example, are irreversibly inactivated with respect to polymerization following exposure to high concentrations of chaotropic agents, reduction/alkylation, or heat.

Affinity Retardation Chromatography. The method (Fig. 10A and B) is a modification of standard affinity chromatography in which a species is bound to protein immobilized on gel-filtration beads[10] but retardation of elution rather than salt displacement is evaluated. The value of the method is that low-affinity interactions can be evaluated with small amounts of soluble ligand and the column with immobilized species, after an initial expenditure of protein, can be reused. One protein species is coupled either to Sepharose CL-4B (Pharmacia) beads by the CNBr method, or to compression-resistant Affi-Prep 10 (Bio-Rad, Richmond, CA) used for HPLC,[57] at a concentration of 2 mg/ml or higher (higher amounts of immobilized protein theoretically increase the sensitivity of the assay but may interfere with the flow rate). The coupled beads are then packed into a long (50 cm) and narrow (6-mm i.d.) column with a thermal jacket to control the reaction temperature. For binding studies with laminin, the intact parental protein is the macromolecule, which is immobilized because, if used in the mobile phase, it becomes trapped in the column matrix at elevated temperatures. A control column of exactly the same size and protein concentration is prepared by coupling albumin or some other nonreacting protein. It is also useful to prepare an inactivated blank column to test for spurious interactions with the beads themselves. A second species (which can be optionally radiolabeled) is then applied as a narrow band to the test and control columns and the elution volumes are determined. An interaction is evidenced by a delay in elution relative to the control column.

Equilibrium Gel Filtration. The method of equilibrium gel filtration[9] as modified for radiolabeled macromolecules[58] holds an advantage over the above technique because all reacting species are maintained under solution conditions and because it is possible to determine both the affinity of interaction and the approximate average size of the resulting complex (Fig. 10C and D). However, the method can be costly and sometimes

[57] J. C. Schittny and C. Schittny, *Eur. J. Biochem.* **216,** 439 (1993).
[58] A. Horwitz, K. Duggan, R. Greggs, C. Decker, and C. Buck, *J. Cell Biol.* **101,** 2134 (1985).

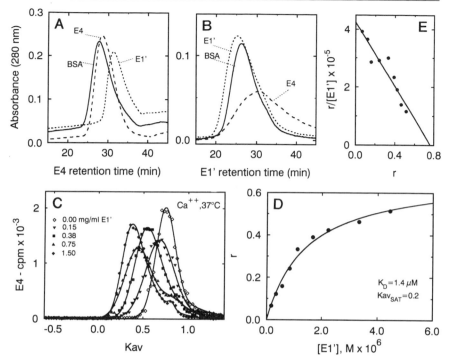

FIG. 10. Affinity retardation chromatography and equilibrium gel filtration. Affinity retardation chromatography (A and B) and equilibrium gel filtration (C–E) analysis of the binding between short arm fragments. (A) Fragment E4 (B1 domains VI–V) was passed as a band down Affi-Prep (Bio-Rad) columns coupled to fragment E1' (dotted line), E4 (dashed line), or BSA (solid line). E4 elution was preferentially retarded on the E1' column, indicating a binding interaction. (B) Reciprocal arrangement of immobilized and mobile species. Fragment E1' (A and B2 short arm complex) was passed down the same columns. E1' was preferentially retarded on the E4 column. (C) Concentration dependency (E1') of binding at 37° in the presence of 1 mM calcium. E4 eluted at earlier K_{av} values as a function of increasing E1' concentration: (◇) no E1'; (▼) 0.15 mg/ml; (■) 0.38 mg/ml; (▲) 0.75 mg/ml; (◆) 1.5 mg/ml. (D) Determination of dissociation constant for E4–E1' binding using a larger number of determinations. Binding was measured from the known concentration of E1' and an elution coefficient r, which is calculated from the binding K_{av} and the nonbinding K_{av}. Apparent binding parameters were determined by curve fitting. (E) Scatchard transform of the same data shown in (D).[5]

even impractical because large quantities of one of the two reacting species are likely to be consumed. The basic procedure is as follows[5]: gel-filtration beads covering the appropriate molecular sieve range (e.g., Sepharose CL-6B for laminin fragments) are packed into 50-cm-long columns (3-mm i.d.) and surrounded by a water jacket whose temperature is maintained by a circulating refrigerator/heater water bath. The column is then permeated with appropriate binding buffer in the presence of carrier albumin (to prevent loss of protein to the column without affecting binding) or with a protein P at a variety of constant concentrations in binding buffer plus carrier protein. The radioiodinated ligand L is applied to the column as a small aliquot and fractions are collected. An internal sodium [^{35}S]sulfate or other low molecular weight standard can be added to correct for small variations in volumes. The elution position of phenol red or radioactive sulfate is used as a marker for the included volume (V_t) and the first Blue dextran 5000 (Pharmacia) peak (or other very large macromolecule) can be used as a measure of the void volume (V_0). The principle of the method is as follows: If a component P is allowed to permeate a gel-filtration column completely, then a putative reacting radiolabeled ligand L, applied as a narrow band, will remain in the presence of constant component P throughout its passage through the column. The ligand L will elute from the column as a larger species at the K_{av} of the complex if strongly bound to component P. If the equilibrium is such that it spends only part of the time bound to component P, then it should elute between the elution volume of the complex and ligand L weighted according to the fraction of time the ligand spends in the bound and free states. The dissociation equilibrium constant can be determined by Scatchard analysis or nonlinear regression analysis of the migration behavior of the ligand at different concentrations of an interacting permeating protein. We can define a relative unitless elution coefficient $r = (K_0 - K_x)/K_0$, where K_0 corresponds to the nonbinding K_{av} of radiolabeled ligand L and K_x corresponds to the K_{av} of ligand L at a constant free molar permeating protein concentration, [P]. Because r at a given protein concentration should be proportional to the amount of ligand bound, then for a single class of interaction, a plot of $r/[P]$ versus r will give a straight line with a slope of $-1/K_D$, a y intercept of r_{sat}/K_D, and an x axis intercept of r_{sat}. The term r_{sat} is the value of the elution coefficient under saturating conditions at infinitely high concentrations of protein P. The K_{av} at saturation [$K_{sat} = K_0 (1 - r_{sat})$] in turn can provide a molecular mass estimate for the complex at saturation if the column is calibrated. Because the free concentration of protein P cannot be measured directly, the concentration of ligand L is maintained much less than protein P, permitting the free concentration to be nearly that of the total concentration of protein P.

Standard and Equilibrium Rate Zonal Ultracentrifugation. Zonal velocity sedimentation (Fig. 11) is a useful procedure to separate large proteins according to sedimentation constant and has been used to demonstrate the polymerization of type IV collagen and laminin and the formation of 7S tetramers of type IV collagen. Protein or other macromolecule (0.2

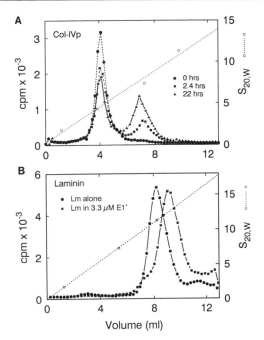

FIG. 11. Rate zonal ultracentrifugation. (A) Standard ultracentrifugation. A pepsin monomeric fragment of type IV collagen lacking the C-terminal globular domain and radioiodinated was incubated at 28° in PBS for the indicated times (●, 0 hr; ■, 2.4 hr; ▲, 22 hr) applied in 0.2-ml aliquots to sucrose gradients, and centrifuged in the cold at 40,000 rpm for 22 hr. Fractions were collected from the top. Note the conversion of the first peak (monomer) to the second peak (tetramer) as a function of time. Standards with known S values were sedimented in parallel (dotted line). (B) Equilibrium rate zonal ultracentrifugation. Radioiodinated laminin (Lm) was applied to a gradient with buffer alone (●) or with buffer containing fragment E1′ (■) and centrifuged at 40,000 rpm for 8 hr at 37°. As a result of an interaction with E1′, laminin sediments at a higher rate.[5]

ml), usually radiolabeled, is applied to the top of 12.6 ml of 5–20% (w/v) linear sucrose gradients in phosphate- or Tris-buffered saline (Fig. 11A). The tubes are centrifuged in an SW40-Ti rotor (Beckman) at 40,000 rpm for an appropriate period of time (depending on the Svedberg constant of the macromolecule under study) and temperature (3, 28, or 35–37°). For laminin we have generally centrifuged for 13 hr at 3° and for 8 hr at 35–37°. For type IV collagen we have centrifuged for 22 hr at 3° and for 13.5 hr at 28°, lowering the speed to 38,000 rpm for the latter condition. Cytochrome c, bovine serum albumin, aldolase, and catalase standards (1.8, 4.3, 7.35, and 11.3S, respectively) are useful standards to determine relative sedimentation values. Like gel filtration, ultracentrifugation can be modified to examine low-affinity interactions (Fig. 11B) between two species by permeating the sucrose gradient with one of the species.[5] This is accomplished by placing the same concentration of unlabeled protein in the two vessels of the mixing chamber used to pour the sucrose gradient. During the run, the initially constant concentration of protein changes such that the upper boundary moves forward leaving a region devoid of protein with excess protein accumulating at the bottom of the tube. This imposes a limitation, that is, the labeled species applied at the top of the gradient must sediment faster than the permeating species. The method has proved especially valuable for the analysis of laminin with its fragments, because we have found that the former becomes trapped in gel-filtration columns warmed to 37°. The procedure has also been used to examine heparin binding to laminin fragment E3.[59]

Polymerization and Polymerization–Inhibition Assays

Both laminin and type IV collagen will polymerize *in vitro,* the former forming a solid gel at higher concentrations. These assembly processes can be evaluated by turbidity and centrifugation.

Turbidity. This approach was adapted from that used to follow the heat gelation of type I and other interstitial collagens. It permits one to follow the development of aggregates over time (Fig. 12). Type IV collagen or laminin extracted from the EHS tumor, previously dialyzed into a thoroughly degassed reaction buffer (phosphate-buffered saline or TBS containing 1 mM calcium chloride in the case of laminin) and subjected to centrifugation to clear preformed aggregates, is adjusted to concentrations ranging from 0.1 to 0.8 mg/ml. The solutions are then placed into clean quartz cuvettes (4-mm diameter) prewarmed to 28° (collagen) or 35–37° (laminin) in a thermally jacketed spectrophotometer. We prefer to use a double-beam instrument because of its greater baseline stability. The turbidity (as absorbance) is measured at 360 nm at the set temperature over a period of one to several hours. Interacting regions of the parent macromolecules can be determined by assaying for inhibition of turbidity using defined fragments.

Sedimentation. Sedimentation is a more reliable technique to quantitate the amount of polymer formed. Different concentrations of laminin or type IV collagen are incubated in small (0.5-ml) aliquots in their respective polymerization buffers for several hours. Type IV collagen samples are centrifuged at 40,000 rpm for 10 min (Ty 65 rotor; Beckman). The amount of polymer is determined either from the pellet or supernatant (Fig. 13).

[59] P. D. Yurchenco, Y.-S. Cheng, and J. C. Schittny, *J. Biol. Chem.* **265,** 3981 (1990).

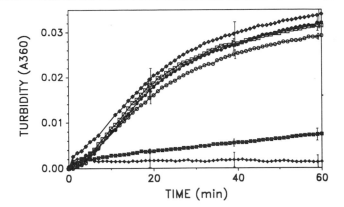

FIG. 12. Turbidity determinations of laminin polymerization. Laminin/entactin, in the presence of different fragments, was incubated at 35° in TBS–1 mM CaCl$_2$ at a concentration of 0.25 mg/ml. Turbidity (measured at 360 nm) was determined at regular time intervals. Elastase fragments E4 (■, 0.125 mg/ml) and E1' (♦, 0.8 mg/ml), which inhibit laminin polymerization in sedimentation assays, blocked the development of turbidity. Elastase fragments E8 (○, 0.48 mg/ml) and P1' (◇, 0.56 mg/ml), and control bovine serum albumin (□, 0.12 mg/ml), which have little or no effect on laminin polymerization in sedimentation assays, did not inhibit the development of turbidity.[10]

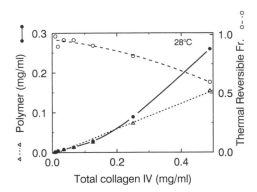

FIG. 13. Sedimentation assay of type IV collagen polymerization. Type IV collagen extracted from EHS tumor (trace radioiodinated; 0.5-ml aliquots) was incubated for 4 hr in neutral salt buffer at different concentrations (●) and then centrifuged at 40,000 rpm at the same temperature for 15 min. To determine thermal reversibility, identical samples were incubated at 28° for 4 hr followed by incubation on ice for 4 hr (fraction, ○). These aliquots were sedimented at 3°. Pelleted radioactivity was determined. Thermally reversible polymer for each concentration is also plotted (△). This fraction decreases at higher concentrations. Because the formation of N-terminal tetramers from pepsin-treated monomers is not found to be thermally reversible, we have concluded that the reversible fraction reflects lateral associations and the nonreversible fraction reflects 7S formation.[7]

For laminin, it is convenient to sediment in Eppendorf tubes (in 0.5-ml Eppendorf tubes in a microcentrifuge) at 10,000 to 12,000 rpm for 10 to 15 min.

Acknowledgment

This work was supported by grants (R01-DK36425 and R01-DK41500) from the NIH and from the American Heart Association.

[24] Extracellular Matrix Assembly

By JOHN A. MCDONALD

Introduction

Extracellular matrix molecules are assembled into highly complex structures with remarkable shape and function in metazoan species. In animals, these include membrane-like basal lamina, bone, tendons, and so on. In many cases the molecules found in these structures exhibit a marked propensity for homophilic, entropy-driven self-assembly into highly ordered arrays *in vitro,* for example, type I collagen.[1] In other cases, assembly involves the formation of complex arrays of two or more components, for example, type IV collagen, laminin, and entactin.[2] Although in many cases proper molecular alignment can be replicated *in vitro,* cells appear also to play a critical role in organizing, shaping, and remodeling extracellular matrices *in vivo.*[3] Virtually all techniques covered in this volume and utilized to study extracellular matrix molecules are also utilized to understand their assembly. Several typical applications of these techniques are outlined in Table I. This chapter illustrates certain examples of their application, but the reader should consult the particular chapter(s) in this volume for detailed protocols.

In the final analysis, understanding extracellular matrix assembly requires a combination of biochemical, biophysical, and molecular genetic approaches. These approaches often begin with knowledge of gene structure and organization and the deduced protein sequence. This in turn may allow the elucidation of secondary and tertiary structure, either by homology of motifs to those with known structure, or by direct analysis.

[1] D. E. Birk, F. H. Silver, and R. L. Trelstad, *in* "Cell Biology of Extracellular Matrix" (E. D. Hay, ed.), 2nd ed., Chapter 7, p. 221. Plenum, New York, 1991.
[2] P. D. Yurchenco and J. C. Schittny, *FASEB J.* **4,** 1577 (1990).
[3] E. D. Hay, ed., "Cell Biology of Extracellular Matrix," 2nd ed. Plenum, New York, 1991.

Copyright © 1994 by Academic Press, Inc.
All rights of reproduction in any form reserved.

TABLE I
GENERAL APPROACHES USED TO STUDY MATRIX ASSEMBLY

Molecular: cDNA cloning
    Expression of full-length polypeptide or domains for direct assembly studies
    Creation of antipeptide or fusion protein antibodies to identify domains important in
       matrix assembly
    Use of combinatorial shape libraries to identify binding motifs
Structural: Biophysical and direct imaging studies
    Optical and electron microscopy of naturally occurring and reconstituted matrices; map-
       ping of epitopes
    Biophysical studies on fragments, intact molecules, or mixtures of purified molecules,
Cellular: *In vitro* matrix assembly system
    Expression of matrix components or receptors in eukaryotic cell lines to test their role
       in matrix assembly
    Testing of binding to cells and assembly using purified matrix components or inhibitors
       (biosynthetic, antibody, domain, or peptide)
    Expression of domains or truncated matrix molecules using plasmid or viral vector systems
    Use of mutant cell lines to study matrix assembly
In vivo
    Definition of normal expression pattern of matrix molecules *in vivo*
    Molecular genetic analysis of human diseases
    Gene targeting in mice
    Transgenic mice (dominant negative mutant approaches)

Expression or purification of specific molecules or domains of molecules implicated in matrix assembly allows binding studies, as well as determination of the role of cell surface components (receptors and other matrix molecules) in their assembly. Finally, phenotypic clues from human or human mutants can be followed by direct sequence analysis, and recreation of a phenotype in mice or other transgenic animals by insertion of a mutant gene or mutating molecules involved in matrix assembly.

Definition of Matrix Assembly

Matrix assembly results from the predictable and orderly homophilic and heterophilic interactions of extracellular matrix molecules with themselves and with other molecules and the cell surface. This process results in a homo- or heteropolymeric complex of defined biochemical composition and architecture. A common theme in matrix assembly is that initial noncovalent interactions are often followed by formation of covalent cross-links, for example, as formed between collagen molecules. The sites on matrix molecules involved in their assembly and the involvement of other components are arguably best elucidated by the application of distinct but complementary approaches performed under conditions as close to physiological as practical. First, a matrix may be reconstituted by mixing purified components or domains of individual molecules, and the resulting matrix compared with that formed *in vivo*.[4] This approach is particularly fruitful when applied to the study of entropy-driven, self-assembling molecules such as laminin or fibrillar collagens. Second, the assembly of matrix may be perturbed or prevented by addition of domains or antibodies to domains implicated in matrix assembly as detailed below. A more elegant approach is to introduce specific mutations in genes encoding matrix molecules or their receptors, mimicking genetic diseases of matrix assembly. For example, Pereira *et al.*[5] created a line of transgenic mice expressing a human gene for the pro-α1(I) chain of type I procollagen that contained an internal deletion. The transgene mimicked a sporadic in-frame deletion of the human gene causing a lethal variant of osteogenesis imperfecta. Some of the resulting transgenic mice were nonviable owing to extensive fractures at birth, whereas 33% had fractures. Third, expression of cell surface receptors or matrix molecules in cells normally lacking these molecules may be used to establish their role in matrix assembly. The need to utilize physiological conditions implies an *in vivo* approach for definitive results, including transgenic expression and gene targeting covered in [19] on gene knockouts.

Differential Extraction of Matrix from Tissue

Differential extraction with chaotropic agents or enzymes has been utilized to study the macromolecular composition of complex structures such as basal lamina. The essence of this approach is that more stable components resist extraction, and exist in an organized, polymeric structure independent of other preferentially extracted matrix components. For example, when the extracellular matrix of the Englebreth–Holm–Swarm (EHS) tumor is extracted with guanidine hydrochloride, a type IV collagen network remains. Degrading the type IV collagen with collagenase leaves a laminin network. Thus, both matrix components can exist as independent networks. Such observations have been used as an argument that each protomer may assemble independently or that one nucleates the other. However, this approach may not identify other molecules that may initiate matrix assembly or modify the biological properties of the final matrix in important ways. Thus, the existence of such a chemically resistant polymer network may not tell us how it was formed initially.

[4] P. D. Yurchenco and Y. S. Cheng, *J. Biol. Chem.* **268**, 17286 (1993).
[5] R. Pereira, J. S. Khillan, H. J. Helminen, E. L. Hume, and D. J. Prockop, *J. Clin. Invest.* **91**, 709 (1993).

Molecular Approaches to Matrix Assembly

Cloning of cDNAs encoding matrix components and receptors has revolutionized the analysis of matrix assembly. Molecular cloning allows the production of recombinant molecules, domains from these molecules, or fusion proteins. These reagents can be used directly in matrix assembly or cell-binding assays. In addition, the deduced sequence may be used to develop additional reagents such as domain-specific synthetic peptide antibodies, or monoclonal antibodies. The promising approach of combinatorial shape libraries using phage display or other techniques has not been specifically applied to the study of matrix assembly, but has been used to identify integrin-binding motifs.[6,7]

Cultured Cells as Models of Matrix Assembly

Cell culture provides a fruitful method by which to study matrix assembly, particularly the initial stages of matrix assembly prior to formation of an insoluble polymeric complex. Accordingly, cell cultures have been particularly helpful in elucidating the role of specific components in the matrix assembly process, both matrix molecules and receptors. Studies with fibronectin provide an informative example, and point out several requirements for successful matrix assembly assays. These include the following:

1. Knowledge of the complexes formed under normal physiological conditions: Although purified macromolecules or fragments thereof may bind to other molecules with high avidity *in vitro,* this binding is generally of interest only if it mimicks binding events leading to formation of a physiological matrix *in vivo.*

2. A reasonable and reproducible assay by which to quantify the assembly process: For sparingly soluble or insoluble matrix components, this may require considerable thought and careful development of reagents and assay conditions. For example, we developed a simple and accurate method for quantifying insoluble fibronectin by enzyme-linked immunosorbent assay (ELISA). This required screening anti-fibronectin monoclonal antibodies until one was identified that bound denatured, reduced, and carboxymethylated fibronectin. The resulting antibody was then used to develop a sensitive, reproducible quantitative capture ELISA for insolu-

[6] D. J. Kenan, D. E. Tsai, and J. D. Keene, *Trends Biochem. Sci.* **19,** 57 (1994).
[7] E. Koivunen, D. A. Gay, and E. Ruoslahti, *J. Biol. Chem.* **268,** 20205 (1993).

ble fibronectin.[8] Highly hydrated matrices rich in hyaluronan may require special techniques for visualization, such as particle exclusion assays.[9]

3. A source of purified matrix components for fragments for binding studies. For certain relatively abundant components (e.g., fibronectin, certain collagens, and laminin), tissue sources (e.g., of EHS matrix) may serve. However, the existence of tissue-specific forms of matrix components, (e.g., laminin or type IV collagen) is an important caveat for matrix assembly studies.[10] For other molecules, recombinant methods including expression in bacteria[11] or insect cells[12] or mammalian cell culture[13] may be used to produce fragments, map monoclonal libraries, introduce epitope tags, or introduce specific mutations.

Use of Purified Domains and Recombinant Molecules to Study Fibronectin Matrix Assembly: A Paradigm for Studies of Matrix Assembly

Under normal extracellular physiological conditions, fibronectin matrices form only on the surface of certain cells *in vitro,* and at specified sites during embryogenesis *in vivo.*[14] To determine the sites involved in this process, we and others have used two complementary approaches.[15] One is based on inhibiting fibronectin matrix assembly, the other on studying its cell-binding properties. These studies were feasible because of the abundance of plasma fibronectin, and the availability of cDNAs. Recombinant approaches for expressing domains and full-length proteins have subsequently enabled similar methods to be utilized for a variety of matrix molecules. It should be noted that fibronectin matrix assembly appears in part to involve specific receptors on the cell surface, and hence approaches utilized for homophilic, self-assembling systems have not been exploited as fully as in studies of other matrix molecules.

The inhibition approach utilizes either antibodies to defined matrix domains or recombinant or proteolytic fragments themselves as competitive inhibitors of the assembly process. The example of fibronectin serves

[8] B. J. Quade and J. A. McDonald, *J. Biol. Chem.* **263,** 19602 (1988).

[9] P. Heldin and H. Pertoft, *Exp. Cell Res.* **208,** 422 (1993).

[10] E. Engvall, D. Earwicker, T. Haaparanta, E. Ruoslahti, and J. R. Sanes, *Cell Regul.* **1,** 731 (1990).

[11] S. L. Young, L.-Y. Chang, and H. P. Erickson, *Dev. Biol.* **161,** 615 (1994).

[12] T. Tsao, J. C. Hsieh, M. E. Durkin, C. Y. Wu, S. Chakravarti, L. J. Dong, M. Lewis, and A. E. Chung, *J. Biol. Chem.* **265,** 5188 (1990).

[13] J. Sottile, J. Schwarzbauer, J. Selegue, and D. F. Mosher, *J. Biol. Chem.* **266,** 12840 (1991).

[14] R. O. Hynes, "Fibronectins." Springer-Verlag, New York, 1989.

[15] J. A. McDonald, *Annu. Rev. Cell Biol.* **4,** 183 (1988).

as a paradigm for studying the assembly of other multidomain, modular matrix proteins. After purifying large quantities of plasma fibronectin, proteolytic fragments representing domains were purified utilizing affinity, ion-exchange, and molecular sieve chromatography.[16] Proteins found in lower abundance or those more difficult to purify could be expressed utilizing a variety of prokaryotic or eukaryotic vectors that create fusion proteins containing unique sequences supporting affinity purification. Domain-specific, polyclonal antibodies were developed by immunization with the purified domain.[17] The resulting domain-specific antibodies were purified by affinity chromatography on the domain immobilized on Sepharose. This step may be important when utilizing antibodies to domains purified from proteolytic digests of the parent molecule (as opposed to recombinant domains), as the immune system may amplify the response to immunogenic fragments contaminating the purified immunogen. For example, rabbits immunized with the 29-kDa amino-terminal domain of fibronectin purified from proteolytic digests also develop a strong antibody response to immunogenic sequences from the carboxyl terminus present at undetectable concentration in preparations of the purified domain. This is a common problem in the development of polyclonal antibodies to many collagens that are poor immunogens.

A library of monoclonal antibodies was developed using intact fibronectin as an immunogen.[18] These reagents were used in simple qualitative and quantitative matrix assembly assays. The qualitative assay utilized immunohistochemistry to visualize extracellular fibronectin.[18] The quantitative assay for matrix assembly was performed by culturing fibroblasts in the presence of purified fragments of fibronectin, and determining the concentration of detergent insoluble and polymeric fibronectin using a capture ELISA.[8] Studies with monoclonal antibodies and fragments of fibronectin revealed a pattern consistent with the involvement of two sites in fibronectin assembly. The first site was located in the I_{1-5} domain of fibronectin, and the second in the RGD-containing domain mediating $\alpha 3\beta 1$, $\alpha 5\beta 1$, and $\alpha v\beta 1$ binding to fibronectin.

These methods further refined the structure–function map of fibronectin, and demonstrated that both fragments and antibodies to sites contained within these fragments inhibited fibronectin assembly. Cell culture has also been used to express truncated forms of fibronectin to determine the features required for assembly.[13] Retroviral vectors are useful for studying

[16] J. A. McDonald and D. G. Kelley, *J. Biol. Chem.* **255**, 8848 (1980).
[17] J. A. McDonald, T. J. Broekelmann, D. G. Kelley, and B. Villiger, *J. Biol. Chem.* **256**, 5583 (1981).
[18] J. A. McDonald, B. J. Quade, T. J. Broekelmann, R. LaChance, K. Forsman, E. Hasegawa, and S. Skiyama, *J. Biol. Chem.* **262**, 2957 (1987).

large molecules such as fibronectin because of their ability to accommodate reasonably large inserts and high infectivity.[19,20] More traditional eukaryotic expression vectors have also been used. Studies with recombinant forms of fibronectin expressed from both eukaryotic expression vectors and retroviruses have further refined and validated the sites involved in fibronectin matrix assembly.

Inhibition of Matrix Assembly Can Be Used to Establish a Role in Deposition of Other Components

The ability to prevent the formation of a pericellular fibronectin matrix also allows us to assess its role as a template or organizer for other matrix components. One candidate for a matrix molecule organized by fibronectin is BM-90 or fibulin.[21-23] BM-90 binds to the carboxyl-terminal region of fibronectin.[24] This domain of fibronectin does not apparently participate directly in cell–fibronectin binding or matrix organization.[8] This suggested that BM-90 might be organized on the cell surface by fibronectin. To test this, the distributions of fibronectin and BM-90 on the surface of human lung fibroblasts were determined by immunofluorescence microscopy.[25] The two molecules were localized in the same fibers at the light microscope level. A functional role for fibronectin in organizing the BM-90 on the cell surface was established by inhibiting fibronectin assembly using antibodies and proteolytic fragments. When fibronectin assembly was prevented, no BM-90 was detected on the cell surface.[25] Thus, interrupting the assembly of one matrix molecule can be used to dissect its role in the organization of other matrix (and nonmatrix) molecules.

Similar approaches have been utilized to study the role of tenascin-C in lung branching morphogenesis.[11] In this case, the authors used polyclonal antibodies as well as recombinant portions of tenascin-C expressed in bacteria to probe the role of tenascin in branching morphogenesis. Interestingly, although several recombinant fragments inhibited this event *in vitro,* antibodies to the recombinant proteins did not. Moreover, lung development apparently proceeds normally in mice lacking tenascin.[26] These re-

[19] F. J. Barkalow and J. E. Schwarzbauer, *J. Biol. Chem.* **266,** 7812 (1991).
[20] J. E. Schwarzbauer, *J. Cell Biol.* **113,** 1463 (1991).
[21] W. S. Argraves, H. Tran, W. H. Burgess, and K. Dickerson, *J. Cell Biol.* **111,** 3155 (1990).
[22] M. Kluge, K. Mann, M. Dziadek, and R. Timpl, *Eur. J. Biochem.* **193,** 651 (1990).
[23] T. C. Pan, M. Kluge, R. Z. Zhang, U. Mayer, R. Timpl, and M. L. Chu, *Eur. J. Biochem.* **215,** 733 (1993).
[24] K. Balbona, H. Tran, S. Godyna, K. C. Ingham, D. K. Strickland, and W. S. Argraves, *J. Biol. Chem.* **267,** 20120 (1992).
[25] J. Roman and J. A. McDonald, *Am. J. Respir. Cell Mol. Biol.* **8,** 538 (1993).
[26] Y. Saga, T. Yagi, Y. Ikawa, T. Sakakura, and S. Aizawa, *Genes Dev.* **6,** 1821 (1992).

sults (and those obtained with $\alpha 5$ integrin-deficient mice) emphasize the need for *in vivo* correlates to *in vitro* matrix assembly or perturbation assays.

Expression of Cell Surface Receptors and Matrix Molecules in Cultured Cells to Study Matrix Assembly

As noted above, fibronectin matrix assembly appears to be a cell surface-mediated event. Cell culture systems have proved to be particularly useful tools with which to explore the role of specific receptors in extracellular matrix assembly. For example, we have utilized the $\alpha 5$ integrin-deficient Chinese hamster ovary (CHO) cell line B2 isolated by Juliano and co-workers to study the role of integrin matrix receptors in fibronectin assembly.[27] This cell line exhibits low baseline adherence to fibronectin and low matrix assembly activity. These cells were transfected with a plasmid containing an insert encoding the full-length human $\alpha 5$ integrin subunit. When supplemented with exogenous fibronectin, the parental cell line assembled fibronectin into the extracellular matrix. The $\alpha 5$ integrin-deficient cell line could not. Restoring $\alpha 5$ integrin expression by transfection restored the ability to assemble a fibronectin matrix (Fig. 1).[27] This demonstrates that the $\alpha 5\beta 1$ integrin functions in this cell line in an early and essential step in fibronectin matrix assembly. Similar studies were performed by Zhang *et al.*,[28] who expressed $\alpha v\beta 1$ integrin in CHO B2 cells. Although conferring the ability to adhere to fibronectin, this integrin did not support fibronectin assembly. The need to verify such studies *in vivo* is demonstrated by the observations of Yang *et al.*,[29] who have shown that fibroblasts derived from mice lacking the $\alpha 5$ integrin gene assemble a normal-appearing fibronectin matrix.

Expression of cell surface receptors implicated in matrix assembly has also been used to study hyaluron-containing pericellular matrices. Expression of the lymphocyte homing receptor CD44 by COS-7 cells, together with exogenous proteoglycan and hyaluronan, results in assembly of a hyaluron-containing pericellular matrix normally absent from these cells.[30]

[27] C. Wu, J. S. Bauer, R. L. Juliano, and J. A. McDonald, *J. Biol. Chem.* **268**, 21883 (1993).
[28] Z. Zhang, A. O. Morla, K. Vuori, J. S. Bauer, R. L. Juliano, and E. Ruoslahti, *J. Cell Biol.* **122**, 235 (1993).
[29] J. T. Yang, H. Rayburn, and R. O. Hynes, *Development* (*Cambridge, UK*) **119**, 1093 (1993).
[30] W. Knudson, E. Bartnik, and C. B. Knudson, *Proc. Natl. Acad. Sci. U.S.A.* **90**, 4003 (1993).

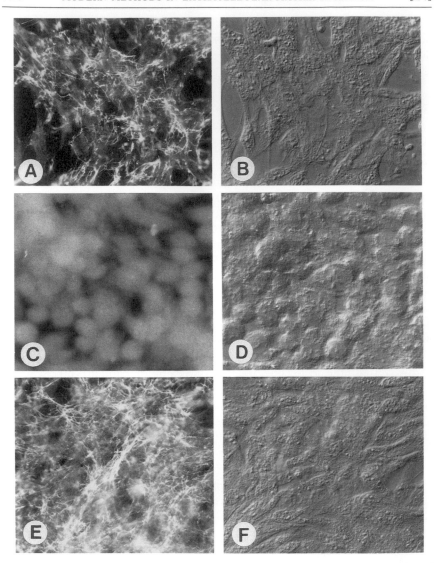

Fig. 1. Use of cell lines transfected with matrix receptors to study matrix assembly and receptor structure–function relationships. The CHO cell line B2, which does not express $\alpha5\beta1$ integrins, adheres poorly to fibronectin and cannot organize a fibronectin-rich pericellular matrix. Transfection of a full-length cDNA encoding the human $\alpha5$ integrin restores the ability to attach and spread on fibronectin, and to assemble a fibronectin-containing matrix. Cells were cultured in the presence of exogenous fibronectin, and studied by differential interference contrast (*right*) or immunofluorescence microscopy (*left*). See Wu *et al.*[27] for complete details. (A and B) CHO wild type assembles a fibronectin matrix. (C and D) $\alpha5$-deficient CHO B2 cells are incapable of assembling fibronectin. (E and F) Expressing $\alpha5\beta1$

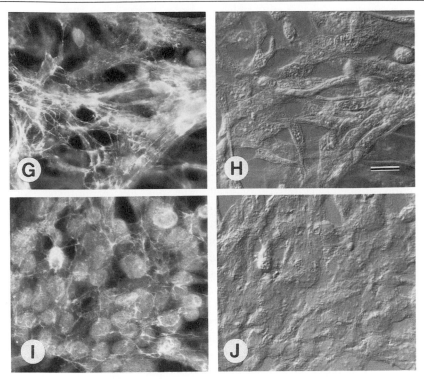

integrins in B2B4 cells using a cDNA encoding the human $\alpha 5$ subunit restores the ability to assemble matrix. (G and H) B2/α1 cells expressing lower levels of the human $\alpha 5$ subunit with a truncated cytoplasmic domain are still capable of assembling a fibronectin matrix. (I and J) 1-23 cells expressing similar levels of $\alpha 5\beta 1$ integrin as the B2/α1 cell line are much less active in matrix assembly. Bar in (H): 200 mm; applies to (A)–(J).

Cell Surface-Binding Assays

Binding studies utilizing proteolytic fragments or recombinant molecules are useful for studying cell–matrix interactions important in matrix assembly. Typical assay conditions involve isolated cells cultured *in vitro* under standard tissue culture conditions. In establishing the assay conditions, the effects of the culture medium, the presence of other matrix molecules, and other factors such as the need for combinations of two tissue types (e.g., mesenchyme and epithelium) should be considered. Typical methods for the detection of bound matrix molecules include labeling via biotinylation, radiolabeling (either metabolic or with tags such as ^{125}I), the use of species-specific antibodies in heterologous systems, or epitope-tagged ligands expressed recombinantly. Microscopy has also been utilized to detect exogenously tagged molecules incorporated into the matrix.[31]

Intracellular Signaling and Matrix Assembly

Cell surface interactions with matrix molecules mediated by integrins involve intracellular signaling.[32] Phosphorylation of integrins and associated molecules localized at focal contacts (e.g., pp125^{fak}) is typically detected in assays involving the plating of suspended cells on specific substrates. After attachment on a specific ligand, the cells are incubated with ^{32}P, and phosphoproteins are detected utilizing immunoprecipitation, sodium dodecyl sulfate–polyacrylamide gel electrophoresis (SDS–PAGE), and autoradiography.[33–35] Using this approach, the phosphorylation of p125^{fak} has been studied in cells interacting with fibronectin. Dedhar *et al.* have extended this analysis of postreceptor ligand events by utilizing antibodies to cluster integrins, as well as β1-activating antibodies. Their reports should be consulted for full details.[36,37]

Specific inhibitors and activators of signaling pathways have also been applied to dissect the interaction of extracellular matrix molecules with receptors. Inhibitors of protein kinase C decrease binding of the 29-kDa amino-terminal domain of fibronectin to cells,[38] and phosphorylation of p125^{fak}. Upregulation of protein kinase C activity augments $\alpha5\beta1$ integrin-mediated cell spreading.[39]

Other Methods of Measuring Interactions of Matrix Molecules with Cells and Other Molecules

A number of methods covered elsewhere in this volume have been utilized to study matrix interactions. Ingham *et al.* have used traditional biophysical approaches including fluorescent polarization to determine K_d

[31] D. M. Peters, L. M. Portz, J. Fullenwider, and D. F. Mosher, *J. Cell Biol.* **111,** 249 (1990).
[32] R. O. Hynes, *Cell (Cambridge, Mass.)* **69,** 11 (1992).
[33] L. J. Kornberg, H. S. Earp, C. E. Turner, C. Prockop, and R. L. Juliano, *Proc. Natl. Acad. Sci. U.S.A.* **88,** 8392 (1991).
[34] L. Kornberg, H. S. Earp, J. T. Parsons, M. Schaller, and R. L. Juliano, *J. Biol. Chem.* **267,** 23439 (1992).
[35] R. L. Juliano and S. Haskill, *J. Cell Biol.* **120,** 577 (1993).
[36] C. Kapron-Bras, L. Fitz-Gibbon, P. Jeevaratham, J. Wilkins, and S. Dedhar, *J. Biol. Chem.* **268,** 20701 (1993).
[37] J. McGlade, B. Brunkhorst, D. Anderson, G. Mbamalu, J. Settleman, S. Dedhar, M. Rozakis-Adcock, L. B. Chen, and T. Pawson, *EMBO J.* **12,** 3073 (1993).
[38] C. E. Somers and D. F. Mosher, *J. Biol. Chem.* **268,** 1 (1993).
[39] K. Vuori and E. Ruoslahti, *J. Biol. Chem.* **268,** 21459 (1993).

values for the interaction of purified fibronectin and fibronectin domains with collagen.[40] This approach, like most other quantitative methods, has the advantage of actually measuring affinities and binding constants, albeit in the simplified milieu utilized by the physical chemist. Expression of specific domains using systems that produce fusion proteins such as glutathione S-transferase allows immobilization of one binding partner and addition of others in the liquid phase. This has been used to advantage to study the interaction of multidomain proteins such as Src and pp125^{fak}.[41] The ELISA has also been utilized for similar purposes,[42] although care must be taken as certain interactions are affected by binding to the plastic substrate. Similar approaches have been used to map binding sites implicated in assembly of basal lamina components including laminin, type IV collagens, and entactin. Proximity detection assays using surface plasmon resonance [Pharmacia (Piscataway, NJ) BiaCore instrument] promise to become useful approaches for probing specific matrix interactions. They offer the advantage of determining k_1 and k_{-1} for binding of one component in a complex mixture to a potential ligand immobilized on a chemically derivatized surface.[43]

Peptides as Probes of Extracellular Matrix Interactions

Synthetic peptides have become powerful tools with which to dissect the phenomena of matrix assembly. They have been used to screen for and mimic recognition sequences in matrix molecules. Peptides have been isolated that inhibit matrix interactions and that stimulate binding or self-assembly. A peptide contained in the homophilic interactive site contained in the first type III repeat in fibronectin has also been used to stimulate fibronectin matrix assembly *in vitro* in the absence of cells.[44] The advent of multiple peptide synthesis has made the creation of overlapping peptide libraries such more affordable. In addition to direct probes of structure–function relationships in matrix assembly, peptides allow the creation of synthetic peptide antibodies directly from sequences deduced from cDNAs.

Antiidiotype or Molecular Mimicry Approaches

Although not yet employed for studies of matrix assembly (to the author's knowledge), antibodies may also serve to identify cognate ligands

[40] K. C. Ingham, S. A. Brew, and B. S. Isaacs, *J. Biol. Chem.* **263**, 4624 (1988).
[41] J. D. Hildebrand, M. D. Schaller, and J. T. Parsons, *J. Cell Biol.* **123**, 993 (1993).
[42] A. Morla and E. Ruoslahti, *J. Cell Biol.* **118**, 421 (1992).
[43] M. Malmqvist, *Nature (London)* **361**, 186 (1993).
[44] A. Morla, Z. Zhang, and E. Ruoslahti, *Nature (London)* **367**, 193 (1994).

via molecular mimicry or antiidiotype approaches. These techniques utilize the information implicit in the hypervariable region of an antibody to determine complementary binding sites in other molecules. This is a specific application of what has been termed the combinatorial shape library method.[6]

In Vivo Approaches

The most common interstitial collagen, type I, is a paradigm for homophilic assembly into highly ordered lateral arrays.[45] The registration required for proper alignment is so precise that single amino acid substitutions can result in disordered fibrillogenesis and subsequent alterations in function.[46] Human mutants in type I collagen that alter fibrillogenesis result in a number of human diseases.[47] Some of these have been replicated in transgenic mouse models, particularly useful in recreating dominant negative mutations.[5]

Direct observation of the ordered complexes formed from purified molecules or precursors by biophysical means and direct imaging by electron microscopy have been particularly useful tools in the study of homophilic self-assembly. These techniques yield information about the arrangement of individual molecules in the polymeric assemblies, as well as the impact of other molecules on the rate of assembly and structure of the final complexes. Prockop has emphasized the susceptibility of these arrays to interruption by mutations, leading to inherited disorders of connective tissue with diverse phenotypes.[46] Despite their intrinsic self-assembly properties, cells may play an important role in organizing collagenous matrices by vectorial secretion.[1] The ultrastructural and image analysis techniques utilized in these studies have been reviewed extensively.[1]

Another large molecule whose structure is disrupted in an inherited human disorder is fibrillin. Mutations in the fibrillin gene are associated with Marfan's syndrome. Fibrillin appears to play a critical role in elastin deposition, although the precise molecular interactions involved in this association are not yet known.[48] In vitro approaches outlined above should help elucidate the mechanisms by which fibrillin mediates elastin assembly in the extracellular matrix.

[45] T. F. Linsenmayer, in "Cell Biology of Extracellular Matrix" (E. D. Hay, ed.), 2nd ed., p. 7. Plenum, New York, 1991.

[46] D. J. Prockop, J. Biol. Chem. 265, 15349 (1990).

[47] D. J. Prockop, N. Engl. J. Med. 326, 540 (1992).

[48] F. Ramirez, L. Pereira, H. Zhang, and B. Lee, BioEssays 15, 589 (1993).

Gene Targeting

Definitive evidence for the role of matrix molecules, specific domains of these molecules, and receptors can now be sought via the powerful technique of gene targeting. The ability of cells derived from $\alpha 5$ integrin-deficient mice to assemble fibronectin has been mentioned above. This raises the question of redundancy in matrix assembly. More precise techniques such as hit-and-run mutations, creation of specific mutations, or use of dominant negative mutation strategies *in vivo* will likely represent the next major advance in studies of matrix assembly.

Section VI

Use of Extracellular Matrix

[25] Cell Differentiation by Extracellular Matrix Components

By RUEDIGER J. BLASCHKE, ANTHONY R. HOWLETT,
PIERRE-YVES DESPREZ, OLE W. PETERSEN, and MINA J. BISSELL

Introduction

The differentiated status of single cells within multicellular organisms and tissues is characterized by their organization into well-defined three-dimensional structures, and the expression of tissue-specific genetic programs. To establish, maintain, and coordinately change these differentiated characteristics during development, cells must constantly exchange information with their microenvironment.

The nature of signals that individual cells receive and send, as well as the mechanisms of signal processing, are extremely complex and not fully understood. A major breakthrough in our perception of the informational network that regulates cell differentiation was made, however, by the fundamental discovery that the extracellular matrix (ECM) takes an active part in this signaling process rather than providing merely a mechanical framework for tissue architecture.[1,2] In addition to, and in many cases synergistically with, a variety of other factors including hormones, vitamins, growth factors, and cell–cell interactions, the ECM participates directly in the control of differential gene expression.[3–6]

The variety of the ECM signaling capacity is reflected by the diversity of its molecular composition. Besides the primary structural components such as collagen, elastin, and several proteoglycans, the ECM contains a tissue-specific configuration of proteins that determines its adhesive and informative properties. Examples of these kinds of components include laminin, type IV collagen, and fibronectin.

---

[1] E. D. Hay, "Cell Biology of Extracellular Matrix." Plenum, New York, 1981.

[2] M. J. Bissell, H. G. Hall, and G. Parry, *J. Theor. Biol.* **99,** 31 (1982).

[3] J. C. Adams and F. M. Watt, *Nature (London)* **340,** 307 (1989).

[4] A. Ben-Ze'ev, G. S. Robinson, N. L. R. Bucher, and S. R. Farmer, *Proc. Natl. Acad. Sci. U.S.A.* **85,** 2161 (1988).

[5] L. M. Reid, *Curr. Opin. Cell Biol.* **2,** 121 (1990).

[6] A. W. Stoker, C. H. Streuli, M. Martins-Green, and M. J. Bissell, *Curr. Opin. Cell Biol.* **2**(5), 864 (1990).

METHODS IN ENZYMOLOGY, VOL. 245

Copyright © 1994 by Academic Press, Inc.
All rights of reproduction in any form reserved.

Additional complexity in ECM signaling results from feedback mechanisms described as "dynamic reciprocity." [2,7] As a result of the signal transduction from the ECM, differentiating cells are able to change their own microenvironment by producing and secreting a different set of ECM components as well as enzyme activities that are required for ECM remodeling.[8,9] These changes in ECM composition, of course, create a novel and modified signaling capacity. Considering the dynamic reciprocity, the ECM appears to be a highly flexible and constantly changing structure that is able to respond perfectly to a broad variety of external impacts and to fulfill the essential requirements of developmental plasticity. This flexibility, together with the enormous informational content of the ECM, poses a challenge to understanding the detailed molecular mechanisms underlying the ECM signaling.

Possible Mechanisms of Extracellular Matrix Signaling

Whereas the active role of the ECM in the regulation of cell differentiation is well recognized, only a few molecular details of the implicated signal transduction pathways have been deciphered to date. It is clear, however, that the initial step of informational transfer across the plasma membrane involves the binding of individual ECM components to suitable cell surface receptors.

The integrins, a widely expressed family of transmembrane proteins, are major binding partners for several ECM components. The members of this family are believed to be the predominant mediators of this cross-membrane information transfer.[10] The fact that all integrins are assembled from α and β subunits and are functional only as $\alpha\beta$ heterodimers is of special interest with regard to their importance in ECM signal transduction. Because the ligand specificity is determined by the $\alpha\beta$ configuration, a given cell can filter the appropriate information presented by its surrounding by expressing a defined assortment of integrin subunits.[11]

Using heterodimeric receptors probably also allows the cell to modulate the information represented by individual ECM components according to its specific needs. Even if two different $\alpha\beta$ integrin heterodimers are

[7] M. J. Bissell and J. Aggler, in "Mechanisms of Signal Transduction by Hormones and Growth Factors" (M. C. Cabot and W. Cabot, eds.), p. 251. Alan R. Liss, New York, 1987.

[8] C. H. Streuli and M. J. Bissell, J. Cell Biol. **110,** 1405 (1990).

[9] Z. Werb, P. M. Tremble, O. Behrendtsen, E. Crowly, and C. H. Damsky, J. Cell Biol. **109,** 877 (1989).

[10] C. H. Damsky and Z. Werb, Curr. Opin. Cell Biol. **4,** 772 (1992).

[11] R. O. Hynes, Cell (Cambridge, Mass.) **69,** 11 (1992).

able to bind the same ligand, they might translate this signal into different intracellular responses.[12]

Currently, a major approach toward a more detailed understanding of the molecular mechanisms involved in the ECM–integrin-mediated signal transduction is based on the hypothesis that such signaling is integrated into biochemical reaction cascades involving protein phosphorylation. An interesting aspect of this hypothesis is the similarity to signal transduction initiated by growth factors. According to this model, the integrins could be considered as analogs of growth factor receptor tyrosine kinases.[13] In fact, these two types of receptors share several similarities. Both are class I transmembrane proteins, they operate only as dimers, and they are positioned at the interface of extracellular–intracellular transmission of information. The predominant difference, however, is the complete absence of any detectable enzymatic activity from the integrin receptors. Conceivably, the first step in the ECM–integrin-triggered signal transduction must include the association of the "ligand-activated" integrin receptor with such enzymatic activities. The resulting ligand–integrin–enzyme complexes would then, in turn, initiate the biochemical reaction cascades. Although these kinds of complexes have not yet been identified, there is substantial evidence for integrin-triggered protein phosphorylation.

Kornberg and colleagues have shown that clustering of integrins on the surface of human carcinoma cells by anti-integrin antibodies leads to an increased phosphorylation of a complex of proteins with molecular masses between 120 and 130 kDa.[14] One component of this protein complex cross-reacts with antibodies against the protein tyrosine kinase pp125^{fak}.[15] This correlation between ECM–integrin interactions and pp125^{fak} phosphorylation has also been observed in other cell systems including mouse fibroblasts[16] and human platelets.[17]

When mouse fibroblasts are plated on fibronectin, pp125^{fak} is attracted to focal adhesion contacts and phosphorylated on tyrosine residues. Human platelets deficient for the major fibronectin receptor, αIIb/β3 integrin, as seen in Glanzman's thrombosthemia patients, show a markedly reduced pp125^{fak} activation in response to thrombin-mediated aggregation. Modu-

[12] R. L. Juliano and S. Haskill, *J. Cell Biol.* **120,** 577 (1993).
[13] A. Ullrich and J. Schlessinger, *Cell (Cambridge, Mass.)* **61,** 203 (1990).
[14] L. Kornberg, H. S. Earp, C. Turner, C. Prokop, and R. L. Juliano, *Proc. Natl. Acad. Sci. U.S.A.* **88,** 8392 (1991).
[15] L. Kornberg, H. S. Earp, J. T. Parsons, M. Schaller, and R. L. Juliano, *J. Biol. Chem.* **267,** 23439 (1992).
[16] K. Burridge, C. E. Turner, and L. Romer, *J. Cell Biol.* **119,** 893 (1992).
[17] L. Lipfert, B. Haimovich, M. D. Schaller, B. S. Cobb, T. J. Parsons, and J. S. Brugge, *J. Cell Biol.* **119,** 905 (1992).

lation of the pp125^{fak} activity seems therefore to be a general and cell type-independent downstream event of integrin-mediated signal transduction.

It has been shown that pp125^{fak} can directly interact with,[18] and is a substrate for, pp60^{src}-like kinases.[19] This might suggest that, rather than being a direct consequence of ECM–integrin interactions, the activation of pp125^{fak} is indirect, and is located further downstream within this pathway. Several lines of evidence argue for an involvement of the cytoskeleton in the activation of pp125^{fak}. It is known, for example, that β1 and β3 integrins can bind directly to cytoskeletal proteins such as α-actinin[20] and talin[21] and that the disruption of the actin-filament organization by cytochalasin D treatment can block the pp125^{fak} phosphorylation.[16] It is therefore possible that one major function of ECM–integrin complexes in the initialization of intracellular signaling is the organization of complex, supramolecular "reaction centers" on the inner surface of the membrane via the cytoskeleton. According to this hypothesis, the establishment of a certain configuration of individual components within such complexes could be the actual trigger for the observed phosphorylation events.

The cytoskeleton seems not only to be involved in the activation of protein tyrosine kinases but might also be one of the major targets of the ECM–integrin-initiated phosphorylation cascades. Paxilin, a cytoskeletal protein that is concentrated in focal adhesions, is rapidly phosphorylated on tyrosine residues in response to ECM adhesion of rat embryonic fibroblasts.[15] Although the consequences of this phosphorylation are still elusive, it is conceivable that among other effects it creates a binding site for Src homology 2 (SH2) domain-containing proteins. The SH2 amino acid motif is commonly found in proteins involved in signal transduction and is known to bind tightly to phosphorylated tyrosine residues.[22] Interestingly, tensin, the second cytoskeletal protein that is tyrosine phosphorylated after rat embryo fibroblasts are plated on ECM substrata,[23] contains such an SH2 domain.[24] Although it is not known how the individual cytoskeletal components are organized within the centers of ECM–integrin

[18] B. S. Cobb, M. D. Schaller, Z. Horng-Len, and J. T. Parsons, *Mol. Cell. Biol.* (in press).
[19] M. D. Schaller, C. A. Borgmann, B. S. Cobb, R. R. Vines, A. B. Reynolds, and J. T. Parsons, *Proc. Natl. Acad. Sci. U.S.A.* **89,** 5192 (1992).
[20] C. A. Otey, F. M. Pavalko, and K. Burridge, *J. Cell Biol.* **111,** 721 (1990).
[21] A. Horwitz, K. Duggan, C. Buck, M. C. Berckerle, and K. Burridge, *Nature (London)* **320,** 531 (1986).
[22] C. A. Koch, D. Anderson, M. F. Moran, C. Ellis, and T. Pawson, *Science* **252,** 668 (1991).
[23] S. M. Bockholt and K. Burridge, *J. Biol. Chem.* **268,** 14565 (1993).
[24] S. Davis, M. L. Lu, S. Lin, J. A. Butler, B. J. Drucker, T. M. Roberts, Q. An, and L. B. Chen, *Science* **252,** 712 (1991).

contacts, it is tempting to deduce from these data that phosphorylation-dependent interactions are involved. Recent results show that in mouse mammary cells, both cytoskeletal changes and protein phosphorylation are necessary for proper signal transduction.[24a]

Unfortunately, we have no indications of the specific biochemical pathways that connect these early events with the downstream effects of integrin-related signaling that finally results in the activation of differentiation-specific gene expression; nor do we know how the implicated reaction cascades are triggered.

It is likely, however, that additional, yet unidentified components including other kinases and/or phosphatases participate in ECM–integrin-initiated intracellular pathways. The identification of such molecules and their integration into the emerging reaction sequences are subjects of an active field of research. Further insights into the biochemistry of these pathways are strongly dependent on the use of model systems that allow the dissection of individual steps of the informational flow from the cell membrane to the nucleus and finally to the suitable target genes.

Mammary Gland as Model for Extracellular
Matrix-Regulated Differentiation

The mammary gland lends itself well as a model system to investigate epithelial cell proliferation, differentiation, and tumorigenesis. Unlike most other organs, the mammary gland develops mostly in the adult animal, reaching its fully functional status only during late stages of pregnancy and lactation. Pregnancy induces a massive proliferation of the epithelial component of the gland, which leads to a branching morphogenesis and development of secretory alveoli from a previously primitive and rather rudimentary duct system. This morphogenesis is accompanied by the production and secretion of milk proteins by the epithelial cells.[25] After the lactation period, this massively developed epithelium is dismantled during involution by a process that appears to involve programmed cell death (apoptosis).[26]

Studies have shown that the extensive ECM remodeling during these developmental processes is not merely a consequence of cell differentiation but, according to the paradigm of "dynamic reciprocity," the ECM itself has a profound influence on the differentiation status of the mammary

[24a] C. D. Rosuelly, P. Y. Desprez, and M. J. Bissell, *Proc. Natl. Acad. Sci. U.S.A.* (in press).
[25] M. C. Nevill and C. W. Daniel, eds., "The Mammary Gland: Development, Regulation and Function." Plenum, New York, 1987.
[26] R. Strange, F. Li, S. Sauer, A. Burkhard, and R. R. Friis, *Development (Cambridge, UK)* **115,** 49 (1992).

epithelium.[2,27-30] Because most of our knowledge about the role of the ECM as a regulator of this differentiation has been obtained by developing and employing mouse cell culture systems, these experiments are now discussed in more detail.

Mouse Mammary Gland Cell Cultures

Primary mouse mammary cells have been successfully isolated from midpregnant mice and maintained under standard cell culture conditions.[31] These cells, however, change their morphology and shut down milk protein synthesis when cultured on tissue culture plastic even in the presence of appropriate lactogenic hormonal stimuli.[32,33] The fully differentiated phenotype including milk protein expression is reestablished only when these cells are cultured on a laminin-rich extracellular matrix isolated from Engelbreth-Holm–Swarm (EHS) tumors.[34,35] Under these conditions the epithelial cells reorganize into alveolar-like structures, initiate milk protein synthesis, and secrete these proteins vectorially into the central lumen of the newly formed structures (Fig. 1). These results clearly demonstrate that the ECM is absolutely required for the functional differentiation of mammary epithelial cells. The findings, however, do not allow discrimination between a direct ECM signaling and an indirect effect of the substratum by simply allowing the cells to form an appropriate three-dimensional structure and to establish proper cell–cell contacts. To rule out cell–cell contacts as the crucial signals for milk protein expression, a "single-cell assay" has been developed and is used in our laboratory.[36]

In these experiments monolayer cultures of primary or secondary, nonfunctional, mammary epithelial cells are dissociated from each other,

[27] M. L. Li, J. Aggler, D. A. Farson, C. Hatier, J. Hassell, and M. J. Bissell, *Proc. Natl. Acad. Sci. U.S.A.* **84**, 136 (1987).

[28] R. S. Talhouk, R. J. Chin, E. N. Unemori, Z. Werb, and M. J. Bissell, *Development (Cambridge, UK)* **112**, 439 (1991).

[29] C. J. Sympson, R. S. Talhouk, C. M. Alexander, J. R. Chin, S. M. Clift, M. J. Bissell, and Z. Werb, *J. Cell Biol.* **125**, 681 (1994).

[30] A. R. Howlett and M. J. Bissell, *Epithelial Cell Biol.* **2**, 79 (1993).

[31] J. T. Emerman, J. Enami, D. R. Pitelka, and S. Nandi, *Proc. Natl. Acad. Sci. U.S.A.* **74**, 4466 (1977).

[32] J. T. Emerman and D. R. Pitelka, *In Vitro* **13**, 316 (1977).

[33] E. Y.-H. Lee, W.-H. Lee, C. S. Kaetzel, G. Parry, and M. J. Bissell, *Proc. Natl. Acad. Sci. U.S.A.* **82**, 149 (1985).

[34] L.-H. Chen and M. J. Bissell, *Cell Regul.* **1**, 45 (1989).

[35] M. H. Barcellos-Hoff, J. Aggler, T. G. Rom, and M. J. Bissell, *Development (Cambridge, UK)* **105**, 223 (1989).

[36] C. H. Streuli, N. Bailey, and M. J. Bissell, *J. Cell Biol.* **115**, 1383 (1991).

embedded into a reconstituted basement membrane substratum, and assayed for β-casein expression by immunofluorescence. The expression of β-casein by single cells, as shown by these assays, allowed for the first time the demonstration that a direct signal transduction from the ECM is necessary for mammary gland epithelial cell differentiation. This finding was supported, and put in even more concrete terms, by experiments in which specific ECM–cell interactions were blocked by anti-β1 integrin antibodies.[36] Studies have shown that laminin is in fact the essential component of the ECM that is required for the differentiation-specific expression of the β-casein gene.[37]

To further characterize the molecular mechanisms underlying this ECM-induced milk protein gene expression, a novel cell strain CID-9 has been developed in our laboratory.[38] This cell strain represents an epithelial cell-enriched fraction of its parental strain COMMA-1-D,[39] which in turn was derived from a mammary gland of a midpregnant mouse. Most important, CID-9 cells maintain their physiological responsiveness to ECM substratum. When cultured on EHS matrix, they not only form structures that resemble the alveoli formed during pregnancy *in vivo* but the epithelial cells of CID-9 also exhibit a strict ECM dependency of milk protein expression. This cell strain therefore provides a functional model system to study ECM-induced differentiation and gene expression *in vitro*.

By using CID-9 cells that were stably transfected with progressive deletions of bovine β-casein promotor-CAT constructs it has been shown that this promotor contains an ECM-response element (BCE 1) that is located approximately 1600 bp upstream of the transcription start point. BCE 1 is functional in a position- and orientation-independent manner, and confers ECM-dependent transcriptional activity to heterologous viral promoters.[40] It therefore completely fulfills the criteria of an enhancer element. This ECM-responsive transcriptional enhancer is an extremely powerful tool that can be used to investigate further the basic mechanisms by which the ECM regulates gene expression.

[36] Deleted in proof.

[37] C. H. Streuli, C. Schmidhauser, N. Bailey, P. Yurchenko, A. P. N. Shubitz, and M. J. Bissell, submitted for publication.

[38] C. Schmidhauser, M. J. Bissell, C. A. Myers, and G. F. Casperson, *Proc. Natl. Acad. Sci. U.S.A.* **87,** 9118 (1990).

[39] K. Danielson, C. J. Oborn, E. M. Durban, J. S. Buetel, and D. Medina, *Proc. Natl. Acad. Sci. U.S.A.* **81,** 3756 (1984).

[40] C. Schmidhauser, G. F. Casperson, C. A. Myers, K. T. Sanzo, S. Bolten, and M. J. Bissell, *Mol. Biol. Cell* **3,** 699 (1992).

Human Mammary Gland Cell Cultures

Human mammary epithelial cells have been isolated from surgical discard tissue (reduction mammoplasty) and from milk and cultured successfully in complex, serum-free medium formulations, in media supplemented with serum, and/or in conditioned media from stromal cells as discussed below. With the exception of epithelial cells isolated from milk, human mammary epithelial cultures are typically established from resting, nonpregnant breast tissue. Similar to the mouse, these cells proliferate continuously in monolayer culture until they senesce but display few differentiated characteristics. Given the evidence provided by studies in the mouse the possibility that human mammary epithelial cell differentiation might also be subject to regulation by external microenvironmental effectors such as extracellular matrix has been raised. Furthermore, there has been an interest in the possibility that microenvironmental regulators of epithelial function, such as ECM, might also influence the process of tumor formation.[41–44] In the mammary gland, deregulation of the interaction of tumor cells with stroma and ECM appear to be of general importance in breast cancer.[45–49] However, the precise mechanisms underlying these interactive processes and the means by which they contribute to the carcinogenesis process remain obscure. To define potential mechanisms of microenvironmental control in human breast cell differentiation and tumorigenesis it is necessary to have a well-defined culture model in which the patterns of normal and tumor cell behavior are preserved. Our group has developed a rapid and versatile assay system that utilizes extracellular matrix to allow accurate discrimination of normal and malignant human breast epithelial cells in culture.[50] Using this assay system we demon-

[41] T. Sakakura, *in* "Understanding Breast Cancer: Clinical Laboratory Concepts" (M. A. Rich, J. C. Hager, and P. Furmanski, eds.), p. 261. Dekker, New York, 1983.

[42] A. Van den Hoof, *Adv. Cancer Res.* **50,** 159 (1988).

[43] R. Chiquet-Ehrismann, P. Kalla, and C. A. Pearson, *Cancer Res.* **49,** 4322 (1989).

[44] P. Bassett, J. P. Bellocq, C. Wolf, I. Stoll, P. Hutin, J. M. Limacher, O. L. Podhajcer, M. P. Cherard, M. C. Rio, and P. Chambon, *Nature (London)* **348,** 699 (1990).

[45] S. L. Schor, J. A. Haggie, P. Durning, A. Howell, L. Smith, R. S. A. Sellwood, and D. Crowther, *Int. J. Cancer* **37,** 831 (1986).

[46] S. Z. Haslam, *in* "Regulatory Mechanisms in Breast Cancer" (M. Lippman and R. Dickson, eds.), p. 401. 1991.

[47] R. Chiquet-Ehrismann, E. J. Mackie, C. A. Pearson, and T. Sakakura, *Cell (Cambridge, Mass.)* **47,** 131 (1986).

[48] E. J. Ormerod, M. J. Warburton, B. Gusterson, C. M. Hughes, and P. S. Rudland, *Histochem. J.* **17,** 1155 (1985).

[49] M. H. Barcellos-Hoff, *Cancer Res.* **53,** 3880 (1993).

[50] O. W. Peterson, L. R. Rønnov-Jessen, A. R. Howlett, and M. J. Bissell, *Proc. Natl. Acad. Sci. U.S.A.* **89,** 9064 (1992).

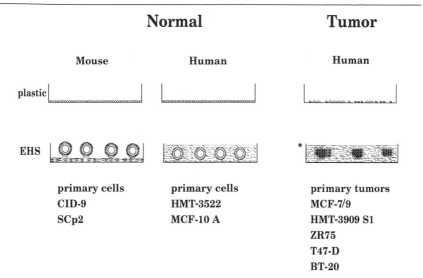

FIG. 1. Morphology of mammary cells in monolayer and EHS-culture. *, not drawn to scale.

strated, for the first time, the capacity of human breast epithelial cells to reexpress their *in vivo* patterns of growth and differentiation when placed within EHS, the laminin-rich reconstituted basement membrane material isolated from EHS tumors. In this assay, normal breast epithelial cells form correctly polarized organotypic spheres resembling acini *in situ*, deposit an endogenous basement membrane, and growth arrest by 7 days in culture. Malignant breast tumor cells on the other hand, form large, dense, unpolarized colonies of cells that fail to deposit basement membrane (BM) and grow continuously (Fig. 1).

The development of such a rapid assay for discriminating normal cells from tumor cells has paved the way for the types of mechanistic studies that until now have been afforded only by rodent models and provides new possibilities for study of the functional significance of microenvironment in breast cancer.

Material and Methods

Preparation of Engelbreth-Holm–Swarm Reconstituted Basement Membrane Matrix[27,36,51]

A basement membrane-secreting tumor, initially described by Engelbreth-Holm and Swarm, appears spontaneously in ST/EH mice. A lami-

[51] D. Medina, M. L. Li, C. J. Oborn, and M. J. Bissell, *Exp. Cell Res.* **172**, 192 (1987).

nin-rich reconstituted basement membrane (EHS matrix) is extracted from this tumor material according to the following method. Ten male C57BL mice are routinely injected intramuscularly with 0.2 ml of tumor homogenate (2 g/ml). Fourteen days after injection, the mice should be made lathyritic to block cross-linking of collagens: 0.1% of BAPN (β-aminopropionitrile fumarate) is added to their water. Seventy to 80 g of tumor can be harvested 4 weeks after injection. The EHS matrix is prepared from the tumor material according to the method described by Kleinman et al.,[52] as detailed below. EHS matrix is also commercially available as Matrigel (Collaborative Research, Bedford, MA).

All operations are performed at 4° or on ice.

Stock Solutions

All stock solutions are to be stored at 4°.

EHS buffer 1
   NaCl (3.4 M)   99.32 g
   Tris-HCl (50 mM), pH 7.4   25 ml of 1 M stock
   Sodium EDTA (4 mM)   10 ml of 0.2 M stock
   Water   to 500 ml
EHS buffer 2
   NaCl (0.2 M)   20 ml of 5 M stock
   Tris-HCl (50 mM), pH 7.4   25 ml of 1 M stock
   Sodium EDTA (4 mM)   10 ml of 0.2 M stock
   Urea (2 M), ultrapure   60 g
   Water   to 500 ml
EHS buffer 3
   NaCl (0.15 M)   90 ml of 5 M stock
   Tris-HCl (50 mM), pH 7.4   150 ml of 1 M stock
   Sodium EDTA (4 mM)   60 ml of 0.2 M stock
   Water   to 3 liters; autoclave

1. *N*-Ethylmaleimide (NEM, 200 mM) is prepared in water, sterilized by filtration, and stored in 10-ml aliquots at $-20°$.

2. Add 1.65 ml of 200 mM NEM to 165 ml of EHS buffer 1.

3. Add 400 μl of 200 mM NEM to 40 ml of EHS buffer 2.

Tumor tissue is weighed and suspended in EHS buffer 1–NEM at 1 g/ml. The tissue is then homogenized with a motorized homogenizer, followed by centrifugation in an SW 27 rotor at 25,000 rpm (112,000 g) for 30 min at 4°. The recovered EHS material in the pellet is dissolved in

[52] H. K. Kleinman, M. L. McCarvey, L. A. Liotta, P. G. Robey, K. Tryggvason, and G. R. Martin, *Biochemistry* **21**, 6188 (1982).

EHS buffer 2–NEM overnight and clarified by centrifugation as above. The EHS matrix-containing supernatants are dialyzed three times against EHS buffer 3–NEM (30 ml of 200 mM NEM in 3 liters of EHS buffer 3) in the presence of 0.1% (v/v) chloroform and then dialyzed against serum-free medium. Total time of dialysis is 48 hr at 4°. The isolated EHS extracellular matrix is frozen at −80° until ready to use.

Each preparation contains 5–10 mg/ml protein, and is routinely assayed for purity and lack of protein degradation on 5% (w/v) reducing sodium dodecyl sulfate (SDS)–polyacrylamide gels followed by silver staining. The EHS extract is thawed slowly on ice and coated dishes are prepared by carefully spreading EHS matrix at 15 μl/cm^2 onto dishes kept on ice. The plates are then incubated for 1–2 hr in a 37° humidified incubator to allow the EHS matrix to gel.

Epithelial Cell Isolation[53–55]

Epithelial cell isolation is carried out according to the procedure of Emerman and Pitelka,[32] except that a different dissociation medium can be used to obtain higher cell yield. This dissociation medium contains collagenase [type II, 0.3% (w/v)], trypsin (0.15%, w/v), fetal calf serum (5%, v/v), F10 medium (0.98%, v/v), sodium bicarbonate (0.12%, w/v), and N-2-hydroxyethylpiperazine-N′-2-ethanesulfonic acid [HEPES, 0.26% (v/v)]. The hormones insulin, hydrocortisone, and prolactin are added (5 μg/ml) just before use and are present throughout culturing. Mammary gland epithelial (MME) cells from pregnant BALB/c mice are isolated by differential centrifugation and/or Percoll. Finely chopped tissue is dissociated at 37° for 50 min. Undissociated tissue is then removed and redigested with fresh dissociation solution for 30 min. The two cell isolates are pooled, and epithelial cells are isolated by differential centrifugation. Cells are counted in a hemocytometer and plated at 3 × 10^5 epithelial cells/cm^2 on 35-mm plastic tissue culture dishes in medium 199 containing gentamicin (50 μg/ml), 11 mM glucose, 10% (v/v) fetal calf serum (FCS), and hormones (insulin, 5 μg/ml; hydrocortisone, 2 μg/ml; prolactin, 3 μg/ml) at 37° in 95% air and 5% CO$_2$. Serum is eliminated from the medium after the first day in culture. The medium is changed every other day. In experiments in which Percoll gradients are used to check the purity of the cells, discontinuous gradients of Percoll (16 ml; 20 to 80%, in 10% steps) in medium 199 are made. Cells (3 × 10^6) are

[53] E. Y. H. Lee, G. Parry, and M. J. Bissell, *J. Cell Biol.* **98**, 146 (1984).
[54] E. Y. H. Lee, M. H. Barcellos-Hoff, L. H. Chen, G. Parry, and M. J. Bissell, *in Vitro Cell. Dev. Biol.* **23**, 221 (1987).
[55] J. T. Emerman and M. J. Bissell, *Adv. Cell Cult.* **6**, 137 (1988).

loaded onto each gradient and then centrifuged at 2350 rpm for 30 min (CFU 500 equipped with a 259 rotor; IEC Centrifuge, Needham Heights, MA). The cells distribute themselves to the top bands. These are removed, the suspensions are diluted with medium 199, and the cells are collected by centrifugation at 800 rpm for 3 min.

Single-Cell Assay from Secondary Cultures[36]

The cells are plated at 4.7×10^5 cells/cm^2 onto various substrata using conditions identical to those for the primaries. For the suspension assays, cells are embedded inside EHS matrix. Freshly harvested cells (10^6) are pelleted for 2 min at 1300 g in sterile microfuge tubes, chilled on ice, immediately resuspended with mild trituration to a density of $2.2–4.5 \times 10^6$ cells/cm^3 in EHS matrix at 0–4°, and 100-μl aliquots are plated onto preformed base layers of EHS matrix. At the latter density the cells are well separated from each other, being on average at least five cell diameters apart. Some cells reaggregate immediately after trypsinization and remain as clusters during the course of the experiment. The cell-containing substrata are gelled (30–60 min, 37°) and then covered with hormone-containing medium. After 6 days of culture in a humidified atmosphere of 5% CO_2 in air, the majority of the cells are still isolated from each other; at 6 days, 44–60% of all the single cells plus clusters are single, 16–31% are doublets, and the remainder exist in larger clusters.

Mouse Mammary Epithelial Cell Strains and Cell Lines

COMMA-1-D. COMMA-1-D[39] is a cell strain derived from midpregnant mammary glands of BALB/c mice. These cells exhibit mammary-specific functional differentiation under appropriate conditions in cell culture. The heterogeneous COMMA-1-D strain, which contains three cell types characterized by antisera to different keratin proteins, is induced by EHS matrix and lactogenic hormones to produce β-casein mRNA. The cell strain maintained as stock cultures is grown in F12 medium supplemented with 5% (v/v) fetal bovine serum (FBS), 15 mM HEPES buffer, gentamicin (50 μg/ml), insulin (5 μg/ml), transferrin (10 μg/ml), and epidermal growth factor (10 ng/ml). For examining the presence and induction of casein, ovine prolactin (3 μg/ml) and hydrocortisone (2 μg/ml) are also added to the medium of the cells plated on EHS as described previously. The cell cultures are incubated at 37° in a humidified atmosphere of 5% CO_2 in air.

CID-9. A subpopulation (designated CID-9[38]) of the mouse mammary epithelial cell strain COMMA-1-D has been developed in which more than 35% of the cells form alveoli-like structures when plated onto EHS matrix and vectorially secrete β-casein. These cells are grown in Dulbecco's modified Eagle's medium (DMEM)–F12 containing 5% (v/v) heat-inactivated FBS, gentamicin (50 μg/ml), and insulin (5 μg/ml). For hormonal and ECM-dependent milk induction, the cells are plated on plastic dishes or EHS matrix at 8×10^4 cells/cm^2 in the presence of 2% (v/v) FBS and insulin. Hydrocortisone (1 μg/ml) and ovine prolactin (3 μg/ml) are also added. Twenty-four hours after plating, the medium is switched to 0% FBS and the cell cultures are incubated at 37° in a humidified atmosphere of 5% CO_2 in air.

SCp2. A clonally derived homogeneous cell line, called SCp2,[56] has been isolated from the functional, but heterogeneous, mouse mammary cell strain CID-9. After the addition of an exogenously supplied basement membrane-derived extracellular matrix, more than 90% of the cells aggregate and form clusters. In the presence of lactogenic hormones (same treatment as for the CID-9 cells), these clusters express β-casein. In general, epithelial cells are plated on top of EHS matrix. An alternative way of inducing differentiation is to seed the cells on glass or plastic, then to overlay the cells with EHS diluted to 1% in culture medium containing the lactogenic hormones.[37] The final concentration of EHS is then about 100 μg/ml. After 1 day of treatment, the cells undergo aggregation and shape changes. After 2–3 days, expression of β-casein can be detected by immunofluorescence.

Initial Culture of Normal Human Breast Epithelial Cells

In the resting human breast the lumenal epithelial cells of the ducts and terminal ductal lobular units (TDLUs) are the major cellular targets for tumor formation. Because of this, the methods described are focused on the culture and characterization of the lumenal epithelial cell fraction.

Reduction Mammoplasties. Primary breast epithelial cell populations are prepared from reduction mammoplasty tissue removed surgically from normal donors. The tissue is mechanically and enzymatically disaggregated to yield organoids free from the majority of contaminating stromal fibroblasts. Pure populations of epithelial cells subsequently develop in monolayer culture using either an undefined medium preparation, MM, as in the first description of prolonged culture of these cells,[57,58] or a serum-free formulation, MCDB-170.[59] Detailed procedures for tissue processing, medium formulation, and primary culture have been described extensively

[56] P. Y. Desprez, C. Roskelley, J. Campisi, and M. J. Bissell, *Mol. Cell. Differ.* **1**, 99 (1993).
[57] M. R. Stampfer, R. C. Hallowes, and A. J. Hackett, *In Vitro* **16**, 415 (1980).
[58] M. R. Stampfer, *In Vitro* **18**, 531 (1982).
[59] S. L. Hammond, R. G. Ham, and M. R. Stampfer, *Proc. Natl. Acad. Sci. U.S.A.* **81**, 5435 (1984).

elsewhere[57,60–62] and are therefore not covered here. This technique yields large quantities of predominantly basal (myoepithelial) cells that senesce at the third passage[61,63] with the exception of a few cells with high proliferative potential that can be passaged a further 10–20 times.[63]

Petersen *et al.* described a modification of this original method that included procedures for greatly enriching the lumenal cell fraction by the use of unpolymerized type I collagen as a substratum for the culture of organoids,[64,65] and by passaging cultures at high density in serum-free CDM3 medium.[50] These latter modified methods are used to prepare primary cells for the assay described here and the collagen substratum and medium formulations are provided below. It is important to note that the lumenal cells are the innermost cells of the gland and as such most sensitive to hypoxic damage imposed by tissue handling. In the operating room, the tissue is cut with sterile scalpels into 1-cm fragments or smaller immediately after excision and transported in 50-ml tubes containing 25 ml of CO_2-equilibrated DMEM–F12 medium. If tissue processing is delayed for more than 1 hr the fragments are cooled to 4°. The tissue is then mechanically and enzymatically disaggregated in either DMEM–F12 or CDM3.[66] The resultant organoids are plated on collagen-coated dishes at a density of 8–10 organoids/cm^2. It may be that the various differences reported in success with cultivation of lumenal cells are not strictly tied to the serum-free medium formulations used but, rather, to the culture substratum, that is, ECM coated or not. To select further for the presence of lumenal cells rather than myoepithelial cells the cultures are subsequently seeded and maintained at high cell densities by plating at 2×10^5/cm^2. This additional modification yields almost pure populations of lumenal cells by taking advantage of the net loss of myoepithelial cells owing to squamous metaplasia.[50]

Milk. An alternative method for obtaining primary epithelial cells is by isolation from human milk samples and culture in a milk–serum medium formulation with macrophage feeder layers for support. This method, which is described in detail elsewhere,[63] yields populations of lumenal cells that can be passaged one or two times before senescence.

[60] R. C. Hallowes, E. J. Bone, and W. A. Jones, *Tissue Cult. Med. Res.* **2,** 213 (1980).
[61] M. R. Stampfer, *J. Tissue Cult. Methods* **9,** 107 (1985).
[62] J. Taylor-Papadimitriou and M. R. Stampfer, *in* "Cell & Tissue Culture: Laboratory Procedures" (J. B. Griffiths, A. Doule, and D. G. Newell, eds.), p. 107. Wiley-Liss, New York, 1992.
[63] J. Taylor-Papadimitriou, M. R. Stampfer, J. Bartek, A. Lewis, M. Boshell, E. B. Lane, and I. M. Leigh, *J. Cell Sci.* **94,** 403 (1989).
[64] O. W. Petersen and B. van Deurs, *Cancer Res.* **47,** 856 (1987).
[65] P. Briand, O. W. Petersen, and B. van Deurs, *In Vitro Cell Dev. Biol.* **23,** 181 (1987).
[66] L. Ronnov-Jessen and O. W. Petersen, *Lab. Invest.* **68,** 696 (1993).

Normal Cell Lines. There are few "normal" human breast epithelial cell lines that can be passaged extensively and retain characteristics of breast epithelial cells *in vivo*.[65,66a–c,67,68] For the purposes of the assay system described here three lines are considered: HMT-3522,[65] MCF-10A,[67] and HBL-100.[68] However, the latter cell line is of questionable value because it contains simian virus 40, is highly aneuploid, and expresses a phenotype in our assay that is intermediate between normal and tumor.[50] Both MCF-10A and HMT-3522 cells can be maintained in serum-free H14 medium (formula provided below) and passaged into new dishes at densities of $2 \times 10^4/cm^2$ or greater.

Initial Culture of Human Breast Carcinoma Cells

Primary and Metastatic Breast Carcinoma. Primary human breast tumor cells are exceptionally difficult to establish routinely in monolayer culture. Although there are a few reports of the growth of cells from primary breast carcinomas,[64,69–72] the cells generally appear phenotypically normal and do not show the profile of keratins expressed by carcinoma cells, suggesting that they develop from normal or benign tissue associated with the tumor. Some success has been obtained with the culture of tumor cells from lymph node or pleural effusion metastases.[73]

For our assay, primary carcinoma biopsy material is collected in CDM3 medium and processed as described above and in the original methods,[57,60,72] with the exception that the digestion is done in CDM3 medium without hyaluronidase but with collagenase at a higher concentration of 600–900 IU/ml.[64] In addition, primary carcinoma organoids are treated differently from normal tissue after enzymatic digestion because the individual carcinoma cells are larger and more fragile than normal cells. Attempts to generate single-cell suspensions from carcinoma organoids almost inevitably result in the loss of all the carcinoma cells and the familiar

[66a] M. R. Stampfer and J. C. Bartley, *Proc. Natl. Acad. Sci. U.S.A.* **82**, 2394 (1985).

[66b] V. Band, J. A. de Caprio, L. Delmolino, V. Kulesa, and R. Sager, *J. Virol.* **12**, 6671 (1991).

[66c] J. Bartek, J. Bartkova, N. Kyprianou, E. Lalani, Z. Staskova, M. Shearer, S. Chang, and J. Taylor-Papadimitrion, *Proc. Natl. Acad. Sci. U.S.A.* **88**, 3520 (1991).

[67] H. D. Soule, T. M. Maloney, S. R. Wolman, W. D. Petersen, R. Brenz, C. M. McGrath, J. Russo, R. J. Pauley, R. F. Jones, and S. C. Brooks, *Cancer Res.* **50**, 6075 (1990).

[68] E. V. Gaffney, *Cell Tissue Res.* **227**, 563 (1982).

[69] O. W. Petersen and B. van Deurs, *Differentiation* (Berlin) **39**, 197 (1988).

[70] V. Band and R. Sager, *Proc. Natl. Acad. Sci. U.S.A.* **86**, 1249 (1989).

[71] H. S. Smith, L. A. Liotta, M. C. Hancock, S. R. Wolman, and A. J. Hackett, *Proc. Natl. Acad. Sci. U.S.A.* **82**, 1805 (1985).

[72] P. S. Rudland, R. C. Hallowes, S. Cox, E. J. Ormerod, and M. J. Warburton, *Cancer Res.* **45**, 3864 (1985).

[73] H. S. Smith, S. R. Wolman, S. H. Dairkee, M. C. Hancock, M. Lippman, A. Leff, and A. J. Hackett, *J. Natl. Cancer Inst.* **78**, 611 (1987).

establishment of normal cell cultures. The period following digestion poses the highest risk of loss of viable carcinoma cells. They must be protected from excess mechanical force from, for example, the use of pipettes and pipette tips with too narrow a bore, from harsh enzymatic treatment from enzymes such as trypsin, and from the toxic effects of endogenous collagenases that are transiently released after tissue processing. Also, contrary to routine practice for the dissagregation of normal breast tissue, the primary carcinoma organoids cannot be frozen for future use for the reasons stated above. Thus, the carcinoma organoids are maintained intact, suspended in 5 ml of CDM3 medium for 2 days using a rotary shaker (40 rpm) to prevent settling and local accumulation of collagenases. After this period no further collagenases are released around the organoids as assayed by collagen gel (O. W. Petersen, unpublished observations). This method allows routine preparation of primary cultures containing carcinoma cells expressing the phenotypic traits of breast carcinoma cells *in vivo.*[50,64] The cells obtained by these methods are identical in terms of morphology, success rate of plating and growth, and ability to produce cell lines from the primary cultures to those previously identified by Rudland *et al.*[72] as "slow adherent colonies" or "nonadherent colonies" (D. Hallowes, personal communication).

Breast Tumor Cell Lines. Several established cell lines from breast carcinomas have been used successfully in our reconstituted basement membrane assay.[50,78] These include HMT-3909 S1,[74] MCF-7/9, ZR75, T47-D, BT-20,[75,76] CAMA-1,[77] and MDA-MB-435.[78] We have determined that monolayer cultures of MCF-7/9 and HMT-3909/S1 can be maintained in serum-free DMEM-F12 (1 : 1) with 2 mM glutamine and DMEM-F12 (1 : 1) with $1.4 \cdot 10^{-6}$ M hydrocortisone respectively on collagen coated plates (method below). MDA-MB-435[78] cells can be maintained on collagen-coated plates (method below) in serum-free CDM3 medium supplemented with 10 times the normal amount of bovine serum albumin (BSA).[78] Such methods may prove suitable for the serum-free culture of the remaining carcinoma cell lines, although this must be established empirically.

Protocol for Reconstituted Basement Membrane Assay

Cell Preparation. Third-passage cultures of normal primary cells, normal cell lines, or carcinoma cell lines are incubated in 1 ml of tryp-

[74] O. W. Petersen, B. van Deurs, K. Vang Nielsen, M. W. Madsen, I. Laursen, I. Balslev, and P. Briand, *Cancer Res.* **50,** 1257 (1990).

[75] L. W. Engel and N. A. Young, *Cancer Res.* **38,** 4327 (1978).

[76] P. Briand and A. Lykkesfelt, *Anticancer Res.* **6,** 85 (1986).

[77] J. Fogh, W. C. Wright, and J. D. Loveless, *Int. Natl. Cancer Inst.* **58,** 209 (1977).

[78] A. R. Howlett, O. W. Petersen, P. S. Steeg, and M. J. Bissell, *J. Natl. Cancer Inst.,* in press (1994).

sin–EDTA per 100-mm dish for 3–10 min at 37°. Tap the dish periodically and monitor the progress of cell detachment microscopically. Resuspend the cells in 10 ml of DMEM–F12 medium and wash out the trypsin by centrifugation at 600 g for 5 min. Resuspend the cells in 5 ml of medium, add 10 μl of soybean trypsin inhibitor (SBTI) per milliliter, and count 10 μl of cell suspension with a hemocytometer. If cell aggregation occurs first add 1 ml of medium, then resuspend, and add the remaining 4 ml followed by SBTI.

Suspensions of carcinoma organoids (10–50 cells/organoid) are counted on regular microscope slides and not with a hemocytometer, to avoid mechanical damage. Organoid densities are determined directly by counting six fields defined by an ocular grid and multiplication of the mean number per field to calculate the total for the culture dish. To make quadruplicate assays adjust the density of cell suspensions to 1×10^6/ml, and carcinoma organoids to approximately 2×10^3 organoids/ml of medium. Microfuge the cell suspension in a sterile Eppendorf tube at 4000 rpm for 2 min. Pellet the carcinoma organoids at approximately 150 g for 5 min. Remove the medium and place the pellets on ice.

Assay Setup and Maintenance. The cell pellets are resuspended, using a cold pipette, in 1.2 ml of ice-cold EHS matrix (7–10 mg/ml) produced by the protocol described above or obtained commercially as Matrigel (Collaborative Research). The carcinoma organoids are gently resuspended in EHS matrix with a cold pipette tip with a wide bore made by cutting the tip. Add 300 μl of the EHS matrix containing either approximately 2.5×10^5 single cells or 500 organoids to each of 4 wells of a 24-well plate (Nunc, Roskilde, Denmark). The wells that receive organoids are precoated with 100 μl of gelled EHS matrix to prevent organoid contact with plastic. The plates are incubated for 1 hr at 37° to allow the EHS to solidify and 1 ml of medium is then added to the resultant gel containing suspended cells or organoids. The medium is changed (1 ml/well) every second or third day and cultures are maintained in a humidified atmosphere of 5% CO_2 in air.

Assay Analysis

Growth. In the early stages of the assay both normal and tumorigenic cells proliferate within the EHS matrix. The growth of primary carcinoma cells is, however, considerably slower than that of carcinoma cell lines. Within 1 or 2 days the growth rate of the normal cells begins to decline and the cells undergo a morphogenesis into small, growth-arrested spheres resembling breast acini *in situ.* Tumor cells, in contrast, continue to grow to form large, disorganized colonies. The growth of cells is measured in three ways: (1) the size of growing spherical structures is measured with an eye piece equipped with a micrometer spindle calibrated with a stage micrometer. Twenty colonies are counted per experimental point; (2) 24-hr [³H]thymidine incorporation ([³H]TdR, 20 Ci/mmol: 2.5 μCi/ml) (NET-027X; New England Nuclear Research Products, Boston, MA) is determined; (3) the final number of cells is counted either in spherical profiles of normal cultures or colony profiles of tumors in sections of EHS gels.

Differentiation. The differentiation of the breast cells can be assessed immunochemically in EHS cultures prefixed in 2% (v/v) paraformaldehyde or −20° methanol–acetone (1 : 1, v/v) and processed for cryostat sectioning or in untreated EHS cultures that have been cryostat sectioned and postfixed in −20° methanol for 5 min. Frozen sections (5 μm) can be stained for milk fat globule membrane antigen (MGFGM-A, sialomucin; Catalog Laboratories, San Francisco), which is a polarity marker specifically expressed at the apical membrane of lumenal cells. In the assay, sialomucin will be localized at the apical lumenal surface and may accumulate within the lumen of spheres formed by normal cells. Tumor cells also express sialomucin but the antigen is expressed in a disorganized random pattern throughout the tumor colony. A second key marker of normal lumenal cell differentiation in our assay is the formation of a type IV collagen and laminin-containing endogenous basement membrane. The detection of human basement membrane material in a mouse matrix is possible through the use of species-specific anti-human antibodies. Sections can be stained with antibodies against either type IV collagen (PHM-12; Accurate Chemicals, Westbury, NJ) or laminin (Life Technologies, Grand Island, NY) to localize a thin, continuous basally deposited basement membrane around spheres formed by normal cells. The basement membrane staining is considerably enhanced if any myoepithelial cells are also present. Tumor cells do not deposit an organized endogenous basement membrane.

Media and Solutions

Trypsin–EDTA

1. Prepare trypsin (0.25%, w/v):

NaCl	8000 mg
KH_2PO_4	200 mg
KCl	200 mg
$Na_2HPO_4 \cdot 2H_2O$	1444 mg
trypsin (Difco, Detroit, MI)	2500 mg
Milli-Q water or equivalent	1000 ml

Adjust the pH to pH 7.6–7.8 and sterile filter.

2. Prepare EDTA (100 mM):

EDTA (tetrasodium salt)	37.2 gm
Milli-Q water or equivalent	1000 ml

Sterile filter.

3. Prepare trypsin–EDTA:

trypsin (0.25%, w/v)	99 ml
EDTA (100 mM)	1 ml

Sterile filter and store at −20°.

Soybean Trypsin Inhibitor

SBTI (T-6522, type 1-S; Sigma)	200 mg
Milli-Q water or equivalent	20 ml

Sterile filter; store at −20° for a maximum of 6 months, or at 4° for a maximum of 14 days.

Collagen-Coated Plates

Solutions
   Phosphate-buffered saline
   Rat tail collagen, ~3 mg/ml (Vitrogen 100, Celtrix Laboratories, Santa Clara, CA). Store at 4°
Procedure: At 0–4° mix 1 ml of Vitrogen with 44 ml of phosphate-buffered saline (PBS). Add 0.12 ml of the Vitrogen–PBS mixture per square centimeter of tissue culture surface, incubate for 24 hr, rinse with 3–5 ml of PBS, add a further 3–5 ml of PBS, and store at 0–4° for up to 4 weeks

CDM3 Medium. The basal medium is DMEM–F12 (1 : 1, v/v), available from Life Technologies (Cat. No. 12400-024) with 15 mM HEPES buffer, with L-glutamine, without sodium bicarbonate. Phenol red-free medium is also available. Powder is dissolved in 1 liter of Milli-Q water or the equivalent. Sodium bicarbonate is added as per manufacturer instructions. Sterile filter (0.2-μm pore size) and store at 4°. Supplement the medium prior to use according to the following schedule.

Additive	Stock solution	Vol (μl)/10 ml of medium	Final concentration
1. Insulin	2 mg/ml	15	3000 ng/ml
2. Transferrin	20 mg/ml	12.5	25 μg/ml
3. Sodium selenite	2.6 μg/ml	10	2.6 ng/ml
4. EGF	20 μg/ml	50	100 ng/ml
5. Estradiol	10^{-7} M	10	10^{-10} M
6. Hydrocortisone	1.4×10^{-3} M	10	1.4×10^{-6} M
7. Cyclic AMP	10^{-5} M	10	10^{-8} M
8. Fibronectin	100 μg/ml	10	0.1 μg/ml
9. Triiodothyronine	6 μg/ml	10	10^{-8} M
10. Ethanolamine	0.1 M	10	10^{-4} M
11. Phosphoethanolamine	0.1 M	10	10^{-4} M
12. Bovine serum albumin	10% (w/v)	10	0.01% (w/v)
13. Ascorbic acid	10 mg/ml	10	10 μg/ml
14. Trace element mix	100$\times$	100	1$\times$
15. Fetuin	2 mg/ml	100	20 μg/ml

CDM3 Medium Stock Solutions

1. Insulin (bovine, Cat. No. 1204718; Boehringer Mannheim, Indianapolis, IN): Dissolve 20 mg in 10 ml of 5 mM HCl. Filter (0.2-μm pore size), aliquot, and store: $-20°$ (6 months), $4°$ (1 month), powder $4°$ (2 years).

2. Transferrin (human, Cat. No. T-2252; Sigma): Dissolve 200 mg in 10 ml of Milli-Q water. Filter (0.2-μm pore size), aliquot, and store: $-20°$ (3 months), $4°$ (1 month), powder $4°$ (1 year).

3. Sodium selenite (Cat. No. 40201; Collaborative Research): Dissolve 100 mg in 5 ml of sterile (0.2-μm filtered) Milli-Q water to give a 20-mg/ml solution. Dilute 25 μl of the 20-mg/ml solution in 1 ml of water to give 500 μg/ml. Dilute 26 μl of the 500-μg/ml solution in 5 ml of water to give a 2.6-μg/ml stock. Do not refilter; aliquot and store concentrated stocks at $-20°$ (1 year), working stock at $-20°$ (1 month), $4°$ (1 week), powder $4°$ (2 years).

4. Epidermal growth factor (EGF) (Cat. No. 40001; Collaborative Research): Dissolve vial contents (100 μg) in 5 ml of sterile (0.2-μm filtered) Milli-Q water. Do not refilter; aliquot and store: $-20°$ (3 months), $4°$ (2 weeks), powder $4°$ (1 year).

5. Estradiol (Cat. No. E-2785; Sigma): In autoclaved glass vials dissolve 25 mg in 3.125 ml of 95% (v/v) ethanol. Serially dilute this stock (8 mg/ml) in 95% ethanol as follows: 50 μl (8 mg/ml) in 1.5 ml (10^{-3} M); 20 μl of 10^{-3} M in 1.98 ml (10^{-5} M); 20 μl of 10^{-5} M in 1.98 ml (10^{-7} M). Store 10^{-3} M at $-20°$ (1 year). Store all other stocks at $-20°$ (6 months) or $4°$ (3 months).

6. Hydrocortisone (Cat. No. 40203, Collaborative Research): Dissolve 50 mg in 10 ml of 95% (v/v) ethanol (1.4×10^{-2} M). Dilute 0.5 ml of the 5-mg/ml solution in 5 ml of ethanol (1.4×10^{-3} M). Store the

$1.4 \times 10^{-2} M$ solution at $-20°$ (1 year) or the $1.4 \times 10^{-3} M$ solution at $-20°$ (6 months).

7. Cyclic AMP (Cat. No. D-0260; Sigma): Dissolve 25 mg in 5 ml of Milli-Q water ($10^{-2} M$). Dilute 10 μl of the $10^{-2} M$ solution into 10 ml of water ($10^{-5} M$). Filter (0.2-μm pore size), aliquot, and store concentrated stock at $-20°$ and working stock at $4°$.

8. Fibronectin (Cat. No. 4008; Collaborative Research): Dissolve one vial (1 mg) in 10 ml of sterile Milli-Q water. Leave for 30 min at room temperature, then aliquot aseptically and store at $-20°$ for 1 month, powder $4°$ (18 months).

9. Triiodothyronine (Cat. No. T-5516; Sigma): Dissolve one vial (1 mg) in 1.67 ml of sterile 0.2-μm filtered Milli-Q water (600 μg/ml). Dilute 100 μl in 9.9 ml of sterile Milli-Q water. Store at $-20°$ (1 year) or $4°$ (1 month).

10. Ethanolamine (Cat. No. E-9508; Sigma): Dilute 60 μl in 10 ml of Milli-Q water. Filter (0.2-μm pore size), aliquot, and store at $4°$.

11. Phosphoethanolamine (Cat. No. P-0503; Sigma): Dissolve 141 mg in 10 ml of Milli-Q water. Filter (0.2-μm pore size), aliquot, and store at $4°$.

12. Bovine serum albumin (Cat. No. A-4919; Sigma): Dissolve 100 mg in 10 ml of Milli-Q water. Filter (0.2-μm pore size), aliquot, and store at $-20°$.

13. Ascorbic acid (Cat. No. A-0278; Sigma): Dissolve 100 mg in 10 ml of Milli-Q water. Filter (0.2-μm pore size), aliquot, and store at $-20°$ (1 month) or $4°$ (1 week).

14. Trace element mix (Cat. No. 062-030124C; Life Technologies): Dissolve 1 vial in 10 ml of sterile (0.2-μm filtered) Milli-Q water. Aliquot and store at $-20°$ (3 months), or as a powder at room temperature (1 year).

15. Fetuin (Cat. No. F-6131; Sigma): Dissolve 40 mg in 20 ml of Milli-Q water. Filter (0.45-μm pore size), aliquot, and store at $-20°$ (3 months).

H14 Medium. The basal medium for H14 is the same as for CDM3 described above. H14 is supplemented with additives 1–6 as described above, except that the final concentration of insulin is 250 ng/ml, transferrin is 10 μg/ml, and EGF is 10 ng/ml. H14 also contains prolactin (5 μg/ml).

Additive	Stock solution	Volume (μl)/10 ml of medium	Final concentration
1. Insulin	100 μg/ml	25	250 ng/ml
2. Transferrin	20 mg/ml	05	10 μg/ml
3. Sodium selenite	2.6 μg/ml	10	2.6 ng/ml
4. EGF	20 μg/ml	05	10 ng/ml
5. Estradiol	$10^{-7} M$	10	$10^{-10} M$
6. Hydrocortisone	$1.4 \times 10^{-3} M$	10	$1.4 \times 10^{-6} M$
7. Prolactin	1 mg/ml (30 U/ml)	50	5 μg/ml

H14 Medium Stock Solutions

Insulin: As per description above, except take 150 μl of the 2-mg/ml CDM3 medium stock solution and dilute in 2.85 ml of Milli-Q water to give a 100-μg/ml stock.

Prolactin (Cat. No. L-6520; Sigma): Dissolve ~5.8 mg (180 IU) in 6 ml of sodium bicarbonate (13.3 mg/ml). Filter (0.2-μm pore size), aliquot, and store at 4° (2 weeks) or as a powder at 4° (1 year).

[26] Therapeutic Application of Matrix Biology

By J. F. Tschopp, W. S. Craig, J. Tolley, J. Blevitt, C. Mazur, and M. D. Pierschbacher

Introduction

The extracellular matrix is a complex network of glycoproteins interwoven into the three-dimensional structure that surrounds all cells in the body. This matrix not only exerts a strong influence on the normal behavior of cells in the body, but also appears to play an important role in many disease states. The discovery of the existence of a family of cell surface receptors, called integrins,[1,2] which are primarily responsible for mediating cell–matrix interactions, has opened up a new opportunity for therapeutic intervention. At least seven, and possibly nine, of the integrins identified to date recognize an epitope in their associated ligands comprising an arginine-glycine-aspartic acid (RGD) sequence. This sequence of amino acids can be incorporated synthetically into small peptides that can then be used as agonists or antagonists of the appropriate integrin to gain a desired cellular response.

One example of the use of RGD-containing proteins as agonists is for the enhancement of the biocompatibility of implantable prosthetic devices. Driven by the significant need to replace components of the body for which transplantation is not practical, there has been a surge during the last 50 years to develop materials that can remain in contact with body tissues and fluids for an extended time without provoking an adverse host response. The search has generally been for inert materials that would be largely ignored by the immune system and other pathways of biological

[1] E. Ruoslahti, *J. Clin. Invest.* **87**, 1 (1991).
[2] R. O. Hynes, *Cell (Cambridge, Mass.)* **42**, 549 (1987).

Copyright © 1994 by Academic Press, Inc.
All rights of reproduction in any form reserved.

degradation. It has become increasingly clear, however, that materials comprising most implants are never completely inert and that the response of the body to their presence will range from the relatively mild reactions that accompany most implants to severe reactions, such as fibrous encapsulation of breast prostheses and surface thrombus formation in small-diameter vascular grafts, which can result in implant failure.[3,4]

Another approach for using RGD-containing peptide therapeutically is for the prevention of unwanted cell–matrix interactions. TP9201 is a highly specific inhibitor of the $\alpha_{IIb}\beta_3$ integrin responsible for platelet aggregation. This peptide has been shown in preclinical studies to inhibit platelet-driven blood clotting at doses that do not increase template bleeding time.[5-8] The unique properties of this peptide may allow its application in therapeutic areas where the separation of bleeding from antithrombotic effects is critical, such as during tissue transfer reconstructive surgery or after ischemic stroke. A challenge in using peptides therapeutically is the maintenance of a constant therapeutic concentration of the drug in the blood over a prolonged period when infusion is not practical.

Enhancement of Biocompatability of Prosthetic Devices by Use of RGD-Containing Peptide

Biomedical scientists and materials engineers have combined efforts in a new approach for improving the biocompatibility of implants. The pathophysiological responses of inflammation, thrombogenesis, infection potential, and neoplasia by the host are lessened or abolished by providing a material surface–tissue interface that encourages a constructive biological interaction. Heparin coatings for surfaces in contact with blood, and calcium phosphate or hydroxylapatite coatings for bone implants, are examples of new approaches that have proved partially successful.[9,10] An

[3] J. A. Ferreira, *Aesthetic Plast. Surg.* **8**, 109 (1984).

[4] C. O. Esquivel and F. W. Blaisdell, *J. Surg. Res.* **41**, 1 (1986).

[5] J. F. Tschopp, E. M. Driscoll, D.-X. Mu, Ss. C. Black, M. D. Pierschbacher, and B. R. Lucchesi, *Coronary Artery Dis.* **4**, 809 (1993).

[6] D. Collen, H. R. Lu, J. M. Stassen, I. Vreys, T. Yasuda, S. Bunting, and H. K. Gold, *Thromb. Haemostasis* **71**, 95 (1994).

[7] C. Mazur, J. F. Tschopp, E. C. Faliakou, K. E. Gould, J. T. Diehl, M. D. Pierschbacher, and R. J. Connolly, *J. Lab. Clin. Med.* (in press) (1994).

[8] S. Cheng, W. S. Craig, D. Mullen, J. F. Tschopp, D. Dixon, and M. D. Pierschbacher, *J. Med. Chem.* **37**, 1 (1994).

[9] S. W. Kim, H. Jacobs, J. Y. Lin, C. Nojori, and T. Skano, *Ann. N.Y. Acad. Sci.* **516**, 116 (1987).

[10] C. A. van Blitterswijk, J. J. Grote, W. Kuypers, C. J. G. Blok-van Hoek, and W. T. Daems, *Biomaterials* **43**, 243 (1985).

RGD-containing peptide sequence used to coat devices is another approach to providing sites on the material surface that will initiate cell interactions and that will allow the natural development of tissue structure at the material–tissue interface. A challenge has been to develop a universal method for coating diverse material surfaces with the RGD epitope in a uniformly effective way.

PepTite-2000 (Telios Pharmaceuticals, Inc., San Diego, Ca.) is a peptide that has been designed to coat a variety of materials noncovalently and present an accessible RGD sequence for cell attachment.[11] The full amino acid sequence of this peptide, acetyl-G(dR)GDSPASSKGGGGS (dR)LLLLLL(dR)-amide, is composed of an amino-terminal presentation sequence, acetyl-G(dR)GDSP-, followed by a spacer sequence, -ASSKGGGGS-, and then, finally, by a carboxy-terminal sequence responsible for attachment of the peptide to material surfaces, -(dR)LLLLL-L(dR)-amide. All of these elements are critical to elicit the proper biological response from the surrounding tissue.

Virtually any of the materials commonly used in prosthetic devices can be coated with PepTite-2000. This includes Dacron, titanium, silicon, Teflon, polyethylene, and polylactic acid. Because the interaction between the molecule and the material surface is noncovalent, any shape or size of device can be coated conveniently.

Method of Coating with PepTite-2000

Reagents/Materials

PepTite-2000 stock solution: Dissolve the peptide in dimethyl sulfoxide (DMSO) or 25% (v/v) ethanol to a final concentration of 5–10 mg/ml. Mix until completely dissolved. This solution is stable for at least 6 months when stored at 25°.

Phosphate-buffered saline (PBS): 50 mM sodium phosphate, pH 7.4, 140 mM sodium chloride.

Coating Procedure

1. Dilute the stock solution to 100 µg/ml immediately before use with sterile tissue culture-grade PBS.

2. Immerse the device in the coating solution at room temperature. Coating is usually complete in several minutes, but immersion may be continued for up to 2 hr to ensure saturation. If there are small spaces or

---

[11] M. D. Pierschbacher, C. J. Honsik, and L. B. Dreisbach, U.S. Pat. 5,120,829 (1992).

cavities in the device, a vacuum with mild agitation will remove trapped gas and improve coating efficiency.

3. Rinse the device with PBS and air dry.

Sterilization

PepTite-2000 activity on the surface is stable to common sterilization methods such as steam autoclaving, γ irradiation, electron beam, and ethylene oxide,[12] providing that the surface of the material itself is not significantly altered by these procedures.

Stability

The activity of the desiccated peptide and the dry, coated device surface is stable for more than 1 year and resistant to mechanical disruption.

When the surface of a material that does not support cell attachment is coated with PepTite-2000, cells attach, adopt a flattened morphology, and spread over the surface. In Fig. 1 MG-63 osteosarcoma cells have been incubated in the absence of serum with silicone and Teflon. There are substantial numbers of cells attached and spread on the PepTite-2000-coated surfaces of both materials, whereas the noncoated surfaces have only a few cells present and these generally possess a rounded morphology. In addition, the attachment on coated surfaces is RGD dependent. The inclusion of the peptide GRGDSP (1 mg/ml) during incubation with the cells reduces cell attachment by greater than 80%, whereas inclusion of the inactive GRGESP peptide fails to reduce attachment.[12]

By providing sites for cell attachment, the PepTite-2000 coating can send a signal to attached cells in a manner similar to extracellular matrix. Immunofluorescence microscopy studies have detected substantial fibronectin biosynthesis and deposition from MG-63 osteosarcoma cells attached to PepTite-2000-coated surfaces after less than 24 hr of culture in serum-free medium.[12] Importantly, with longer incubation the fibronectin receptor on these cells, the integrin $\alpha_5\beta_1$, binds to the deposited matrix and localizes in areas of focal contacts. However, after only 4 hr of culture it is the $\alpha_v\beta_3$ integrin, that is localized at areas of focal contact to the peptide sequence.[12] In addition, actin stress fibers form that terminate at the same focal contacts. This suggests that the integrin $\alpha_v\beta_3$ is used by the cell to attach to PepTite-2000-coated surfaces and that other integrins can be recruited to interact with newly formed matrix at the surface.

---

[12] J. R. Glass, W. S. Craig, K. Dickerson, and M. D. Pierschbacher, *Mater. Res. Soc. Smp. Proc.* **252,** 331 (1992).

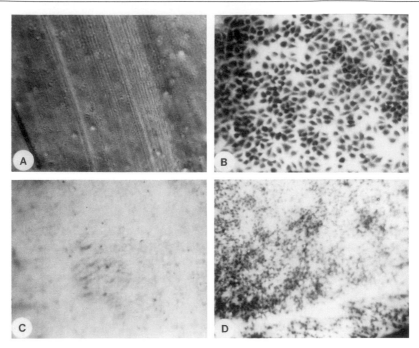

FIG. 1. Cell attachment on PepTite-2000-coated silicone and Teflon. The attachment of MG-63 osteosarcoma cells after a 2-hr incubation at 37°. Very few cells attach to the noncoated silicon (A) and Teflon (C). On the other hand, there is definite cell attachment and the cells display a well-spread morphology on the respective coated surfaces (B and D).

Soft tissues respond to PepTite-2000-coated surfaces by exhibiting faster tissue ingrowth (Fig. 2). This response can be observed when a woven Dacron material coated with PepTite-2000 is implanted into the back of a rabbit. After 7 days the coated Dacron shows nearly complete

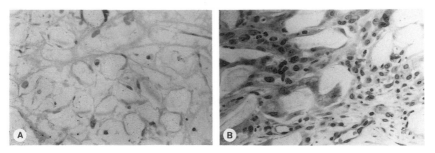

FIG. 2. Cell growth into PepTite-2000-coated Dacron. Noncoated (A) and coated (B) Dacron mesh after subcutaneous implantation for 7 days.

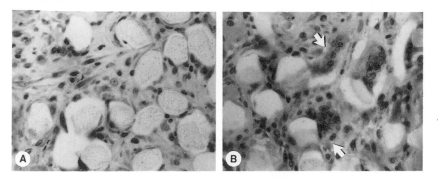

FIG. 3. Response of soft tissue to PepTite-2000-coated Dacron. Noncoated and coated Dacron mesh after subcutaneous implantation for 14 days. Giant cells are minimal in the coated Dacron (A). In the noncoated implant (B), however, a majority of the fibers are surrounded by these inflammatory cells (arrow).

ingrowth in the interfiber spaces (Fig. 2B). In contrast, an uncoated Dacron subcutaneous implant shows little ingrowth at this time point (Fig. 2A). Furthermore, the region is characterized by a higher and more uniform deposition of collagen as evidenced by staining with Masson's trichrome. In these same experiments, the PepTite-2000-coated Dacron elicited a much weaker inflammatory response from the tissue. When, after longer implantation time, tissue ingrowth finally occurred in the control Dacron, there were large numbers of giant cells associated with the fibers (Fig. 3B), indicative of the inflammatory nature of Dacron. In contrast, very few giant cells were present in the coated samples (Fig. 3A), even after 3 months of implantation.

PepTite-2000 can easily coat virtually any material that is used in prosthetic implants without the need for any other chemical modification of the surface. It directs cell attachment using the normal mechanism of integrin–RGD interactions. These interactions elicit a normal physiological response, with the synthesis and deposition of new matrix components. The consequence is faster tissue infiltration and a reduction in pathophysiological responses such as excessive inflammation and fibrous scar formation. These qualities suggest that PepTite-2000 can be used to substantially increase the biocompatibility of prosthetic implants.

Assays for Determination of Cyclic RGD-Containing Peptide
    Concentration in Plasma

Natural and many synthetic peptides behave as ideal targeted drugs. These molecules are, however, difficult to administer extravascularly because they are water-soluble macromolecules that do not cross mucosal surfaces readily, and they have short half-lives. Therefore the number of successful applications has remained small in comparison to the number of available peptides. As a result, administration is commonly by injection or infusion. An additional problem can be the rapid whole-body clearance of many peptides. In the therapeutic context, a longer duration of action may be needed, and this will require the administration of large doses, depot forms, or continuous infusions.

Peptides can be modified in a number of ways to improve their properties, such as size reduction, replacement of amino acid side chains, conformational restrictions, modification of the peptide backbone to stabilize the molecule, or chemical modification of the N-terminal/C-terminal ends to improve proteolytic stability.[13,14] These approaches can increase potency and greatly prolong duration of action but they do not usually lead to an orally bioavailable product except in cases of very small molecules. An example of an application is TP9201. Modifications have resulted in increased potency, selectivity, and plasma half-lives, but do not overcome the bioavailability problems associated with nonparenteral administration.

One of the requirements to study potency, duration of action, and to overcome bioavailability problems is to study the pharmacokinetics and disposition of the peptides. Factors contributing to the potential difficulty of studying and describing pharmacokinetics of peptides include the following: (1) analytical methods for unchanged substance and its metabolites often lack specificity; (2) multiple metabolites may be formed rapidly; (3) clearance from the plasma is usually relatively rapid; (4) fast degradation of primary metabolites to small fragments and amino acids removes evidence of the primary inactivating step; and (5) degradation may occur at many sites in the body.

Because peptides are complex molecules, it requires considerable effort to identify their metabolites. Even measuring the concentration of the unchanged molecule can be problematic because small changes in structure may be undetected when using a radioimmunoassay or bioassay. Some assurance of reliability can be obtained by comparing results of complementary methods such as high-performance liquid chromatography (HPLC) and enzyme-linked immunosorbent assay (ELISA).

Here the analysis of a synthetic RGD-containing peptide, TP9201 [Ac-CNPRGD(Y-OMe)RC-NH$_2$], a potent inhibitor of platelet aggregation,[5-8] in plasma by two independent methods, HPLC and ELISA, is presented. Plasma samples were collected from the animal at short times after injec-

[13] M. Faucliere, M. L. Fosling, J. J. Jones, and T. S. Pryor, *Adv. Drug Res.* **15,** 29 (1986).
[14] M. T. Humphrey and P. S. Ringrose, *Drug Metab. Rev.* **17,** 283 (1986).

tion. Injections can be given by intravenous (iv), subcutaneous (sc), or intramuscular (im) routes. Administration by subcutaneous and intramuscular routes results in a prolonged pharmacological half-life as compared to the intravenous route.[15] The pattern of peptide release into and elimination from the blood can be analyzed using the methods described below. These methods allow for simple precipitation of plasma proteins, dilution of the remaining plasma, and direct analysis without prior purification of the peptide. Because the formation of early metabolites of TP9201 is slow compared to its rate of excretion, and no accumulation of TP9201 in tissues has been observed, analysis of TP9201 pharmacokinetics turned out to be relatively straightforward.

Analysis of TP9201 Plasma Levels by HPLC

Reagents and Materials

Sodium citrate, 3.8% (w/v) in H_2O; heparin, 50 U/ml in H_2O
Trichloroacetic acid, 6% (w/v)
Milli-Q water (Millipore, Bedford, MA)
Trifluoroacetic acid
C_4 protein column (Cat. No. 214TP54; Vydac, Hesperia, CA)
HPLC apparatus with automatic injector
Online data collection system with software (ChromePerfect, Mountain View, CA) for analysis
Table-top microcentrifuge
Centrifuge tubes, pipetting tools

Procedure

1. Obtain anticoagulated blood (use either citrate or heparin as anticoagulant). Put whole blood into a 1/10 volume of either 3.8% (w/v) sodium citrate or heparin (50 U/ml). Normal plasma obtained from animals or human subjects before administration of TP9201 should be analyzed by this method so that any plasma peaks present after acid extraction can be accounted for.

2. Administer TP9201 via intravenous or extravascular administration.

3. Centrifuge the blood in a microfuge to obtain plasma. Remove the plasma and freeze it immediately for long-term storage.

4. Extract the large plasma proteins by adding 75 μl of plasma to 225 μl of trichloroacetic acid (6%, w/v).

---

[15] C. Mazur, J. F. Tschopp, and M. D. Pierschbacher, *Abst. Assoc. Pharm. Sci. Am. West. Reg. Meet. 1994* (1994).

5. Vortex briefly and incubate the acidified mixture on ice for 10 min.
6. Centrifuge the sample at 14,000 g for 4 min.
7. Recover the supernatant and discard the pellet.
8. Analyze the supernatant by reversed-phase HPLC.
9. Use a C_4 protein column (Cat. No. 214TP54; Vydac) and elute with a mobile phase acetonitrile gradient as follows:

Mobile phase A: Milli-Q water plus 0.1% (v/v) trifluoroacetic acid
Mobile phase B: 80% (v/v) acetonitrile in Milli-Q water plus 0.1% (v/v) trifluoroacetic acid

0–10 min	0% mobile phase B
10–25 min	0–25% mobile phase B
25–27 min	25–100% mobile phase B
27–32 min	100% mobile phase B
32–34 min	100–0% mobile phase B
34–55 min	0% mobile phase B

10. Monitor absorbance at 215 nm for 25 min.

Quantitation. TP9201 can be quantitated by creating a standard curve of the peptide in normal plasma, analyzing these samples using the protocol described above, and creating a calibration file within the ChromePerfect software (Fig. 4). Quantitation in the experimental plasma samples will automatically be done by the software.

Analysis of TP9201 Plasma Levels by Receptor-Based ELISA

Reagents and Materials

Trichloroacetic acid, 6% (w/v) in water
Neutralization buffer

HEPES	50 mM
NaCl	300 mM
Octylglucoside	40 mM
CaCl$_2$	1 mM
MgCl$_2$	2 mM
NaOH (added fresh just before use)	240 mM

Tris-buffered saline (TBS)

Tris	10 mM
NaCl	150 mM

Adjust the pH to 7.3

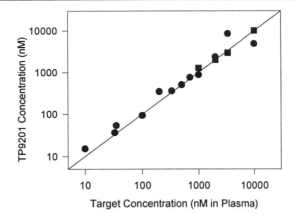

FIG. 4. Validation of TP9201 analysis in human plasma. Human blood was spiked with TP9201 to obtain varying final concentrations. Plasma was prepared, acid precipitated, and analyzed by a second operator using both methods [(●) ELISA; (■) HPLC]. The actual TP9201 concentration in plasma is in good agreement with the measured concentration.

TBS–Tween: 0.05% (v/v) Tween 20 in TBS

TBS–OG: 20 mM octylglucoside in TBS

Fibrinogen: Fibrinogen [plasminogen-free, human plasma fibrinogen (Cat. No. 341578; Calbiochem, La Jolla, CA)], 10 μg/ml in TBS.

Receptor: Each preparation of RGD peptide affinity purified $\alpha_{IIb}\beta_3$[16] must be titrated to determine the maximum and minimum detectable binding to the fibrinogen-coated wells. For the assay, the receptor should be diluted to yield a final concentration in the well of approximately 80% of the maximum binding asymptote in TBS–OG, 1 mM CaCl$_2$, 1 mM MgCl$_2$

Primary antibody: Fibrinogen-adsorbed rabbit anti-$\alpha_{IIb}\beta_3$ serum or monoclonal anti-$\alpha_{IIb}\beta_3$ IgG diluted to a nonlimiting concentration for maximal receptor binding in TBS–Tween

Secondary antibody: Peroxidase-conjugated goat anti-(primary antibody species) IgG diluted to a nonlimiting concentration in TBS–Tween

Developer

NaH$_2$PO$_4$ 120 mM

Citric acid   50 mM

Adjust the pH to 5.0. Add fresh just before use:

[16] R. Pytela, M. D. Pierschbacher, M. H. Ginsberg, E. F. Plow, and E. Ruoslahti, *Science* **231,** 1159 (1986).

o-Phenylenediamine hydrochloride   400 μg/ml
H$_2$O$_2$ (30%) (w/w)   0.24 μl/ml

Procedure

1. Ligand plate coating: 200 μl of the fibrinogen solution is added to each well of Nunc (Roskilde, Denmark) MaxiSorp microtiter plates. Each plate is sealed to prevent evaporation, and incubated overnight at room temperature on a gyratory shaker. The coated plates may be stored at 4° for up to 10 days before use.

2. Standard curve preparation: Normal plasma treated like the samples (same anticoagulant and storage) is used to create standards of known peptide concentrations. Make serial dilutions in one-half log steps of the peptide in plasma across a range known to show minimal and maximal inhibition of receptor binding in the assay. (This range is from 33 ng/ml to 33 μg/ml for most preparations of TP9201.) Use the same peptide stock that was used for the *in vivo* study.

3. Sample preparation: 1 : 1, 1 : 3, and 1 : 7 dilutions (twofold serial dilutions) of each sample should be made with the normal plasma, to ensure that the final concentration of the peptide in the well of at least one of the diluted samples will be in the range of the standards. This dilution scheme yields one-eighth, one-forth, one-half diluted, and an undiluted allocation of each sample.

4. Sample precipitation: One part of each of the dilutions of the samples and standards is mixed with three parts ice-cold trichloroacetic acid (TCA) (75 μl of plasma and 325 μl of TCA) and allowed to precipitate on ice for 10 min. The solution is then centrifuged at ~14,000 relative centrifugal force (RCF; 14,000 rpm in a microfuge) for 4 min at room temperature. The supernatant is removed and reserved. This procedure removes most of the plasma proteins from the samples.

5. Sample neutralization: The sodium hydroxide is added to the neutralization buffer just before use. A cloudy precipitate will form. Vortex the buffer well to keep the precipitate in suspension prior to mixing the buffer with the samples. One part of the TCA supernatants is mixed with one part neutralization buffer (100 μl plus 100 μl). The pH of this mixture should be between pH 7 and pH 8.

6. Plate washing: The solution in the plate is removed and the plate tapped dry on paper towels. TBS–Tween (300 μl) is added to each well of the plate and incubated on a gyratory shaker at room temperature for 5 min. This procedure is repeated twice more for a total of three washes, then the plate is tapped dry in preparation for the next step of the assay.

7. Receptor incubation: Add 75 μl of the neutralized sample and 75 μl of standard to the washed, fibrinogen-coated well of the plate in duplicate.

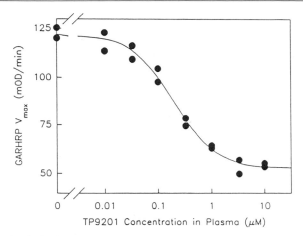

FIG. 5. TP9201 inhibition of $\alpha_{IIb}\beta_3$–fibrinogen binding in the ELISA (GARHRP, horseradish peroxidase conjugated goat anti-rabbit IgG).

Twenty-five microliters of the receptor solution (with four times the final receptor concentration desired) is added to each well. The plate is sealed / to prevent evaporation and incubated at room temperature on a gyratory shaker for 3 hr to overnight.

8. Wash the plate, using the same technique as described above.

9. Primary antibody incubation: Add 100 μl of the primary antibody to each well. The plate is sealed to prevent evaporation and incubated at room temperature on a gyratory shaker for 3 hr to overnight.

10. Wash the plate.

11. Secondary antibody incubation: Add 100 μl of the secondary antibody to each well. The plate is incubated at room temperature on a gyratory shaker for 1 to 3 hr.

12. Wash the plate.

13. Developing and reading the plate: The microtiter plate reader should be set up to perform a kinetic reading of the plates measuring the OD at 450 nm every 12 sec. Add 100 μl of freshly prepared developer to each well and immediately begin a kinetic reading of the plate. Collect data for 5 to 10 min. Determine the maximum velocity (V_{max}) of the enzymatic reaction in each well over an approximately 2-min period of the reaction (10 data points).

Quantitation: Peptide Concentration Determination. Fit a logistic equation to the values measured for the standards. A logistic function is

$$y = [(a - d)/(1 + (x/c)^b)] + d$$

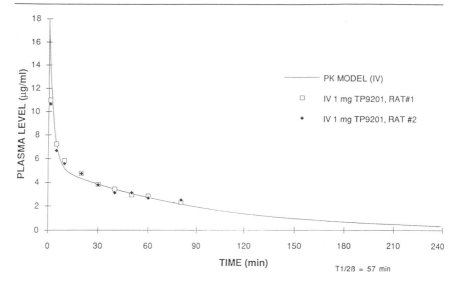

FIG. 6. Example of TP9201 pharmacokinetics in rats after iv injection and HPLC plasma analysis. (—) Pharmacokinetic model (iv); (□) TP9201, 1 mg, iv, rat 1; (◆) TP9201, 1 mg, iv, rat 2.

where a is the asymptotic maximum, d is the asymptotic minimum, b is the slope parameter, and c is the inflection point or IC_{50}. For these studies x is the peptide concentration and y is the V_{max} value. A concentration for each of the samples with a V_{max} value within the range of the standards

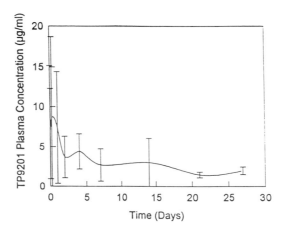

FIG. 7. Example of TP9201 pharmacokinetics of TP9201 infusion by minipumps in rat over a 4 week period and an analysis of plasma samples by ELISA.

can be determined from this equation. The final calculation of the concentration of the peptide in plasma is made by correcting for the dilution factors of the samples. A typical TP9201 inhibition of $\alpha_{IIb}\beta_3$ fibrinogen is shown in Fig. 5.

Discussion

Both methods, HPLC and $\alpha_{IIb}\beta_3$/fibrinogen-based ELISA, are useful assays to detect TP9201 levels in plasma from animals or humans. Figure 6 shows the pharmacokinetic profile of TP9201 in the rat after a 1-mg intravenous injection. The HPLC method using ultraviolet (UV) detection shows a lower detection limit of 1 μM TP9201 (1.14 μg/ml plasma). By contrast, the ELISA assay shows a lower detection limit of 20 nM and a upper limit of 5 μM. Figure 7 shows the pharmacokinetic profile of TP9201 infusion over a 4-week period in the rat, using Alzet miniosmotic pumps (ALZA Corp., Palo Alto, CA). TP9201 plasma levels were detected by the ELISA method. Detection of 5 μM TP9201 or higher with the ELISA method is not practical because of potential increases in errors from the multiple dilutions of plasma.

The levels of ruggedness of the two assays are different. The HPLC assay is operator insensitive, reproducible, and requires little time to perform. By contrast the ELISA procedure requires careful operator technique and more operator time per sample analyzed. The ELISA is sensitive to reagent variability and the reagents needed are valuable, especially the receptor preparations ($\alpha_{IIb}\beta_3$) and antibodies used.

Author Index

Numbers in parentheses are footnote reference numbers and indicate that an author's work is referred to although the name is not cited in the text.

A

Aaronson, S. A., 178
Aarts, P. A. M. M., 167, 170(36)
Abbott, L. A., 292
Abbott, M., 23
Abbott, M. H., 24
Abbott-Brown, D., 80
Abe, N., 5(145), 26, 27(145), 28(145)
Abedin, M. Z., 10
Abelson, J., 425, 426(27)
Abraham, J. A., 232
Achari, A., 378
Ackerman, M. S., 356, 367
Ackerman, R., 17
Adachi, E., 18, 481, 482(54)
Adams, J. C., 63, 70, 72, 72(35), 422(13), 423, 535
Adams, S. P., 87, 159, 173, 347, 369
Adey, N. B., 376
Adler, S., 252
Adra, C. N., 391, 393(17)
Aebersold, R., 275
Aeschlimann, D., 63, 71(18)
Aggeler, J., 174, 175(63)
Aggler, J., 536, 540, 543(27)
Agin, P. P., 185
Aguilera, A., 14
Aizawa, S., 54, 388, 409(8), 417(8), 524
Akiyama, S. K., 365, 368
Akkerman, J. W. N., 171
Albelda, S., 84
Albelda, S. M., 144, 145(70), 297, 298(8), 309(8)
Albini, A., 178
Albrechtsen, R., 85, 87, 91(4), 93(8), 94(8), 97(8), 101(8), 104, 134(40), 135
Alexander, C. M., 540

Alexander, P., 378
Ali, F., 366, 369, 382
Allen, E. R., 43
Allen, J., 302
Allen, J. B., 242
Allen, L., 38
Allen, N. D., 418
Altieri, D. C., 347, 366
Alvial, G., 120
Amagai, M., 441
Ambrogio, C., 366, 368
Ambros, V., 263
Amesi-Impiombato, F. S., 21
Amiot, M., 196, 198, 206(14, 27), 208(14, 27), 209(27)
An, Q., 538
Anderson, D., 528, 538
Anderson, D. C., 308, 446, 447(109, 113)
Anderson, P. S., 385
Andersson-Fisone, C., 61
Andrade, R., 120
Andreasen, A. P., 242
Andres, G., 202
Andrews, D., 108
Andrieux, A., 347, 368
Andrulis, I., 313
Angello, J., 258
Angerer, L. M., 77
Angerer, R. C., 77
Anselmetti, D., 470, 481(16), 482(16), 483(16)
Antignac, C., 4
Antonsson, P., 63, 71, 71(18), 72, 72(38)
Aoyama, T., 34, 39(21)
Apfelroth, S. D., 37
Appel, J. R., 353, 367
Appella, E., 68
Appleby, M. W., 419

H

Herstburger, E., 273
Hese, K. M., 168, 173
Hessle, H., 29
Heuser, J. E., 479, 507
Hewett, D., 39
Hewett, D. R., 38
Hibbs, M. L., 308–309
Hickey, R., 419
Higashiyama, S., 232
Higley, A., 382
Higuchi, R., 424
Hildebrand, A., 349, 369
Hildebrand, B., 34
Hildebrand, J. D., 529
Hildreth, J. E. K., 427(31), 429
Hilgenberg, L., 110, 112(30), 113(30), 114(30), 116(30)
Hill, J. E., 441
Hilton, D. J., 394
Hindriks, G., 167
Hinek, A., 349, 367
Hinkes, M. T., 222, 223(12)
Hinuma, S., 391, 392(19), 393(19)
Hiraiwa, H., 58–59
Hirano, S., 293
Hirano, T., 445(102), 446
Hirayama, Y., 446, 447(111)
Hirosawa, K., 22, 23(121)
Hirose, T., 84
Hirschmann, R., 361, 369
Hirvonen, H., 25, 26(141), 86–87, 472
Hishino, T., 470
Hizawa, K., 84
Hjerpe, A., 63, 71(18)
Hoare, K., 196
Hockfield, S., 105
Hodgkin, J., 263, 268(24)
Hodson, E., 184
Hoess, R. H., 351, 353, 368, 370, 378(28), 385–386
Hof, L., 287
Hofer, U., 54, 55(25), 59
Hoffman, S., 52–54, 60, 60(24), 123–125, 125(58)
Hoffström, B. G., 87, 316
Hofmann, M., 208
Hofschneider, P. H., 305
Hogan, B., 258, 260(12), 263(12), 394, 405(32), 408(32), 411(32), 418(32)
Hogan, B. L., 88

Hogan, B. L. M., 74, 77, 91, 258
Hogarth, P. M., 431, 433(49)
Hogervorst, F., 136, 139, 141(60), 185, 429, 431(34)
Hogg, P. J., 65–66
Hojima, Y., 481, 482(54)
Holland, P. W. H., 74, 77, 258
Hollander, D., 431, 433(52)
Hollenbach, P. W., 418
Hollenbaugh, D., 196, 198(17), 201(17), 205(17), 207(17)
Holley, B., 131(22), 132
Hollinger, M. A., 242
Hollister, D. W., 29, 31, 37, 37(13), 38(13), 39(13, 29), 42(29)
Holloway, P. W., 488
Holly, F. W., 361, 369
Holm, M., 305
Holmes, D., 474
Holmes, D. A., 470, 479(13)
Holmes, D. F., 479, 505
Holn, H. P., 196
Holt, G. D., 68
Holton, J. L., 441, 444(89)
Holund, B., 344
Holzmann, B., 439
Hong, X., 104
Honkanen, N., 26
Honsik, C. J., 558
Höök, M., 55, 70, 81, 81(27), 132, 159, 161, 196, 220, 243, 248(24)
Hooper, M., 391
Hopkinson, S. B., 7, 9(23)
Hori, H., 31, 37(13), 38(13), 39(13)
Horie, K., 14, 15(73)
Horigan, E., 497
Horigan, E. A., 91
Horn, G. T., 424
Horne, T. J., 454
Horng-Len, Z., 538
Horodniceanu, F., 303
Hortsch, M., 271, 275(2)
Horvitz, H. R., 257
Horwitz, A., 512, 538
Horwitz, A. F., 179, 297–298, 298(6, 9), 309(4, 9), 365, 369, 427, 433(28)
Hoshino, T., 14, 15(73)
Hostikka, S. L., 4
Hotta, Y., 274
Houde, M., 439

Subject Index